Circuits and Applications Using Silicon Heterostructure Devices

Circuits and Applications Using Silicon Heterostructure Devices

Edited by
John D. Cressler

CRC Press
Taylor & Francis Group
Boca Raton London New York

CRC Press is an imprint of the
Taylor & Francis Group, an **informa** business

The material was previously published in *Silicon Heterostructure Handbook: Materials, Fabrication, Devices, Circuits and Applications of SiGe and Si Strained-Layer Epitaxy* © Taylor and Francis 2005.

CRC Press
Taylor & Francis Group
6000 Broken Sound Parkway NW, Suite 300
Boca Raton, FL 33487-2742

Printed in the United States of America on acid-free paper
10 9 8 7 6 5 4 3 2 1

International Standard Book Number-13: 978-1-4200-6694-4 (Hardcover)

Library of Congress Cataloging-in-Publication Data

Circuits and applications using silicon heterostructure devices / editor, John D. Cressler.
p. cm.
Includes bibliographical references and index.
ISBN 978-1-4200-6694-4 (alk. paper)
1. Bipolar transistors. 2. Heterostructures. 3. Bipolar integrated circuits--Design and construction. 4. Semiconductors. I. Cressler, John D. II. Title.

TK7871.96.B55C73 2008
621.3815'28--dc22

2007030347

Visit the Taylor & Francis Web site at
http://www.taylorandfrancis.com

and the CRC Press Web site at
http://www.crcpress.com

For the tireless efforts
Of the many dedicated scientists and engineers
Who helped create this field and make it a success.
I tip my hat, and offer sincere thanks from all of us
Who have benefitted from your keen insights and imaginings.

And ...

For Maria:
My beautiful wife, best friend, and soul mate for these 25 years.
For Matthew John, Christina Elizabeth, and Joanna Marie:
God's awesome creations, and our precious gifts.
May your journey of discovery never end.

He Whose Heart Has Been Set
On The Love Of Learning And True Wisdom
And Has Exercised This Part of Himself,
That Man Must Without Fail Have Thoughts
That Are Immortal And Divine,
If He Lay Hold On Truth.

Plato

Εκείνος που έχει δώσει την
ψυχή του στην Αγάπη για Μάθηση
και Αληθινή Σοφία,
και έχει Ασκηθεί για τούτο,
Ένας τέτοιος μόνο Άνδρας μπορεί
το δίχως άλλο να κάνει σκέψεις
Αθάνατες και Θείες,
Εάν στηριχθεί στην Αλήθεια.

Πλάτωνας

Foreword

Progress in a given field of technology is both desired and expected to follow a stable and predictable long-term trajectory. Semilog plots of technology trends spanning decades in time and orders of magnitude in value abound. Perhaps the most famous exemplar of such a technology trajectory is the trend line associated with Moore's law, where technology density has doubled every 12 to 18 months for several decades. One must not, however, be lulled into extrapolating such predictability to other aspects of semiconductor technology, such as device performance, or even to the long-term prospects for the continuance of device density scaling itself. New physical phenomena assert themselves as one approaches the limits of a physical system, as when device layers approach atomic dimensions, and thus, no extrapolation goes on indefinitely.

Technology density and performance trends, though individually constant over many years, are the result of an enormously complex interaction between a series of decisions made as to the layout of a given device, the physics behind its operation, manufacturability considerations, and its extensibility into the future. This complexity poses a fundamental challenge to the device physics and engineering community, which must delve as far forward into the future as possible to understand when physical law precludes further progress down a given technology path. The early identification of such impending technological discontinuities, thus providing time to ameliorate their consequences, is in fact vital to the health of the semiconductor industry. Recently disrupted trends in CMOS microprocessor performance, where the "value" of processor-operating frequency was suddenly subordinated to that of integration, demonstrate the challenges remaining in accurately assessing the behavior of future technologies. However, current challenges faced in scaling deep submicron CMOS technology are far from unique in the history of semiconductors.

Bipolar junction transistor (BJT) technology, dominant in high-end computing applications during the mid-1980s, was being aggressively scaled to provide the requisite performance for future systems. By the virtue of bipolar transistors being vertical devices rather than lateral (as CMOS is), the length scale of bipolar transistors is set by the ability to control layer thicknesses rather than lateral dimensions. This allowed the definition of critical device dimensions, such as base width, to values far below the limits of optical lithography of the day. Although great strides in device performance had been made by 1985, with unity gain cutoff frequencies (f_T) in the range 20–30 GHz seemingly feasible, device scaling was approaching limits at which new physical phenomena became significant. Highly scaled silicon BJTs, having base widths below 1000 Å, demonstrated inordinately high reverse junction leakage. This was due to the onset of band-to-band tunneling between heavily doped emitter and base regions, rendering such devices unreliable. This and other observations presaged one of the seminal technology discontinuities of the past decade, silicon–germanium (SiGe) heterojunction bipolar transistor (HBT) technology being the direct consequence.

Begun as a program to develop bipolar technology with performance capabilities well beyond those possible via the continued scaling of conventional Si BJTs, SiGe HBT technology has found a wealth of applications beyond the realm of computing. A revolution in bipolar fabrication methodology, moving from device definition by implantation to device deposition and definition by epitaxy, accompanied by the exploitation of bandgap tailoring, took silicon-based bipolar transistor performance to levels never anticipated. It is now common to find SiGe HBTs with performance figures in excess of 300 GHz for both f_T and f_{max}, and circuits operable at frequencies in excess of 100 GHz.

A key observation is that none of this progress occurred in a vacuum, other than perhaps in the field of materials deposition. The creation of a generation of transistor technology having tenfold improved performance would of itself have produced far less ultimate value in the absence of an adequate eco-system to enable its effective creation and utilization. This text is meant to describe the eco-system that developed around SiGe technology as context for the extraordinary achievement its commercial rollout represented.

Early SiGe materials, of excellent quality in the context of fundamental physical studies, proved near useless in later device endeavors, forcing dramatic improvements in layer control and quality to then enable further development. Rapid device progress that followed drove silicon-based technology (recall that SiGe technology is still a silicon-based derivative) to unanticipated performance levels, demanding the development of new characterization and device modeling techniques. As materials work was further proven SiGe applications expanded to leverage newly available structural and chemical control.

Devices employing ever more sophisticated extensions of SiGe HBT bandgap tailoring have emerged, utilizing band offsets and the tailoring thereof to create SiGe-based HEMTs, tunneling devices, mobility-enhanced CMOS, optical detectors, and more to come. Progress in these diverse areas of device design is timely, as I have already noted the now asymptotic nature of performance gains to be had from continued classical device scaling, leading to a new industry focus on innovation rather than pure scaling. Devices now emerging in SiGe are not only to be valued for their performance, but rather their variety of functionality, where, for example, optically active components open up the prospect of the seamless integration of broadband communication functionality at the chip level.

Access to high-performance SiGe technology has spurred a rich diversity of exploratory and commercial circuit applications, many elaborated in this text. Communications applications have been most significantly impacted from a commercial perspective, leveraging the ability of SiGe technologies to produce extremely high-performance circuits while using back level, and thus far less costly, fabricators than alternative materials such as InP, GaAs, or in some instances advanced CMOS.

These achievements did not occur without tremendous effort on the part of many workers in the field, and the chapters in this volume represent examples of such contributions. In its transition from scientific curiosity to pervasive technology, SiGe-based device work has matured greatly, and I hope you find this text illuminating as to the path that maturation followed.

Bernard S. Meyerson
IBM Systems and Technology Group

Preface

While the idea of cleverly using silicon–germanium (SiGe) and silicon (Si) strained-layer epitaxy to practice bandgap engineering of semiconductor devices in the highly manufacturable Si material system is an old one, only in the past decade has this concept become a practical reality. The final success of creating novel Si heterostructure transistors with performance far superior to their Si-only homojunction cousins, while maintaining strict compatibility with the massive economy-of-scale of conventional Si integrated circuit manufacturing, proved challenging and represents the sustained efforts of literally thousands of physicists, electrical engineers, material scientists, chemists, and technicians across the world.

In the electronics domain, the fruit of that global effort is SiGe heterojunction bipolar transistor (SiGe HBT) BiCMOS technology, and strained Si/SiGe CMOS technology, both of which are at present in commercial manufacturing worldwide and are rapidly finding a number of important circuit and system applications. As with any new integrated circuit technology, the industry is still actively exploring device performance and scaling limits (at present well above 300 GHz in frequency response, and rising), new circuit applications and potential new markets, as well as a host of novel device and structural innovations. This commercial success in the electronics arena is also spawning successful forays into the optoelectronics and even nanoelectronics fields. The Si heterostructure field is both exciting and dynamic in its scope.

The implications of the Si heterostructure success story contained in this book are far-ranging and will be both lasting and influential in determining the future course of the electronics and optoelectronics infrastructure, fueling the miraculous communications explosion of the twenty-first century. While several excellent books on specific aspects of the Si heterostructures field currently exist (for example, on SiGe HBTs), this is the first reference book of its kind that "brings-it-all-together," effectively presenting a comprehensive perspective by providing very broad topical coverage ranging from materials, to fabrication, to devices (HBT, FET, optoelectronic, and nanostructure), to CAD, to circuits, to applications. Each chapter is written by a leading international expert, ensuring adequate depth of coverage, up-to-date research results, and a comprehensive list of seminal references. A novel aspect of this book is that it also contains "snap-shot" views of the industrial "state-of-the-art," for both devices and circuits, and is designed to provide the reader with a useful basis of comparison for the current status and future course of the global Si heterostructure industry.

This book is intended for a number of different audiences and venues. It should prove to be a useful resource as:

1. A hands-on reference for practicing engineers and scientists working on various aspects of Si heterostructure integrated circuit technology (both HBT, FET, and optoelectronic), including materials, fabrication, device physics, transistor optimization, measurement, compact modeling and device simulation, circuit design, and applications
2. A hands-on research resource for graduate students in electrical and computer engineering, physics, or materials science who require information on cutting-edge integrated circuit technologies

3. A textbook for use in graduate-level instruction in this field
4. A reference for technical managers and even technical support/technical sales personnel in the semiconductor industry

It is assumed that the reader has some modest background in semiconductor physics and semiconductor devices (at the advanced undergraduate level), but each chapter is self-contained in its treatment.

In this age of extreme activity, in which we are all seriously pressed for time and overworked, my success in getting such a large collection of rather famous people to commit their precious time to my vision for this project was immensely satisfying. I am happy to say that my authors made the process quite painless, and I am extremely grateful for their help. The list of contributors to this book actually reads like a global "who's who" of the silicon heterostructure field, and is impressive by any standard. I would like to formally thank each of my colleagues for their hard work and dedication to executing my vision of producing a lasting Si heterostructure "bible." In order of appearance, the "gurus" of our field include:

Lawrence E. Larson, University of California at San Diego, USA
Donald Y.C. Lie, Dynamic Research Corporation and University of California at San Diego, USA
Qingqing Liang, Georgia Institute of Technology, USA
Leo C.N. de Vreede, Delft University of Technology, The Netherlands
Mark P. van der Heijden, Delft University of Technology, The Netherlands
Hermann Schumacher, University of Ulm, Germany
Johann-Friedrich Luy, DaimlerChrysler Research, Germany
John R. Long, Delft University of Technology, The Netherlands
Sudipto Chakraborty, Georgia Institute of Technology, USA
Joy Laskar, Georgia Institute of Technology, USA
John Papapolymerou, Georgia Institute of Technology, USA
Manos Tentzeris, Georgia Institute of Technology, USA
Rong Lin Li, Georgia Institute of Technology, USA
Kyutae Lim, Georgia Institute of Technology, USA
Stephane Pinel, Georgia Institute of Technology, USA
Daniel A. Friedman, IBM Thomas J. Watson Research Center, USA
Mounir Meghelli, IBM Thomas J. Watson Research Center, USA
Katsuyoshi Washio, Hitachi, Ltd., Japan
Didier Belot, ST Microelectronics, France
Michael Schröter, University of California at San Diego, USA
Ramana M. Malladi, IBM Microelectronics, USA

I would also like to thank my graduate students and post-docs, past and present, for their dedication and tireless work in this fascinating field. I rest on their shoulders. They include: David Richey, Alvin Joseph, Bill Ansley, Juan Roldán, Stacey Salmon, Lakshmi Vempati, Jeff Babcock, Suraj Mathew, Kartik Jayanaraynan, Greg Bradford, Usha Gogineni, Gaurab Banerjee, Shiming Zhang, Krish Shivaram, Dave Sheridan, Gang Zhang, Ying Li, Zhenrong Jin, Qingqing Liang, Ram Krithivasan, Yun Luo, Tianbing Chen, Enhai Zhao, Yuan Lu, Chendong Zhu, Jon Comeau, Jarle Johansen, Joel Andrews, Lance Kuo, Xiangtao Li, Bhaskar Banerjee, Curtis Grens, Akil Sutton, Adnan Ahmed, Becca Haugerud, Mustayeen Nayeem, Mustansir Pratapgarhwala, Guofu Niu, Emery Chen, Jongsoo Lee, and Gnana Prakash.

Finally, I am grateful to Tai Soda at Taylor & Francis for talking me into this project, and supporting me along the way. I would also like to thank the production team at Taylor & Francis for their able assistance (and patience!), especially Jessica Vakili.

The many nuances of the Si heterostructure field make for some fascinating subject matter, but this is no mere academic pursuit. In the grand scheme of things, the Si heterostructure industry is already reshaping the global communications infrastructure, which is in turn dramatically reshaping the way life on planet Earth will transpire in the twenty-first century and beyond. The world would do well to pay attention. It has been immensely satisfying to see both the dream of Si/SiGe bandgap engineering, and this book, come to fruition. I hope our efforts please you. Enjoy!

John D. Cressler
Editor

Editor

John D. Cressler received a B.S. in physics from the Georgia Institute of Technology (Georgia Tech), Atlanta, Georgia, in 1984, and an M.S. and Ph.D. in applied physics from Columbia University, New York, in 1987 and 1990. From 1984 to 1992 he was on the research staff at the IBM Thomas J. Watson Research Center in Yorktown Heights, New York, working on high-speed Si and SiGe bipolar devices and technology. In 1992 he left IBM Research to join the faculty at Auburn University, Auburn, Alabama, where he served until 2002. When he left Auburn University, he was Philpott–Westpoint Stevens Distinguished Professor of Electrical and Computer Engineering and director of the Alabama Microelectronics Science and Technology Center.

In 2002, Dr. Cressler joined the faculty at Georgia Tech, where he is currently Ken Byers Professor of Electrical and Computer Engineering. His research interests include SiGe devices and technology; Si-based RF/microwave/millimeter-wave mixed-signal devices and circuits; radiation effects; device-circuit interactions; noise and linearity; reliability physics; extreme environment electronics, 2-D/3-D device-level simulation; and compact circuit modeling. He has published more than 350 technical papers related to his research, and is author of the books *Silicon-Germanium Heterojunction Bipolar Transistors*, Artech House, 2003 (with Guofu Niu), and *Reinventing Teenagers: The Gentle Art of Instilling Character in Our Young People*, Xlibris, 2004 (a slightly different genre!).

Dr. Cressler was Associate Editor of the *IEEE Journal of Solid-State Circuits* (1998–2001), Guest Editor of the *IEEE Transactions on Nuclear Science* (2003–2006), and Associate Editor of the *IEEE Transactions on Electron Devices* (2005–present). He served on the technical program committees of the IEEE International Solid-State Circuits Conference (1992–1998, 1999–2001), the IEEE Bipolar/BiCMOS Circuits and Technology Meeting (1995–1999, 2005–present), the IEEE International Electron Devices Meeting (1996–1997), and the IEEE Nuclear and Space Radiation Effects Conference (1999–2000, 2002–2007). He currently serves on the executive steering committee for the IEEE Topical Meeting on Silicon Monolithic Integrated Circuits in RF Systems, as international program advisor for the IEEE European Workshop on Low-Temperature Electronics, on the technical program committee for the IEEE International SiGe Technology and Device Meeting, and as subcommittee chair of the 2004 Electrochemical Society Symposium of SiGe: Materials, Processing, and Devices. He was the Technical Program Chair of the 1998 IEEE International Solid-State Circuits Conference, the Conference Co-Chair of the 2004 IEEE Topical Meeting on Silicon Monolithic Integrated Circuits in RF Systems, and the Technical Program Chair of the 2007 IEEE Nuclear and Space Radiation Effects Conference.

Dr. Cressler was appointed an IEEE Electron Device Society Distinguished Lecturer in 1994, an IEEE Nuclear and Plasma Sciences Distinguished Lecturer in 2006, and was awarded the 1994 Office of Naval Research Young Investigator Award for his SiGe research program. He received the 1996 C. Holmes

MacDonald National Outstanding Teacher Award by Eta Kappa Nu, the 1996 Auburn University Alumni Engineering Council Research Award, the 1998 Auburn University Birdsong Merit Teaching Award, the 1999 Auburn University Alumni Undergraduate Teaching Excellence Award, an IEEE Third Millennium Medal in 2000, and the 2007 Georgia Tech Outstanding Faculty Leadership in the Development of Graduate Students Award. He is an IEEE Fellow.

On a more personal note, John's hobbies include hiking, gardening, bonsai, all things Italian, collecting (and drinking!) fine wines, cooking, history, and carving walking sticks, not necessarily in that order. He considers teaching to be his vocation. John has been married to Maria, his best friend and soul-mate, for 25 years, and is the proud father of three budding scholars: Matt, Christina, and Jo-Jo.

Dr. Cressler can be reached at School of Electrical and Computer Engineering, 777 Atlantic Drive, N.W., Georgia Institute of Technology, Atlanta, GA 30332-0250 U.S.A. or cressler@ece.gatech.edu http://users.ece.gatech.edu/~cressler/

Contents

1 **The Big Picture** 1-1
John D. Cressler

2 **A Brief History of the Field** 2-1
John D. Cressler

3 **Overview: Circuits and Applications** 3-1
John D. Cressler

4 **SiGe as an Enabler for Wireless Communications Systems** 4-1
Lawrence E. Larson and Donald Y.C. Lie

5 **LNA Optimization Strategies** 5-1
Qingqing Liang

6 **Linearization Techniques** 6-1
Leo C.N. de Vreede and Mark P. van der Heijden

7 **SiGe MMICs** 7-1
Hermann Schumacher

8 **SiGe Millimeter-Wave ICs** 8-1
Johann-Friedrich Luy

9 **Wireless Building Blocks Using SiGe HBTs** 9-1
John R. Long

10 **Direct Conversion Architectures for SiGe Radios** 10-1
Sudipto Chakraborty and Joy Laskar

11 **RF MEMS Techniques in Si/SiGe** 11-1
John Papapolymerou

12 **Wideband Antennas on Silicon** 12-1
Manos Tentzeris and Rong Lin Li

13 **Packaging Issues for SiGe Circuits** 13-1
Kyutae Lim, Stephane Pinel, and Joy Laskar

14 **Industry Examples of the State-of-the-Art: IBM—High-Speed Circuits for Data Communications Applications** 14-1
Daniel A. Friedman and Mounir Meghelli

15 **Industry Examples at the State-of-the-Art: Hitachi** 15-1
Katsuyoshi Washio

16 **Industry Examples at the State-of-the-Art: ST** 16-1
Didier Belot

A.1 **Properties of Silicon and Germanium** A.1-1
John D. Cressler

A.2 **The Generalized Moll–Ross Relations** A.2-1
John D. Cressler

A.3 **Integral Charge-Control Relations** A.3-1
Michael Schröter

A.4 **Sample SiGe HBT Compact Model Parameters** A.4-1
Ramana M. Malladi

Index I-1

1

The Big Picture

1.1 The Communications Revolution 1-1
1.2 Bandgap Engineering in the Silicon Material System 1-3
1.3 Terminology and Definitions 1-4
1.4 The Application Space 1-5
1.5 Performance Limits and Future Directions 1-9

John D. Cressler
Georgia Institute of Technology

1.1 The Communications Revolution

We are at a unique juncture in the history of humankind, a juncture that amazingly we engineers and scientists have dreamed up and essentially created on our own. This pivotal event can be aptly termed the "Communications Revolution," and the twenty-first century, our century, will be the era of human history in which this revolution plays itself out.

This communications revolution can be functionally defined and characterized by the pervasive acquisition, manipulation, storage, transformation, and transmission of "information" on a global scale. This information, or more generally, knowledge, in its infinitely varied forms and levels of complexity, is gathered from our analog sensory world, transformed in very clever ways into logical "1"s and "0"s for ease of manipulation, storage, and transmission, and subsequently regenerated into analog sensory output for our use and appreciation. In 2005, this planetary communication of information is occurring at a truly mind-numbing rate, estimates of which are on the order of 80 Tera-bits/sec (10^{12}) of data transfer across the globe in 2005 solely in wired and wireless voice and data transmission, 24 hours a day, 7 days a week, and growing exponentially. The world is quite literally abuzz with information flow—communication.* It is for the birth of the Communications Revolution that we humans likely will be remembered for 1000 years hence. Given that this revolution is happening during the working careers of most of us, I find it a wonderful time to be alive, a fact of which I remind my students often.

Here is my point. No matter how one slices it, at the most fundamental level, it is semiconductor devices that are powering this communications revolution. Skeptical? Imagine for a moment that one could flip a switch and instantly remove all of the integrated circuits (ICs) from planet Earth. A moment's reflection will convince you that there is not a single field of human endeavor that would not come to a grinding halt, be it commerce, or agriculture, or education, or medicine, or entertainment. Life as we in the first world know it in 2005 would simply cease to exist. And yet, remarkably, the same result would not have been true 50 years ago; even 20 years ago. Given the fact that we humans have been on planet Earth in our present form for at least 1 million years, and within communities

*I have often joked with my students that it would be truly entertaining if the human retina was sensitive to longer wavelengths of electromagnetic radiation, such that we could "see" all the wireless communications signals constantly bathing the planet (say, in greens and blues!). It might change our feelings regarding our ubiquitous

having entrenched cultural traditions for at least 15,000 years, this is truly a remarkable fact of history. A unique juncture indeed.

Okay, hold on tight. It is an easy case to make that the semiconductor silicon (Si) has single-handedly enabled this communications revolution.* I have previously extolled at length the remarkable virtues of this rather unglamorous looking silver-grey element [1], and I will not repeat that discussion here, but suffice it to say that Si represents an extremely unique material system that has, almost on its own, enabled the conception and evolving execution of this communications revolution. The most compelling attribute, by far, of Si lies in the economy-of-scale it facilitates, culminating in the modern IC fabrication facility, effectively enabling the production of gazillions of low-cost, very highly integrated, remarkably powerful ICs, each containing millions of transistors; ICs that can then be affordably placed into widgets of remarkably varied form and function.†

So what does this have to do with the book you hold in your hands? To feed the emerging infrastructure required to support this communications revolution, IC designers must work tirelessly to support increasingly higher data rates, at increasingly higher carrier frequencies, all in the design space of decreasing form factor, exponentially increasing functionality, and at ever-decreasing cost. And by the way, the world is going portable and wireless, using the same old wimpy batteries. Clearly, satisfying the near-insatiable appetite of the requisite communications infrastructure is no small task. Think of it as job security!

For long-term success, this quest for more powerful ICs must be conducted within the confines of conventional Si IC fabrication, so that the massive economy-of-scale of the global Si IC industry can be brought to bear. Therein lies the fundamental motivation for the field of Si heterostructures, and thus this book. Can one use clever nanoscale engineering techniques to custom-tailor the energy bandgap of fairly conventional Si-based transistors to: (a) improve their performance dramatically and thereby ease the circuit and system design constraints facing IC designers, while (b) performing this feat without throwing away all the compelling economy-of-scale virtues of Si manufacturing? The answer to this important question is a resounding "YES!" That said, getting there took time, vision, as well as dedication and hard work of literally thousands of scientists and engineers across the globe.

In the electronics domain, the fruit of that global effort is silicon–germanium heterojunction bipolar transistor (SiGe HBT) bipolar complementary metal oxide semiconductor (BiCMOS) technology, and is in commercial manufacturing worldwide and is rapidly finding a number of important circuit and system applications. In 2004, the SiGe ICs, by themselves, are expected to generate US$1 billion in revenue globally, with perhaps US$30 billion in downstream products. This US$1 billion figure is projected to rise to US$2.09 billion by 2006 [2], representing a growth rate of roughly 42% per year, a remarkable figure by any economic standard. The biggest single market driver remains the cellular industry, but applications in optical networking, hard disk drives for storage, and automotive collision-avoidance radar systems are expected to represent future high growth areas for SiGe. And yet, in the beginning of 1987, only 18 years ago, there was no such thing as a SiGe HBT. It had not been demonstrated as a viable concept. An amazing fact.

In parallel with the highly successful development of SiGe HBT technology, a wide class of "transport enhanced" field effect transistor topologies (e.g., strained Si CMOS) have been developed as a means to boost the performance of the CMOS side of Si IC coin, and such technologies have also recently begun

*The lone exception to this bold claim lies in the generation and detection of coherent light, which requires direct bandgap III–V semiconductor devices (e.g., GaAs of InP), and without which long-haul fiber communications systems would not be viable, at least for the moment.

†Consider: it has been estimated that in 2005 there are roughly 20,000,000,000,000,000,000 (2×10^{19}) transistors on planet Earth. While this sounds like a large number, let us compare it to some other large numbers: (1) the universe is roughly 4.2×10^{17}sec old (13.7 billion years), (2) there are about 1×10^{21} stars in the universe, and (3) the universe is about 4×10^{23} miles across (15 billion light-years)! Given the fact that all 2×10^{20} of these transistors have been produced since December 23, 1947 (following the invention of the point-contact transistor by Bardeen, Brattain, and Shockley), this is a truly remarkable feat of human ingenuity.

to enter the marketplace as enhancements to conventional core CMOS technologies. The commercial success enjoyed in the electronics arena has very naturally also spawned successful forays into the optoelectronics and even nanoelectronics fields, with potential for a host of important downstream applications.

The Si heterostructure field is both exciting and dynamic in its scope. The implications of the Si heterostructure success story contained in this book are far-ranging and will be both lasting and influential in determining the future course of the electronics and optoelectronics infrastructure, fueling the miraculous communications explosion of our twenty-first century. The many nuances of the Si heterostructure field make for some fascinating subject matter, but this is no mere academic pursuit. As I have argued, in the grand scheme of things, the Si heterostructure industry is already reshaping the global communications infrastructure, which is in turn dramatically reshaping the way life of planet Earth will transpire in the twenty-first century and beyond. The world would do well to pay close attention.

1.2 Bandgap Engineering in the Silicon Material System

As wonderful as Si is from a fabrication viewpoint, from a device or circuit designer's perspective, it is hardly the ideal semiconductor. The carrier mobility for both electrons and holes in Si is comparatively small compared to their III–V cousins, and the maximum velocity that these carriers can attain under high electric fields is limited to about 1×10^7 cm/sec under normal conditions, relatively "slow." Since the speed of a transistor ultimately depends on how fast the carriers can be transported through the device under sustainable operating voltages, Si can thus be regarded as a somewhat "meager" semiconductor. In addition, because Si is an indirect gap semiconductor, light emission is fairly inefficient, making active optical devices such as diode lasers impractical (at least for the present). Many of the III–V compound semiconductors (e.g., GaAs or InP), on the other hand, enjoy far higher mobilities and saturation velocities, and because of their direct gap nature, generally make efficient optical generation and detection devices. In addition, III–V devices, by virtue of the way they are grown, can be compositionally altered for a specific need or application (e.g., to tune the light output of a diode laser to a specific wavelength). This atomic-level custom tailoring of a semiconductor is called bandgap engineering, and yields a large performance advantage for III–V technologies over Si [3]. Unfortunately, these benefits commonly associated with III–V semiconductors pale in comparison to the practical deficiencies associated with making highly integrated, low-cost ICs from these materials. There is no robust thermally grown oxide for GaAs or InP, for instance, and wafers are smaller with much higher defect densities, are more prone to breakage, and are poorer heat conductors (the list could go on). These deficiencies translate into generally lower levels of integration, more difficult fabrication, lower yield, and ultimately higher cost. In truth, of course, III–V materials such as GaAs and InP fill important niche markets today (e.g., GaAs metal semiconductor field effect transistor (MESFETs) and HBTs for cell phone power amplifiers, AlGaAs- or InP-based lasers, efficient long wavelength photodetectors, etc.), and will for the foreseeable future, but III–V semiconductor technologies will never become mainstream in the infrastructure of the communications revolution if Si-based technologies can do the job.

While Si ICs are well suited to high-transistor-count, high-volume microprocessors and memory applications, RF, microwave, and even millimeter-wave (mm-wave) electronic circuit applications, which by definition operate at significantly higher frequencies, generally place much more restrictive performance demands on the transistor building blocks. In this regime, the poorer intrinsic speed of Si devices becomes problematic. That is, even if Si ICs are cheap, they must deliver the required device and circuit performance to produce a competitive system at a given frequency. If not, the higher-priced but faster III–V technologies will dominate (as they indeed have until very recently in the RF and microwave markets).

The fundamental question then becomes simple and eminently practical: is it possible to improve the performance of Si transistors enough to be competitive with III–V devices for high-performance applications, while preserving the enormous yield, cost, and manufacturing advantages associated with conventional Si fabrication? The answer is clearly "yes," and this book addresses the many nuances associated with using SiGe and Si-strained layer epitaxy to practice bandgap engineering in the Si material system, a process culminating in, among other things, the SiGe HBT and strained Si CMOS, as well as a variety of other interesting electronic and optoelectronic devices built from these materials. This totality can be termed the "Si heterostructures" field.

1.3 Terminology and Definitions

A few notes on modern usage and pronunciation in this field are in order (really!). It is technically correct to refer to silicon–germanium alloys according to their chemical composition, $Si_{1-x}Ge_x$, where x is the Ge mole fraction. Following standard usage, such alloys are generally referred to as "SiGe" alloys. Note, however, that it is common in the material science community to also refer to such materials as "Ge:Si" alloys.

A SiGe film that is carbon doped (e.g., less than 0.20% C) in an attempt to suppress subsequent boron out-diffusion (e.g., in HBTs) is properly referred to as a SiGe:C alloy, or simply SiGeC (pronounced "silicon germanium carbon," *not* "silicon germanium carbide"). This class of SiGe alloys should be viewed as optimized SiGe alloys, and are distinct from SiGe films with a much higher C content (e.g., 2% to 3% C) that might be used, for instance, to lattice-match SiGeC alloys to Si.

Believe it or not, this field also has its own set of slang pronunciations. The colloquial usage of the pronunciation \'sig-ee\ to refer to "silicon–germanium" (begun at IBM in the late 1990s) has come into vogue (heck, it may make it to the dictionary soon!), and has even entered the mainstream IC engineers's slang; pervasively.*

In the electronics domain, it is important to be able to distinguish between the various SiGe technologies as they evolve, both for CMOS (strained Si) and bipolar (SiGe HBT). Relevant questions in this context include: Is company X's SiGe technology more advanced than company Y's SiGe technology? For physical as well as historical reasons, one almost universally defines CMOS technology (Si, strained Si, or SiGe), a lateral transport device, by the drawn lithographic gate length (the CMOS technology "node"), regardless of the resultant intrinsic device performance. Thus, a "90-nm" CMOS node has a drawn gate length of roughly 90 nm. For bipolar devices (i.e., the SiGe HBT), however, this is not so straightforward, since it is a vertical transport device whose speed is not nearly as closely linked to lithographic dimensions.

In the case of the SiGe HBT it is useful to distinguish between different technology generations according to their resultant *ac* performance (e.g., peak common-emitter, unity gain cutoff frequency (f_T), which is (a) easily measured and unambiguously compared technology to technology, and yet is (b) a very strong function of the transistor vertical doping and Ge profile and hence nicely reflects the degree of sophistication in device structural design, overall thermal cycle, epi growth, etc.) [1]. The peak f_T generally nicely reflects the "aggressiveness," if you will, of the transistor scaling which has been applied to a given SiGe technology. A higher level of comparative sophistication can be attained by also invoking the maximum oscillation frequency (f_{max}), a parameter which is well correlated to both intrinsic profile and device parasitics, and hence a bit higher on the ladder of device performance metrics, and thus more representative of actual large-scale circuit performance. The difficulty in this case is that f_{max} is far more ambiguous than f_T, in the sense that it can be inferred from various gain definitions (e.g., U vs. MAG), and in practice power gain data are often far less ideal in its behavior over frequency, more sensitive to accurate deembedding, and ripe with extraction "issues."

We thus term a SiGe technology having a SiGe HBT with a peak f_T in the range of 50 GHz as "first generation;" that with a peak f_T in the range of 100 GHz as "second generation;" that with a peak f_T in the range of 200 GHz as "third generation;" and that with a peak f_T in the range of 300 GHz as "fourth generation." These are loose definitions to be sure, but nonetheless useful for comparison purposes.

*I remain a stalwart holdout against this snowballing trend and stubbornly cling to the longer but far more satisfying "silicon–germanium."

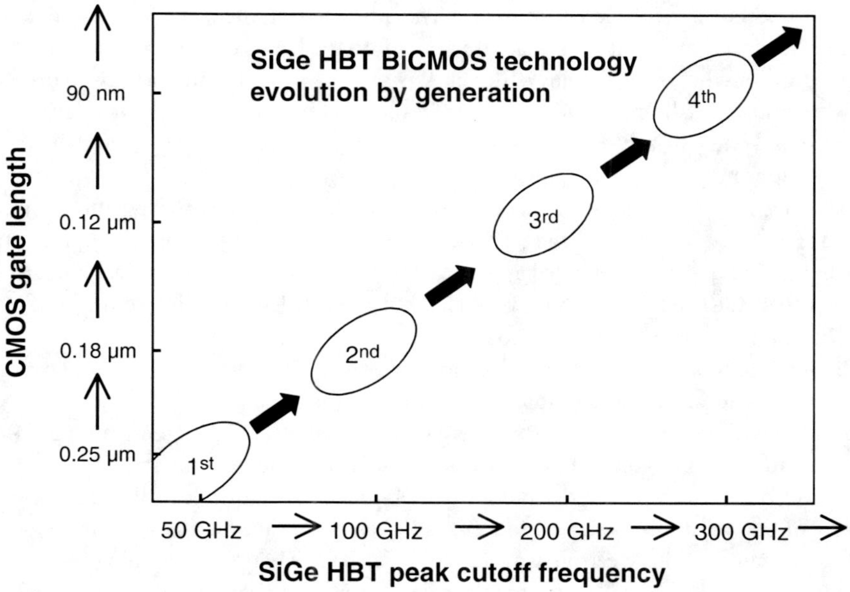

FIGURE 1.1 Evolution of SiGe HBT BiCMOS technology generations, as measured by the peak cutoff frequency of the SiGe HBT, and the CMOS gate length.

A complicating factor in SiGe technology terminology results from the fact that most, if not all, commercial SiGe HBT technologies today also contain standard Si CMOS devices (i.e., SiGe HBT BiCMOS technology) to realize high levels of integration and functionality on a single die (e.g., single-chip radios complete with RF front-end, data converters, and DSP). One can then speak of a given generation of SiGe HBT BiCMOS technology as the most appropriate intersection of both the SiGe HBT peak f_T and the CMOS technology node (Figure 1.1). For example, for several commercially important SiGe HBT technologies available via foundry services, we have:

- IBM SiGe 5HP—50 GHz peak f_T SiGe HBT + 0.35 μm Si CMOS (first generation)
- IBM SiGe 7HP—120 GHz peak f_T SiGe HBT + 0.18 μm Si CMOS (second generation)
- IBM SiGe 8HP—200 GHz peak f_T SiGe HBT + 0.13 μm Si CMOS (third generation)
- Jazz SiGe 60—60 GHz peak f_T SiGe HBT + 0.35 μm Si CMOS (first generation)
- Jazz SiGe 120—150 GHz peak f_T SiGe HBT + 0.18 μm Si CMOS (second generation)
- IHP SiGe SGC25B—120 GHz peak f_T SiGe HBT + 0.25 μm Si CMOS (second generation)

All SiGe HBT BiCMOS technologies can thus be roughly classified in this manner. It should also be understood that multiple transistor design points typically exist in such BiCMOS technologies (multiple breakdown voltages for the SiGe HBT and multiple threshold or breakdown voltages for the CMOS), and hence the reference to a given technology generation implicitly refers to the most aggressively scaled device within that specific technology platform.

1.4 The Application Space

It goes without saying in our field of semiconductor IC technology that no matter how clever or cool a new idea appears at first glance, its long-term impact will ultimately be judged by its marketplace "legs" (sad, but true). That is, was the idea good for a few journal papers and an award or two, or did someone actually build something and sell some useful derivative products from it? The sad reality is that the semiconductor field (and we are by no means exceptional) is rife with examples of cool new devices that never made it past the pages of the IEDM digest! The ultimate test, then, is one of stamina. And sweat. Did the idea make it out of the research laboratory and into the hands of the manufacturing lines? Did it pass the qualification-checkered flag, have design kits built around it, and get delivered to real circuit designers who built ICs, fabricated them, and tested them? Ultimately, were the derivative ICs inserted into real systems—widgets—to garner leverage in this or that system metric, and hence make the products more appealing in the marketplace?

Given the extremely wide scope of the semiconductor infrastructure fueling the communications revolution, and the sheer volume of widget possibilities, electronic to photonic to optoelectronic, it is useful here to briefly explore the intended application space of Si heterostructure technologies as we peer out into the future. Clearly I possess no crystal ball, but nevertheless some interesting and likely lasting themes are beginning to emerge from the fog.

SiGe HBT BiCMOS is the obvious ground-breaker of the Si heterostructures application space in terms of moving the ideas of our field into viable products for the marketplace. The field is young, but the signs are very encouraging. As can be seen in Figure 1.2, there are at present count 25 + SiGe HBT industrial fabrication facilities on line in 2005 around the world, and growing steadily. This trend points to an obvious recognition that SiGe technology will play an important role in the emerging electronics infrastructure of the twenty-first century. Indeed, as I often point out, the fact that virtually every major player in the communications electronics field either: (a) has SiGe up and running in-house, or (b) is using someone else's SiGe fab as foundry for their designers, is a remarkable fact, and very encouraging in the grand scheme of things. As indicated above, projections put SiGe ICs at a US$2.0 billion level by 2006, small by percentage perhaps compared to the near trillion dollar global electronics market, but growing rapidly.

The intended application target? That obviously depends on the company, but the simple answer is, gulp, a little bit of everything! As depicted in Figure 1.3 and Figure 1.4, the global communications landscape is exceptionally diverse, ranging from low-frequency wireless (2.4 GHz cellular) to the fastest high-speed wireline systems (10 and 40 Gbit/sec synchronous optical network (SONET)). Core CMOS technologies are increasingly being pushed into the lower frequency wireless space, but the compelling drive to higher carrier frequencies over time will increasingly favor SiGe technologies.

At present, SiGe ICs are making inroads into: the cellular industry for handsets [global system for mobile communications—GSM, code division multiple access (CDMA), wideband CDMA (W-CDMA), etc.], even for power amplifiers; various wireless local area networks (WLAN) building blocks, from components to fully integrated systems ranging from 2.4 to 60 GHz and up; ultrawide band (UWB) components; global positioning systems (GPS); wireless base stations; a variety of wireline networking products, from 2.5 to 40 Gbit/sec (and higher); data converters (D/A and A/D); high-speed memories; a variety of instrumentation electronics; read-channel memory storage products; core analog functions (op amps, etc.); high-speed digital circuits of various flavors; radiation detector

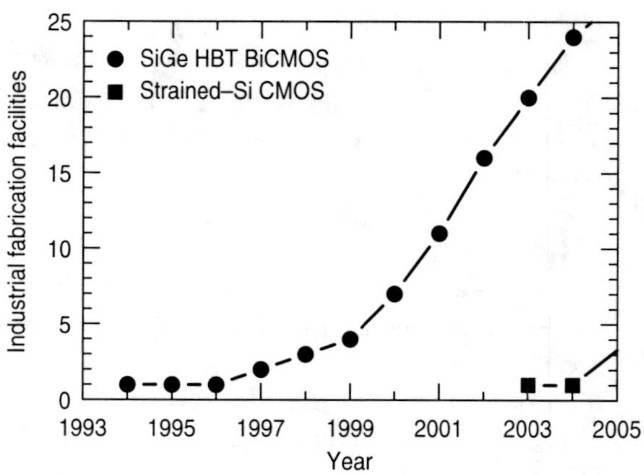

FIGURE 1.2 Number of industrial SiGe and strained Si fabrication facilities.

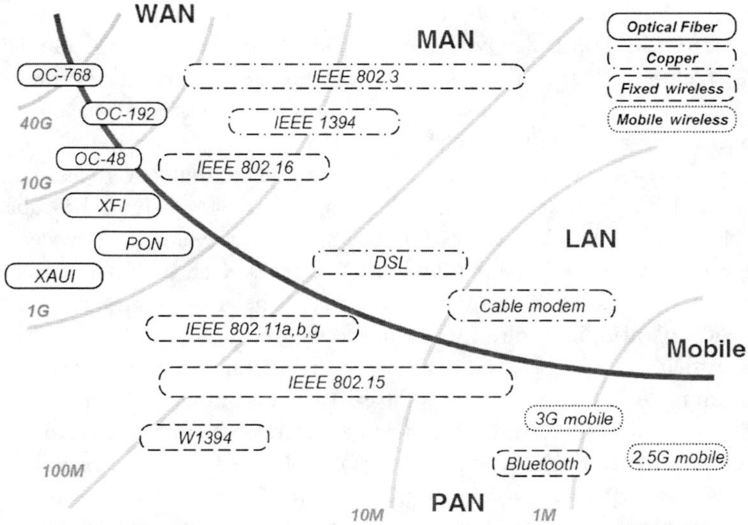

FIGURE 1.3 The global communications landscape, broken down by the various communications standards, and spanning the range of: wireless to wireline; fixed to mobile; copper to fiber; low data rate to broadband; and local area to wide area networks. WAN is *wide area network*, MAN is *metropolitan area network*, the so-called "last mile" access network, LAN is *local area network*, and PAN is *personal area network*, the emerging in-home network. (Used with the permission of Kyutae Lim.)

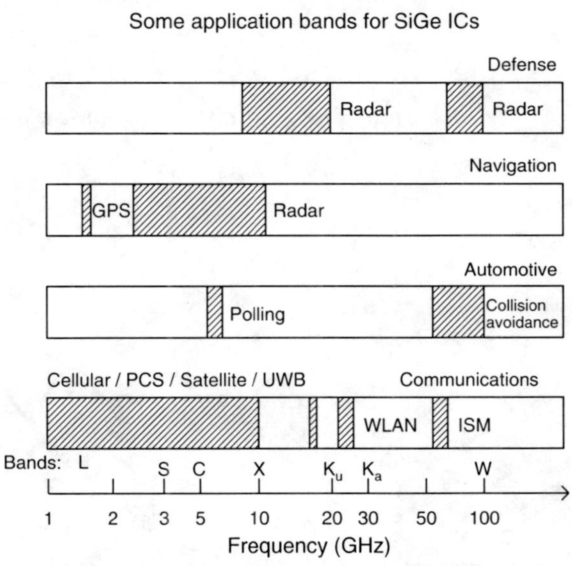

FIGURE 1.4 Some application frequency bands for SiGe integrated circuits.

electronics; radar systems (from 3 to 77 GHz and up); a variety space-based electronics components; and various niche extreme environment components (e.g., cryogenic (77 K) hybrid superconductor–semiconductor systems). The list is long and exceptionally varied—this is encouraging. Clearly, however, some of these components of "everything" are more important than others, and this will take time to shake out.

The strength of the BiCMOS twist to SiGe ICs cannot be overemphasized. Having both the high-speed SiGe HBT together on-chip with aggressively scaled CMOS allows one great flexibility in system design, the depths of which is just beginning to be plumbed. While debates still rage with respect to the most cost-effective partitioning at the chip and package level (system-on-a-chip versus system-in-a-package, etc.), clearly increased integration is viewed as a good thing in most camps (it is just a question of how much), and SiGe HBT BiCMOS is well positioned to address such needs across a broad market sector.

The envisioned high-growth areas for SiGe ICs over the new few years include: the cellular industry, optical networking, disk drives, and radar systems. In addition, potential high-payoff market areas span the emerging mm-wave space (e.g., the 60 GHz ISM band WLAN) for short range, but very high data rate (Gbit/sec) wireless systems. A SiGe 60 GHz single-chip/package transceiver (see Figure 1.5 for IBM's vision of such a beast) could prove to be the "killer app" for the emerging broadband multimedia market. Laughable? No. The building blocks for such systems have already been demonstrated using third-generation SiGe technology [4], and fully integrated transceivers are under development.

The rest of the potential market opportunities within the Si heterostructures field can be leveraged by successes in the SiGe IC field, both directly and indirectly. On the strained Si CMOS front, there are existent proofs now that strained Si is likely to become a mainstream component of conventional CMOS scaling at the 90-nm node and beyond (witness the early success of Intel's 90-nm logic technology built around uniaxially strained Si CMOS; other companies are close behind). Strained Si would seem to represent yet another clever technology twist that CMOS device technologists are pulling from their bag of tricks to keep the industry on a Moore's law growth path. This was not an obvious development (to me anyway) even a couple of years back. A wide variety of "transport enhanced" Si-heterostructure-based FETs have been demonstrated (SiGe-channel FETs, Si-based high electron mobility transistors (HEMTs), as well as both uniaxially and biaxially strained FETs, etc). Most of these devices, however, require complex substrate engineering that would have seemed to preclude giga-scale integration level needs for microprocessor-level integration. Apparently not so. The notion of using Si heterostructures (either

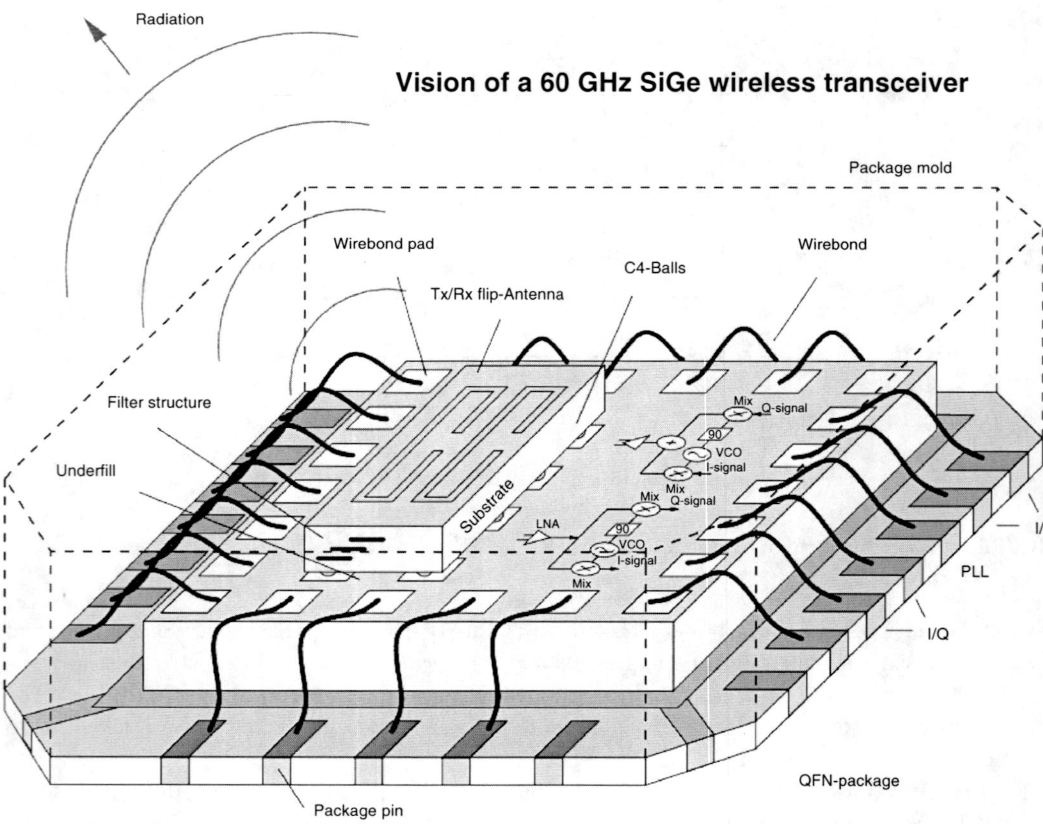

FIGURE 1.5 Vision for a single-chip SiGe mm-wave transceiver system. (Used with the permission of Ullrich Pfeiffer.)

uniaxial or biaxial strain or both) to boost conventional CMOS performance appears to be an appealing path for the future, a natural merging point I suspect for SiGe strained layers found in SiGe HBT BiCMOS (which to date contains only conventional Si CMOS) and strained Si CMOS.

From the optoelectronics camp, things are clearly far less evolved, but no less interesting. A number of functional optoelectronic devices have been demonstrated in research laboratories. Near-term successes in the short wavelength detector arena and light emitting diodes (LEDs) are beginning to be realized. The achievement of successful coherent light emission in the Si heterostructure system (e.g., via quantum cascade techniques perhaps) would appear to be the "killer app" in this arena, and research in this area is in progress. More work is needed.

1.5 Performance Limits and Future Directions

We begin with device performance limits. Just how fast will SiGe HBTs be 5 years from now? Transistor-level performance in SiGe HBTs continues to rise at a truly dizzying pace, and each major conference seems to bear witness to a new performance record (Figure 1.6). Both first- and second-generation SiGe HBT BiCMOS technology is widely available in 2005 (who would have thought even 3 years ago that fully integrated 100+ GHz Si-based devices would be "routine" on 200 mm wafers?), and even at the 200 GHz (third-generation) performance level, six companies (at last count) have achieved initial technology demonstrations, including IBM (Chapter 7), Jazz (Chapter 8), IHP (Chapter 11), ST Microelectronics (Chapter 12), Hitachi (Chapter 9), and Infineon (Chapter 10). (See *Fabrication of SiGe HBT BiCMOS Technology* for these chapters.) Several are now either available in manufacturing, or are very close (e.g., [5]). At press time, the most impressive new stake-in-the-ground is the report (June 2004) of the newly optimized "SiGe 9T" technology, which simultaneously achieves 302 GHz peak f_T and 306 GHz peak f_{max}, a clear record for any Si-based transistor, from IBM (Figure 1.7) [6]. This level of ac performance was achieved at a BV_{CEO} of 1.6 V, a BV_{CBO} of 5.5 V, and a current gain of 660. Noise measurements on these devices yielded NF_{min}/G_{assoc} of 0.45 dB/14 dB and 1.4 dB/8 dB at 10 and 25 GHz, respectively. Measurements of earlier (unoptimized) fourth-generation IBM SiGe HBTs have yielded record values of 375 GHz peak f_T [7] at 300 K and above 500 GHz peak f_T at 85 K. Simulations suggest that THz-level (1000 GHz) intrinsic transistor performance is not a laughable proposition in SiGe HBTs (Chapter 16, see *Silicon Heterostructure Devices*). This fact still amazes even me, the eternal optimist of SiGe performance! I, for one, firmly believe that we will see SiGe HBTs above-500 GHz peak

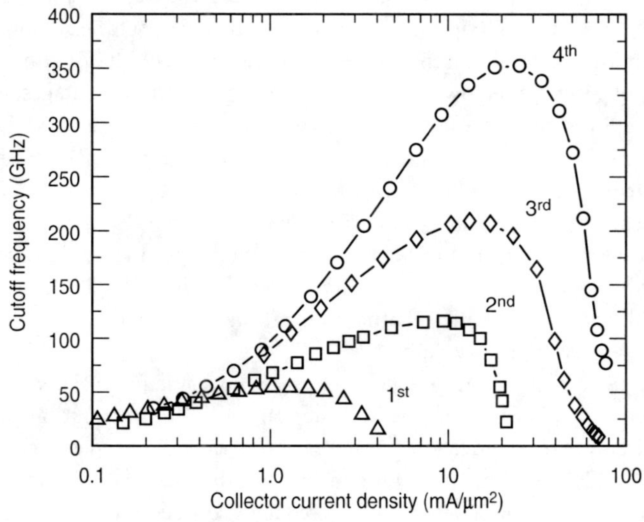

FIGURE 1.6 Measured cutoff frequency as a function of bias current density for four different SiGe HBT technology generations.

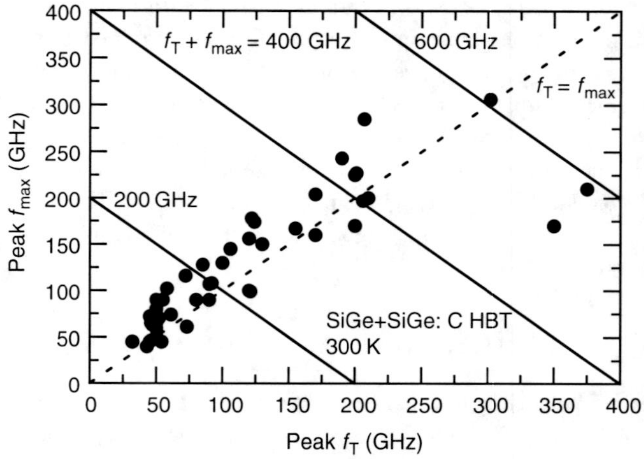

FIGURE 1.7 Measured maximum oscillation frequency versus cutoff frequency for a variety of generations of SiGe HBT BiCMOS technology shown in Figure 1.1.

f_T and f_{max} fully integrated with nanometer-scale (90 nm and below) Si CMOS (possibly strained Si CMOS) within the next 3 to 5 years.

One might logically ask, particularly within the confines of the above discussion on ultimate market relevance, why one would even attempt to build 500 GHz SiGe HBTs, other than to win a best-paper award, or to trumpet that "because-it's-there" Mount Everest mentality we engineers and scientists love so dearly. This said, if the future "killer app" turns out to be single-chip mm-wave transceiver systems with on-board DSP for broadband multimedia, radar, etc., then the ability of highly scaled, highly integrated, very high performance SiGe HBTs to dramatically enlarge the circuit/system design space of the requisite mm-wave building blocks may well prove to be a fruitful (and marketable) path.

Other interesting themes are emerging in the SiGe HBT BiCMOS technology space. One is the very recent emergence of complementary SiGe (C-SiGe) HBT processes (npn + pnp SiGe HBTs). While very early pnp SiGe HBT prototypes were demonstrated in the early 1990s, only in the last 2 years or so have fully complementary SiGe processes been developed, the most mature of which to date is the IHP SGC25C process, which has 200 GHz npn SiGe HBTs and 80 GHz pnp SiGe HBTs (Chapter 11, see *Fabrication of SiGe HBT BiCMOS Technology*). Having very high-speed pnp SiGe HBTs on-board presents a fascinating array of design opportunities aimed particularly at the analog/mixed-signal circuit space. In fact, an additional emerging trend in the SiGe field, particularly for companies with historical pure analog circuit roots, is to target lower peak f_T, but higher breakdown voltages, while simultaneously optimizing the device for core analog applications (e.g., op amps, line drivers, data converters, etc.), designs which might, for instance, target better noise performance, and higher current gain-Early voltage product than mainstream SiGe technologies. One might even choose to park that SiGe HBT platform on top of thick film SOI for better isolation properties (Chapter 13, see *Fabrication of SiGe HBT BiCMOS Technology*). Another interesting option is the migration of high-speed vertical SiGe HBTs with very thin film CMOS-compatible SOI (Chapter 5, see *Fabrication of SiGe HBT BiCMOS Technology*). This technology path would clearly favor the eventual integration of SiGe HBTs with strained Si CMOS, all on SOI, a seemingly natural migratory path.

If one accepts the tenet that integration is a good thing from a system-level perspective, the Holy Grail in the Si heterostructure field would, in the end, appear to be the integration of SiGe HBTs for RF through mm-wave circuitry (e.g., single-chip mm-wave transceivers complete with on-chip antennae), strained Si CMOS for all DSP and memory functionality, both perhaps on SOI, Si-based light emitters, SiGe HBT modulator electronics, and detectors for such light sources, together with on-chip waveguides to steer the light, realized all on one Si wafer to produce a "Si-based optoelectronic superchip" [8], that could do-it-all. These diverse blocks would be optional plug-in modules around a core SiGe

HBT + strained Si CMOS IC technology platform, perhaps with flip-chip (or other) packaging techniques to join different sub-die to the main superchip (e.g., for a Si-based detector or laser).

I know, I know. It is not obvious that even if each of these blocks could be realized, that it would make economic sense to do so for real systems. I have no quarrel with that. I think such a Si-based superchip is a useful paradigm, however, to bind together all of the clever objects we wish to ultimately build with Si heterostructures, from electronic to photonic, and maintain the vision of the one overarching constraint that guides us as we look forward—keep whatever you do compatible with high-volume manufacturing in Si fabrication facilities if you want to shape the path of the ensuing communications revolution. This Si-based superchip clearly remains a dream at present. A realizable dream? And if realizable, commercially viable? Who knows? Only time will tell. But it is fun to think about.

As you peruse this book you hold in your hands, which spans the whole Si heterostructure research and development space, from materials, to devices, to circuit and system applications, I think you will be amazed at both the vision, cleverness, and smashing successes of the many scientists and engineers who make up our field. Do not count us out! We are the new architects of an oh-so-very-interesting future.

References

1. JD Cressler and G Niu. *Silicon–Germanium Heterojunction Bipolar Transistors.* Boston, MA: Artech House, 2003.
2. "SiGe devices market to hit $2 billion in 2006," article featured on CompoundSemicoductor.net, http://compoundsemiconductor.net/articles/news/8/3/22/1.
3. F Capasso. Band-gap engineering: from physics and materials to new semiconductor devices. *Science*, 235:172–176, 1987.
4. S Reynolds, B Floyd, U Pfeiffer, and T Zwick. 60 GHz transciever circuits in SiGe bipolar technology. Technical Digest of the IEEE International Solid-State Circuits Conference, San Francisco, 2004, pp 442–443.
5. AJ Joseph, D Coolbaugh, D Harame, G Freeman, S Subbanna, M Doherty, J Dunn, C Dickey, D Greenberg, R Groves, M Meghelli, A Rylyakov, M Sorna, O Schreiber, D Herman, and T Tanji. 0.13 μm 210 GHz f_T SiGe HBTs—expanding the horizons of SiGe BiCMOS. Technical Digest of the IEEE International Solid-State Circuits Conference, San Francisco, 2002, pp 180–182.
6. J-S Rieh, D Greenberg, M Khater, KT Schonenberg, J-J Jeng, F Pagette, T Adam, A Chinthakindi, J Florkey, B Jagannathan, J Johnson, R Krishnasamy, D Sanderson, C Schnabel, P Smith, A Stricker, S Sweeney, K Vaed, T Yanagisawa, D Ahlgren, K Stein, and G Freeman. SiGe HBTs for millimeter-wave applications with simultaneously optimized f_T and f_{max}. Proceedings of the IEEE Radio Frequency Integrated Circuits (RFIC) Symposium, Fort Worth, 2004, pp 395–398.
7. JS Rieh, B Jagannathan, H Chen, KT Schonenberg, D Angell, A Chinthakindi, J Florkey, F Golan, D Greenberg, S-J Jeng, M Khater, F Pagette, C Schnabel, P Smith, A Stricker, K Vaed, R Volant, D Ahlgren, G Freeman, K Stein, and S Subbanna. SiGe HBTs with cutoff frequency of 350 GHz. Technical Digest of the IEEE International Electron Devices Meeting, San Francisco, 2002, pp 771–774.
8. R Soref. Silicon-based photonic devices. Technical Digest of the IEEE International Solid-State Circuits Conference, 1995, pp 66–67.

2

A Brief History of the Field

John D. Cressler
Georgia Institute of Technology

2.1 Si–SiGe Strained Layer Epitaxy 2-1
2.2 SiGe HBTs 2-3
2.3 SiGe–Strained Si FETs and Other SiGe Devices 2-6

In the historical record of any field of human endeavor, being "first" is everything. It is often said that "hindsight is 20–20," and it is tempting in many cases to ascribe this or that pivotal event as "obvious" or "easy" once the answer is known. Anyone intimately involved in a creative enterprise knows, however, that it is never easy being first, and often requires more than a little luck and maneuvering. Thus the triumphs of human creativity, the "firsts," should be appropriately celebrated. Still, later chroniclers often gloss over, and then eventually ignore, important (and sometimes very interesting) twists and turns, starts and stops, of the winners as well as the second and third place finishers, who in the end may in fact have influenced the paths of the winners, sometimes dramatically. The history of our field, for instance, is replete with interesting competitive battles, unusual personalities and egos, no small amount of luck, and various other fascinating historical nuances.

There is no concise history of our field available, and while the present chapter is not intended to be either exhaustive or definitive, it represents my firm conviction that the history of any field is both instructive and important for those who follow in the footsteps of the pioneers. Hopefully this brief history does not contain too many oversights or errors, and is offered as a step in the right direction for a history of pivotal events that helped shape the Si heterostructures field.

2.1 Si–SiGe Strained Layer Epitaxy

The field of Si-based heterostructures solidly rests on the shoulders of materials scientists and crystal growers, those purveyors of the semiconductor "black arts" associated with the deposition of pristine films of nanoscale dimensionality onto enormous Si wafers with near infinite precision. What may seem routine today was not always so. The Si heterostructure story necessarily begins with materials, and circuit designers would do well to remember that much of what they take for granted in transistor performance owes a great debt to the smelters of the crystalline world. Table 2.1 summarizes the key steps in the development of SiGe–Si strained layer epitaxy.

Given that Ge was the earliest and predominant semiconductor pursued by the Bell Laboratories transistor team, with a focus on the more difficult to purify Si to come slightly later, it is perhaps not surprising that the first study of SiGe alloys, albeit unstrained bulk alloys, occurred as early as 1958 [1]. It was recognized around 1960 [2] that semiconductor epitaxy* would enable more robust and controllable transistor fabrication. Once the move to Si-based processing occurred, the field of Si epitaxy was

*The word "epitaxy" (or just "epi") is derived from the Greek word *epi*, meaning "upon" or "over."

TABLE 2.1 Milestones in the Development of SiGe–Si Strained Layer Epitaxy

Historical Event	Year	Ref.
First investigation of the bandgap of unstrained SiGe alloys	1958	[1]
First epitaxially grown layer to be used in a transistor	1960	[2]
First investigation of high-temperature Si epitaxy	1963	[3]
Concept of critical thickness for epitaxial strained layers	1963	[4]
Energy minimization approach for critical thickness	1963	[5]
Force-balance approach for critical thickness	1974	[6]
First growth of SiGe strained layers	1975	[7]
First growth of SiGe epitaxy by MBE	1984	[8]
First stability calculations of SiGe strained layers	1985	[9]
First measurements of energy bandgap in SiGe strained layers	1985	[10,11]
First growth of Si epitaxy by LRP-CVD	1985	[12]
First 2D electron gas in the SiGe system	1985	[13]
First growth of Si epitaxy by UHV/CVD	1986	[14]
First measurements of band alignments in SiGe–Si	1986	[15]
First growth of SiGe epitaxy by UHV/CVD	1988	[16]
First step-graded relaxed SiGe substrate	1988	[16]
First growth of SiGe epitaxy by LRP-CVD	1989	[17]
First growth of Si epitaxy by AP-CVD	1989	[18]
First 2D hole gas in the SiGe system	1989	[19]
First growth of SiGe epitaxy by AP-CVD	1991	[20]
First majority hole mobility measurements in SiGe	1991	[21]
First minority electron mobility measurements in SiGe	1992	[22]
First growth of lattice-matched SiGeC alloys	1992	[23]
First growth of SiGe layers with carbon doping	1994	[24]
First stability calculations to include a Si cap layer	2000	[25]

launched, the first serious investigation of which was reported in 1963 [3]. Early Si epitaxy was exclusively conducted under high-temperature processing conditions, in the range of 1100°C, a temperature required to obtain a chemically pure and pristine growth interface on the Si host substrate for the soon-to-be-grown crystalline Si epi. High-temperature Si epi has been routinely used in basically this same form for over 40 years now, and represents a mature fabrication technique that is still widely practiced for many types of Si devices (e.g., high-speed bipolar transistors and various power devices).

Device engineers have long recognized the benefits of marrying the many virtues of Si as a host material for manufacturing electronic devices, with the bandgap engineering principles routinely practiced in the III–V system. Ultimately this requires a means by which one can perform epitaxial deposition of thin Si layers on large Si substrates, for both p- and n-type doping of arbitrary abruptness, with very high precision, across large wafers, and doping control at high dynamic range. Only a moment's reflection is required to appreciate that this means the deposition of the Si epi must occur at very low growth temperatures, say 500°C to 600°C (not "low" per se, but low compared to the requisite temperatures needed for solid-state diffusion of dopants in Si). Such a low-temperature Si epi would then facilitate the effective marriage of Si and Ge, two chemically compatible elements with differing bandgaps, and enable the doping of such layers with high precision, just what is needed for device realizations. Clearly the key to Si-based bandgap engineering, Si-heterostructures, our field, is the realization of device quality, low-temperature Si epi (and hence SiGe epi), grown pseudomorphically* on large Si host substrates. Conquering this task proved to be remarkably elusive and time consuming.

In the III–V semiconductor world, where very low processing temperatures are much easier to attain, and hence more common than for Si, the deposition of multiple semiconductors on top of one another proved quite feasible (e.g., GaAs on InP), as needed to practice bandgap engineering, for instance,

*The word "pseudo" is derived from the Greek word *pseudēs*, meaning "false," and the word "morphic" is derived from the Greek word *morphē*, meaning "form." Hence, pseudomorphic literally means *false-form*.

resulting in complex material composites having differing lattice constants in intimate physical contact. To accommodate the differing lattice constants while maintaining the crystallinity of the underlying films, strain is necessarily induced in the composite film, and the notion of a film "critical thickness," beyond which strain relaxation occurs via fundamental thermodynamic driving forces, was defined as early as 1963 [4], as were the energy minimization techniques needed for calculating such critical thicknesses [5]. Alternative "force-balance" techniques for addressing the so-called stability issues in strained layer epitaxy came from the III–V world in 1974, and were applied to SiGe strained layer epitaxy in 1985 [9]. Interestingly, however, research continues today on stability in complicated (e.g., compositionally graded) SiGe films, and only very recently have reasonably complete theories been offered which seem to match well with experiment [25].

The first reported growth of SiGe strained layers was in 1975 in Germany [7], but the field did not begin to seriously heat up until the early 1980s, when several teams pioneered the application of molecular beam epitaxy (MBE) to facilitate materials studies of device-quality strained SiGe on Si in 1984 [8]. Optical studies on these films resulted in encouraging findings concerning the beneficial effects of strain on the band-edge properties of SiGe [10,11], paving the way for serious contemplation of devices built from such materials. Parallel paths toward other low-temperature Si epi growth techniques centered on the ubiquitous chemical vapor deposition (CVD) approach were simultaneously pursued, culminating in the so-called limited-reaction-processing CVD (LRP-CVD) technique (Si epi in 1985 [12], and SiGe epi in 1989 [17]), the ultrahigh-vacuum CVD (UHV/CVD) technique (Si epi in 1986 [14] and SiGe epi in 1988 [16]), and various atmospheric pressure CVD (AP-CVD) techniques (e.g., Si epi in 1989 [18], and SiGe epi in 1991 [20]). These latter two techniques, in particular, survive to this day, and are widely used in the SiGe heterojunction bipolar transistor (HBT) industry.

Device-quality SiGe–Si films enabled a host of important discoveries to occur, which have important bearing on device derivatives, including the demonstration of both two-dimensional electron and hole gases [13,19], and the fortuitous observation that step-graded SiGe buffer layers could be used to produce device-quality strained Si on SiGe, with its consequent conduction band offsets [16]. This latter discovery proved important in the development of SiGe–Si heterostructure-based FETs. Both majority and minority carrier mobility measurements occurred in the early 1990s [21,22], although reliable data, particularly involving minority carriers, remain sparse in the literature. Also in the early 1990s, experiments using high C content as a means to relieve strain in SiGe and potentially broaden the bandgap engineering space by lattice-matching SiGe:C materials to Si substrates (a path that has to date not borne much fruit, unfortunately), while others began studying efficacy of C-doping of SiGe, a result that ultimately culminated in the wide use today of C-doping for dopant diffusion suppression in SiGe:C HBTs [23,24].

The Si–SiGe materials field continues to evolve. Commercial single wafer (AP-CVD) and batch wafer (UHV/CVD) Si–SiGe epi growth tools compatible with 200 mm (and soon 300 mm) Si wafers exist in literally dozens of industrial fabrication facilities around the world, and SiGe growth can almost be considered routine today in the ease in which it can be integrated into CMOS-compatible fabrication processes. It was clearly of paramount importance in the ultimate success of our field that some of the "black magic" associated with robust SiGe film growth be removed, and this, thankfully, is the case in 2005.

2.2 SiGe HBTs

Transistor action was first demonstrated by Bardeen and Brattain in late December of 1947 using a point contact device [26]. Given all that has transpired since, culminating in the Communications Revolution, which defines our modern world (refer to the discussion in Chapter 1), this pivotal event surely ranks as one of the most significant in the course of human history—bold words, but nevertheless true. This demonstration of a solid-state device exhibiting the key property of amplification (power gain) is also unique in the historical record for the precision with which we can locate it in time—December 23,

1947, at about 5 p.m. Not to be outdone, Shockley rapidly developed a theoretical basis for explaining how this clever object worked, and went on to demonstrate the first true bipolar junction transistor (BJT) in 1951 [27]. The first BJT was made, ironically in the present context, from Ge. The first silicon BJT was made by Teal in 1954 using grown junction techniques. The first diffused silicon BJT was demonstrated in 1956 [28], and the first epitaxially grown silicon BJT was reported in 1960, see Ref. [2].

The concept of the HBT is surprisingly an old one, dating in fact to the fundamental BJT patents filed by Shockley in 1948 [29]. Given that the first bipolar transistor was built from Ge, and III–V semiconductors were not yet on the scene, it seems clear that Shockley envisioned the combination of Si (wide bandgap emitter) and Ge (narrow bandgap base) to form a SiGe HBT. The basic formulation and operational theory of the HBT, for both the traditional wide bandgap emitter plus narrow bandgap base approach found in most III–V HBTs, as well as the drift-base (graded) approach used in SiGe HBTs today, was pioneered by Kroemer, and was largely in place by 1957 [30–32]. It is ironic that Kroemer in fact worked hard early on to realize a SiGe HBT, without success, ultimately pushing him toward the III–V material systems for his heterostructure studies, a path that proved in the end to be quite fruitful for him, since he shared the Nobel Prize in physics in 2000 for his work in (III–V) bandgap engineering for electronic and photonic applications [33]. While III–V HBT (e.g., AlGaAs–GaAs) demonstrations began appearing in the 1970s, driven largely by the needs for active microwave components in the defense industry, reducing the SiGe HBT to practical reality took 30 years after the basic theory was in place due to material growth limitations. As pointed out [34] the semiconductor device field is quite unique in the scope of human history because "science" (theoretical understanding) preceded the "art" (engineering and subsequent technological advancement). Once device-quality SiGe films were finally achieved in the mid-1980s, however, progress was quite rapid. Table 2.2 summarizes the key steps in the evolution of SiGe HBTs.

The first functional SiGe HBT was demonstrated by an IBM team in December 1987 at the IEDM [35]. The pioneering result showed a SiGe HBT with functional, albeit leaky, dc characteristics; but it was a SiGe HBT, it worked (barely), and it was the first.* It is an often overlooked historical point, however, that at least four independent groups were simultaneously racing to demonstrate the first functional SiGe HBT, all using the MBE growth technique: the IBM team [35], a Japanese team [62], a Bell Laboratories team [63], and a Linköping University team [64]. The IBM team is fairly credited with the victory, since it presented (and published) its results in early December of 1987 at the IEDM (it would have been submitted to the conference for review in the summer 1987) [35]. Even for the published journal articles, the IBM team was the first to submit its paper for review (on November 17, 1987) [65]. All four papers appeared in print in the spring of 1988. Other groups soon followed with more SiGe HBT demonstrations.

The first SiGe HBT demonstrated using (the ultimately more manufacturable) CVD growth technique followed shortly thereafter, in 1989, first using LRP-CVD [17], and then with UHV/CVD [36]. Worldwide attention became squarely focused on SiGe technology, however, in June 1990 at the IEEE VLSI Technology Symposium with the demonstration of a non-self-aligned UHV/CVD SiGe HBT with a peak cutoff frequency of 75 GHz [37,38]. At that time, this SiGe HBT result was roughly twice the performance of state-of-the-art Si BJTs, and clearly demonstrated the future performance potential of the technology (doubling of transistor performance is a rare enough event that it does not escape significant attention!). Eyebrows were raised, and work to develop SiGe HBTs for practical circuit applications began in earnest in a large number of industrial and university laboratories around the world.†

The feasibility of implementing pnp SiGe HBTs was also demonstrated in June 1990 [40]. In December 1990, the simplest digital circuit, an emitter-coupled-logic (ECL) ring oscillator, using

---

*An interesting historical perspective of early SiGe HBT development at IBM is contained in Ref. [61].

†A variety of zero-Dt, mesa-isolated, III–V-like high-speed SiGe HBTs were reported in the early 1990s (e.g., Ref. [66]), but we focus here on fully integrated, CMOS-compatible SiGe HBT technologies, because they are inherently more manufacturable, and hence they are the only ones left standing today, for obvious reasons.

TABLE 2.2 Milestones in the Development of SiGe HBTs

Historical Event	Year	Ref.
First demonstration of transistor action	1947	[26]
Basic HBT concept	1948	[29]
First demonstration of a bipolar junction transistor	1951	[27]
First demonstration of a silicon bipolar transistor	1956	[28]
Drift-base HBT concept	1954	[30]
Fundamental HBT theory	1957	[31,32]
First epitaxial silicon transistors	1960	[2]
First SiGe HBT	1987	[35]
First ideal SiGe HBT grown by CVD	1989	[17]
First SiGe HBT grown by UHV/CVD	1989	[36]
First high-performance SiGe HBT	1990	[37,38]
First self-aligned SiGe HBT	1990	[39]
First SiGe HBT ECL ring oscillator	1990	[39]
First pnp SiGe HBT	1990	[40]
First operation of SiGe HBTs at cryogenic temperatures	1990	[41]
First SiGe HBT BiCMOS technology	1992	[42]
First LSI SiGe HBT integrated circuit	1993	[43]
First SiGe HBT with peak f_T above 100 GHz	1993	[44,45]
First SiGe HBT technology in 200-mm manufacturing	1994	[46]
First SiGe HBT technology optimized for 77 K	1994	[47]
First radiation tolerance investigation of SiGe HBTs	1995	[48]
First report of low-frequency noise in SiGe HBTs	1995	[49]
First SiGe:C HBT	1996	[50]
First high-power SiGe HBTs	1996	[51,52]
First sub-10 psec SiGe HBT ECL circuits	1997	[53]
First high-performance SiGe:C HBT technology	1999	[54]
First SiGe HBT with peak f_T above 200 GHz	2001	[55]
First SiGe HBT with peak f_T above 300 GHz	2002	[56]
First complementary (npn + pnp) SiGe HBT technology	2003	[57]
First C-SiGe technology with npn and pnp f_T above 100 GHz	2003	[58]
First vertical SiGe HBT on thin film (CMOS compatible) SOI	2003	[59]
First SiGe HBT with both f_T and f_{max} above 300 GHz	2004	[60]

self-aligned, fully integrated SiGe HBTs was produced [39]. The first SiGe BiCMOS technology (SiGe HBT + Si CMOS) was reported in December 1992 [42]. Theoretical predictions of the inherent ability of SiGe HBTs to operate successfully at cryogenic temperatures (in contrast to Si BJTs) were first confirmed in 1990 [41], and SiGe HBT profiles optimized for the liquid nitrogen temperature environment (77 K) were reported in 1994 [48]. The first LSI SiGe HBT circuit (a 1.2 Gsample/sec 12-bit digital-to-analog converter—DAC) was demonstrated in December 1993 [43]. The first SiGe HBTs with frequency response greater than 100 GHz were described in December 1993 by two independent teams [44,45], and the first SiGe HBT technology entered commercial production on 200-mm wafers in December 1994 [46].

The first report of the effects of ionizing radiation on advanced SiGe HBTs was made in 1995 [48]. Due to the natural tolerance of epitaxial-base bipolar structures to conventional radiation-induced damage mechanisms without any additional radiation-hardening process changes, SiGe HBTs are potentially very important for space-based and planetary communication systems applications, spawning an important new sub-discipline for SiGe technology. The first demonstration that epitaxial SiGe strained layers do not degrade the superior low-frequency noise performance of bipolar transistors occurred in 1995, opening the way for very low-phase noise frequency sources [49].

Carbon-doping of epitaxial SiGe layers as a means to effectively suppress boron out-diffusion during fabrication has rapidly become the preferred approach for commercial SiGe technologies, particularly those above first-generation performance levels. Carbon-doping of SiGe HBTs has its own interesting history, dating back to the serendipitous discovery [50] in 1996 that incorporating small amounts of C into a SiGe epi layer strongly retards (by an order of magnitude) the diffusion of the boron (B) base layer during subsequent thermal cycles. Given that maintaining a thin base profile during fabrication is perhaps the most challenging aspect of building a manufacturable SiGe technology, it is somewhat surprising that it took so long for the general adoption of C-doping as a key technology element. I think it is fair to say that most SiGe practitioners at that time viewed C-doping with more than a small amount of skepticism, given that C can act as a deep trap in Si, and C contamination is generally avoided at all costs in Si epi processes, particularly for minority carrier devices such as the HBT. At the time of the discovery of C-doping of SiGe in 1996, most companies were focused on simply bringing up a SiGe process and qualifying it, relegating the potential use of C to the back burner. In fairness, most felt that C-doping was not necessary to achieve first-generation SiGe HBT performance levels. The lone visionary group to solidly embrace C-doping of SiGe HBTs at the onset was the IHP team in Germany, whose pioneering work eventually paid off and began to convince the skeptics of the merits of C-doping. The minimum required C concentration for effective out-diffusion suppression of B was empirically established to be in the vicinity of 0.1% to 0.2% C (i.e., around $1 \times 10^{20}\ \text{cm}^{-3}$). Early on, much debate ensued on the physical mechanism of *how* C impedes the B diffusion process, but general agreement for the most part now exists and is discussed in Chapter 11 (see *SiGe and Si Strained-Layer Epitaxy for Silicon Heterostructure Devices*). The first high-performance, fully integrated SiGe:C HBT technology was reported in 1999 [54].

The first "high-power" SiGe HBTs (S band, with multiwatt output power) were reported in 1996 using thick collector doping profiles [51,52]. The 10-psec ECL circuit performance barrier was broken in 1997 [53]. The 200-GHz peak f_T performance barrier was broken in November 2001 for a non-self-aligned device [55], and for a self-aligned device in February 2002 [67]. By 2004, a total of six industrial laboratories had achieved 200 GHz performance levels. A SiGe HBT technology with a peak f_T of 350 GHz (375 GHz values were reported in the IEDM presentation) was presented in December 2002 [56], and this 375 GHz f_T value remains a record for room temperature operation (it is above 500 GHz at cryogenic temperatures), and an optimized version with both f_T and f_{max} above 300 GHz was achieved in June 2004 [60]. This combined level of 300+ GHz for both f_T and f_{max} remains a solid record for any Si-based semiconductor device.

Other recent and interesting developments in the SiGe HBT field include the first report of a complementary (npn + pnp) SiGe HBT (C-SiGe) technology in 2003 [57], rapidly followed by a C-SiGe technology with f_T for both the npn and pnp SiGe HBTs above 100 GHz [58]. In addition, a novel vertical npn SiGe HBT has been implemented in thin-film (120 nm) CMOS-compatible SOI [59]. Besides further transistor performance enhancements, other logical developments to anticipate in this field include the integration of SiGe HBTs with strained-Si CMOS for a true all-Si-heterostructure technology.

Not surprisingly, research and development activity involving SiGe HBTs, circuits built from these devices, and various SiGe HBT technologies, in both industry and at universities worldwide, has grown very rapidly since the first demonstration of a functional SiGe HBT in 1987, only 18 years in the past.

2.3 SiGe–Strained Si FETs and Other SiGe Devices

The basic idea of using an electric field to modify the surface properties of materials, and hence construct a "field-effect" device, is remarkably old (1926 and 1935), predating even the quest for a solid-state amplifier [68]. Given the sweeping dominance of CMOS technology in the grand scheme of the electronics industry today, it is ironic that the practical demonstration of the BJT preceded that of the MOSFET by 9 years. This time lag from idea to realization was largely a matter of dealing with the many perils associated with obtaining decent dielectric materials in the Si system—doubly ironic given that Si has such a huge natural advantage over all other semiconductors in this regard. Bread-and-butter notions of ionic contamination, de-ionized water, fixed oxide charge, surface state passivation, and clean-room techniques in semiconductor fabrication had to be learned the hard way. Once device-quality SiO_2 was obtained in the late 1950s, and a robust gate dielectric could thus be fabricated, it was not long until the first functional MOSFET was demonstrated in 1960 [69]. The seemingly trivial (remember, however, that hindsight is 20–20!) connection of n-channel and p-channel MOSFETs to form low-power CMOS in 1963 [70] paved the way (eventually) to the high-volume, low-cost, highly integrated microprocessor, and the enormous variety of computational engines that exist today as a result.

Like their cousin, the SiGe HBT, SiGe–strained Si FETs did not get off the ground until the means for accomplishing the low-temperature growth of Si epitaxy could be realized. Once that occurred in the mid-1980s the field literally exploded. Table 2.3 summarizes the milestones in the evolution of SiGe–strained Si FETs, as well as a veritable menagerie of other electronic and optoelectronic components built from SiGe–strained Si epitaxy.

It was discovered as early as 1971 that direct oxidation of SiGe was a bad idea for building gate dielectrics [71]. Given that gate oxide quality, low-temperature deposited oxides, did not exist in the mid-1980s, the earliest FET demonstrations were modulation-doped, Schottky-gated, FETs, and both n-channel and p-channel SiGe MODFETs were pioneered as early as 1986 using MBE-grown material [72,73]. Before the SiGe MOSFET field got into high gear in the 1990s, a variety of other novel device demonstrations occurred, including: the first SiGe superlattice photodetector [74], the first SiGe Schottky barrier diodes (SBD) in 1988 [75], the first SiGe hole-transport resonant tunneling diode (RTD) in 1988 [76], and the first SiGe bipolar inversion channel FET (BiCFET) in 1989, a now-extinct dinosaur [77]. Meanwhile, early studies using SiGe in conventional CMOS gate stacks to minimize dopant depletion effects and tailor work functions, a fairly common practice in CMOS today, occurred in 1990 [78], and the first SiGe waveguides on Si substrates were produced in 1990 [79].

The first functional SiGe channel pMOSFET was published in 1991, and shortly thereafter, a wide variety of other approaches aimed at obtaining the best SiGe pMOSFETs (see, for instance, Refs. [93–95]). The first electron-transport RTD was demonstrated in 1991 [81], and the first LED in SiGe

TABLE 2.3 Milestones in the Development of SiGe–Strained Si FETs and Other Devices

Historical Event	Year	Ref.
Field effect device concept	1926	[68]
First Si MOSFET	1960	[69]
First Si CMOS	1963	[70]
First oxidation study of SiGe	1971	[71]
First SiGe nMODFET	1986	[72]
First SiGe pMODFET	1986	[73]
First SiGe photodetector	1986	[74]
First SiGe SBD	1988	[75]
First SiGe hole RTD	1988	[76]
First SiGe BiCFET	1989	[77]
First SiGe gate CMOS technology	1990	[78]
First SiGe waveguide	1990	[79]
First SiGe pMOSFET	1991	[80]
First SiGe electron RTD	1991	[81]
First SiGe LED	1991	[82]
First SiGe solar cell	1992	[83]
First a-SiGe phototransistor	1993	[84]
First SiGe pMOSFET on SOI	1993	[85]
First strained Si pMOSFET	1993	[86]
First strained Si nMOSFET	1994	[87]
First SiGe:C pMOSFET	1996	[88]
First SiGe pFET on SOS	1997	[89]
First submicron strained Si MOSFET	1998	[90]
First vertical SiGe pFET	1998	[91]
First strained Si CMOS technology	2002	[92]

also in 1991 (a busy year for our field). In 1992, the first a-SiGe solar cell was discussed [83], and in 1993, the first high-gain a-SiGe phototransistor [84]. The first SiGe pMOSFETs using alternate substrate materials were demonstrated, first in SOI in 1993 [85], and then on sapphire in 1997 [88], the first SiGe:C channel pMOSFET was demonstrated in 1996 [89], and the first vertical SiGe FET was published in 1998 [92].

Because of the desire to use Si-based bandgap engineering to improve not only the p-channel MOSFET, but also the n-channel MOSFET, research in the early- to mid-1990s in the FET field began to focus on strained Si MOSFETs on relaxed SiGe layers, with its consequent improvement in both electron and hole transport properties. This work culminated in the first strained Si pMOSFET in 1993 [87], and the first stained Si nMOSFET in 1994 [88], and remains an intensely active research field today. Key to the eventual success of strained Si CMOS approaches was that significant mobility enhancement could be achieved in both nFETs and pFETs down to very short (sub-micron) gate lengths, and this was first demonstrated in 1998 [90]. Strained Si CMOS at the 90-nm node and below is rapidly becoming mainstream for most serious CMOS companies, and the first commercial 90 nm strained Si CMOS technology platform was demonstrated by Intel in 2002 [91]. At last count, there were upwards of a half-dozen companies (e.g., Texas Instruments and IBM) also rapidly pushing toward 90 nm (and below) strained Si CMOS technologies, utilizing a variety of straining techniques, and thus it would appear that strained Si CMOS will be a mainstream IC technology in the near future, joining SiGe HBT BiCMOS technology. This is clearly outstanding news for our field. The merger of SiGe HBTs with strained Si CMOS would be a near-term logical extension.

References

1. R Braunstein, AR Moore, and F Herman. Intrinsic optical absorption in germanium–silicon alloys. *Physical Review B* 32:1405–1408, 1958.
2. HC Theuerer, JJ Kleimack, HH Loar, and H Christensen. Epitaxial diffused transistors. *Proceedings of the IRE* 48:1642–1643, 1960.
3. BA Joyce and RR Bradley. Epitaxial growth of silicon from the pyrolysis of monosilane on silicon substrates. *Journal of the Electrochemical Society* 110:1235–1240, 1963.
4. JH van der Merwe. Crystal interfaces. Part I. Semi-infinite crystals. *Journal of Applied Physics* 34:117–125, 1963.
5. JH van der Merwe. Crystal interfaces. Part II. Finite overgrowths. *Journal of Applied Physics* 34:123–127, 1963.
6. JW Matthews and AE Blakeslee. Defects in epitaxial multilayers: I. Misfit dislocations in layers. *Journal of Crystal Growth* 27:118–125, 1974.
7. E Kasper, HJ Herzog, and H Kibbel. A one-dimensional SiGe superlattice grown by UHV epitaxy. *Journal of Applied Physics* 8:1541–1548, 1975.
8. JC Bean, TT Sheng, LC Feldman, AT Fiory, and RT Lynch. Pseudomorphic growth of Ge_xSi_{1-x} on silicon by molecular beam epitaxy. *Applied Physics Letters* 44:102–104, 1984.
9. R People and JC Bean. Calculation of critical layer thickness versus lattice mismatch for Ge_xSi_{1-x}/Si strained layer heterostructures. *Applied Physics Letters* 47:322–324, 1985.
10. R People. Indirect bandgap of coherently strained $Si_{1-x}Ge_x$ bulk alloys on $\langle 0\ 0\ 1\rangle$ silicon substrates. *Physical Review B* 32:1405–1408, 1985.
11. DV Lang, R People, JC Bean, and AM Sergent. Measurement of the bandgap of Ge_xSi_{1-x}/Si strained-layer heterostructures. *Applied Physics Letters* 47:1333–1335, 1985.
12. JF Gibbons, CM Gronet, and KE Williams. Limited reaction processing: silicon epitaxy. *Applied Physics Letters* 47:721–723, 1985.
13. G Abstreiter, H Brugger, T Wolf, H Joke, and HJ Kerzog. Strain-induced two-dimensional electron gas in selectively doped Si/Si_xGe_{1-x} superlattices. *Physical Review* 54:2441–2444, 1985.
14. BS Meyerson. Low-temperature silicon epitaxy by ultrahigh vacuum/chemical vapor deposition. *Applied Physics Letters* 48:797–799, 1986.

15. R People and JC Bean. Band alignments of coherently strained Ge_xSi_{1-x}/Si heterostructures on ⟨0 0 1⟩ Ge_ySi_{1-y} substrates. *Applied Physics Letters* 48:538–540, 1986.
16. BS Meyerson, KJ Uram, and FK LeGoues. Cooperative phenomena is silicon/germanium low temperature epitaxy. *Applied Physics Letters* 53:2555–2557, 1988.
17. CA King, JL Hoyt, CM Gronet, JF Gibbons, MP Scott, and J Turner. Si/Si_{1-x}/Ge_x heterojunction bipolar transistors produced by limited reaction processing. *IEEE Electron Device Letters* 10:52–54, 1989.
18. TO Sedgwick, M Berkenbilt, and TS Kuan. Low-temperature selective epitaxial growth of silicon at atmospheric pressure. *Applied Physics Letters* 54:2689–2691, 1989.
19. PJ Wang, FF Fang, BS Meyerson, J Mocera, and B Parker. Two-dimensional hole gas in Si/$Si_{0.85}Ge_{0.15}$ modulation doped heterostructures. *Applied Physics Letters* 54:2701–2703, 1989.
20. P Agnello, TO Sedgwick, MS Goorsky, J Ott, TS Kuan, and G Scilla. Selective growth of silicon–germanium alloys by atmospheric-pressure chemical vapor deposition at low temperatures. *Applied Physics Letters* 59:1479–1481, 1991.
21. T Manku and A Nathan. Lattice mobility of holes in strained and unstrained $Si_{1-x}Ge_x$ alloys. *IEEE Electron Device Letters* 12:704–706, 1991.
22. T Manku and A Nathan. Electron drift mobility model for devices based on unstrained and coherently strained $Si_{1-x}Ge_x$ grown on ⟨0 0 1⟩ silicon subtrate. *IEEE Transactions on Electron Devices* 39:2082–2089, 1992.
23. K Erbel, SS Iyer, S Zollner, JC Tsang, and FK LeGoues. Growth and strain compensation effects in the ternary $Si_{1-x-y}Ge_xC_y$ alloy system. *Applied Physics Letters* 60:3033–3035, 1992.
24. HJ Osten, E Bugiel, and P Zaumseil. Growth of inverse tetragonal distorted SiGe layer on Si(0 0 1) by adding small amounts of carbon. *Applied Physics Letters* 64:3440–3442, 1994.
25. A Fischer, H-J Osten, and H Richter. An equilibrium model for buried SiGe strained layers. *Solid-State Electronics* 44:869–873, 2000.
26. J Bardeen and WH Brattain. The transistor, a semi-conductor triode. *Physical Review* 71:230–231, 1947.
27. W Shockley, M Sparks, and GK Teal. p–n junction transistors. *Physical Review* 83:151–162, 1951.
28. M Tanenbaum and DE Thomas. Diffused emitter and base silicon transistors. *Bell System Technical Journal* 35:23–34, 1956.
29. See, for instance, W Shockley. U.S. Patents 2,502,488, 2,524,035, and 2,569,347.
30. H Kroemer. Zur theorie des diffusions und des drifttransistors. Part III. *Archiv der Elektrischen Ubertragungstechnik* 8:499–504, 1954.
31. H Kroemer. Quasielectric and quasimagnetic fields in nonuniform semiconductors. *RCA Review* 18:332–342, 1957.
32. H Kroemer. Theory of a wide-gap emitter for transistors. *Proceedings of the IRE* 45:1535–1537, 1957.
33. B Brar, GJ Sullivan, and PM Asbeck. Herb's bipolar transistors. *IEEE Transactions on Electron Devices* 48:2473–2476, 2001.
34. RM Warner. Microelectronics: Its unusual origin and personality. *IEEE Transactions on Electron Devices* 48:2457–2467, 2001.
35. SS Iyer, GL Patton, SL Delage, S Tiwari, and J.M.C. Stork. Silicon–germanium base heterojunction bipolar transistors by molecular beam epitaxy. Technical Digest of the IEEE International Electron Devices Meeting, San Francisco, 1987, pp. 874–876.
36. GL Patton, DL Harame, JMC Stork, BS Meyerson, GJ Scilla, and E Ganin. Graded-SiGe-base, poly-emitter heterojunction bipolar transistors. *IEEE Electron Device Letters* 10:534–536, 1989.
37. GL Patton, JH Comfort, BS Meyerson, EF Crabbé, E de Frésart, JMC Stork, JY-C Sun, DL Harame, and J Burghartz. 63-75 GHz f_T SiGe-base heterojunction-bipolar technology. Technical Digest IEEE Symposium on VLSI Technology, Honolulu, 1990, pp. 49–50.
38. GL Patton, JH Comfort, BS Meyerson, EF Crabbé, GJ Scilla, E de Frésart, JMC Stork, JY-C Sun, DL Harame, and J Burghartz. 75 GHz f_T SiGe base heterojunction bipolar transistors. *IEEE Electron Device Letters* 11:171–173, 1990.

39. JH Comfort, GL Patton, JD Cressler, W Lee, EF Crabbé, BS Meyerson, JY-C Sun, JMC Stork, P-F Lu, JN Burghartz, J Warnock, K Kenkins, K-Y Toh, M D'Agostino, and G Scilla. Profile leverage in a self-aligned epitaxial Si or SiGe-base bipolar technology. Technical Digest IEEE International Electron Devices Meeting, Washington, 1990, pp. 21–24.
40. DL Harame, JMC Stork, BS Meyerson, EF Crabbé, GL Patton, GJ Scilla, E de Frésart, AA Bright, C Stanis, AC Megdanis, MP Manny, EJ Petrillo, M Dimeo, RC Mclntosh, and KK Chan. SiGe-base PNP transistors fabrication with n-type UHV/CVD LTE in a "NO DT" process. Technical Digest IEEE Symposium on VLSI Technology, Honolulu, 1990, pp. 47–48.
41. EF Crabbeé, GL Patton, JMC Stork, BS Meyerson, and JY-C Sun. Low temperature operation of Si and SiGe bipolar transistors. Technical Digest IEEE International Electron Devices Meeting, Washington, 1990, pp. 17–20.
42. DL Harame, EF Crabbé, JD Cressler, JH Comfort, JY-C Sun, SR Stiffler, E Kobeda, JN Burghartz, MM Gilbert, J Malinowski, and AJ Dally. A high-performance epitaxial SiGe-base ECL BiCMOS technology. Technical Digest IEEE International Electron Devices Meeting, Washington, 1992, pp. 19–22.
43. DL Harame, JMC Stork, BS Meyerson, KY-J Hsu, J Cotte, KA Jenkins, JD Cressler, P Restle, EF Crabbé, S Subbanna, TE Tice, BW Scharf, and JA Yasaitis. Optimization of SiGe HBT technology for high speed analog and mixed-signal applications. Technical Digest IEEE International Electron Devices Meeting, San Francisco, 1993, pp. 71–74.
44. E Kasper, A Gruhle, and H Kibbel. High speed SiGe-HBT with very low base sheet resistivity. Techncial Digest IEEE International Electron Devices Meeting, San Francisco, 1993, pp. 79–81.
45. EF Crabbé, BS Meyerson, JMC Stork, and DL Harame. Vertical profile optimization of very high frequency epitaxial Si- and SiGe-base bipolar transistors. Technical Digest IEEE International Electron Devices Meeting, Washington, 1993, pp. 83–86.
46. DL Harame, K Schonenberg, M Gilbert, D Nguyen-Ngoc, J Malinowski, S-J Jeng, BS Meyerson, JD Cressler, R Groves, G Berg, K Tallman, K Stein, G Hueckel, C Kermarrec, T Tice, G Fitzgibbons, K Walter, D Colavito, T Houghton, N Greco, T Kebede, B Cunningham, S Subbanna, JH Comfort, and EF Crabbé. A 200 mm SiGe-HBT technology for wireless and mixed-signal applications. Technical Digest IEEE International Electron Devices Meeting, Washington, 1994, pp. 437–440.
47. JD Cressler, EF Crabbé, JH Comfort, JY-C Sun, and JMC Stork. An epitaxial emitter cap SiGe-base bipolar technology for liquid nitrogen temperature operation. *IEEE Electron Device Letters* 15:472–474, 1994.
48. JA Babcock, JD Cressler, LS Vempati, SD Clark, RC Jaeger, and DL Harame. Ionizing radiation tolerance of high performance SiGe HBTs grown by UHV/CVD. *IEEE Transactions on Nuclear Science* 42:1558–1566, 1995.
49. LS Vempati, JD Cressler, RC Jaeger, and DL Harame. Low-frequency noise in UHV/CVD Si- and SiGe-base bipolar transistors. Proceedings of the IEEE Bipolar/BiCMOS Circuits and Technology Meeting, Minnneapolis, 1995, pp. 129–132.
50. L Lanzerotti, A St Amour, CW Liu, JC Sturm, JK Watanabe, and ND Theodore. $Si/Si_{1-x-y}Ge_xC_y/Si$ heterojunction bipolar transistors. *IEEE Electron Device Letters* 17:334–337, 1996.
51. A Schüppen, S Gerlach, H Dietrich, D Wandrei, U Seiler, and U König. 1-W SiGe power HBTs for mobile communications. *IEEE Microwave and Guided Wave Letters* 6:341–343, 1996.
52. PA Potyraj, KJ Petrosky, KD Hobart, FJ Kub, and PE Thompson. A 230-Watt S-band SiGe hetero-junction junction bipolar transistor. *IEEE Transactions on Microwave Theory and Techniques* 44:2392–2397, 1996.
53. K Washio, E Ohue, K Oda, M Tanabe, H Shimamoto, and T Onai. A selective-epitaxial SiGe HBT with SMI electrodes featuring 9.3-ps ECL-Gate Delay. Technical Digest IEEE International Electron Devices Meeting, San Francisco, 1997, pp. 795–798.
54. HJ Osten, D Knoll, B Heinemann, H Rücker, and B Tillack. Carbon doped SiGe heterojunction bipolar transistors for high frequency applications. Proceedings of the IEEE Bipolar/BiCMOS Circuits and Technology Meeting, Minneapolis, 1999, pp. 109–116.

55. SJ Jeng, B Jagannathan, J-S Rieh, J Johnson, KT Schonenberg, D Greenberg, A Stricker, H Chen, M Khater, D Ahlgren, G Freeman, K Stein, and S Subbanna. A 210-GHz f_T SiGe HBT with non-self-aligned structure. *IEEE Electron Device Letters* 22:542–544, 2001.
56. JS Rieh, B Jagannathan, H Chen, KT Schonenberg, D Angell, A Chinthakindi, J Florkey, F Golan, D Greenberg, S-J Jeng, M Khater, F Pagette, C Schnabel, P Smith, A Stricker, K Vaed, R Volant, D Ahlgren, G Freeman, K Stein, and S Subbanna. SiGe HBTs with cut-off frequency of 350 GHz.Technical Digest of the IEEE International Electron Devices Meeting, San Francisco, 2002, pp. 771–774.
57. B El-Kareh, S Balster, W Leitz, P Steinmann, H Yasuda, M Corsi, K Dawoodi, C Dirnecker, P Foglietti, A Haeusler, P Menz, M Ramin, T Scharnagl, M Schiekofer, M Schober, U Schulz, L Swanson, D Tatman, M. Waitschull, JW Weijtmans, and C Willis. A 5V complementary SiGe BiCMOS technology for high-speed precision analog circuits. Proceedings of the IEEE Bipolar/BiCMOS Circuits and Technology Meeting, Toulouse, 2003, pp. 211–214.
58. B Heinemann, R Barth, D Bolze, J Drews, P Formanek, O Fursenko, M Glante, K Glowatzki, A Gregor, U Haak, W Höppner, D Knoll, R Kurps, S Marschmeyer, S Orlowski, H Rücker, P Schley, D Schmidt, R Scholz, W Winkler, and Y Yamamoto. A complementary BiCMOS technology with high speed npn and pnp SiGe:C HBTs. Technical Digest of the IEEE International Electron Devices Meeting, Washington, 2003, pp. 117–120.
59. J Cai, M Kumar, M Steigerwalt, H Ko, K Schonenberg, K Stein, H Chen, K Jenkins, Q Ouyang, P Oldiges, and T Ning. Vertical SiGe-base bipolar transistors on CMOS-compatible SOI substrate. Proceedings of the IEEE Bipolar/BiCMOS Circuits and Technology Meeting, Toulouse, 2003, pp. 215–218.
60. J-S Rieh, D Greenberg, M Khater, KT Schonenberg, J-J Jeng, F Pagette, T Adam, A Chinthakindi, J Florkey, B Jagannathan, J Johnson, R Krishnasamy, D Sanderson, C Schnabel, P Smith, A Stricker, S Sweeney, K Vaed, T Yanagisawa, D Ahlgren, K Stein, and G Freeman. SiGe HBTs for millimeter-wave applications with simultaneously optimized f_T and f_{max}. Proceedings of the IEEE Radio Frequency Integrated Circuits (RFIC) Symposium, Fort Worth, 2004, pp. 395–398.
61. DL Harame and BS Meyerson. The early history of IBM's SiGe mixed signal technology. *IEEE Transactions on Electron Devices* 48:2555–2567, 2001.
62. T Tatsumi, H Hirayama, and N Aizaki. $Si/Ge_{0.3}Si_{0.7}$ heterojunction bipolar transistor made with Si molecular beam epitaxy. *Applied Physics Letters* 52:895–897, 1988.
63. H Temkin, JC Bean, A Antreasyan, and R Leibenguth. Ge_xSi_{1-x} strained-layer heterostructure bipolar transistors. *Applied Physics Letters* 52:1089–1091, 1988.
64. D-X Xu, G-D Shen, M Willander, W-X Ni, and GV Hansson. n-Si/p-Si_{1-x}/n-Si double-heterojunction bipolar transistors. *Applied Physics Letters* 52:2239–2241, 1988.
65. GL Patton, SS Iyer, SL Delage, S Tiwari, and JMC Stork. Silicon–germanium-base heterojunction bipolar transistors by molecular beam epitaxy. *IEEE Electron Device Letters* 9:165–167, 1988.
66. A Gruhle, H Kibbel, U König, U Erben, and E Kasper. MBE-Grown Si/SiGe HBTs with high β, f_T, and f_{max}. *IEEE Electron Device Letters* 13:206–208, 1992.
67. AJ Joseph, D Coolbaugh, D Harame, G Freeman, S Subbanna, M Doherty, J Dunn, C Dickey, D Greenberg, R Groves, M Meghelli, A Rylyakov, M Sorna, O Schreiber, D Herman, and T Tanji. 0.13 μm 210 GHz f_T SiGe HBTs—expanding the horizons of SiGe BiCMOS. Technical Digest IEEE International Solid-State Circuits Conference, San Francisco, 2002, pp. 180–182.
68. H. Lilienfeld Patent, 1926; O. Heil, British patent number 439,457, 1935.
69. D Khang and MM Atalla. Silicon–silicon dioxide field induced surface devices. Solid State Research Conference, Pittsburgh, 1960.
70. FM Wanlass and CT Sah. Nanowatt logic using field-effect metal-oxide-semiconductor triodes (MOSTs). IEEE International Solid-State Circuits Conference, Philadelphia, 1963, pp. 32–33.
71. P Balk. Surface properties of oxidized germanium-doped silicon. *Journal of the Electrochemical Society* 118:494–495, 1971.
72. H Daembkes, H-J Herzog, H Jorke, H. Kibbel, and E Kasper. The n-channel SiGe/Si modulation doped field-effect transistor. *IEEE Transactions on Electron Devices* 33:633–638, 1986.

73. TP Pearsall and JC Bean. Enhancement and depletion-mode p-channel Ge_xSi_{1-x} modulation-doped field effect transistor. *IEEE Electron Device Letters* 7:308–310, 1986.
74. H Temkin, TP Pearsall, JC Bean, RA Logan, and S. Luryi. Ge_xSi_{1-x} strained-layer superlattice waveguide photodetectors operating near 1.3 μm. *Applied Physics Letters* 48:963–965, 1986.
75. RD Thompson, KN Tu, J Angillelo, S Delage, and SS Iyer. Interfacial reaction between Ni and MBE grown SiGe alloys. *Journal of the Electrochemical Society* 135:3161–3163, 1988.
76. HC Liu, D Landheer, M Buchmann, and DC Houghton. Resonant tunneling diode in the $Si_{1-x}Ge_x$ system. *Applied Physics Letters* 52:1809–1811, 1988.
77. RC Taft, JD Plummer, and SS Iyer. Demonstration of a p-channel BiCFET in the Ge_xSi_{1-x}/Si system. *IEEE Electron Device Letters* 10:14–16, 1989.
78. TJ King, JR Pfriester, JD Scott, JP McVittie, and KC Saraswat. A polycrystalline SiGe gate CMOS technology. Technical Digest of the IEEE International Electron Devices Meeting, Washington, 1990, pp. 253–256.
79. RA Soref, F Namavar, and JP Lorenzo. Optical waveguiding in a single-crystal layer of germanium–silicon grown on silicon. *Optics Letters* 15:270–272, 1990.
80. DK Nayak, JCS Woo, JS Park, KL Wang, and KP MacWilliams. Enhancement-mode quantum-well Ge_xSi_{1-x} PMOS. *IEEE Electron Device Letters* 12:154–156, 1991.
81. K Ismail, BS Meyerson, and PJ Wang. Electron resonant tunneling in Si/SiGe double barrier diodes. *Applied Physics Letters* 59:973–975, 1991.
82. DC Houghton, JP Noel, and NL Rowell. Electroluminescence and photoluminesence from SiGe alloys grown on (1 0 0) silicon by MBE. *Materials Science and Engineering B* 9:237–244, 1991.
83. DS Chen, JP Conde, V Chu, S Aljishi, JZ Liu, and S Wagner. Amorphous silicon–germanium thin-film photodetector array. *IEEE Electron Device Letters* 13:5–7, 1992.
84. S-B Hwang, YK Fang, K-H Chen, C-R Liu, J-D Hwang, and M-H Chou. An a-Si:H/a-Si, Ge:H bulk barrier phototransistor with a-SiC:H barrier enhancement layer for high-gain IR optical detector. *IEEE Transactions on Electron Devices* 40:721–726, 1993.
85. DK Nayak, JCS Woo, GK Yabiku, KP MacWilliams, JS Park, and KL Wang. High mobility GeSi PMOS on SIMOX. *IEEE Electron Device Letters* 14:520–522, 1993.
86. DK Nayak, JCS Woo, JS Park, KL Wang, and KP MacWilliams. High-mobility p-channel metal-oxide semiconductor field-effect transistor on strained Si. *Applied Physics Letters* 62:2853–2855, 1993.
87. J Welser, JL Hoyt, and JF Gibbons. Electron mobility enhancement in strained-Si n-type metal-oxide semiconductor field-effect transistors. *IEEE Electron Device Letters* 15:100–102, 1994.
88. SK Ray, S John, S Oswal, and SK Banerjee. Novel SiGeC channel heterojunction pMOSFET. Technical Digest of the IEEE International Electron Devices Meeting, Washington, 1996, pp. 261–264.
89. SJ Mathew, WE Ansley, WB Dubbelday, JD Cressler, JA Ott, JO Chu, PM Mooney, KL Vavanagh, BS Meyerson, and I Lagnado. Effect of Ge profile on the frequency response of a SiGe pFET on sapphire technology. Technical Digest of the IEEE Device Research Conference, Boulder, 1997, pp. 130–131.
90. K Rim, JL Hoyt, and JF Gibbons. Transconductance enhancement in deep submicron strained-Si n-MOSFETs. Technical Digest of the IEEE International Electron Devices Meeting, Washington, 1998, pp. 707–710.
91. KC Liu, SK Ray, SK Oswal, and SK Banerjee. $Si_{1-x}Ge_x$/Si vertical pMOSFET fabricated by Ge ion implantation. *IEEE Electron Device Letters* 19:13–15, 1998.
92. S Thompson, N. Anand, M Armstrong, C Auth, B Arcot, M Alavi, P Bai, J Bielefeld, R Bigwood, J Brandenburg, M Buehler, S Cea, V Chikarmane, C Choi, R Frankovic, T Ghani, G Glass, W Han, T Hoffmann, M Hussein, P Jacob, A Jain, C Jan, S Joshi, C Kenyon, J Klaus, S Klopcic, J Luce, Z Ma, B McIntyre, K Mistry, A Murthy, P Nguyen, H Pearson, T Sandford, R Schweinfurth, R Shaheed, S Sivakumar, M Taylor, B Tufts, C Wallace, P Wang, C Weber, and M Bohr. A 90 nm logic technology featuring 50 nm strained silicon channel transistors, 7 layers of Cu interconnects, low k ILD, and 1 μm^2 SRAM Cell. Technical Digest of the IEEE International Electron Devices Meeting, Washington, 2002, pp. 61–64.

93. VP Kesan, S Subbanna, PJ Restle, MJ Tejwani, JM Aitken, SS Iyer, and JA Ott. High performance 0.25 μm p-MOSFETs with silicon–germanium channels for 300 K and 77 K operation. Technical Digest of the IEEE International Electron Devices Meeting, San Francisco, 1991, pp. 25–28, 1991.
94. S Verdonckt-Vanderbroek, E Crabbé, BS Meyerson, DL Harame, PJ Restle, JMC Stork, AC Megdanis, CL Stanis, AA Bright, GMW Kroesen, and AC Warren. High-mobility modulation-doped, graded SiGe-channel p-MOSFETs. *IEEE Electron Device Letters* 12:447–449, 1991.
95. S Verdonckt-Vanderbroek, E Crabbé, BS Meyerson, DL Harame, PJ Restle, JMC Stork, and JB Johnson. SiGe-channel heterojunction p-MOSFETs. *IEEE Transactions on Electron Devices* 41:90–102, 1994.

3

Overview: Circuits and Applications

John D. Cressler
Georgia Institute of Technology

One of the unique merits of this book lies in its extreme breadth. What begins with materials, must not end with devices, but rather must also span the circuit and system application space. This is particularly so in our field because its success is intimately tied to the uncanny ability of Si-based devices to be integrated, enabling the construction of large systems from many diverse components in a very small space at low cost. This final section provides coverage of this application space; the real world, if you will. In Chapter 4, "SiGe as an Enabler of Wireless Communications Systems," L. Larson of the University of California at San Diego gives a broad view of the merits of SiGe for emerging wireless communications systems. In Chapter 5, "LNA Optimization Strategies," by Q. Liang of Georgia Tech, new techniques for improved LNA design are addressed, and Chapter 6, "Linearization Techniques," by L. de Vreede of the Delft University of Technology, presents a comprehensive view of linearization techniques in devices and circuits. The next three chapters span the RF to millimeter-wave IC space, beginning with Chapter 7, "SiGe MMICs," by H. Schumacher of the University of Ulm, and then moving up in frequency in Chapter 8, "SiGe Millimeter-Wave ICs," by J. Luy of DaimlerChrysler, and then down in frequency in Chapter 9, "Wireless Building Blocks Using SiGe HBTs," by J. Long of the Delft University of Technology. New wireless radio architectures are covered in Chapter 10, "Direct Conversion Architectures for SiGe Radios," by S. Chakraborty of Georgia Tech. MEMS processing represents an important emerging area in silicon fabrication and applications, and is addressed in Chapter 11, "RF MEMS Techniques in Si/SiGe," by J. Papapolymerou of Georgia Tech. Future system integration approaches necessarily require robust packaging techniques and even on-board antennae for signal transmission, as addressed in Chapter 12, "Wideband Antennas on Silicon," by M. Tentzeris of Georgia Tech and Chapter 13, "Packaging Issues for SiGe Circuits," by K. Lim of Georgia Tech. Finally, in the last three chapters, we take a snapshot of the state-of-the-art in the IC application space. While by definition this view holds only for 2005, a blink of the eye in this dynamic field, it nonetheless provides a nice glimpse of the future, as envisioned by several industry leaders: Chapter 14, "Industry Examples at the State-of-the-Art: IBM," by D. Friedman of IBM Research; Chapter 15, "Industry Examples at the State-of-the-Art: Hitachi," by K. Washio of Hitachi, and Chapter 16, "Industry Examples at the State-of-the-Art: ST Microelectronics," by D. Belot of ST Microlectronics.

4

SiGe as an Enabler for Wireless Communications Systems*

4.1 Introduction 4-1
4.2 Architectural Design Considerations for Wireless Receivers and Transmitters 4-3
4.3 SiGe HBT Transistor Performance for Wireless Transceivers 4-8
4.4 SiGe HBT-Based RF Circuit Considerations 4-12
Si–SiGe HBT Low-Noise Amplifier Fundamentals • Si–SiGe HBT Voltage-Controlled Oscillators and Frequency Synthesizers
4.5 Examples of SiGe HBT-Based RF Wireless Transceivers 4-15
4.6 Summary 4-19

Lawrence E. Larson
University of California at San Diego

Donald Y.C. Lie
Dynamic Research Corporation and University of California at San Diego

4.1 Introduction

The desire to communicate quickly and reliably with our family, friends, and colleagues is one of the most widespread of human needs, and wireless telephony has exploded in the last decade as a ubiquitous tool to fulfill that desire. Over 400 million cellular handsets were sold in the years 2002 and 2003, and the market is expected to grow to nearly a billion phones per year within the next decade. At the same time, the market for the now-ubiquitous wireless local area network 802.11a/b/g is expected to exceed 100 million per year. The cellular telephone and wireless LAN card have become so common and widespread, such an integral part of modern existence, that it is easy to forget that they were considered to be expensive novelties just 15 years ago.

This revolution in communications has resulted from the confluence of a variety of technological factors: advances in communications theory, networking architectures, semiconductor technology, and transceiver design. The wireless "revolution" would not have happened without the advances in each of these areas, and no one technology can plausibly lay claim to be the dominant technology driving us forward. However, the combination of stunning advances in semiconductor technology, e.g., Moore's law, combined with improved approaches to transceiver design, has enabled the size, cost, and battery life of the wireless transceiver to be shrunk to that of a typical consumer item, within plausible reach of

*Portions of this chapter first appeared in the *IEEE Transactions on Electron Devices* and *International Journal on Wireless and Optical Communications*.

half the population on the planet. It is our contention that SiGe BiCMOS technology is an ideal candidate for implementation of these advanced wireless devices of the future.

The communications medium that a wireless transceiver typically finds itself in is often referred to as "hostile," since the path—or channel—from the transmitter to the receiver is subject to time-varying obstructions and multipath fading, as well as Doppler effects. This is in contrast to "point-to-point" communications links—either wireless, fiber-optic or free-space optical—where the channel is essentially nontime-varying or "stationary." This hostile channel affects both the design of the transmitter and receiver in profound ways, and the next section will summarize some of the key challenges associated with the wireless transceiver, as well as some of the new approaches that are developed to produce fully monolithic versions of wireless transceivers.

A related challenge for next generation wireless devices is the necessity to deliver multiband as well as multistandard functionality in the future. A typical functional block diagram of the electronics of a cellular telephone that might accomplish this is shown in Figure 4.1, where a next generation European handset might support GSM (at 900 MHz) for voice applications, as well as W-CDMA (at 2000 MHz) for data applications, Bluetooth (at 2.4 GHz) for piconet applications, GPS (at 1.5 GHz) for position location, and 802.11 (at 2.4 GHz) for wireless local area network functionality. Of course this seems impractical today, but it is expected to be well within the reach of semiconductor and system technology within the next few years.

In a manner similar to what Moore's law has achieved for digital integrated circuits, we expect that RF microwave circuits will benefit from the same lithographic scaling advances, and the entire industry can significantly reduce the costs and the form factors of communication products by achieving a higher level of RF system integration on a single integrated circuit. However, the RF portion of a typical high-performance wireless communication system remains a mixture of devices made with different technologies. For example, III–V compound (GaAs or InP-based) low-noise amplifiers (LNAs), switches, or power amplifiers (PAs) are often used in cellular telephones or wireless LAN products today, along with

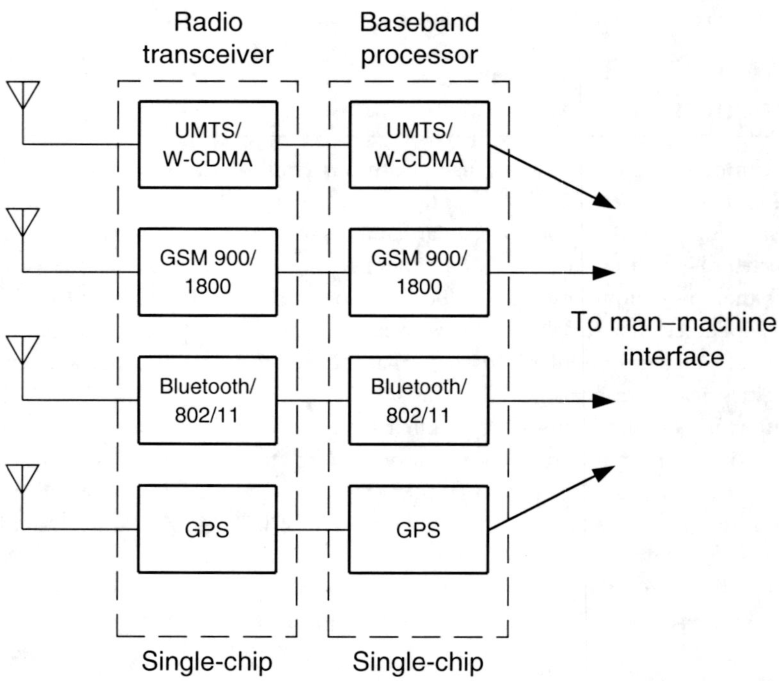

FIGURE 4.1 Multimode and multiband cellular handset architecture. These features will be common in future handset designs.

lower-cost silicon components for other functions, preventing the high integration levels and ultralow cost of a "single-chip radio."

This is due in large part to the differing performance between III–V-based and Si-based transistors, where the superior electron transport and breakdown characteristics provide the III–V-based devices with superior performance. While GaAs FETs, PHEMTs, and heterojunction bipolar transistors (HBTs) can be inexpensive and highly efficient at gigahertz frequencies and cellular handset power levels, they offer a limited ability to integrate with CMOS baseband chips. Although CMOS technology has achieved impressive levels of RF integration lately—especially for cost-sensitive applications like Bluetooth—its performance lags behind that of III–V-based devices and circuits.

As a result, Si–SiGe heterostructures—and specifically Si–SiGe HBTs—are under extensive investigation, since they can provide nearly III–V levels of performance with the low cost of a Si-based technology [1]. HBTs that utilize Si–SiGe heterolayers extended the high-frequency limit of Si-based bipolar technology to cutoff frequencies f_T well above 200 GHz, a frequency range that has been historically dominated by GaAs-based devices [2,3]. In addition, fabrication processes for Si–SiGe devices are compatible with those routinely used for CMOS ICs, enabling the manufacture of Si–SiGe BiCMOS technology. When combined with the state-of-the-art digital CMOS devices, the SiGe BiCMOS technology offers a unique suite of devices to enable dramatic improvements in RF–analog–mixed-signal IC integration. This technology truly has become an "enabler" for single-chip implementation of high-performance wireless communications systems.

4.2 Architectural Design Considerations for Wireless Receivers and Transmitters

The standard receiver architecture for wireless systems—the venerable heterodyne—is shown in Figure 4.2. Since its initial development by Edwin Armstrong in the 1910s, the heterodyne architecture has remained the preferred approach for the implementation of the vast majority of radio-frequency applications in the world. Its perennial popularity is due to its ability to reproducibly pick out

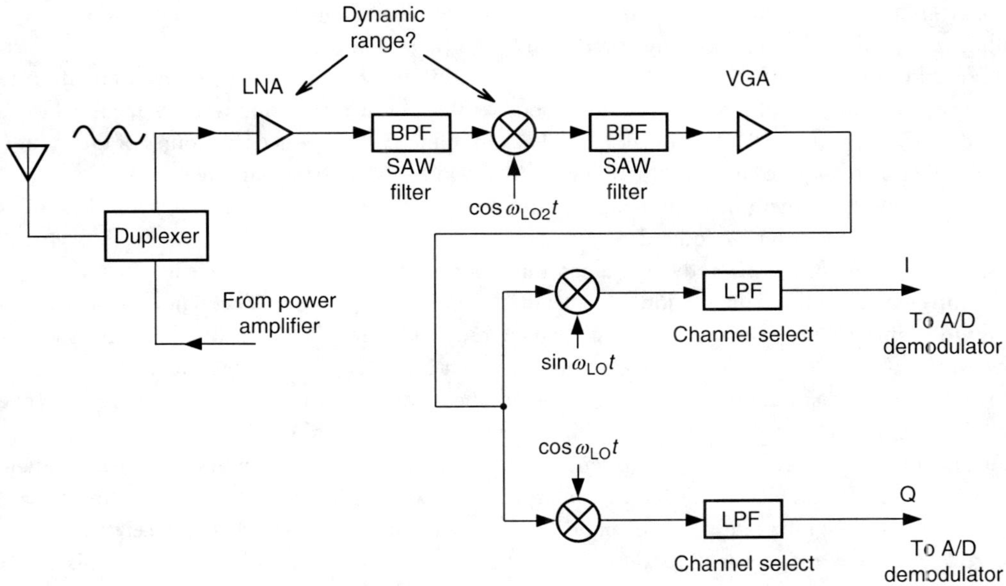

FIGURE 4.2 Typical heterodyne receiver used in cellular handsets.

narrow-bandwidth high-frequency signals from the surrounding background clutter of signals outside the frequency range of interest.

One way to view these constraints of background clutter is through the "football field" metaphor, first proposed by Paul C. Davis of Bell Laboratories [4]. The popular European GSM system has a minimum received signal sensitivity level (the smallest level of the desired received signal) at the output of the antenna of -102 dB m ($10^{-13.2}$ W), but the largest interferer also received by the antenna has a level of 0 dB m (10^{-3} W). If you imagine the desired signal power to be normalized to the head of a pin, roughly 1 mm in diameter, then the largest interferer is roughly the size of two football fields 100 m $\times$ 100 m. Receiving a GSM signal is analogous to the problem finding the head of a pin in a football field without being able to actually going onto the field to look for it. In addition, this has to be accomplished in less than 100 msec, which is typically the time it takes for the cellular handset to receive a call. Viewed through this lens, the modern cellular handset is truly a technological marvel.

The heterodyne receiver accomplishes this through a combination of filtering and careful frequency planning. In the heterodyne receiver, the radio signal is sent from the receiving antenna to an LNA, whose purpose is to boost the signal level without reducing the signal-to-noise ratio significantly. The signal level at the antenna can range between 1 μV rms and nearly 100 mV rms—over a 100-dB variation. At the low end of the signal range, the LNA performance is fundamentally limited by thermodynamic issues, while at the high end of the signal range, the challenge is to minimize the effects of nonlinearities on receiver performance. These diverse requirements are often referred to as the "LNA Bottleneck" [5]. As a result, the high-frequency LNA must exhibit excellent performance over both small-signal and large-signal conditions.

Following the LNA, the signal is typically passed through a mixer, which essentially multiplies the input signal by a local oscillator signal of constant frequency, producing an output signal whose frequency is the difference between the two inputs—the so-called intermediate frequency (IF)—and whose amplitude is proportional to the original input signal. Preceding the mixer, an analog filter eliminates the response to an undesired input signal at $(2f_{lo} - f_{rf})$ that would also downconvert to the intermediate frequency. This image-reject filter is typically implemented with a physically large surface acoustic wave (SAW) filter. In addition to their size, these filters have extremely unforgiving sensitivities to variations in source impedance, ground loops, etc. The dilemma of image rejection and its elimination in heterodyne receivers is one of the fundamental limitations on performance and power reduction in radio-frequency systems. A highly integrated transceiver will allow for these filters to be dispensed with, significantly reducing power dissipation and physical size in the transceiver.

A second limitation of traditional frequency translating mixers and the heterodyne architecture is their sensitivity to a menagerie of spurious responses that result from nonlinearities in the amplifiers preceding the mixer, as well as the mixer itself. These nonlinearities produce harmonics of the input and local oscillator frequencies that can themselves mix down to the intermediate frequency. The potential range of frequencies where this unfortunate set of circumstances can occur is nearly limitless, so very high linearity in the mixer is required.

The architecture of the heterodyne also has a number of problems, which make it very poorly suited for completely monolithic integration—the key to lower power operation. The major problems are the ubiquitous image and spurious responses, which must be carefully controlled through bulky and expensive off-chip filters. These filters represent the major impediment to raising the level of integration of wireless radios, since they cannot be easily implemented monolithically. Therefore, alternative architectures that do not suffer from these limitations are actively explored.

Substantial progress has been made recently in the area of direct downconversion—or homodyne—approaches for wireless receivers, which also eliminate the need for image rejection filters, and are better suited to monolithic integration. A schematic diagram of a typical direct conversion receiver is shown in Figure 4.3. An excellent review of recent research in this field is presented in Refs. [6,7]. In this case, the IF is at dc, and the in-phase and quadrature (I and Q) paths of the mixer contain the positive and negative frequency components of the desired signal. The advantages of this particular architecture are

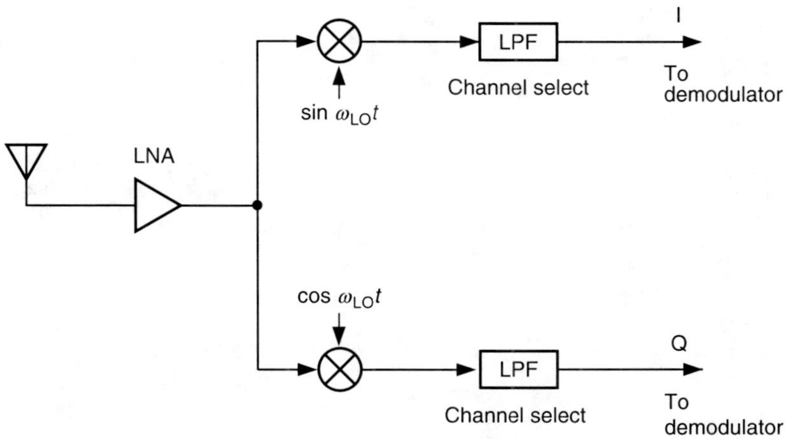

FIGURE 4.3 The direct downconversion—or homodyne—receiver. The absence of the image response makes this approach particularly attractive for monolithic integration.

that it is uniquely well suited to monolithic integration, due to its lack of complex filtering, and its intrinsically simple architecture.

However, although it is actively researched, the direct conversion receiver has not gained completely widespread acceptance to date, especially in high-performance wireless receivers, due to its intrinsic sensitivity to dc offset problems, even-order harmonics of the input signal that interfere with the desired signal, and local-oscillator leakage problems back to the antenna. These issues are all actively pursued by a variety of worldwide research groups, and it is anticipated that they will gradually become solved with further design maturity. The LO feedthrough problem has been addressed through the use of sub-harmonic mixer approaches—a technique borrowed from millimeter-wave radio astronomy—where the mixer is driven at half the desired frequency [8]. This reduces the problem of LO feedthrough and frequency "pulling" of the local oscillator at the expense of a higher local oscillator drive power.

The limitations of the direct conversion approach can once again be seen by examining a specific example—in this case in the GSM environment. In this case, the signal level at the antenna at the minimum sensitivity level is roughly 1 μV rms. The maximum gain from the antenna to the low-frequency output of the mixer is at most 20 dB, due to dynamic range considerations. So, the remaining low-frequency gain in the system, from the output of the mixer to the input of the A–D converters, must be in the order of 80 dB (10,000). This presents two problems. First, any small offset voltage in the subsequent amplifiers will saturate the remaining stages; an input offset voltage of 1 mV will be amplified to 10 V. Second, the input referred noise voltage of the amplifier itself must be in the order of 5 nV/Hz$^{1/2}$ or less. This is a challenging goal for a low-current baseband amplifier. The residual dc offset problem is the largest concern in most practical systems, and it is typically addressed through a variety of background calibration algorithms.

Despite these inherent problems, the homodyne architecture has become very common in GSM handsets recently. For example, recent GSM phones—such as the Ericsson A1018 and T28—have employed the direct conversion approach very successfully, and the Ericsson R520m triband GSM/GPRS phone employs the homodyne approach in a multimode multiband architecture [9].

Transmitter architectures are also the focus of intense international development. The main role of the transmitter section is to convert the digitally generated baseband signals—typically in-phase (I) and quadrature (Q)—to amplitude and phase modulated RF signals superimposed on a fixed carrier frequency. The classic approach for this process is known as the indirect upconversion approach, which is shown in Figure 4.4. A fixed intermediate frequency (or IF) is modulated with the I/Q baseband signal. The resulting signal is filtered and then further upconverted with a second mixer to the desired frequency. In CDMA-oriented systems—such as IS95 or W-CDMA—the power level coming out of the

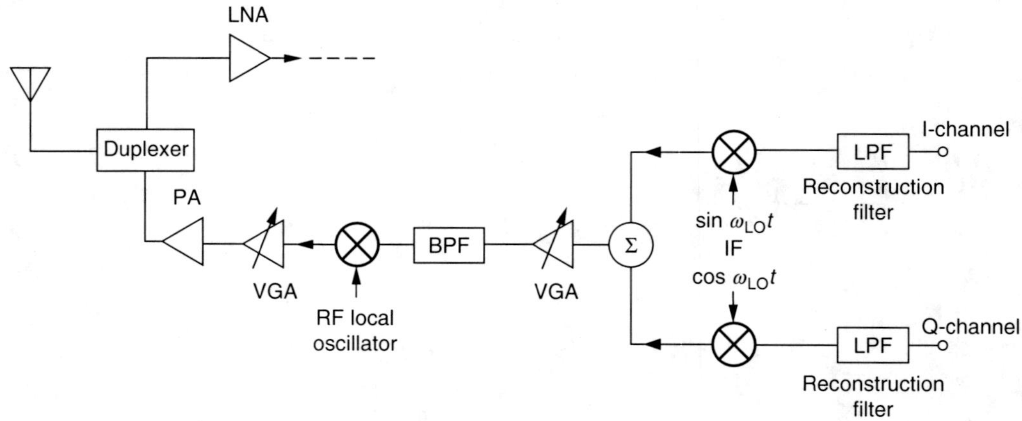

FIGURE 4.4 Indirect upconversion transmitter architecture. This is the most common approach for transmitters requiring both amplitude and phase modulation of the carrier.

antenna must be strictly controlled over a very wide range (roughly 80 dB) to prevent the "near–far" problem common to direct sequence spread spectrum multiuser systems.

The indirect approach has several advantages, which have made it popular for a practical system. The variation in gain can be at both the IF and final RF, so the gain can be distributed throughout the upconversion chain without concern for isolation issues. The IF is at a relatively low frequency, so the first I/Q modulator can be made very accurate, which improves the performance of the modulator and reduces the dc current consumption. Also, since the first IF is at a fixed frequency, a filter with a sharp rolloff frequency can be used to eliminate any spurious noise resulting from the DAC output driving the I/Q modulator.

Despite its intrinsic advantages, there are several disadvantages with the indirect upconversion approach. These include the need for two frequency synthesizers to generate the necessary frequencies, and the fact that the final upconversion mixer generates an output at two frequencies (LO + IF and LO − IF). Since only one of these two frequencies is desired, the mixer has to dissipate extra dc power to accommodate the extra (unwanted) signal. Finally, harmonics of the LO and IF frequencies (nLO ± mIF) can multiply (or intermodulate) together to create a spurious in-band signal. This problem can be minimized through a careful choice of the IF and LO frequencies.

The direct upconversion approach is an alternative to the indirect approach and is shown in Figure 4.5. Its simplicity is reminiscent of the homodyne receiver approach—the I/Q baseband signal is directly upconverted to the RF. The main advantage here is the elimination of the second synthesizer, and the attendant spurious responses this and the first upconversion mixer create. However, now there are two high-power mixers operating at the high upconversion frequency (instead of one) and the accuracy of the I/Q modulator at the higher frequency is greatly diminished. Furthermore, the local oscillator frequency and the transmit frequency are now one and the same, so the output of the PA can alter—or "pull"—the local oscillator with the same I/Q modulation as the transmitted signal. This problem can be eliminated by utilizing a subharmonic mixer, as discussed previously, or using a local oscillator at twice the desired frequency, and then dividing the resulting LO frequency by two on-chip. This eliminates the frequency pulling problem, although several other problems remain with the approach.

One remaining problem with the direct upconversion approach is that the 80-dB output power variation must now be achieved by varying the gain either at the high RF frequency or at low frequencies in the separate I/Q paths. At microwave frequencies, on-chip isolation is typically 40 to 60 dB at best, so achieving a full 80-dB change in gain entirely at high frequency becomes very difficult. At the same time, it is difficult to vary the gain of the circuit at baseband, since any I/Q gain mismatch at lower frequencies

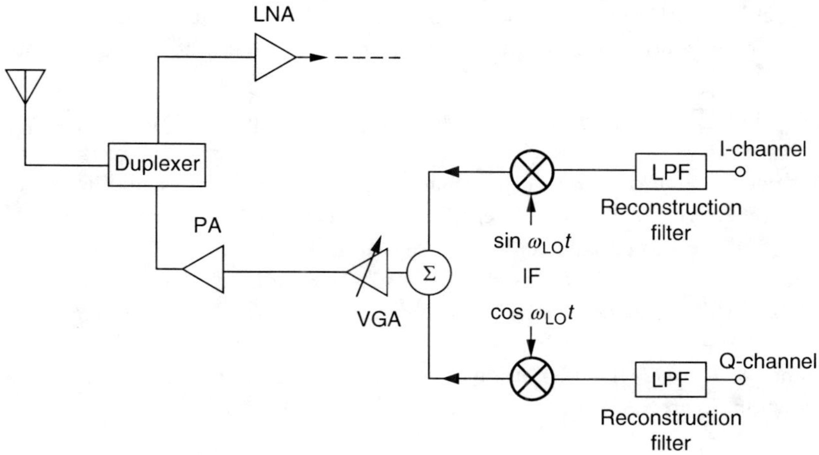

FIGURE 4.5 Direct upconversion architecture. In this case, the baseband signal is directly converted to RF.

translates directly into errors in the transmitted signal, and the effects of local oscillator feedthrough on output power become more pronounced at the low gain levels. The problems of direct upconversion systems remain a challenge, and they are the subject of much active worldwide research.

This discussion of transmitter and receiver architectures naturally leads us to considerations of the interactions between the transmitter and receiver of a typical wireless handset. These issues will become increasingly important in the future, as multimode and multiband handsets become more prevalent. As an example, a multimode 2G/3G GSM/W-CDMA handset will have to contend with differing frequencies, bandwidths, modulation formats, and access methods. Note that there are several possibilities for compromised performance with this proposed system and frequency plan. As an example, consider the case where the GSM portion of the receiver is "on," listening to a signal at 959.9 MHz, the W-CDMA receiver is "off," and the GSM receiver uses the direct conversion approach discussed previously. Furthermore, there is a nearby W-CDMA handset transmitting at 1922.5 MHz. In this case, the unlucky GSM receiver would "receive" the nearby transmitted W-CDMA signal, the mixer would mix it with the second harmonic of the GSM local oscillator (at 1919.8 MHz), and downconvert the resulting spurious signal to 3 MHz, partially corrupting the desired signal that is also in the same frequency range. Although this chain of events seems implausible at first glance, the very small level of the desired signal (less than 100 dB m) and the large value of the potential interferer (up to 0 dB m) make this scenario an unfortunate reality. Considerations of co-interference of different frequency plans must be carefully considered as we move into a more complex environment in the future.

The reference frequencies required for upconversion and downconversion of the transmitted and received signals are generated by a frequency synthesizer, which uses a precise reference (usually produced by crystal oscillator) to synthesize the necessary local oscillator frequencies. In this case, the phase noise of the synthesized signal must be as low as possible to accurately modulate and demodulate the signal. Furthermore, the synthesizer itself is a complex RF–analog–digital circuit, which generates copious amounts of digital switching noise and harmonics. Historically, the synthesizer circuit was contained on a separate integrated circuit but, with system-on-chip implementations, this noise must be isolated from the sensitive receiver circuits despite the fact that they share a common substrate and package environment. This presents a fundamental challenge to the integration level of these complex circuits.

The digital portion of the communication system performs the key functions of modulation and demodulation (the so-called "modem"), carrier recovery, timing recovery, symbol recovery, equalization, channel coding, power detection, and calibration, among others. Separate digital controllers also perform media access control (MAC) functions as well as a variety of other control functions.

The eventual goal is to include all these digital functions on the same integrated circuit substrate as the RF and analog circuits in order to realize a true "single-chip" communications system implementation.

4.3 SiGe HBT Transistor Performance for Wireless Transceivers

The key active device parameters for enhanced circuit performance of noise and linearity for most RF applications are the short-circuit unity current gain frequency (f_T) and the maximum unity power gain frequency (f_{MAX}). These two parameters have made astonishing progress in recent years in SiGe HBT technology, with recently reported values in excess of 200 GHz [10]. The next most important issue is breakdown voltage which, together with noise considerations, sets the dynamic range limitation of most circuits.

If we examine the Si–SiGe HBT first, using the physical cross section and equivalent circuit model of the device shown in Figure 4.6, the f_T is given by

$$\frac{1}{2\pi f_T} = \tau_B + \tau_C + \frac{kT}{q}(C_{je} + C_{cb}) + (R_{ex} + R_c)C_{cb} \qquad (4.1)$$

where R_{ex} and R_c are the parasitic emitter and collector resistances, C_{cb} is the collector–base junction capacitance, C_{je} is the emitter–base junction capacitance, τ_B is the base transit time, and τ_C is the collector transit time.

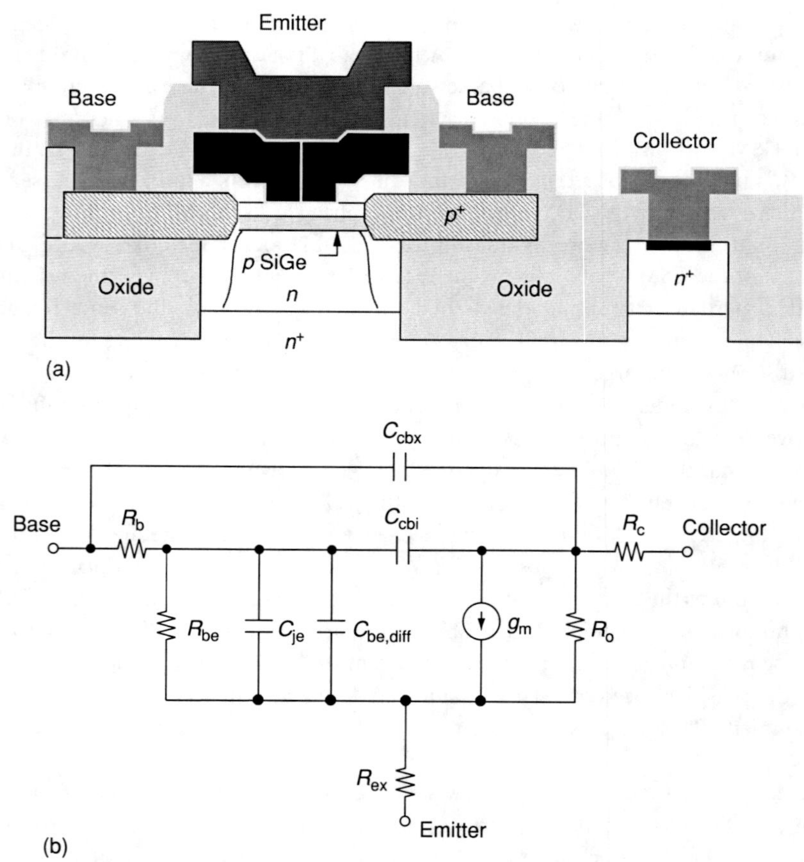

FIGURE 4.6 Si–SiGe HBT. (a) Cross section of device. (b) Equivalent circuit model of transistor.

In most high-frequency applications, the base and collector transit times dominate the f_T, and the other parasitic-related terms have a secondary effect. For this same physical structure and equivalent circuit model, the f_{MAX} of the transistor is given by [11]

$$f_{MAX} \approx \sqrt{\frac{f_T}{8\pi\tau_{cb}}} \tag{4.2}$$

where τ_{CB} is approximately $R_b C_{cb}$ but can be more accurately described as a weighted average of the distributed base resistance and base-collector capacitance.

These expressions highlight the critical role of vertical scaling to improve the f_T and f_{MAX} for bipolar device performance. At the same time lateral scaling of the devices is equally critical, to further reduce extrinsic base resistance and collector–base capacitance. Most scaling efforts with HBT structures aim to keep the f_{MAX} equal to or slightly larger than the f_T.

The dependence of transistor f_T and f_{MAX} on base width can be seen clearly from the plots of measured devices in Figure 4.7, where the clear dependence of transit time on base width has a significant effect on f_T [12]. The effect of base width on f_{MAX} is less pronounced, due to the additional necessity to keep base resistance equally low.

The other absolutely key issue for RF applications of scaled transistors is the breakdown voltage of the device, which influences the dynamic range of operation. The breakdown voltage of a transistor is mostly an issue for the implementation of PAs in the transmitter section, although other circuit areas can benefit from a high breakdown voltage as well. The breakdown voltage issue is detemined by the physics of the device at high electric fields, the varied physical mechanisms that lead to device failure, and the interaction of the breakdown mechanisms with the external circuit.

The bipolar device is fundamentally limited by avalanche multiplication in the collector–base region. This breakdown effect is traded-off against the increasing f_T of the transistor, and the $BV \cdot f_T$ product is the key consideration for most high-frequency applications and is a material-related constant known as the Johnson limit [14]. In the bipolar device, the collector–base junction typically experiences avalanche breakdown first, and the device can be characterized by the collector–emitter breakdown voltage when the emitter is open-circuited (BVCBO) or when the base is open-circuited (BVCEO). The former is usually larger than the latter, due to current gain in the emitter–base region, and can be approximated by

$$BV_{CEO} \approx \frac{BV_{CBO}}{\beta^{1/n}} \tag{4.3}$$

where β is the dc current gain of the transistor and n is a constant that varies from 2 to 5, depending on a variety of physical factors. When the devices have very shallow doping (as in the high f_T case), the transistors exhibit nonlocal avalanche, and the $BV \cdot f_T$ of the device can exceed its value seen for lower frequency devices [15]. Figure 4.8 plots the BVCEO and BVCBO for modern bipolar devices, and the effects of nonlocal avalanching on breakdown voltage can clearly be seen at the higher f_T values, where the breakdown voltage does not change significantly as the f_T increases. In the operation of a PA circuit, the device can typically operate at peak voltages in excess of BVCEO, but less than BVCBO, due to the time-dependent nature of the carrier multiplication process [16] and the impedances presented at each terminal.

This last issue of terminal impedances is crucial in the operation bipolar devices for PAs, since the current gain at the emitter–base junction influences the breakdown characteristics. The collector–base avalanche current can be modeled by

$$i_{AV} = C_{av} v_{CB}^{m} i_C' \tag{4.4}$$

where C_{av} is a technology-dependent avalanche breakdown constant.

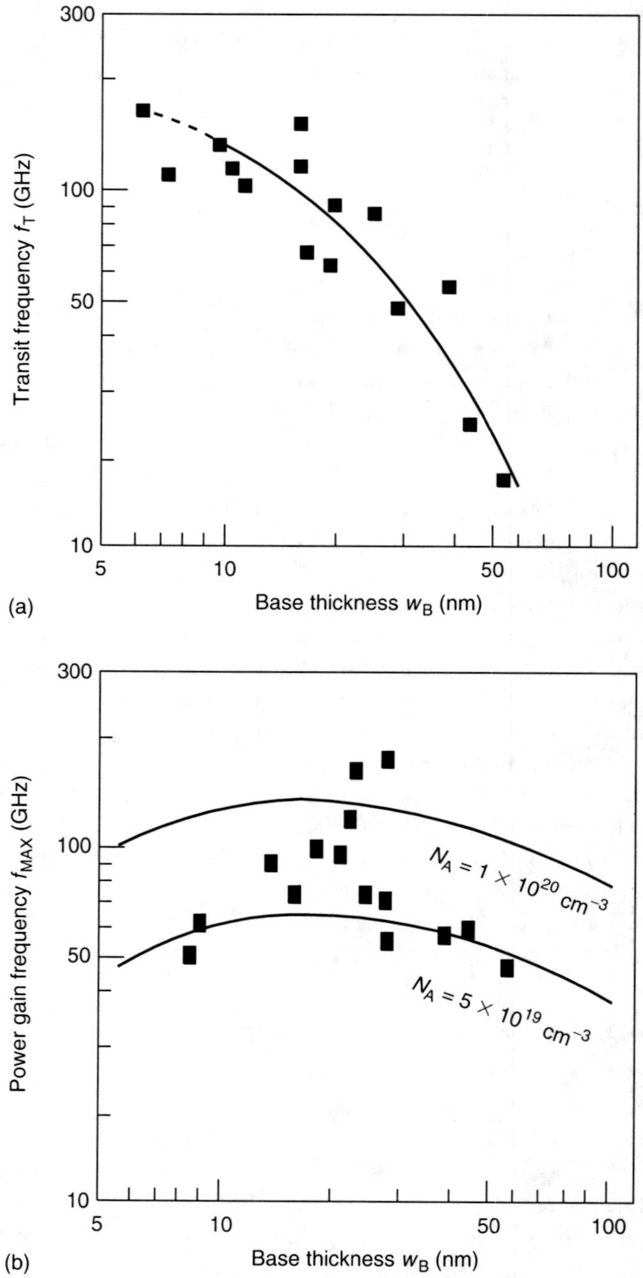

FIGURE 4.7 Si–SiGe HBT speed as a function of base width. (From U. Konig, A. Gruhle, and A. Schuppen. Proc. 1995 IEEE GaAs IC Symposium, pp. 14–18. With permission). (a) The f_T demonstrates a clear base width dependence, and (b) the device f_{MAX} is also affected by the base resistance but the improvement with decreasing base thickness is much less pronounced. (From L. Larson. *IEEE Trans. Electron Dev.* 50(3), 2003, 683–699. With permission.)

The transistor exhibits breakdown when

$$\partial i_C' / \partial v_{CE} \rightarrow \infty \tag{4.5}$$

which can be rewritten as

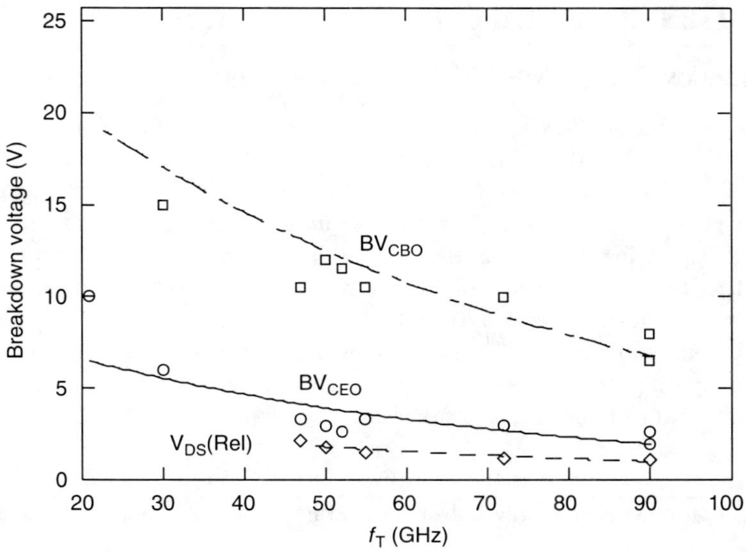

FIGURE 4.8 Comparison of voltage limitations of MOSFETS and HBTs as a function of f_T. (From L. Larson. *IEEE Trans. Electron Dev.* 50(3), 2003, 683–699. With permission.) The V_{DS}(Rel) of the MOSFET is the recommended operating voltage to minimize long-term degradation of the transistor. The Si–SiGe HBT BVCEO and BVCBO maintain a roughly 1:3 relationship from 20 to 90 GHz.

$$\frac{\partial i_C'}{\partial v_{CE}} = \frac{\partial i_C'}{\partial v_B}\frac{\partial v_B}{\partial i_{AV}}\frac{\partial i_{AV}}{\partial v_{CE}} \tag{4.6a}$$

$$= \left(\frac{g_m}{1+g_m r_e}\right)\left(r_b \,\middle\|\, \frac{\beta(1+g_m r_e)}{g_m}\right)\frac{\partial i_{AV}}{\partial v_{CE}} \tag{4.6b}$$

$$\approx g_m' r_{in}\frac{\partial i_{AV}}{\partial v_{CB}} \tag{4.6c}$$

where g_m' is the effective transconductance of the device (including the feedback effects of any extrinsic emitter impedance) and r_{in} is the input impedance consisting of the parallel combination of the extrinsic source impedance (including the base resistance R_b) and the input impedance. In the limiting case of a low-source impedance, r_{in} is simply the transistor base resistance R_b.

Then Equation 4.6a to Equation 4.6c result in

$$BV_{CBO} = \left(\frac{1}{C_{av}g_m' r_b}\right)^{1/m} \tag{4.7}$$

which illustrates the dependence of breakdown voltage on base resistance; as the base resistance increases, the internal feedback shunts more and more of the avalanche current to the emitter, increasing the positive feedback that leads to breakdown.

In the limit of a high source impedance (BVCEO), r_{in} increases to approximately β/g_m and

$$BV_{CEO} \approx BV_{CBO}\left(\frac{g_m' r_b}{\beta}\right)^{1/m} \tag{4.8}$$

which illustrates the well-known relationship between BVCBO and BVCEO in the bipolar transistor. The dependence of bipolar breakdown voltage on source impedance can be exploited in PA design to significantly increase the safe operating voltage range.

4.4 SiGe HBT-Based RF Circuit Considerations

Si–SiGe HBT Low-Noise Amplifier Fundamentals

The front-end LNA is one of the key determiners of the performance of a complete receiver, since the overall signal-to-noise ratio of the final received signal is set by the noise performance of this particular amplifier. Typical wireless application frequencies today are in the 1 to 5 GHz range. Fortunately, the microwave noise performance of SiGe HBTs bipolar has improved dramatically in recent years, thanks to aggressive technology scaling that was largely designed to improve digital circuit performance.

The input referred noise performance of a radio receiver determines the minimum signal level that can be reliably demodulated. As a result, it is a key factor in determining the range and power dissipation of the entire communications system. The noise factor (F), defined as the degradation of the signal-to-noise ratio of an input signal as it passes through the amplifier, is the standard metric for determining the noise performance of a radio-frequency receiver and is given by

$$F = \frac{\text{total noise power delivered to the load impedance}}{\text{total noise power delivered to the load impedance due only to the source}} \tag{4.9}$$

and the noise figure (NF) is defined as F in decibels [i.e., NF $=$ 10 log 10(F)].

Given this noise model of the bipolar transistor, the transistor will exhibit the following minimum noise factor as a function of source impedance [17]

$$\mathrm{F}_{\mathrm{min}} = 1 + \frac{R_\mathrm{b}}{R_\mathrm{s}} + \frac{g_\mathrm{m}}{2}\frac{f^2}{f_\mathrm{T}^2}\frac{(R_\mathrm{b}+R_\mathrm{s})^2}{R_\mathrm{s}} + \frac{g_\mathrm{m}}{2\beta}\frac{f^2}{f_\mathrm{T}^2}\frac{\left[(R_\mathrm{b}+R_\mathrm{s})^2 + X_\mathrm{s}^2\right]}{R_\mathrm{s}} \tag{4.10}$$

So, the keys to lowering the noise figure of the bipolar device are the reduction in R_b or an increase in the f_T. In this respect, the bipolar device has achieved "near-ideal" performance at sub-5 GHz frequencies recently, as state of the art the f_T of the devices exceeds 100 GHz.

Circuit linearity affects the performance of the receiver as well. An RF receiver is typically operated well below its 1-dB compression point, and therefore small-signal linearity is the key performance metric. As an example, in the GSM receiver case, the circuit must be able to amplify a signal of roughly 10^{-13} W while simultaneously receiving an undesired signal many orders of magnitude larger. The key figures-of-merit here are the input intercept point and cross-modulation sensitivity. Transmitters are typically operated at high levels of output power, and so their large-signal linearity is the key consideration.

From the perspective of receiver design, which encompasses the low-noise amplification stages as well as the downconversion mixer, circuit nonlinearity arises from weak nonlinearities both in the dependent sources (principally the transconductance) and charge storage elements (capacitors) within the transistor; at low frequencies, the former consideration dominates.

The standard small-signal linearity figure-of-merit for a receiver amplifier is the third-order input-referred intercept point (IIP3). This is defined as the input power level of two input signals (at frequencies f_1 and f_2), where the extrapolated undesired third-order output nonlinear response intersects the desired first-order linear response. Although this figure-of-merit has many limitations in practical situations, its ease of measurement and calculation make it a perennial favorite among microwave engineers. The second-order input-referred intercept point (IIP2)—the input power level where the extrapolated second-order response intersects the desired first-order response—is also sometimes specified, although it is usually less important than the IIP3.

The nonlinearities introduce frequency dependence to the nonlinearity, which considerably complicates the analysis. The situation can be simplified if we consider resistive terminations only at each terminal of the transistor. In this case, the work of Vaidyanathan et al. [18] employing a Volterra-series analysis clarifies the relationship between the high-frequency linearity of the bipolar transistor and its physical design, particularly the relationship between the high-frequency linearity and the behavior of its "loaded" unity current-gain frequency f_T, where the loaded unity current-gain frequency is defined as the frequency where the current-gain drops to unity with the appropriate terminating impedances.

As an example, at sufficiently high frequencies, and without avalanche breakdown occurring, the OIP2 of a bipolar transistor is given by the relatively simple relationship [18]

$$\mathrm{OIP2}(2f) \approx \frac{4f_\mathrm{T}}{|f'_\mathrm{T}|} \tag{4.11}$$

where f'_T is the derivative of the loaded f_T with respect to collector current. To minimize the second-order intermodulation distortion, the transistor should be designed to have as constant an f_T as possible, and the device will have the highest OIP2 near the peak of the f_T versus i_C curve.

The important OIP3 behavior is more complicated than in the OIP2 case, but some important generalizations can be derived from the analysis of device operation. At sufficiently high frequencies, and when the device is operated at the peak of the i_C–f_T curve, the OIP3 of the bipolar transistor can be approximated by [18]

$$\mathrm{OIP3}(2f_2 - f_1)\Big|_{f'_\mathrm{T}=0} \approx \left|\frac{8f_\mathrm{T}}{f''_\mathrm{T}}\right|^{1/2} \tag{4.12}$$

where f''_T is the second derivative of the loaded f_T with respect to collector current. These results imply that, when the device is operated at the peak of its f_T versus collector current curve, the best distortion performance is obtained when the device has a high f_T and when the f_T curve is as "flat" as possible. Both the OIP2 and OIP3 results mentioned above demonstrate that the "ideal" bipolar transistor is defined as one with very low junction capacitances; and hence, nearly constant f_T will have outstanding high-frequency linearity, and that this intrinsic linearity can improve with future device scaling.

Si–SiGe HBT Voltage-Controlled Oscillators and Frequency Synthesizers

The voltage-controlled oscillator (VCO) provides the frequency reference for the upconversion of the transmitted signal or downconversion of the received signal. The VCO frequency is usually not accurate enough by itself to provide the correct downconversion or upconversion frequency, and so it is usually phase-locked to a more precise reference frequency. The key performance issues with this circuit are phase noise, power dissipation, and frequency tuning range. Unlike many other circuits, the performance of the passive devices can have a significant impact on the performance of this circuit.

The phase noise of the oscillator is the ratio of the power in the desired output (the carrier) to the output power in a 1 Hz bandwidth at a given frequency offset from the carrier, when the amplitude variation on the carrier has been removed through a limiting process. So, the phase noise is expressed in units of dB c/Hz at a specified offset frequency. Ideally, the spectrum of the VCO output is a delta-function in the frequency domain, so the ideal VCO phase noise would be infinite dBc/Hz at all offset frequencies. Phase noise contributes to a variety of deleterious effects in radio systems, including a rise in the receiver noise floor and reciprocal mixing. A simplified schematic of a bipolar transistor monolithic differential LC-tuned VCO, along with its most significant noise sources is shown in Figure 4.9. The cross-coupled differential transistor pair presents a negative impedance to the resonator, canceling the resistive losses in the resonator and enabling sustained oscillation. Frequency variation is achieved with a reverse-biased pn-junction diode or accumulation-mode MOS varactor, which changes the resonant frequency of the circuit. The close-in phase noise behavior at an offset f_m from the carrier frequency f_0 in the differential LC-tuned VCO is determined from the well-known Leeson's [19] model to be

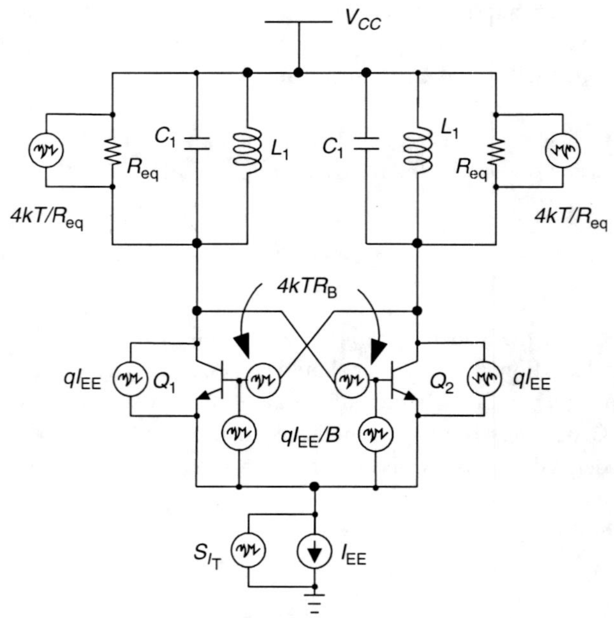

FIGURE 4.9 Simplified schematic of a monolithic SiGe HBT transistor LC-tuned VCO with noise sources.

$$L(f_m) = \frac{2kTR_{eq}F}{A_0^2}\left(\frac{f_0}{2Qf_m}\right)^2 \tag{4.13}$$

where k is Boltzman's constant, T is the absolute temperature, A_0 is the amplitude of oscillation, Q is the resonator-loaded quality factor, and F is the excess noise factor. Leeson's model shows that phase noise is reduced as the amplitude of oscillation is increased. However, once the amplitude of oscillation drives the transistors in the cross-coupled differential pair into saturation, the loaded quality factor of the resonator is lowered and phase noise degrades significantly. It also illustrates the tradeoff between the power dissipation and phase noise, since a large amplitude will lead to both lowered phase noise and higher power dissipation. Leeson's equation clearly shows the importance of maximizing the quality factor of the resonating circuit. The excess noise factor F is determined by the wideband noise from the cross-coupled differential transistor pair and the dc current noise source, taking the nonlinear operation of the oscillator into account. In the case of a bipolar VCO, the excess noise factor F can be approximated to be [20]

$$F \approx 1 + \frac{R_{bb}}{2R_{eq}}\left(\frac{f_T}{f_0}\right) + \frac{qI_T R_{eq}}{2kT}\left(\frac{\Delta V}{\pi A_0}\right)^2\left(1 + \mathrm{sinc}^2\left(\frac{\Delta V}{2A_0}\right)\right) + \frac{S_{I_T} R_{eq}}{8kT} \tag{4.14}$$

where ΔV is the signal level required to make the cross-coupled differential transistor pair switch completely to one side, R_{eq} is the parallel equivalent impedance of the resonator, I_T is the dc current, and S_{I_T} is the mean square current noise power spectral density. This illustrates the importance of minimizing base resistance for low-phase noise operation as well as the slight penalty incurred through the use of a high f_T device. Low-frequency noise from the dc current source results in amplitude modulation of the carrier, and therefore little phase noise contribution from this source. However, dc current source noise at frequencies near the even harmonics of the oscillator creates both amplitude and phase noise.

Due to the absence of a high-quality on-chip varactor diode, the fully integrated VCOs that employ on-chip LC tanks typically suffer from limited tuning range, which need to cover the practical (approximately $\pm 10\%$) lot-to-lot variation in capacitance values. If the tuning range of integrated VCOs can be made significantly larger than 20%, it would be ideal to have one single VCO covering the multiple frequency bands required by several different wireless standards (e.g., W-CDMA/GSM/PCS) and for both the transmit and receive modes. Such a large bandwidth requirement makes the fully integrated VCO design more challenging to meet the low-phase noise and low variation on VCO gain (i.e., K_{VCO}) requirements for wireless applications. For example, a 3.4 to 4.6 GHz fully monolithic SiGe VCO designed for the GSM/DCS/PCS applications was recently reported [21]. The VCO provides a tuning range of 33% and with a very good phase noise of −138 dBc/Hz at 3 MHz offset at 3.4 GHz. After a divide-by-4, the VCO phase noise performance at 900 MHz is approximately −147 dBc/Hz at 3 MHz offset, which meets the challenging GSM TX VCO spec of −143 dBc/Hz at 3 MHz offset with a margin. The large tuning range and low-phase noise are achieved by using a combination of coarse tuning via digital selection of MOS capacitors and analog fine-tuning using P–N junction varactors. The chip is fabricated in a 50/65-GHz f_T/f_{max} SiGe 0.25 μm BiCMOS production process [22]. Figure 4.10 and Figure 4.11 show the schematic and the die photograph of this fully monolithic VCO, respectively.

4.5 Examples of SiGe HBT-Based RF Wireless Transceivers

SiGe HBT technology has become so ubiquitous recently, that a complete listing of all the wireless transceivers implemented in the technology would complete an entire book. Here, we will just focus on some of the 2.5G/3G cellular implementations of SiGe transceivers. In the 3G UMTS/W-CDMA system, the signal bandwidth is adjustable, so the direct-conversion receiver architecture is particularly suitable

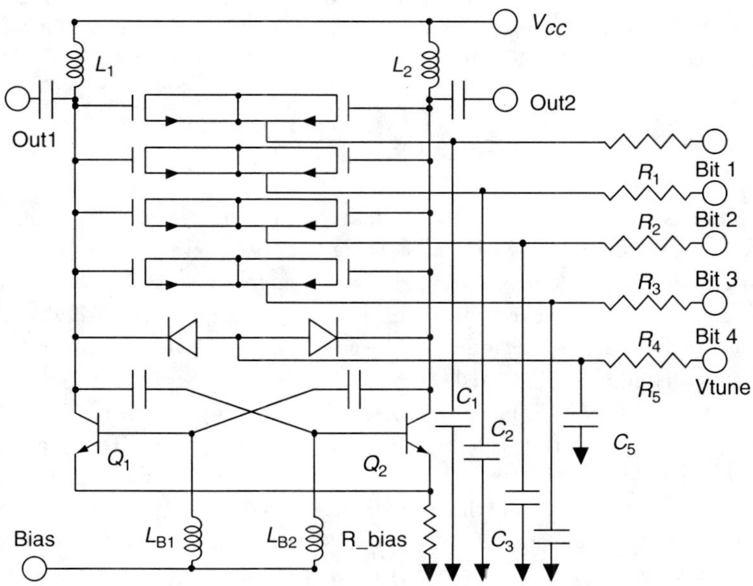

FIGURE 4.10 Schematic for a fully monolithic multiband VCO with 33% tuning range. (From D.Y.C. Lie, X. Yuan, L.E. Larson, T. Robinson, A. Senior, X. Wang, J. Mecke, and M. Case. IEEE Bipolar/BICMOS Circuits and Technology Meeting (BCTM), 2002, Monterey, CA, pp. 65–68. With permission.)

FIGURE 4.11 Die photograph of the fully monolithic VCO shown in Figure 4.10. (From D.Y.C. Lie, X. Yuan, L.E. Larson, T. Robinson, A. Senior, X. Wang, J. Mecke, and M. Case. Proc. IEEE Bipolar/BICMOS Circuits and Technology Meeting (BCTM), 2002, Monterey, CA, pp. 65–68. With permission.)

for W-CDMA implementation where the bandwidth of the receiver is determined by the cutoff frequency of the low-pass filters in the baseband. There have been several recent reports that demonstrated the feasibility of W-CDMA direct-conversion receivers in SiGe [23–26]. Besides the issue of dc offsets and AM-detection, it is challenging to meet the overall receiver sensitivity, selectivity, and dynamic range, as the overall NF needs to be $\leq$9 dB (including the 3.5 to 4 dB loss from the RF system switch and the diplexer) and the IIP3 needs to be high enough to pass the intermodulation/blocker tests while consuming the lowest dc power possible. The strong signal leakage from the transmitter through the duplexer can also interact with the LO signals leaked from the mixers to generate dc offsets. Lie et al. [27] have recently reported a W-CDMA direct-conversion front-end receiver chip consisting of an LNA, a dual-gain RF variable-gain amplifier (RF-VGA), two direct-down-conversion mixers, an I/Q quadrature generator, and a base-band five-gain-stage VGA that was designed and manufactured in a 0.25-μm SiGe BiCMOS production process. A very low LO-induced dc offset value of less than 300 μV (uncalibrated) was measured at the output of the mixers, as the LO signal was fed into the chip at twice the RF frequency. The measured cascaded noise figure for the chip (including the SAW filter) was 4.3 dB at the maximum gain mode, and the IIP2 and IIP3 were +37 and -16.5 dBm, respectively. The reported

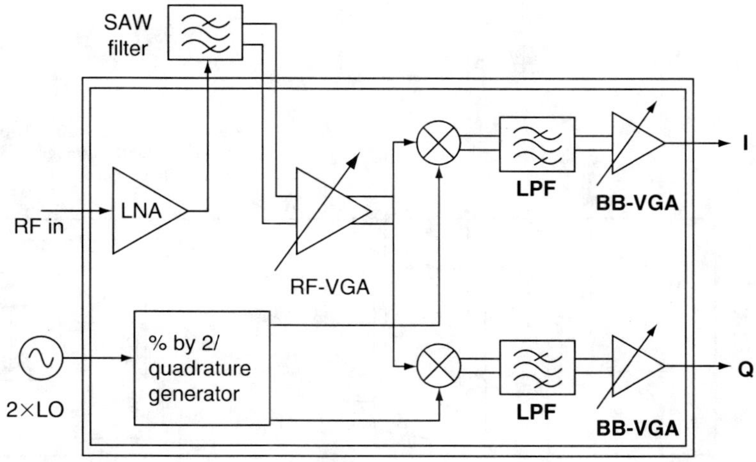

FIGURE 4.12 Block diagram of W-CDMA SiGe direct-conversion front-end receiver chip. (From D.Y.C. Lie, J. Kennedy, D. Livezey, B. Yang, T. Robinson, N. Sornin, T. Beukema, L.E. Larson, A. Senior, C. Saint, J. Blonski, N. Swanberg, P. Pawlowski, D. Gonya, X. Yuan, and H. Zamat. Proc. IEEE Radio Frequency Integrated Circuits Conference (RFIC), 2002, pp. 31–35. With permission.)

LO leakage and the dc offset values are among the best in the literature for W-CDMA direct-conversion receiver chips, with an uncalibrated I/Q amplitude/phase mismatch of less than 0.1 dB and 1.5°, respectively. The overall chip performance meets all the essential parameters and BER requirements of W-CDMA front-end receiver specs. Figure 4.12 shows the block diagram of the W-CDMA SiGe direct-conversion front-end receiver chip as an example of direct-receiver architecture [27]. A BER $= 0.1\%$ is reached at a signal level of approximately -122.9 dBm at the LNA input, which meets the 3GPP receiver sensitivity spec of -117 dBm at antenna or approximately -121 dBm at LNA input at the worst-case (assuming a 4-dB loss from the system switch and TX SAW filter loss).

A transmitter creates a modulated carrier at the desired output frequency by upconversion and power amplification. Many of today's CDMA and TDMA mobile phones use the two-step transmitter approach because this architecture has proven to be less problematic, but it requires the addition of high-quality filters at both the RF and IF stages. The IF filter is required to attenuate the wideband noise floor and the higher harmonics of the IF generated by the quadrature (I/Q) modulator. The RF filter is required to attenuate the unwanted sidebands and spurs generated from the upconverting mixing process. Another issue with a two-step transmitter is that two LO signals must be generated, which may require a second phase-locked loop (PLL). The primary advantage of a two-step transmitter is that narrowband filtering and gain control can be implemented efficiently at the IF stage, which improves the dynamic range of the transmitter. This architecture also benefits from the local oscillators operating at different frequencies from the transmit frequency, which avoids the injection pulling caused by the PA.

Figure 4.13 shows an example of a highly linear W-CDMA SiGe W-CDMA two-step transmitter IC with high dynamic range and on-chip transformers [28]. It meets or exceeds all necessary specifications for 3GPP compliance transmitter IC for UE Output Power Class-3. This TX IC consists of a wideband I/Q modulator, a narrowband IFVGA, a variable-gain mixer, an RFVGA, a differential class A/B driver, a LO driver, and on-chip low-pass and band-pass filters. The transmitter achieves carrier suppression >35 dBc; side-band suppression >40 dB; and RMS error-vector-magnitude (EVM) $<3\%$ at 3 dBm output power. The measured adjacent-channel-leakage ratio (ACLR) is less than -49 dBc (i.e., at ± 5 MHz offset: -49 dBc/3.84 MHz; at ± 10 MHz offset: -59 dBc/3.84 MHz). The TX IC maximum output power is 7 dBm and with overall 100 dB dynamic range (i.e., ± 13 dB margin compared with

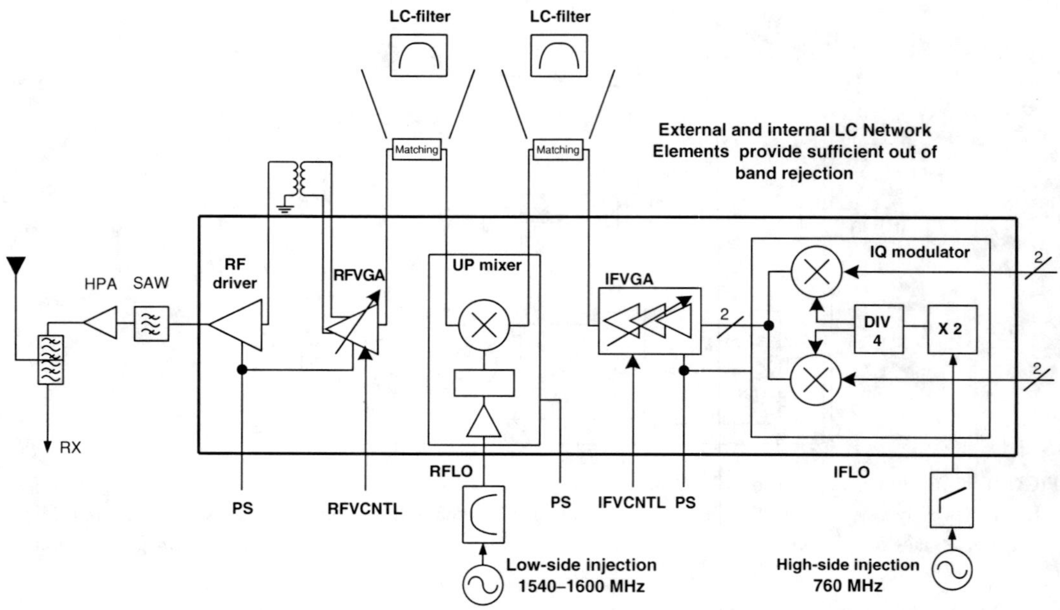

FIGURE 4.13 Block diagram of W-CDMA SiGe heterodyne transmitter chip. (From D. Malhi, L. Larson, D. Wang, C. Demirdag, and V. Pereira. *IEEE J. Solid-State Circ.* 38(9), 2003, 1570–1574. With permission.)

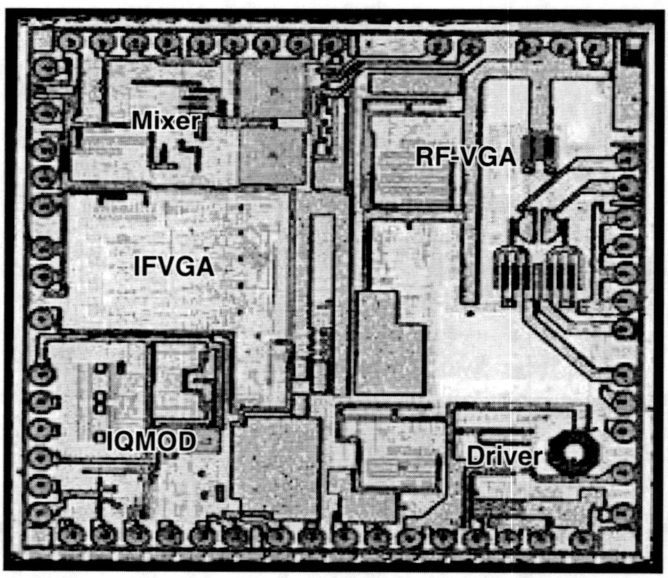

FIGURE 4.14 Die photograph of the W-CDMA SiGe heterodyne transmitter chip shown in Figure 4.13.

3GPP spec to allow for process–temperature–environment variations). The out-of-band noise and spurs for the DCS band (1805 to 1880 MHz) is −121 dBm/Hz, for the TDD band (1893.5 to 1991.6 MHz) is −95.8 dBm/Hz, and for the Rx band (2110 to 2170 MHz) is −120 dBm/3.84 MHz. Figure 4.14 shows a die picture of this TX IC.

Recently, Molnar et al. [29] have reported a single-chip BiCMOS quad-band direct-conversion GSM/GPRS RF transceiver with fully integrated VCOs and fractional-N synthesizer. Subharmonic mixers are used in the direct-conversion receiver to ensure the LO to RF isolation, where the receive frequency is 2/3 and 4/3 of the UHF frequency for GSM versus DCS/PCS mode, respectively. The LO leakage at the receive band is measured approximately −110 dB m, which, together with a high receiver IIP2, reduces the dc offsets induced by large interferers coupling to the LO signal. A proprietary on-chip calibration helps the receiver IIP2 to be above 65 dB m for all modes. An on-chip transformer is used to do the single-ended to differential-signal conversion from LNA to drive a two-stage polyphase filter. The receiver AGC provides programmable gain from 9 to 125 dB in 2-dB steps, with eight poles of baseband filtering and a three-stage dc offset calibration correction. The NF is roughly below 4 dB for all modes under nominal operating condition. The fractional-N synthesizer includes a fully integrated wideband UHF VCO (1.2 to 1.7 GHz), where a band-switching VCO coarse-tuning technique is used (same as the one shown in Figure 4.10). The transmitter uses a translational loop with in-loop vector modulation. The transmit VCOs cover all four bands and meet the GSM out-of-band noise spec.

We show in Figure 4.15 a single-chip transceiver in Si BiCMOS technology that successfully meets the DECT specifications [30]. With an advanced SiGe BiCMOS technology, the performance of the RF products can be significantly improved, enabling IC designers to build multistandard multiband single-chip RF systems for cellular applications. Similar to what has been recently reported by Molnar et al. in 0.35-μm BiCMOS technology, Figure 4.16 shows a die picture of a multiband single-chip direct-conversion SiGe BiCMOS transceiver that supports GSM/PCS/DCS multistandards as a highly integrated RF-SoC product [31]. Note that the IC in Figure 4.16 contains a 3.6-GHz VCO, transceiver, and synthesizer. However, the PA is still not integrated on-chip.

4.6 Summary

To achieve the highest level of RFIC integration for single-chip, multiband, multistandard wireless devices of the future, we expect SiGe-based HBT technology to play a significant role. This is due to its

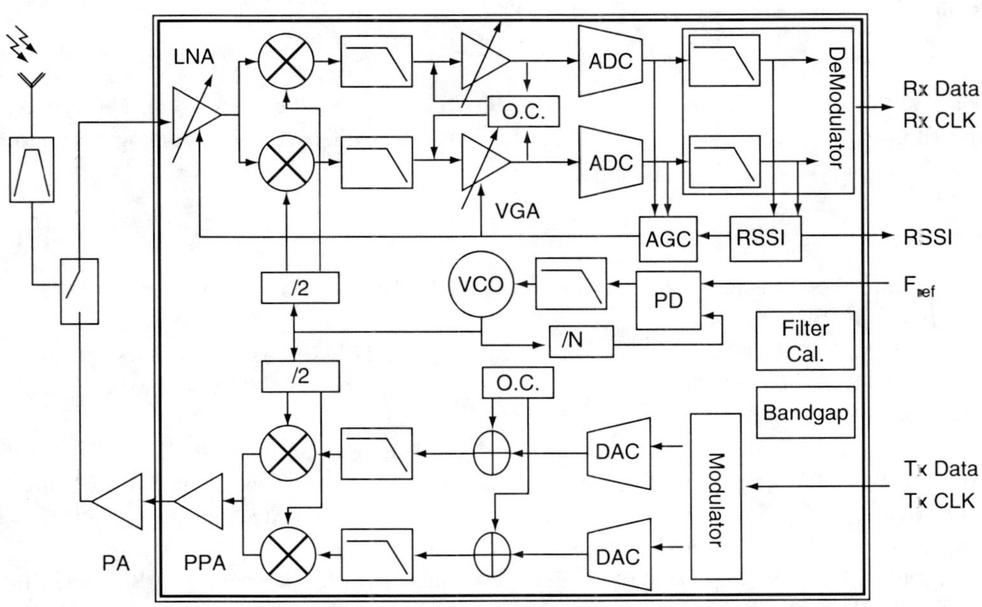

FIGURE 4.15 Block diagram of a single-chip direct-conversion BiCMOS DECT transceiver. (From F.O. Eynde, J. Craninckx, and P. Goetschalckx. IEEE Int. Solid-State Circuits Conf., Dig., 2000, p. 138. With permission.) Note only the PA, antenna switch, and filter are external to the chip.

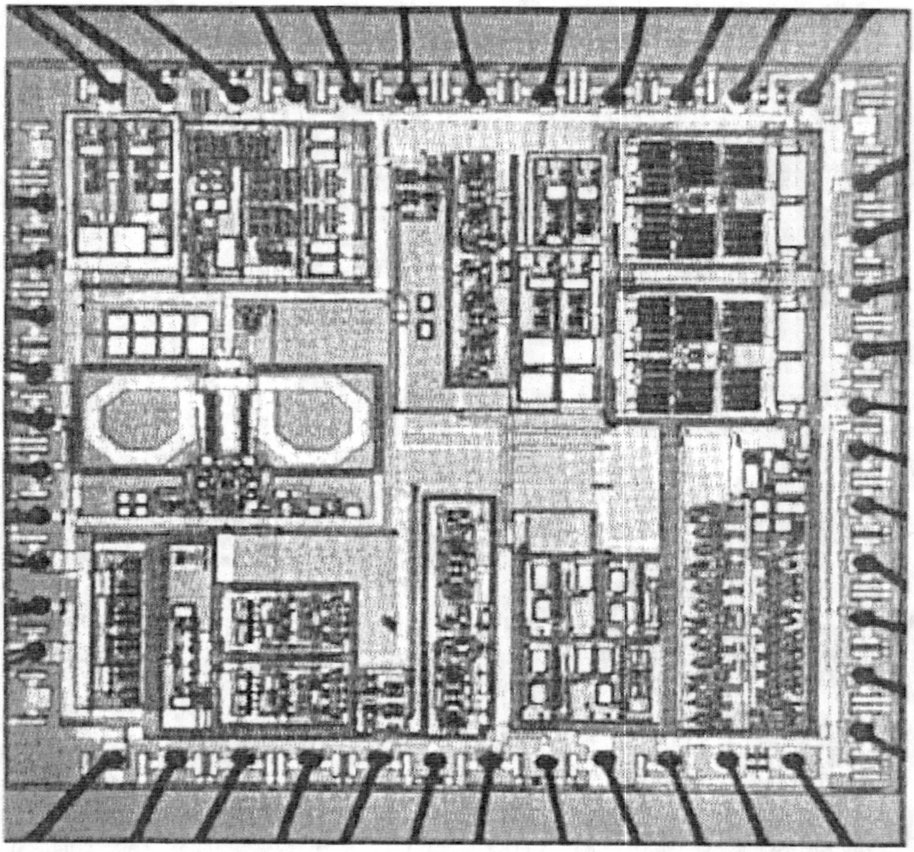

FIGURE 4.16 Die photograph of a multiband single-chip direct-conversion SiGe BiCMOS transceiver that supports GSM/PCS/DCS multistandards as a RF-SoC product. (From J. Sevenhas, F. Op't Eynde, and P. Reusens. *IEEE Trans. Microwave Theory Tech.* 50, 2002, 235–244. With permission.)

superior performance, ease of integration with CMOS, and low overall cost structure. This chapter has illustrated many of the key device, circuit, and system issues that will make this technology an important part of the wireless future.

References

1. D. Harame. Si/SiGe epitaxial-base transistors—Part I: Materials, physics, and circuits. *IEEE Trans. Electron Dev.*, 1995, 455–468.
2. D.Y.C. Lie and K.L. Wang. Si/SiGe heterostructures for Si-based nanoelectronics. *Handbook of Advanced Electronic and Photonic Devices and Materials*, Chapter 1, H.S. Nalwa, ed., Academic Press, San Diego, 2000, pp. 1–69.
3. D. Nguyen-Ngoc, D.L. Harame, J.C. Malinowski, S.J. Jeng, K.T. Schonenberg, M.M. Gilbert, G.D. Berg, S. Wu, M. Soyuer, K.A. Tallman, K.J. Stein, R.A. Groves, S. Subbanna, D.B. Colavito, D.A. Sunderland, and B.S. Meyerson. A 200 mm SiGe-HBT BiCMOS technology for mixed signal applications. *IEEE Proc. BCTM*, 1995, 89–92.
4. P.C. Davis. Merits and requirements of a few RF architectures. Proc. 1999 Bipolar/BiCMOS Circuits and Technology Meeting, Minneapolis, MN, September 1999, pp. 62–66.
5. B. Gilbert. The Design of Si/SiGe LNAs from the Ground Up. IEEE BCTM Short Course Notes, 1996.

6. A. Abidi. Direct-conversion radio transceivers for digital communications. *IEEE JSSSC* 30(12), 1995, 1399–1410.
7. F. Aschwanden. Direct conversion—how to make it work in TV tuners. *IEEE Trans. Consumer Electron.* 42(3), 1996, 729–751.
8. L. Sheng, J.C. Jensen, and L. Larson. A wide-bandwidth Si/SiGe HBT direct conversion sub-harmonic mixer/downconverter. *IEEE J. Solid-State Circ.* 35(9), 2000, 1329–1337.
9. S. Mattison. Architecture and technology for Multistandard transceivers. Proceedings of Bipolar/BiCMOS Circuits and Technology Meeting, 2001, pp. 82–85.
10. B. Jagannathan, M. Khater, F. Pagette, J. Rieh, D. Angell, H. Chen, H.J. Florkey, F. Golan, D. Greenberg, R. Groves, R.S. Jeng, J. Johnson, E. Mengistu, K. Schonenberg, C. Schnabel, P. Smith, A. Stricker, D. Ahlgren, G. Freeman, K. Stein, and S. Subbanna. Self-aligned site NPN transistors with 285 GHz f_{MAX} and 207 GHz f_T in a manufacturable technology. *IEEE Electron Dev. Lett.* 23(5), 2002, 258–260.
11. M. Vaidyanathan and D. Pulfrey. Extrapolated f_{MAX} of heterojunction bipolar transistors. *IEEE Trans. Electron Dev.* 46(2), 1999, 301–309.
12. U. Konig, A. Gruhle, and A. Schuppen. SiGe devices and circuits: where are the advantages over III–V. Proc. 1995 IEEE GaAs IC Symposium, pp. 14–18.
13. L. Larson. Silicon technology tradeoffs for radio-frequency/mixed-signal "systems-on-a-chip". *IEEE Trans. Electron Dev.* 50(3), 2003, 683–699.
14. E.O. Johnson. Physical limitation on frequency and power parameters of transistors. *IEEE Intern. Conv. Record* 5, 1965, 27.
15. R. Jos. Future developments and technology options in cellular phone power amplifiers: From power amplifier to integrated RF front-end module. Proceedings of the 2000 Bipolar/BiCMOS Circuits and Technology Meeting, Minneapolis, MN, pp. 118–125.
16. A. Inoue, S. Nakatsuka, R. Hattori, and Y. Matsuda. The maximum operating region in SiGe HBTs for RF power amplifiers. Proceedings of 2002 IEEE MTT-S International Microwave Symposium, Seattle, WA, USA, June 2002, pp. 1023–1026.
17. G. Niu, J. Cressler, Z. Shiming, W. Ansley, C. Webster, and D. Harame. A unified approach to RF and microwave noise parameter modeling in bipolar transistors. *IEEE Trans. Electron Dev.* 48(11), 2001, 2568–2574.
18. M. Vaidyanathan, L. Larson, M. Iwamoto, P. Asbeck, and P. Gudem. A basic theory of high frequency distortion in bipolar transistors. *IEEE Trans. Microwave Theory Tech.* 51(2), 2003, 448–461.
19. D. Leeson. Simple model of a feedback oscillator noise spectrum. *Proc. IEEE* 54, 1956.
20. T.K. Johansen and L.E. Larson. IEEE Radio Frequency Integrated Circuits (RFIC) Symposium, June 2003, pp. 273–276.
21. D.Y.C. Lie, X. Yuan, L.E. Larson, T. Robinson, A. Senior, X. Wang, J. Mecke, and M. Case. Phase noise analysis of fully-integrated digitally-tuned wideband Si/SiGe BiCMOS VCOs. Proc. IEEE Bipolar/BICMOS Circuits and Technology Meeting (BCTM), 2002, Monterey, CA, pp. 65–68.
22. D. Nguyen-Ngoc, D.L. Harame, J.C. Malinowski, S.J. Jeng, K.T. Schonenberg, M.M. Gilbert, G.D. Berg, S. Wu, M. Soyuer, K.A. Tallman, K.J. Stein, R.A. Groves, S. Subbanna, D.B. Colavito, D.A. Sunderland, and B.S. Meyerson. A 200 mm SiGe-HBT technology for wireless and mixed-signal applications. *IEEE Proc. BCTM*, 1995, 89–92.
23. A. Pärssinen, J. Jussila, J. Ryynänen, L. Sumanen, and K.A.I. Halonen. A 2-GHz wide-band direct conversion receiver for WCDMA applications. *IEEE J. Solid-State Circ.* 34(12), 1999, 1893.
24. J. Ryynänen, K. Kivekäs, J. Jussila, A. Pärssinen, and K. Halonen. A dual-band RF front-end for WCDMA and GSM applications. Proc. IEEE Custom Integrated Circuits Conference (CICC), 2000, p. 175.
25. K. Itoh, T. Katsura, H. Nagano, T. Yamaguchi, Y. Hamade, M. Shimozawa, N. Suematsu, R. Hayashi, W. Palmer, and M. Goldfarb. 2 GHz band even harmonic type direct conversion receiver with ABB-IC for W-CDMA mobile terminal. IEEE MTT-S Digest, 2000, p. 1957.

26. H. Pretl, W. Schelmbauer, L. Maurer, H. Westermayr, R. Weigel, B.-U. Klepser, B. Adler, and J. Fenk. A W-CDMA zero-IF front-end for UMTS in a 75 GHz SiGe BiCMOS technology. IEEE RFIC Symp. Dig., 2001, pp. 9–12.
27. D.Y.C. Lie, J. Kennedy, D. Livezey, B. Yang, T. Robinson, N. Sornin, T. Beukema, L.E. Larson, A. Senior, C. Saint, J. Blonski, N. Swanberg, P. Pawlowski, D. Gonya, X. Yuan, and H. Zamat. A direct-conversion W-CDMA front-end SiGe receiver chip. Proc. IEEE Radio Frequency Integrated Circuits Conference (RFIC), 2002, pp. 31–35.
28. D. Malhi, L. Larson, D. Wang, C. Demirdag, and V. Pereira. SiGe W-CDMA transmitter for mobile terminal application. *IEEE J. Solid-State Circ.* 38(9), 2003, 1570–1574.
29. A. Molnar, R. Magoon, G. Hatcher, J. Zachan, W. Rhee, M. Damgaard, W. Domino, and N. Vakilian. A single-chip quad-band (850/900/1800/1900 MHz) direct conversion GSM/GPRS RF transceiver with integrated VCOs and fractional-n synthesizer. IEEE Int. Solid-State Circuits Conf., Dig., 2002, pp. 232–233.
30. F.O. Eynde, J. Craninckx, and P. Goetschalckx. A fully-integrated zero-IF DECT transceiver. IEEE Int. Solid-State Circuits Conf., Dig., 2000, p. 138.
31. J. Sevenhas, F. Op't Eynde, and P. Reusens. The silicon radio decade. *IEEE Trans. Microwave Theory Tech.* 50, 2002, 235–244.

5

LNA Optimization Strategies

5.1 Introduction 5-1
5.2 RF LNA Design Concerns 5-2
Noise Factor/Figure • Gain and Matching • Linearity • Power Consumption and Stability
5.3 SiGe Cascode LNA Design Example 5-5
Optimization Strategies • Device Models • Simulation and Optimization • Analytical Expressions and Design Rules of Thumb
5.4 Summary 5-13

Qingqing Liang
Georgia Institute of Technology

5.1 Introduction

In an RF receiver, one of the most important features is its noise characteristics; that is, the attribute of collecting the desired input signal while producing less noise. The noise factor F and noise figure NF are the parameters that measure the noise characteristics of RF receivers,

$$F = \frac{SN_{in}}{SN_{out}}, \tag{5.1}$$

$$NF = 10 \log_{10}(F), \tag{5.2}$$

where SN_{in} and SN_{out} are the signal-to-noise ratios at the input and output of the amplifier, respectively.

Using the definitions above, one can calculate the minimum detectable input power level $P_{in,min}$ of a receiver:

$$P_{in,min} = F \times SN_{req} \times P_{nf}, \tag{5.3}$$

where the SN_{req} is the required signal-to-noise power ratio at the output of the receiver, P_{nf} is the input noise floor. $P_{in,min}$ is a fundamental parameter of a receiver: the smaller $P_{in,min}$, the higher sensitivity, and the wider space over which the receiver is functional. Note that both SN_{req} and P_{rf} are not determined by the receiver. Therefore, to improve the sensitivity of the receiver (i.e., a smaller $P_{in,min}$), F should be minimized.

In a linear cascaded system (Figure 5.1), the overall noise factor F_{total} can be written as [1]

$$F_{total} = F_1 + \frac{F_2 - 1}{G_1} + \frac{F_3 - 1}{G_1 G_2} + \cdots, \tag{5.4}$$

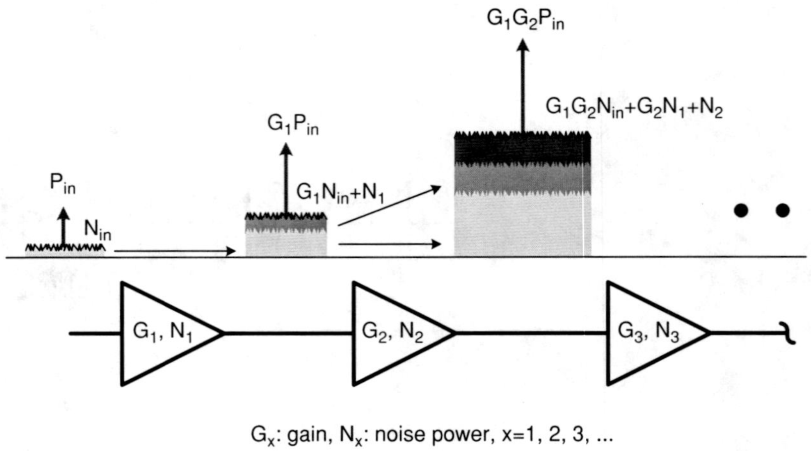

FIGURE 5.1 Signal and noise gain in cascade amplifiers.

where F_1, F_2, and F_3 are noise factors of stages 1, 2, and 3, G_1, G_2, and G_3 are gains* of stages 1, 2, and 3, and so forth. Observe that if the gain of the first stage (G_1) is high enough, the overall noise factor is dominated by the noise factor of the first stage (F_1). Therefore, a low-noise amplifier (LNA), which provides a decent gain with a smaller F, is required and attached at the front of a receiver to obtain low F_{total}.

Compared with other RF building blocks (e.g., mixers, oscillators), the circuit structure of LNA is kept simple, since more devices produce more noise. There are three main circuit topologies of LNA: single-transistor, cascade, and cascode [2–4]. Figure 5.2 shows the three different LNA equivalent circuit structures. Compared with single-transistor LNAs, the cascade and cascode LNAs provide higher gain and isolation. However, the single-transistor LNAs require less dc power consumption and can achieve lower noise factor. Different topology is then used depending on the application requirement.

From a device perspective, the SiGe HBT is a better contender for Si-based RF LNAs because of its high speed, high gain, low noise, and good linearity characteristics [5]. The improvement of the device technology broadens the design space for LNAs. On the other hand, it raises the demand of general and elaborated optimization on LNAs to achieve better performances, indicating that the design and optimization is not only on a circuit-level (e.g., different circuit structures, biasing current) but also on a device-level (e.g., device geometry, device characteristics). In this chapter, we will focus on general design concerns and optimization strategies of LNAs. An example of a inductively degenerated cascode SiGe LNA optimization is then demonstrated.

5.2 RF LNA Design Concerns

Noise Factor/Figure

According to linear two-port noise theory [6], the noise factor (F) or noise figure (NF) is a function of the source impedance (Y_S) and the noise parameters of the two-port system (F_{min}, R_n, and Y_{opt}):

$$F = F_{min} + \frac{R_n}{G_S}\left|Y_S - Y_{opt}\right|^2, \tag{5.5}$$

where G_S is the real part of Y_S, F_{min} is the minimum noise factor, R_n is the noise impedance, and Y_{opt} is the optimum admittance of the system. The noise parameters are functions of the biasing current and

*The available gain is used in this situation, therefore Equation 5.4 assumes that the input and output of stages 1, 2, and 3 are matched.

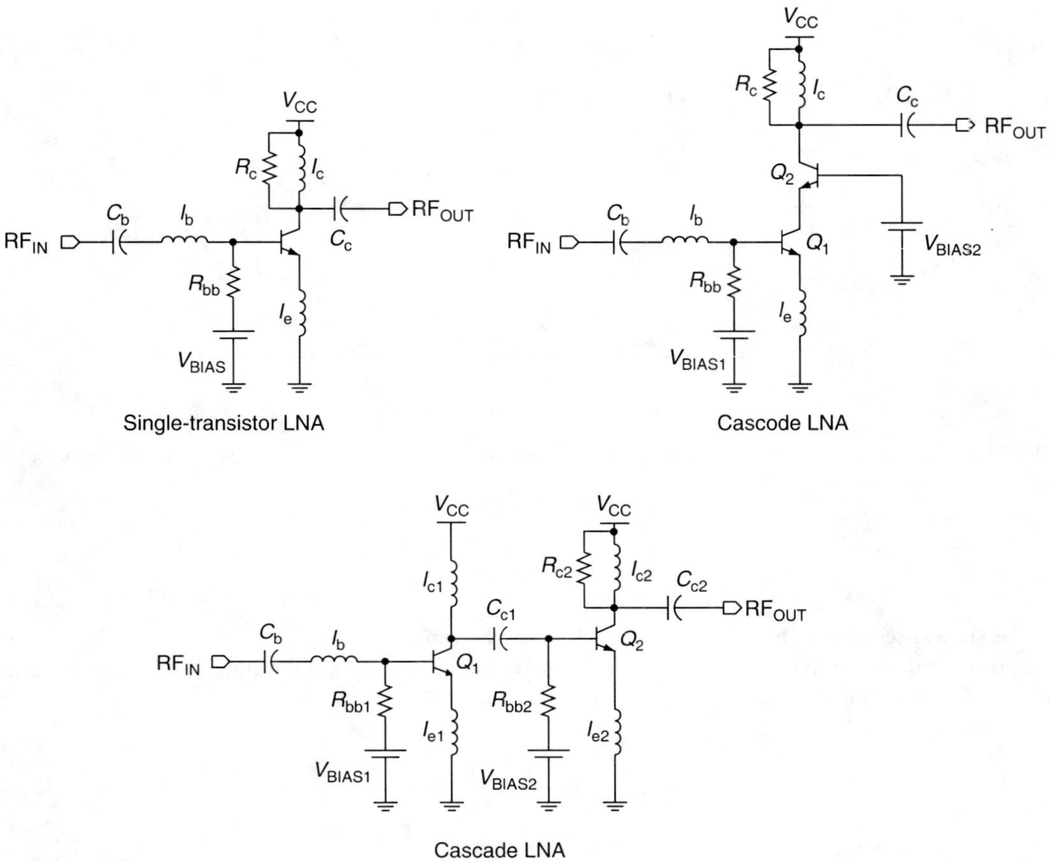

FIGURE 5.2 The three different LNA equivalent circuit structures.

device properties (both active and passive). Figure 5.3 shows the noise factor and minimum noise factor as a function of the collector current (I_C) of a single-transistor SiGe HBT LNA.

In RF LNA, the source impedance is fixed (usually 50 Ω). Then the optimization goal is to engineer the noise parameters of the system for a lower F. One method is to find out the minimum F_{min} in design space and then tune the Y_{opt} (without changing F_{min}) close to Y_S. This approach produces the lowest F of the amplifier. As shown in Figure 5.3, for example, the minimum F_{min} is at $I_C = 3.0$ mA. Therefore, to obtain the lowest NF (i.e., 0.93 dB), I_C should equal 3 mA and $Y_{opt} = Y_S$.

In reality, however, other specifications (e.g., gain, linearity) of the LNA need to be considered, and trade-offs between the noise factor and these specifications are unavoidable. In this case, the noise figure F does not need to be tuned to F_{min} because F_{min} also depends on design variables (e.g., device geometry and biasing current [7]). As shown in Figure 5.3, the actual NF at $I_C = 6$ mA is lower than NF_{min} at $I_C = 10$ mA. It implies that losing the noise factor requirement can either trade for a larger biasing current range while $Y_{opt} = Y_S$ holds, or a smaller I_C range but more flexible noise matching requirement (i.e., $Y_{opt} \neq Y_S$, hence there is a larger device geometry range to choose). Due to these concerns, general optimization strategies will be developed in the next section.

Gain and Matching

The gain of LNA also affects the overall noise factor of the system. Among various gain definitions, the transducer gain (G_T) and available gain (G_A) are used as the figure-of-merit since the source impedance of LNA is fixed.

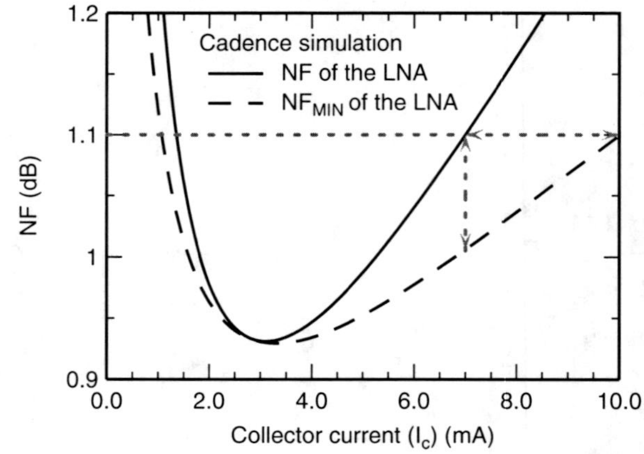

FIGURE 5.3 The actual noise factor and minimum noise factor as a function of the collector current (I_C) of a single-transistor SiGe HBT LNA.

The transducer gain is the ratio of the actual power delivered by the amplifier to an arbitrary load to the power available from the source. It is a function of the *s*-parameters of the amplifier (s_{11}, s_{12}, s_{21}, and s_{22}) and the source and load reflection coefficients (Γ_S and Γ_L):

$$G_T = |s_{21}|^2 \frac{(1-|\Gamma_S|^2)(1-|\Gamma_L|^2)}{|1-\Gamma_S s_{11}|^2|1-\Gamma_L \Gamma_2|^2}, \tag{5.6}$$

where

$$\Gamma_2 = s_{22} + \frac{s_{12}s_{21}\Gamma_S}{1-s_{11}\Gamma_S}. \tag{5.7}$$

The available gain is the ratio of the power available at the output of the amplifier to the power available from the source. It is actually the maximum transducer gain (by sweeping the load impedance) and is a function of the *s*-parameters and the source reflection coefficients:

$$G_A = |s_{21}|^2 \frac{1-|\Gamma_S|^2}{|1-\Gamma_S s_{11}|^2(1-|\Gamma_2|^2)}. \tag{5.8}$$

The source impedance is usually 50 Ω. Therefore $G_A = |s_{21}|^2/(1-|s_{22}|^2)$.

When an LNA is designed exclusively as an RF building block, low voltage-standing-wave-ratio (VSWR) at the input and output of the LNA is also required. This requirement offers good matching between the LNA and other blocks (e.g., antenna, mixer, filter). Therefore, the $|s_{11}|$ and $|s_{22}|$ of the LNA should be minimized. In ideal conditions, $s_{11}=s_{22}=0$ and $G_A=|s_{21}|^2$. Then the optimization goals concerning the gain and matching are to maximize $|s_{21}|$ and minimize $|s_{11}|$ and $|s_{22}|$.

Linearity

Another concern in LNA is linearity. It confines the maximum input power of the LNA. For circuits such as LNA operating at small-signal range, the figure-of-merit of linearity is the input third-order intercept point (IIP3).

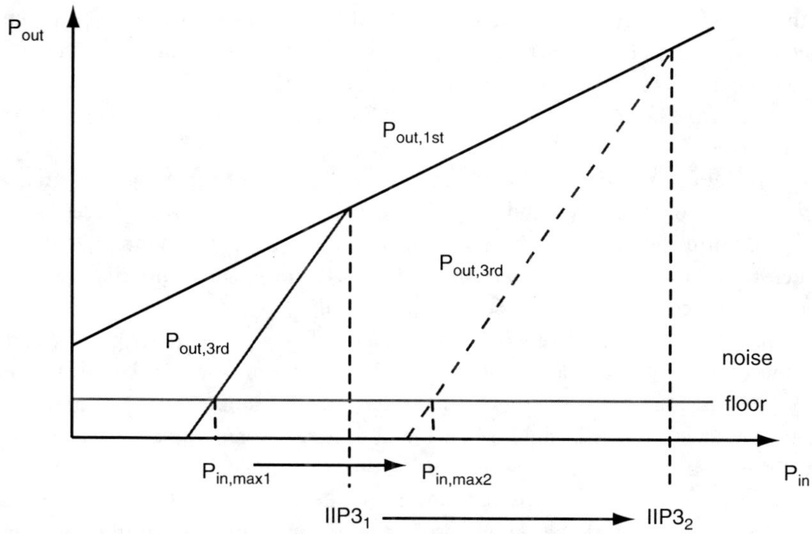

FIGURE 5.4 The first- and third-order output powers as a function of input power of the LNAs with different IIP3s.

Figure 5.4 shows the first- and third-order output powers as a function of input power of the LNAs with different IIP3s. In the RF receiver, the third-order distortion of the output power is an unexpected signal, and should be lower than a required level (e.g., the noise floor) to maintain the output signal quality. Correspondingly, the input power should be lower than $P_{\text{in, max}}$

$$P_{\text{in,max}} = \text{IIP3} - (\text{IIP3} + \text{gain-P}_{\text{nf}})/3 \tag{5.9}$$

Where P_{nf} is Power level of the noise floor. Observe that the LNA with higher IIP3 exhibits a higher $P_{\text{in,max}}$, indicating a wider range for acceptable input power. Therefore, maximizing IIP3 is another goal of LNA optimization.

Power Consumption and Stability

The total dc power is approximately equal to the product of the biasing current and supply voltage:

$$P_{\text{DC}} = \sum_i I_{\text{bias},i} V_{\text{bias},i}, \tag{5.10}$$

where $I_{\text{bias},i}$ and $V_{\text{bias},i}$ are the dc current and voltage supply at pad *i*, respectively. The voltage supply is typically fixed, therefore smaller biasing current is desired for low-power LNA design. Additionally, a robust LNA should be unconditionally stable at any frequency range. These requirements are considered as design constraints rather than optimization goals.

5.3 SiGe Cascode LNA Design Example

Optimization Strategies

Without loss of generality, the optimization goals for RF LNAs are low noise (NF), high gain ($|s_{21}|^2$), low VSWR ($|s_{11}|$ and $|s_{22}|$), and high linearity (IIP3) with the low power consumption and stability constraints; and the design variables are the values of inductors (L_x), capacitors (C_y), resistors (R_z), the device geometry (L_E)*, and biasing current (I_C)†. The optimization goals can be written as functions of design variables [e.g., NF(L_x, C_y, R_z, L_E, I_C), IIP3(L_x, C_y, R_z, L_E, I_C)]. Therefore, the problem can be generalized as optimizing a group of goal-functions in an N-dimensional space, where N is the number of the design variables.

For a better evaluation of the overall performance, we first choose the goal-functions that are not considered in the trade-off (e.g., $|s_{11}|$ and $|s_{22}|$). The dimension of the design space decreases as these goal-functions are optimized. In inductively degenerated cascode LNAs (Figure 5.2), for example, the values of inductors and capacitors are determined when $|s_{11}|$ and $|s_{22}|$ are minimized, and consequently the design space is reduced to two-dimensional (L_E and I_C).

Second, the values of those goal-functions, which are of interest to us (e.g., NF, gain, IIP3), are calculated in the design subspace, and contours of these functions are drawn. In this example, the contour-lines of NF, gain, and IIP3 are plotted in the two-dimensional ($L_E - I_C$) space.

Moreover, the design subspace shrinks when the additional design constraints are considered. For instance, the power consumption constraint requires that I_C should be smaller than a required level ($I_{C,max}$). Therefore, the design $L_E - I_C$ subspace diminishes from $(0,\infty) \times (0,\infty)$ to $(0,\infty) \times (0, I_{C,max})$. The optimum design point, as shown later, can be easily found in this contour plot on the design subspace.

Device Models

In this chapter, a simplified device model is used in analysis to derive analytical expressions and design rules of thumb. In actual design experience, delicate compact models, such as VBIC, HICUM, or MEXTRAM, are used so that the parasitics and second-order effects are considered.

IBM 5HP SiGe HBTs are used in the design. Table 5.1 summarizes the measured transistor parameters of a unit SiGe HBT. Figure 5.5 shows the large-signal nonlinear model of the SiGe HBT. Here I_{CE} represents the collector current transported from the emitter, I_{BE} represents the hole injection into the emitter, and I_{CB} represents the avalanche multiplication current:

$$I_{CB} = I_{CE}(M-1) = I_{C0}(V_{BE})F_{Early}(M-1), \tag{5.11}$$

where I_{C0} (V_{BE}) is I_C measured at zero V_{CB}, M is the avalanche multiplication factor, and F_{Early} is the Early effect factor. $M-1$ and F_{Early} were experimentally extracted using a technique proposed in Ref. [8].

$M-1$ is often modeled only as a function of V_{CB}. In the SiGe HBTs used in this work, however, $M-1$ is also a strong function of J_C [9]. The following equation was used to describe the V_{CB} and J_C dependence of $M-1$:

TABLE 5.1 Measured Parameters of a SiGe HBT, the Device Size is $0.5 \times 20 \times 2\,\mu m^2$

Peak β	110
Peak f_T	51 GHz
τ_{ec}	2.7 psec
r_{bb} at $I_C = 10$ mA	8.9 Ω
BV_{ceo}	3.3 V

*For brevity, the effect of device geometry is viewed as the effect of laying out multiple unit devices in parallel. Here the emitter length L_E is used and is equivalent to a device scaling factor.

†I_C denotes the collector biasing current. For MOSFETs, I_{DS} is used to denote the drain-source biasing current.

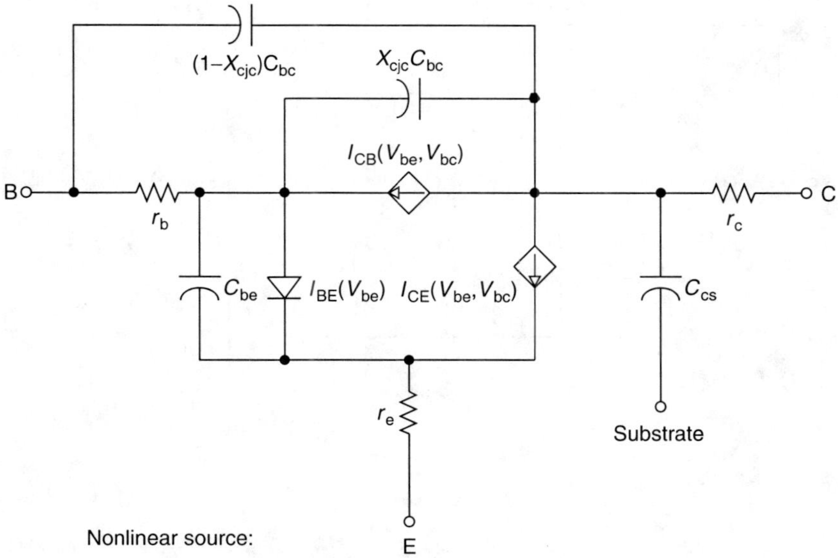

FIGURE 5.5 The large-signal nonlinear model of the SiGe HBT.

$$M-1=\frac{V_{\mathrm{CB}}}{aV_{\mathrm{CB0}}}\exp\left(-\frac{V_{\mathrm{m}}^{b}}{a^{1-b}V_{\mathrm{CB}}^{b}}\right), \tag{5.12}$$

where a is an empirical coefficient that accounts for Kirk effect

$$a=1-\tanh\left[\frac{I_{\mathrm{C}}}{I_{\mathrm{C0}}}\exp\left(\frac{V_{\mathrm{CB}}}{V_{\mathrm{R}}}\right)\right], \tag{5.13}$$

where V_{CB0}, V_{m}, V_{R}, I_{C0}, and b are model parameters, and b is typically 0.33 to 0.5. These parameters were determined by fitting the measured $M-1$ data. In addition, C_{cb} is the collector base junction depletion capacitance, C_{cs} is the collector substrate depletion capacitance, and C_{be} consists of diffusion capacitance C_{de} and depletion capacitance C_{te} according to

$$C_{\mathrm{be}}=C_{\mathrm{de}}+C_{\mathrm{te}}, \tag{5.14}$$

$$C_{\mathrm{de}}=g_{\mathrm{m}}\tau_{\mathrm{f}}. \tag{5.15}$$

In this model, six nonlinear sources are considered: nonlinear transconductance (g_{m}), nonlinear EB conductance (g_{be}), nonlinear capacitances (C_{be}, C_{bc}, and C_{cs}), and avalanche multiplication (I_{CB}).

In the RF LNA noise simulation, only broadband noise is considered.* Figure 5.6 shows the noise model of the SiGe HBT, where $\overline{i_{\mathrm{b}}^2}$ and $\overline{i_{\mathrm{c}}^2}$ are the base and collector current shot noise sources

*Since the devices in LNA are operating at small signal range, the device-induced low-frequency noise is negligible at radio frequencies.

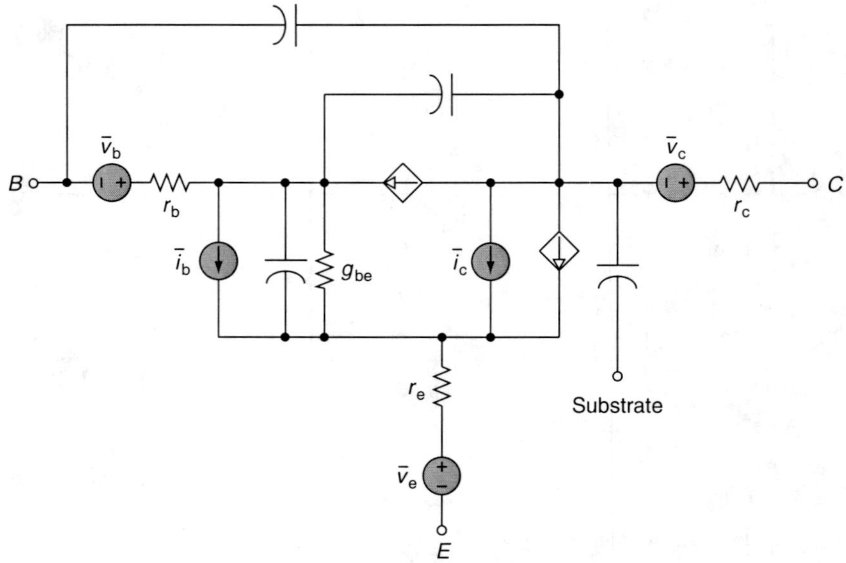

FIGURE 5.6 The noise model of the SiGe HBT.

$$\overline{i_b^2} = 2qI_b\ \Delta f, \tag{5.16}$$

$$\overline{i_c^2} = 2qI_c\ \Delta f. \tag{5.17}$$

and $\overline{v_{rb}^2}$, $\overline{v_{re}^2}$, and $\overline{v_{rc}^2}$ are the thermal noise sources produced by the base, emitter, and collector resistances, respectively:

$$\overline{v_{rx}^2} = 4kTR_x\ \Delta f, \tag{5.18}$$

where "x" refers to "b," "e," or "c" for the base, emitter, or collector resistances.

Simulation and Optimization

The general, Volterra-series method [10] is used to derive analytical expression of IIP3. As mentioned, the optimization process is adopted in the presented design methodology: first, for any given L_E and I_C, L_B, L_E, L_1, R_1, and C_1 are calculated to obtain 50 Ω input and output impedance; second, the circuit netlist is updated with the calculated values and then gain ($|s_{21}|^2$), NF, and IIP3 are simulated. The two-dimensional gain, NF, and IIP3 contours are thus calculated by sweeping the two design variables. Note that the matching requirements are satisfied at each $L_E - I_C$ point in the contour plot.

A determination of the optimum L_E and I_C that balances NF, gain ($|s_{21}|^2$), and IIP3 can be obtained. Figure 5.7 shows the NF = 1.2 dB, gain = 15 dB, and IIP3 contours as a function of L_E and I_C. The design space for NF $\leq$ 1.2 dB is within the dashed line, and the design space for gain >15 dB is above the dash-dotted line. Within the design space that meets both NF $\leq$ 1.2 dB and gain >15 dB, observe that IIP3 changes dramatically, from approximately −5 to 15 dB m. The optimum design point for a maximum IIP3 is thus L_E = 80 μm and I_C = 7.5 mA. The maximum IIP3 is then above 15 dBm, with a resultant noise figure of 1.15 dB.

In contrast, IIP3 at the design point optimum for noise figure (L_E = 60 μm, I_C = 4 mA) is only 0 dBm, with NF = 1.04 dB. The LNA design point optimum for IIP3 is a better overall choice because noise figure is only degraded about 0.11 dB, while IIP3 is significantly higher (by 15 dB). The disadvantage of this design point, however, is that the required bias current is 3.5 mA higher.

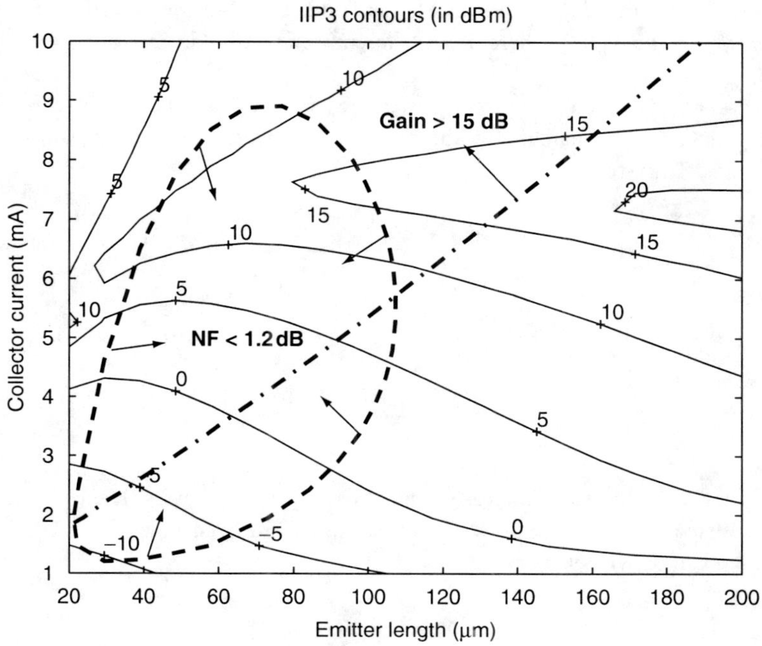

FIGURE 5.7 The NF = 1.2 dB, gain = 15 dB, and IIP3 contours as a function of L_E and I_C.

Therefore, for an optimized design (80 μm and 7.5 mA), IIP3 = 15.8 dB m, gain is 18 dB, NF = 1.15 dB, and $|s_{11}|,|s_{22}| < -30$ dB. Furthermore, if the power consumption constraint shrinks to $I_C \leq$ 5.5 mA, an IIP3 of 5 dB m can be obtained at L_E = 50 μm and I_C = 5.5 mA, with a near-minimum noise figure of 1.08 dB.

Analytical Expressions and Design Rules of Thumb

The procedure introduced above is a numerical methodology that produces accurate simulation results. Furthermore, analytical expressions of the required inductors value, gain, NF, and IIP3 can also be derived by simplifying the device model [11]. These analytical expressions provide the design rules of thumb.

To derive the analytical expressions, we neglect the emitter and collector resistance, base–collector and collector–substrate capacitance, and avalanche multiplication current. The analytical expressions of L_E and L_B can be written as:

$$l_e \simeq \frac{R_s}{2\pi f_T} - \frac{2\pi f_T(1/g_{be} - R_s)}{\omega^2 \beta^2}, \tag{5.19}$$

$$l_b \simeq \frac{2\pi f_T(1/g_{be} - R_s)}{\omega^2 \beta} - l_e \simeq \frac{2\pi f_T(1/g_{be} - R_s)}{\omega^2 \beta} - \frac{R_s}{2\pi f_T}, \tag{5.20}$$

where $\omega = 2\pi f$ is the circuit operating frequency, $R_s = 50\,\Omega$ is the source impedance, and f_T is the cutoff frequency at the chosen $L_E - I_C$ point. Note that g_{be} was neglected in the traditional input impedance matching equations, assuming $1 \ll \omega\beta/2\pi f_T$ and $R_s \ll 1/g_{be}$ [12]. However, the assumption of $2\pi f_T \ll \omega\beta$ is no longer valid when f_T is much higher than the operational frequencies for SiGe HBT LNAs. In RFIC fabrication, smaller l_b is preferred* because that the Q-factor of typical on-wafer inductors is small and hence will degrade noise performance. From the equations above,

*The value l_e is usually much smaller than l_b, then the effect of the parasitics of l_e on noise is negligible.

when I_C is fixed, the required l_b becomes smaller as f_T decreases. At low injection, f_T decreases as the current density ($J_C = I_C/W_E/L_E$, where W_E is the emitter width) decreases, then larger L_E requires smaller l_b.

Assuming that $\omega \ll 2\pi f_T$, then the current gain of the common-base stage is close to unity. Moreover, since the input impedance is matched to the source resistance R_s, the current injected into the EB junction is constant. Thus, the current gain equals to

$$\frac{\beta g_{be}}{g_{be} + j\omega C_{be}} = \frac{\beta}{1 + j\omega\beta/2\pi f_T}, \tag{5.21}$$

and the gain of power can be written as

$$G = \frac{\beta^2 R_{load}}{[1 + (\omega\beta/2\pi f_T)]^2 R_s}. \tag{5.22}$$

At low injection, as J_C increases, f_T increases, and thus the gain increases. Hence, gain increases with increasing I_C and fixed L_E, and decreases with increasing L_E and fixed I_C, as expected.

The noise figure can be written as:

$$\mathrm{NF} = 10\log_{10}(1 + n_{ib} + n_{ic} + n_{vb}), \tag{5.23}$$

$$n_{ib} = \frac{(g_{be}R_s)^2 + [B(1 - g_{be}R_s)]^2}{2g_{be}R_s}, \tag{5.24}$$

$$n_{ic} = \frac{4(g_{be}R_s)^2 + [g_{be}R_s/B + B(1 - g_{be}R_s)]^2}{2g_m R_s}, \tag{5.25}$$

$$n_{vb} = \frac{r_b}{R_s}, \tag{5.26}$$

where $B = 2\pi f_T/\omega\beta$. Under an overall power consumption constraint, I_C normally is less than 10 mA, meaning $g_{be}R_s \ll 1$. Rewriting the above equations, one obtains

$$n_{ib} = \frac{(g_{be}R_s)^2 + B^2}{2g_{be}R_s}, \tag{5.27}$$

$$n_{ic} = \frac{4(g_{be}R_s)^2 + [g_{be}R_s/B + B]^2}{2g_m R_s}, \tag{5.28}$$

$$n_{vb} = \frac{r_b}{R_s}. \tag{5.29}$$

According to these equations, when L_E is fixed, at lower I_C, the collector current shot noise (n_{ic}) is dominant. The NF increases as I_C decreases. At higher I_C, the base current shot noise (n_{ib}), which is proportional to I_C, dominates NF. Thus, there is an optimum value of I_C that minimizes NF.

When I_C is fixed, at a small L_E, the thermal noise caused by the base resistance (n_{rb}) is dominant. At a large L_E, however, the collector current shot noise (n_{ic}) contribution dominates NF. Therefore, there is an optimum L_E that balances the noise caused by r_b and I_C.

Applying Volterra series, one can derive

$$\mathrm{IM3} \simeq \left|\frac{3}{4}C(\omega_1,\omega_2)L(\omega)(1 - G(2\omega_1) - 2G(\omega_1 - \omega_2))\right|, \tag{5.30}$$

where

$$C(\omega_1,\omega_2) = \frac{1}{6V_t^2} v_{\text{in},1}(\omega_1) v_{\text{in},1}(-\omega_2) \simeq \frac{1}{6V_t^2} |v_{\text{in},1}^2(\omega_1)|, \tag{5.31}$$

and $v_{\text{in},1}$ is the first-order ac voltage on the EB junction. In addition,

$$L(\omega) = \frac{V_t K(\omega) G(\omega)}{I_C}, \tag{5.32}$$

where

$$G(\omega) = \frac{A(\omega) I_C}{B(\omega) + A(\omega) I_C}, \tag{5.33}$$

$$K(\omega) = \frac{B(\omega)}{A(\omega) V_t}, \tag{5.34}$$

$$A(\omega) = \left(j\omega\tau_f + \frac{1}{\beta}\right)(Z_b(\omega) + Z_e(\omega)) + Z_e(\omega), \tag{5.35}$$

$$B(\omega) = V_t[1 + j\omega C_{te}(Z_b(\omega) + Z_e(\omega))], \tag{5.36}$$

$$Z_b(\omega) = j\omega l_b + R_s, \tag{5.37}$$

$$Z_e(\omega) = j\omega l_e, \tag{5.38}$$

where l_e and l_b are determined by Equation 5.19 and Equation 5.20.

$C(\omega_1, \omega_2)$ is the square of the magnitude of the first-order ac voltage across the EB junction. As I_C increases, the voltage drop across the EB junction decreases, and thus this term decreases. $L(\omega)$ is proportional to the current gain divided by I_C at the operating frequency, and also decreases as I_C increases. The third term of IM3, $|(1-G(2\omega_1)-2G(\omega_1-\omega_2))|$, determines the nonlinearity cancellation and is called cancellation term. Figure 5.8 shows the $|1-G(2\omega)|$, $|2G(\Delta\omega)|$, and $|1-G(2\omega) - 2G(\Delta\omega)|$ as a function of I_C. At $I_C = 7$ mA, $|1 - G(2\omega)|$ is equal to $|2G(\Delta\omega)|$, and they are both in phase. Thus, $|1-G(2\omega)-2G(\Delta\omega)|$ is minimized at this point.

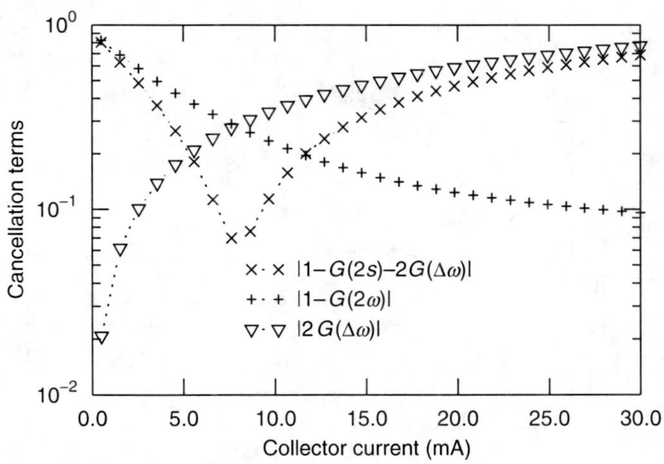

FIGURE 5.8 Cancellation term and its two components as a function of I_C. Note that the minimum value of this term responds to the maximum cancellation.

Fixing L_E, the cancellation term can be maximized at a specific I_C. An IM3 valley or an IIP3 peak is observed when the effect of cancellation dominates. Figure 5.9 shows the three terms, the total IM3, and the IIP3 for the input-impedance matched amplifier. The minimum value of the cancellation term, IM3, and the maximum IIP3 occur at the same value of I_C, proving that the cancellation term dominates IM3.

The IM3 is a similar function of L_E. However, at some I_Cs, the other two terms dominate IM3. Thus, when I_C is fixed at these values, no IIP3 peak can be observed. Figure 5.10 and Figure 5.11 show the three terms, the IM3, and the IIP3 as a function of L_E at $I_C = 6$ mA and $I_C = 7.5$ mA. An IIP3 peak can be observed at $I_C = 6$ mA, but no IIP3 peak is observed at $I_C = 7.5$ mA. The cancellation is pushed to higher current and weakened as I_C increases, then the other two terms dominate the IM3 trend.

The derived behavior of gain, NF, and IIP3 as varying L_E and I_C is identical to the simulations, although they offer better intuitive insight into the optimum LNA design space. Therefore, one can at the first-order locate the optimum design range using the equations above.

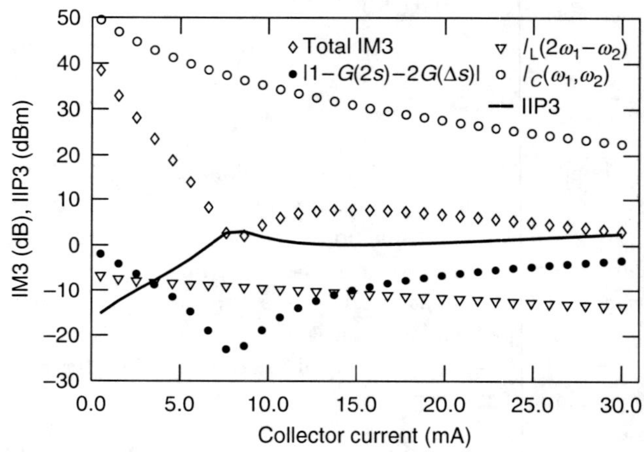

FIGURE 5.9 The three terms, the total IM3, and the IIP3 for the input-impedance matched amplifier.

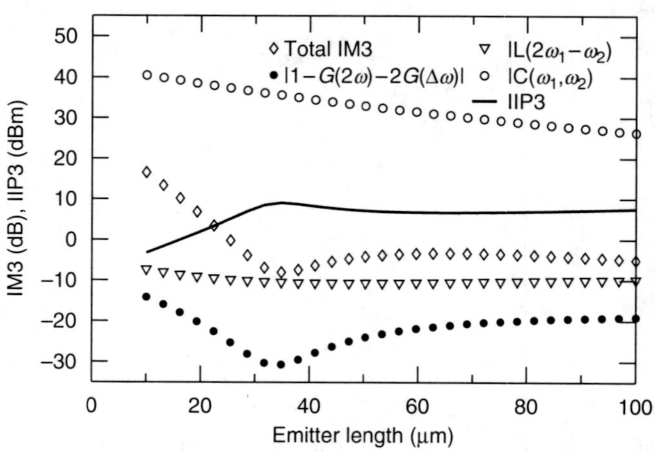

FIGURE 5.10 IM3 terms and the IIP3 as a function of L_E at $I_C = 6$ mA.

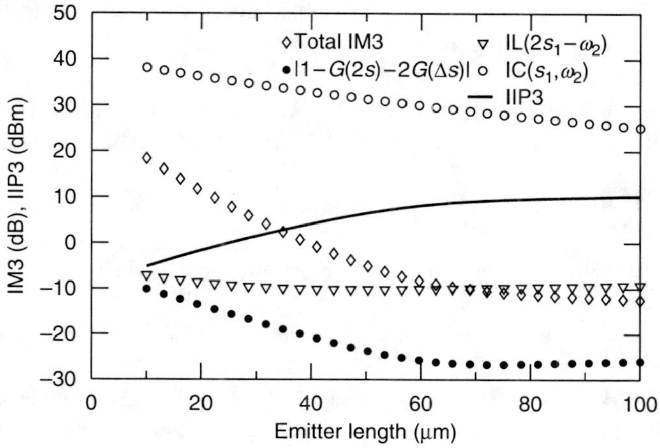

FIGURE 5.11 IM3 terms and the IIP3 as a function of L_E at $I_C = 7.5$ mA.

5.4 Summary

In this chapter, we proposed a generalized LNA design problem: how to trade-off various requirements of LNA in an extended design space, including all design variables. Then we presented optimization strategies and demonstrated a design example of an inductively degenerated cascade LNA using IBM 5HP technology. As shown in the example, close to the valley of NF, the noise figure changes slightly in a decently wide design-space, while IIP3 drops dramatically from 15 to −5 dB m. The results indicate that the optimization methodology, which considers not only NF but also IIP3 and gain, is necessary.

The presented optimization strategies provide a convenient approach for an LNA design that balances all specifications. Furthermore, we derived the analytical expressions and design rules of thumb, which facilitate the determination of the optimum design range.

Acknowledgments

We would like to thank D. Herman, A. Joseph, G. Freeman, D. Ahlgren, J. Dunn, B. Meyerson, and the IBM SiGe team for their support and contributions. This work was supported by IBM, the Semiconductor Research Corporation, and the Georgia Electronic Design Center at Georgia Tech.

References

1. B Razavi. *RF Microelectronics.* Indianapolis, IN: Prentice-Hall, 1997.
2. P Crippa, S Orcioni, F Ricciardi, and C Turchetti. Design of a 4.4 to 5 GHz LNA in 0.25-μm SiGe BiCMOS technology. Proceedings of the International Symposium on Circuits and Systems, 2003, pp. 333–336.
3. Q Liang, G Niu, JD Cressler, S Taylor, and DL Harame. Geometry and bias current optimization for SiGe HBT cascode low-noise amplifiers. Technical Digest of the IEEE Radio Frequency Integrated Circuits Symposium, 2002, pp. 407–410.
4. M Soyuer, J-O Plouchart, H Ainspan, and J Burghartz. A 5.8-GHz 1-V low-noise amplifier in SiGe bipolar technology. Technical Digest of the IEEE Radio Frequency Integrated Circuits Symposium, 1997, pp. 19–22.
5. JD Cressler. SiGe HBT technology: a new contender for Si-based RF and microwave circuit applications. *IEEE Trans. Micro. Theory Tech.* 46:572–589, 1998.
6. HA Haus, WR Atkinson, WH Fonger, WW Mcleod, GM Branch, WA harris, EK Stodola, WB Davenport Jr, SW Harrison, and TE Talpey. Representation of noise in linear twoports. *Proc. IRE* 48:67–74, 1960.

7. S Zhang, G Niu, JD Cressler, AJ Joseph, G Freeman, and DL Harame. The effects of geometrical scaling on the frequency response and noise performance of SiGe HBTs. *IEEE Trans. Electron Dev.* 49:429–435, 2002.
8. G Niu, JD Cressler, S Zhang, U Gogineni, and DC Ahlgren. Measurement of collector–base junction avalanche multiplication effects in advanced UHV/CVD SiGe HBTs. *IEEE Trans. Electron Dev.* 46:1007–1015, 1999.
9. G Niu, Q Liang, JD Cressler, CS Webster, and DL Harame. *IEEE Trans. Micro. Theory Tech.* 49:1558–1565, 2002.
10. P Wambacq and W Sansen. *Distortion Analysis of Analog Integrated Circuits.* New York, NY: Kluwer Academic, 1998.
11. JD Cressler and G Niu. *Silicon–Germanium Heterojunction Bipolar Transistors.* Boston, MA: Artech House, 2003.
12. H Schumacher, U Erben, W Dürr, and K-B Schad. Low-noise, low-power wireless front-end MMICs using SiGe HBTs. *IEICE Trans. Electron E* 82-C:1943–1950, 1999.

6

Linearization Techniques

6.1 Introduction 6-1
6.2 Nonlinear Distortion Concepts 6-2
Power Series Approach • Single-Tone Excitation • Two-Tone Excitation • Cascaded Systems • Feedback Configurations • Hard Nonlinearities and Clipping
6.3 Circuit Interaction of Device Nonlinearities 6-9
Dominant Device Nonlinearities • Low Current/Power Operation • High Current/Power Operation
6.4 RF Design Techniques for Linearity 6-13
Current-Mode Operation • Negative Feedback Techniques • Shaping the Transfer Function • Harmonic Matching Techniques

Leo C.N. de Vreede and
Mark P. van der Heijden
Delft University of Technology

6.1 Introduction

Currently Si–SiGe BiCMOS process technology is the workhorse for the implementation of wireless building blocks. Depending on the application, circuit function, and communication standard, a wide variety of requirements and specifications have to be fulfilled. A general trend in wireless communication systems is the use of increased data rates within confined frequency bands. For this purpose, frequency bandwidth-efficient schemes for the digitally modulated signals (e.g. EDGE-GSM, WCDMA) have been introduced. One of the common characteristics of these communication standards is the large amplitude modulation component, which puts high linearity demands on the circuit blocks involved.

For this reason linearity is currently a key requirement in modern RF front-end design, since the transceiver linearity mainly determines the proper reception or transmission of RF signals. This increased linearity requirement is in conflict with the general demand to lower the DC power consumption of new product generations. Consequently, the designer is confronted with the challenge to develop low DC power consuming, highly linear circuit solutions for the basic RF circuit functions like the low-noise amplifier (LNA), the up/down converting mixer, and the transmitter power amplifier (PA).

This chapter is intended to provide the reader with basic understanding of nonlinear distortion in bipolar circuits and provide solutions towards the design of more power efficient and linear circuit implementations. In order to address this complicated task in a unified manner, we start in Section 6.2 with the review of basic distortion concepts relevant to RF design. Then, in Section 6.3 we study the linearity of a bipolar transistor for a simple but practical test circuit and identify the dominant distortion phenomena for a particular bias condition. Based on the knowledge of the dominant nonlinearities, we can distinguish various design techniques for improved circuit linearity. Finally, Section 6.4 introduces the basics of these design methods and gives various design considerations and constraints by comparing the different circuit implementations for their performance based on a 70 GHz SiGe reference device. All linearity considerations and conclusions are supported analytically as well as by harmonic balance simulations using the Mextram 504 model [1]. Especially, emphasis is placed on clarifying the dominant distortion phenomena and distortion cancellation conditions in today's most popular and promising circuit topologies.

6.2 Nonlinear Distortion Concepts

All practical active devices have nonlinear electrical characteristics; this is especially true for bipolar transistors, which have an exponential relationship between output current and input voltage. For this reason, circuits based on active devices will exhibit nonlinear distortion. Although there are some circuits that only exist by the grace of nonlinear device operation, like oscillators or frequency multipliers. In most cases, however, nonlinear circuit behavior is undesired, since it causes signal corruption and channel-to-channel interference.

Power Series Approach

Depending on the circuit specification under consideration, linearity is expressed in terms like gain compression, intermodulation distortion, cross-modulation, blocking, and desensitization [2]. In order to gain more insight in nonlinear distortion phenomena, we consider for the moment the basic nonlinear properties of two widely used bipolar building blocks, namely the common-emitter (CE) stage in Figure 6.1 and the differential pair in Figure 6.2. These building blocks are often favored in LNA or mixer designs, due to their inherent high gain and low noise behavior. In our analytical considerations we assume for now a *memoryless* system. This basically means that we neglect all charge-storage elements and their associate nonlinearities in the active device, which is of course a strong simplification. Later in Section 6.4, we deal with these cases by using the more rigorous Volterra series approach for the analysis of weak nonlinear circuits [3,4]. For now, we model the transfer functions of the CE-stage and the differential pair with a power series representation, as given by

$$y(t) = a_1 x(t) + a_2 x^2(t) + a_3 x^3(t) + \cdots \tag{6.1}$$

in which the a_n are the Taylor coefficients of the transfer function at the desired operating point X_0, defined as

$$a_n = \frac{1}{n!}\frac{\mathrm{d}^n y(X_0)}{\mathrm{d}x^n} \tag{6.2}$$

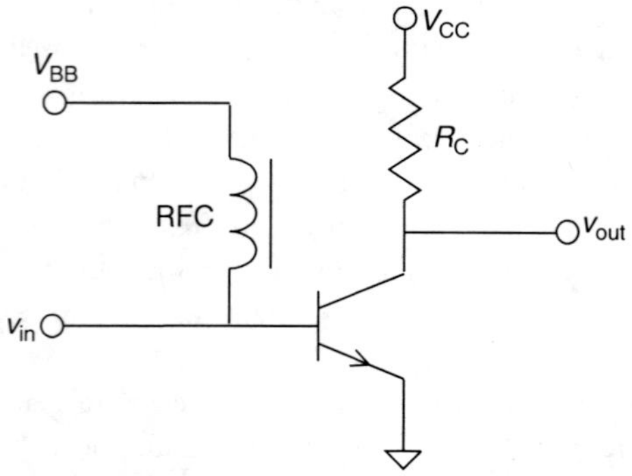

FIGURE 6.1 The common-emitter configuration.

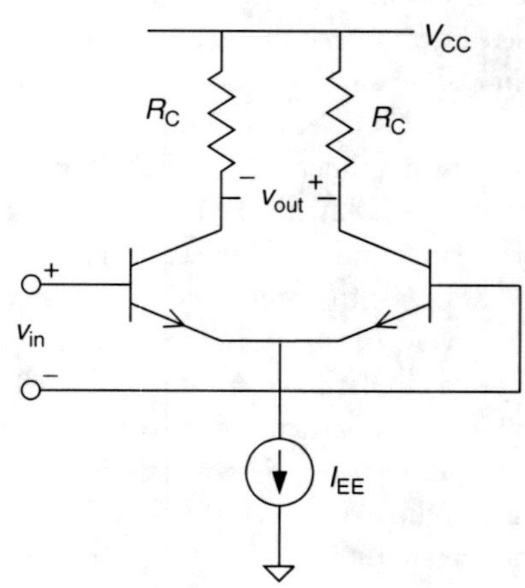

FIGURE 6.2 The differential pair.

The voltage transfer function of the CE-stage with ideal DC- voltage biasing and voltage drive at the base can now be expressed as

$$V_{\mathrm{OUT}} = V_{\mathrm{CC}} - R_{\mathrm{C}}I_{\mathrm{C}} = V_{\mathrm{CC}} - R_{\mathrm{C}}I_{\mathrm{S}}\left[\exp\left(\frac{V_{\mathrm{IN}}}{V_{\mathrm{T}}}\right) - 1\right] \tag{6.3}$$

So, the nonlinear Taylor coefficients of V_{OUT} are given by

$$a_1 = -\frac{R_{\mathrm{C}}I_{\mathrm{C}}}{V_{\mathrm{T}}}, \quad a_2 = -\frac{R_{\mathrm{C}}I_{\mathrm{C}}}{2V_{\mathrm{T}}^2}, \quad a_3 = -\frac{R_{\mathrm{C}}I_{\mathrm{C}}}{6V_{\mathrm{T}}^3} \tag{6.4}$$

Similarly, we can write the well-known voltage transfer function of the differential pair as [5]:

$$V_{\mathrm{OUT}} = R_{\mathrm{C}}I_{\mathrm{EE}}\tan\mathrm{h}\left(\frac{V_{\mathrm{IN}}}{2V_{\mathrm{T}}}\right) \tag{6.5}$$

and the nonlinear coefficients of V_{OUT} are

$$a_1 = \frac{I_{\mathrm{EE}}R_{\mathrm{C}}}{2V_{\mathrm{T}}}, \quad a_2 = 0, \quad a_3 = -\frac{I_{\mathrm{EE}}R_{\mathrm{C}}}{24V_{\mathrm{T}}^3} \tag{6.6}$$

Note that the differential pair has an odd-transfer function, consequently all a_n are zero if n is even. We will now use the previously introduced power series models to explain some distortion phenomena under single and two-tone excitation.

Single-Tone Excitation

If we excite the nonlinear system in (6.1) with a sinusoidal signal $x(t) = A\cos(\omega t)$, harmonic signals will be generated at various frequencies:

$$y(t) = \overbrace{\frac{1}{2}a_2A^2}^{\text{DC}} + \overbrace{\left(a_1A + \frac{3}{4}a_3A^3\right)\cos\omega t}^{\text{Fundamental}} + \overbrace{\frac{1}{2}a_2A^2\cos 2\omega t}^{\text{2nd-Harmonic}} + \overbrace{\frac{1}{4}a_3A^3\cos 3\omega t}^{\text{3rd-Harmonic}} + \cdots \tag{6.7}$$

This type of distortion is called harmonic distortion and *is especially harmful in broadband amplifiers* such as in cable television (CATV) or audio amplifiers. In order to quantify harmonic distortion, usually the ratio is taken of the amplitude at the specific harmonic frequencies with the amplitude of the fundamental frequency, which is a_1A. In this way the second-harmonic distortion (HD2) is expressed as

$$\mathrm{HD2} = \frac{1}{2}\left|\frac{a_2}{a_1}\right|A, \tag{6.8}$$

and the third-harmonic distortion (HD3) is expressed as

$$\mathrm{HD3} = \frac{1}{4}\left|\frac{a_3}{a_1}\right|A^2 \tag{6.9}$$

In RF circuit design, these harmonics are almost always automatically filtered out, due to *the narrowband nature* of the circuits involved.

Two other nonlinear phenomena that we can quantify using a single-tone excitation is the DC-shift due to self-biasing and gain compression or expansion. As we can see from (6.7), the self-biasing ($\frac{1}{2}a_2A^2$) is a second-order nonlinear effect and yields an increase in the DC-bias current. Gain compression or expansion can be explained by inspecting the fundamental amplitude in (6.7). We see that as the input signal amplitude A rises, the factor $\frac{3}{4}a_3A^3$ (depending on the sign of the cubic term a_3) may cause an increase or decrease of the fundamental signal (i.e. gain expansion or compression, respectively). For example, the cubic term of the CE configuration in (6.4) has the same sign as the linear term, leading to some gain expansion. On the other hand, the cubic term of the differential pair in (6.6) has an opposite sign with respect to the linear term, leading to gain compression at high drive levels. In fact, also the CE configuration will suffer from gain compression at some point. However, this will be due to other, strong nonlinear effects such as signal clipping to the supply voltage or saturation of the collector current (see "Hard Nonlinearities and Clipping" section), which are generally not described properly by a power series expansion. In general, gain compression is quantified by the 1 dB gain compression point ($P_{1\text{dB}}$) and is defined as the input or output power level where the gain has dropped 1 dB with respect to the small-signal gain.

Two-Tone Excitation

Two-tone testing is quite common for wireless systems, since it yields information about various undesired distortion phenomena, such as intermodulation distortion (IMD), crossmodulation distortion, blocking, and desensitization [2]. Here we only quantify intermodulation distortion, since in general it gives sufficient information about the other distortion types. In general, a two-tone test is performed by exciting a system with two sinusoidal signals with equal amplitude, namely: $x(t) = A\cos(\omega_1 t) + A\cos(\omega_2 t)$. By substituting this two-tone input signal into (6.1), we obtain the following expression for the output signal, containing the distortion products up the third harmonic and intermodulation frequencies:

$$y(t) = \overbrace{a_2A^2}^{\text{DC}} + \overbrace{\left(a_1A + \frac{9}{4}a_3A^3\right)\cos\omega_{1,2}t}^{\text{Fundamental}} + \overbrace{\frac{1}{2}a_2A^2\cos 2\omega_{1,2}t}^{\text{2nd-Harmonic}} + \overbrace{\frac{1}{4}a_3A^3\cos 3\omega_{1,2}t}^{\text{3rd-Harmonic}}$$
$$+ \underbrace{a_2A^2\cos(\omega_2 \pm \omega_1)}_{\text{2nd-orderIMD}} + \underbrace{\frac{3}{4}a_3A^3\cos(2\omega_{1,2} \pm \omega_{2,1})}_{\text{3rd-orderIMD}} + \cdots \quad (6.10)$$

In which $\omega_{1,2}$ refers to the two tones at ω_1 and ω_2, respectively. Besides the DC-shift and the harmonics, second-order intermodulation products (second-order IMD) are generated at the envelope (or baseband) frequency $\omega_2 - \omega_1$ and sum frequency $\omega_2 + \omega_1$. Furthermore, third-order intermodulation products (third-order *IMD*) are generated at $2\omega_{1,2} - \omega_{2,1}$ adjacent to the fundamentals and at $2\omega_{1,2} + \omega_{2,1}$ adjacent to the third harmonic. In general, the third-order IMD products at $2\omega_1 - \omega_2$ and $2\omega_2 - \omega_1$ are the most troublesome in RF circuit design, since these products are close to the desired channels and therefore cannot be filtered out, yielding signal corruption. The second-order IMD product at $\omega_2 - \omega_1$ is equally important, when considering homodyne or zero-IF receivers. This is because, unwanted signals in the receive band (e.g., caused by channels that are unwanted but yet close in frequency to the wanted signal) can down convert through second-order nonlinearities, corrupting the desired baseband signal [2].

To illustrate some of the distortion figure of merits, we plot in Figure 6.3 the amplitude of the fundamental, the second- and third-order IMD as function of input signal amplitude as expressed in (6.10). Like HD2 and HD3, we now introduce IM2 and IM3, which are defined as the ratio of the intermodulation distortion amplitude with respect to the fundamental signal amplitude:

$$\text{IM2} = \left|\frac{a_2}{a_1}\right| A \quad (6.11)$$

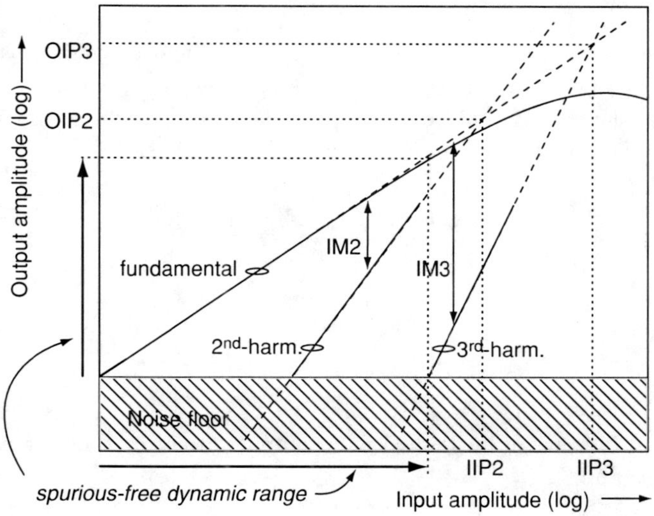

FIGURE 6.3 The fundamental, the second-order, and third-order intermodulation amplitude as function of input amplitude.

$$\mathrm{IM3} = \frac{3}{4}\left|\frac{a_3}{a_1}\right| A^2 \tag{6.12}$$

Since we only consider a memoryless system (power series), we find that IM2 = $2\dot{\mathrm{H}}\mathrm{D2}$ and IM3 = $3\dot{\mathrm{H}}\mathrm{D3}$, which in general does not hold for systems with memory. Another figure-of-merit, which is commonly used in low-power or weakly nonlinear circuits, is the input second-order intercept point (IIP2) and the input third-order intercept point (IIP3). Both quantities are defined as the input amplitude A, where the extrapolated fundamental amplitude intersects with the extrapolated amplitude of second and third-order intermodulation products as depicted in Figure 6.3

$$\mathrm{IIP2} = \left|\frac{a_1}{a_2}\right| \tag{6.13}$$

$$\mathrm{IIP3} = \sqrt{\frac{4}{3}\left|\frac{a_1}{a_3}\right|} \tag{6.14}$$

If we calculate all these distortion figures for the voltage-driven differential pair and the CE stage using the coefficients in (6.4) and (6.6), the IM2 and IM3 become

$$\mathrm{IM2_{CE}} = \frac{1}{2}\frac{A}{V_\mathrm{T}}, \quad \mathrm{IM2_{DP}} = 0 \tag{6.15}$$

$$\mathrm{IM3_{CE}} = \frac{1}{8}\frac{A^2}{V_\mathrm{T}^2}, \quad \mathrm{IM3_{DP}} = \frac{1}{16}\frac{A^2}{V_\mathrm{T}^2} \tag{6.16}$$

and the IIP2 and IIP3 become

$$\mathrm{IIP2_{CE}} = 2V_\mathrm{T}, \quad \mathrm{IIP2_{DP}} = \infty \tag{6.17}$$

$$\mathrm{IIP3_{CE}} = \sqrt{8}\cdot V_\mathrm{T}, \quad \mathrm{IIP3_{DP}} = 4V_\mathrm{T} \tag{6.18}$$

Discussion

The previous analysis showed that the input intercept point of a bipolar transistor is independent of the DC bias current. This can be better understood by looking at the output current of a bipolar transistor as function of input voltage. This can be evaluated with the following analysis, where we use the power series of the exponential collector-current nonlinearity:

$$i_c(t) = g_m v_{be}(t) + g_{m,2} v_{be}^2(t) + g_{m,3} v_{be}^3(t) \tag{6.19}$$

in which

$$g_m = \frac{I_C}{V_T}, \quad g_{m,2} = \frac{I_C}{2V_T^2}, \quad g_{m,3} = \frac{I_C}{6V_T^3} \tag{6.20}$$

If we now write $v_{be}(t)$ in terms of the desired linear output current $i_{c1}(t)$ and transconductance g_m as $v_{be}(t) = i_{c1}(t)/g_m$, we find

$$i_c(t) = i_{c1}(t) + B_2 i_{c1}^2(t) + B_3 i_{c1}^3(t), \tag{6.21}$$

in which $B_n = g_{m,n}/ g_m{}^n$. Normalizing the above for I_C yields

$$m(t) = \frac{i_c(t)}{I_C} = m_1(t) + I_C B_2 m_1^2(t) + I_C^2 B_3 m_1^3(t) \tag{6.22}$$

with $m_1(t) = i_{c1}(t)/I_C$, being the relative linear current swing. Substituting the nonlinear coefficients of (6.20) in (6.22), we obtain [6]

$$m(t) = m_1(t) + \frac{1}{2} m_1^2(t) + \frac{1}{6} m_1^3(t) \tag{6.23}$$

Consequently, the IM3 for a CE stage can be written as

$$\mathrm{IM3} = \frac{3}{4} I_C^2 B_3 |m_1|^2 = \frac{1}{8} \left| \frac{i_{c1}}{I_C} \right|^2 \tag{6.24}$$

Note that $m_1(t) = i_{c1}(t)/I_C = v_{be}(t)/V_T$, which for $v_{be}(t) = A\cos(\omega_1 t) + A\cos(\omega_2 t)$ leads to the same IM3-expression for the CE stage as found in (6.16). From (6.24) we observe that IM3 is a function of the relative output current swing. Therefore, linearity can be improved for a required output current swing by increasing the DC bias current. In the same way we can argue that the linearity improves by decreasing the relative voltage swing at the input with respect to the thermal voltage V_T for a given DC bias current.* This is why in practice it is always possible to improve the output linearity of a transistor stage at the expense of DC power consumption. This becomes more evident when we consider the output third-order intercept point (OIP3) of a bipolar transistor. The OIP3 is defined as the amplitude of the fundamental at the output at which the extrapolated IM3 amplitude intercepts the extrapolated linear part of the fundamental amplitude (see Figure 6.3). Alternatively, the OIP3 can also directly be calculated by multiplying IIP3 with the linear gain factor (i.e., $g_m = I_C/V_T$ in case of a voltage driven CE-stage, where we monitor the output current), consequently

$$\mathrm{OIP3} = \sqrt{8} I_C = g_m \mathrm{IIP3} \tag{6.25}$$

*Decreasing the input voltage swing is commonly referred to as back-off.

From this we can conclude that for a purely voltage driven bipolar transistor, the OIP3 is proportional to the collector current I_C, as long as the nonlinear distortion of the bipolar transistor is dominated by its exponential nonlinearity.

Cascaded Systems

In RF transceiver design, we are normally dealing with a cascade of nonlinear circuits blocks (e.g., the cascade of the LNA and the downconverting mixer). In order to calculate the (spurious-free) dynamic range of the receiver (see Figure 6.3), we need to know how the nonlinearities of the individual blocks are referred back to the input. The easiest way to do this, is by calculating an overall input third-order intercept point in terms of the individual intercept points of all stages. Therefore, we consider the two cascaded nonlinear stages in Figure 6.4 of which the nonlinearity of stage A is given by (1) and the nonlinearity of stage B by

$$z(t) = b_1 y(t) + b_2 y^2(t) + b_3 y^3(t) \tag{6.26}$$

We can calculate the overall IIP3 of the cascaded system by calculating the intersect of the fundamental response with the resulting overall third-order nonlinear response. We can do this by substituting (1) into (26) and regroup the first- and third-order terms of the overall response, yielding [2]:

$$\mathrm{IIP3}_{\mathrm{cas}} = \sqrt{\frac{4}{3}\left|\frac{a_1 b_1}{a_1^3 b_3 + 2a_1 a_2 b_2 + a_3 b_1}\right|} \tag{6.27}$$

In a worst-case scenario, the terms of the denominator under the square-root of (6.27) will add linearly, yielding the following approximation:

$$\mathrm{IIP3}^2_{\mathrm{cas}} = \left[\frac{1}{\mathrm{IIP3}^2_{\mathrm{A}}} + \frac{a_1^2}{\mathrm{IIP3}^2_{\mathrm{B}}} + \frac{3a_2 b_2}{2b_1}\right]^{-1} \tag{6.28}$$

where $\mathrm{IIP3}_A$ and $\mathrm{IIP3}_B$ represent the input IP3 of the first and second nonlinear circuit blocks in Figure 6.4. The last term in (6.28) represents the secondary mixing in stage B of the second-harmonic and the fundamental provided by stage A. Although, in general, people tend to ignore this term, in practical situations it can influence the overall linearity significantly. For the moment, however, we can conclude from (6.28) that by increasing the gain (a_1) of stage A, the linearity of stage B becomes relatively more important and tends to form a natural limit for the overall linearity of the system.

Feedback Configurations

Negative feedback plays an important role in RF circuit design in both a beneficial and harmful way. Undesired linear (or nonlinear) feedback through the base–collector capacitance and emitter resistance are examples of the harmful aspects of feedback, since it basically limits the high-frequency gain of the device. A positive aspect is the possibility to create an accurate overall linear transfer function with "linear" passive components. This is why it is interesting to study the effect of negative feedback on the overall linearity of a system in more detail. For this purpose we consider a nonlinear amplifier with a

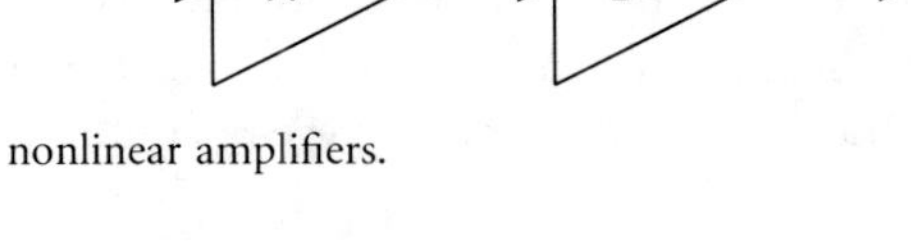

FIGURE 6.4 A cascade of two nonlinear amplifiers.

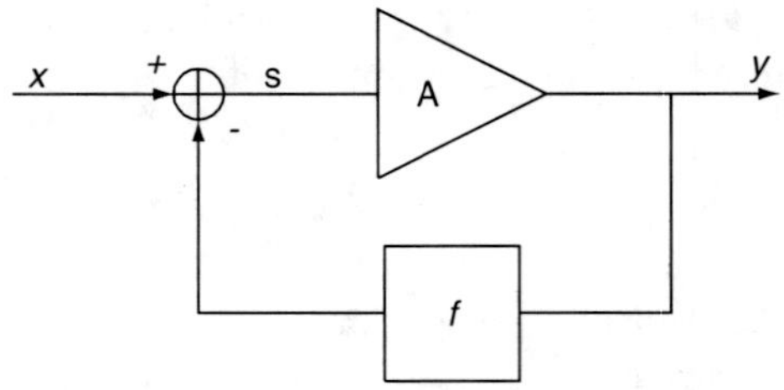

FIGURE 6.5 A nonlinear amplifying element with linear feedback.

linear negative feedback loop as given in Figure 6.5 for which the transfer of the nonlinear amplifier core is given by

$$y(t) = a_1 s(t) + a_2 s^2(t) + a_3 s^3(t) \tag{6.29}$$

We can develop the expression for the IM2 and IM3 of the overall feedback amplifier in terms of the Taylor coefficients of the nonlinear core and the negative feedback element *f*, given the fact that for Figure 6.5 we can write

$$s(t) = x(t) - fy(t) \tag{6.30}$$

Substitution of (6.30) in (6.29) yields

$$y(t) = a_1[x(t) - fy(t)] + a_2[x(t) - fy(t)]^2 + a_3[x(t) - fy(t)]^3 \tag{6.31}$$

By now substituting

$$y(t) = n_1 x(t) + n_2 x^2(t) + n_3 x^3(t) \tag{6.32}$$

in both left and right-hand side of (6.31), we can solve for n_1, n_2, and n_3 by isolating the terms with the corresponding power of $x(t)$, yielding:

$$n_1 = \frac{a_1}{1 + a_1 f}, \quad n_2 = \frac{a_2}{(1 + a_1 f)^3}, \quad n_3 = \frac{a_3(1 + a_1 f) - 2a_2^2 f}{(1 + a_1 f)^5} \tag{6.33}$$

With these nonlinear Taylor coefficients of the overall negative feedback amplifier, we can compare the second- and third-order IMD of the feedback amplifier $\mathrm{IM2_{FB}}$ and $\mathrm{IM3_{FB}}$ to the second- and third-order IMD of the nonlinear amplifier core alone ($\mathrm{IM2_{Amp}}$ and $\mathrm{IM3_{Amp}}$), which are expressed as

$$\mathrm{IM2_{FB}} = \frac{n_2}{n_1} A = \frac{a_2}{a_1(1 + a_1 f)^2} A = \frac{\mathrm{IM2_{Amp}}}{(1 + a_1 f)^2} \tag{6.34}$$

$$\mathrm{IM3_{FB}} = \frac{3}{4}\frac{n_3}{n_1} A^2 = \frac{3}{4}\frac{a_3}{a_1(1 + a_1 f)^3} A^2 = \frac{\mathrm{IM3_{Amp}}}{(1 + a_1 f)^3}, \tag{6.35}$$

where we assumed for simplicity that $a_2 = 0$ in the calculation of IM3. In that case, we see from (6.34) that the feedback action reduces the second-order intermodulation distortion of the original amplifier ($\mathrm{IM2_{Amp}}$) by a factor of $(1 + a_1 f)^2$ and the third-order intermodulation distortion of the original amplifier ($\mathrm{IM3_{Amp}}$) is reduced by a factor $(1 + a_1 f)^3$. Note that there is another contribution to $\mathrm{IM3_{FB}}$, when ($a_2 \neq 0$). In that case, secondary mixing of even harmonics with the fundamental occurs due to the feedback action, yielding an additional contribution to $IM3_{FB}$, as can can be observed from (6.33).

Hard Nonlinearities and Clipping

When the amplitude of the input signal is increased, at some point the output signal that can be generated by the active device(s) will be limited by the bias circuit and are no longer able to follow the input signal over its complete cycle. This is called clipping, yielding a dramatic loss in linearity performance. For this reason clipping conditions should be avoided at all times. An effective way to identify potential clipping problems in a design is to plot the load lines of all active devices in the circuit for a time domain signal with increasing amplitude. In this way voltage or current clipping conditions for each device can be easily recognized and encountered for.

6.3 Circuit Interaction of Device Nonlinearities

Until now we have addressed the basics of nonlinear distortion using ideal bipolar devices with no memory effects, driven by an ideal voltage source. As a consequence, the interaction of the nonlinearities with the surrounding circuitry was more or less omitted. However when considering RF circuit design with real devices, we have to extend our scope in order to identify the dominant sources of distortion at a given bias and frequency point of operation.

One of the most commonly used configurations to test the device behavior for single-tone excitation is the common-emitter stage configuration with 50 Ohm loading conditions at input and output [7–10]. The motivation of this choice is mainly based on the wide availability of 50-Ohm "small-signal" device characterization equipment and consequently, the ease of measurement. One should be aware, however, that device linearity is the result of a strong interaction between active device and external (harmonic) loading conditions provided by the surrounding circuitry. For this reason one must be very careful to qualify device linearity for modulated signals under these simple, far from ideal loading conditions. Note for example that even the presents of simple bias Tees in combination with the DC bias sources in the test setup, can yield ill-defined impedances at the base-band or envelope frequency ($\omega_2 - \omega_1$) for a two-tone excitation. As will be discussed in Section 6.4, this influences the device linearity measurements to a great extent. Based on these considerations, semiconductor manufactures prefer to use practical but rather simple and therefore well-defined test configurations to quantify the linearity performance of their devices. These test configurations are often simplified situations of a meaningful application of their devices, and are adapted to some degree to the desired impedance level and biasing condition of the active device under test (DUT). Although not providing the ultimate linearity performance of the DUT, these benchmark circuits are very useful to identify the influence regions of the various distortion sources [12].

Dominant Device Nonlinearities

A typical benchmark circuit is shown in Figure 6.6, which represents a two-tone test for a dual-loop feedback single-stage transistor network. Note that this configuration can provide simultaneous matched impedance conditions at input- and output by setting the resistors R_1 and R_2 to their proper values [11]. In this section we use such benchmark circuit to investigate the linearity of a typical high-speed SiGe transistor, like the 0.5 μm $\times$ 20.3 μm Philips QUBiC4G BNA style transistor with a peak-f_T of 70 GHz @ 10 mA. In the remainder of this chapter, we will use this device as a reference transistor to analyze its nonlinear behavior for various circuit conditions. To introduce this device, Figure 6.7 and Figure 6.8 show the $I_C(V_{CE})$, $f_T(I_C)$, and F_{min} (I_C) characteristics. As can be observed from this figure, the device exhibits a peak-f_T of approximately 70 GHz, an F_{min} of 0.8 dB, while its BV_{CEO} is 2.5 V. To identify the dominant distortion characteristics of this device under different bias conditions, we consider the circuit topology of Figure 6.6. A two-tone test with small amplitude was performed on

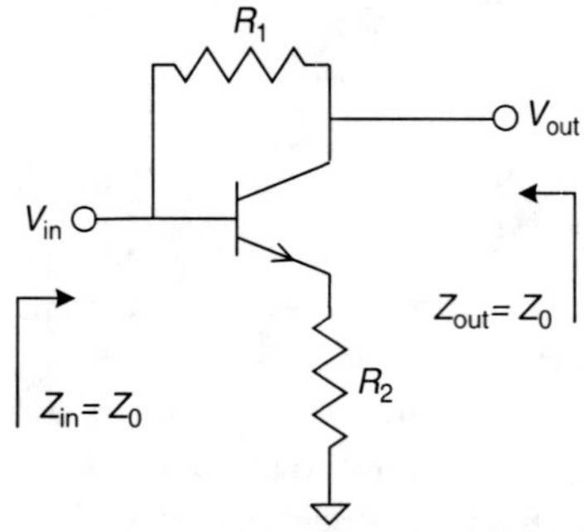

FIGURE 6.6 AC-schematic of a typical benchmark circuit for two-tone testing of an active device for linearity.

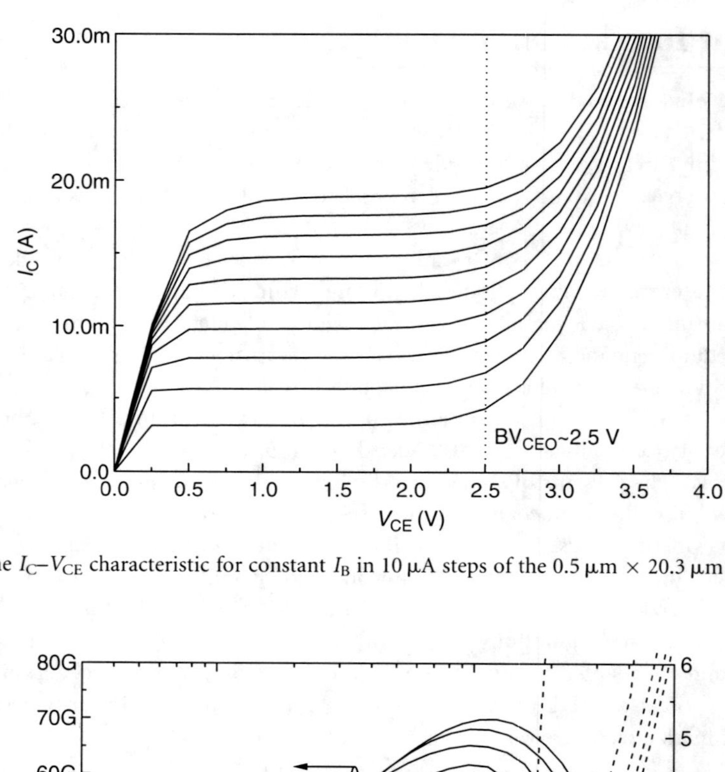

FIGURE 6.7 The I_C–V_{CE} characteristic for constant I_B in 10 μA steps of the 0.5 μm × 20.3 μm Philips QUBIC4G SiGe BJT.

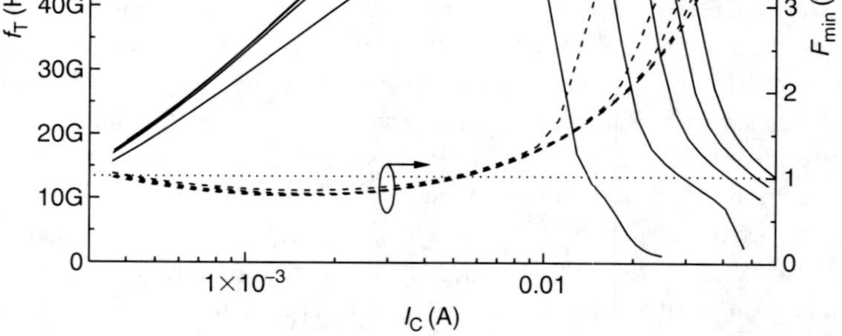

FIGURE 6.8 The f_T and F_{min} versus I_C characteristics derived at 5 GHz for V_{CE} = 0.5, 1, 1.5, 2, and 2.5 V of the 0.5 μm × 20.3 μm Philips QUBIC4G SiGe BJT.

this circuit in order to compute the OIP3 in each bias point of the I_C–V_{CE} plane, as shown in Figure 6.9. By plotting the lines of constant OIP3 in the I_C–V_{CE} plane of the bipolar device, we can identify three regions where different nonlinearities dominate. In support of this identification, Figure 6.9 also indicates the bias conditions where "weak"-avalanche sets in ($V_{CE} \sim 2.5\,\text{V}$), the peak-$f_T$ occurs, as well as the points where the f_T of the device has been reduced to half of its peak value due to high current effects.

The "Low" Current Region

In the region where $I_C < I_C @ f_{T_{peak}}$ the distortion is dominated by the exponential dependency of I_B, I_C, and the diffusion term of C_{be} ($C_{DE} = \tau_f \cdot g_m$) on the input voltage V_{BE}. This can be easily verified by observing there is almost no variation of OIP3 with V_{CE} but only with I_C. Consequently, the DC bias current sets the output linearity, what is in line with the conclusions found for the "bare" CE-stage in Section 4.2.

The Base–Collector/Nonlinear Transit Time Dominated Region

At higher DC-current levels the influence of the exponential distortion on the OIP3 is reduced, therefore other nonlinearities become visible. One of the most important effects is the nonlinear feedback through the base–collector capacitance C_{bc} and the variation of τ_f [13]. Both effects result from the base-charge modulation by the collector voltage and current swing, and become more dominant at a lower V_{CE} or at higher current levels close to quasisaturation. This is due to the strong increase of C_{bc} and τ_f at these bias conditions. Since in general, feedback defines the transfer of a network [11], nonlinear feedback through C_{bc} and τ_f will lead to a nonlinear transfer function and consequently to a nonlinear circuit operation.

The Avalanche Region

At "very" high collector voltages "weak" avalanche effects occur [14], resulting in an avalanche current between the base and the collector. Consequently, again a nonlinear feedback appears which interacts with the other device non-linearities. At higher voltages, this interaction can degrade or improve the device linearity depending on the phase relations of the individual distortion components. Note, for example, that at low current levels the IIP3 should be constant as function of V_{CE}. However, we observe in Figure 6.9 that there is some improvement of the IIP3 for very high V_{CE} due to the influence of

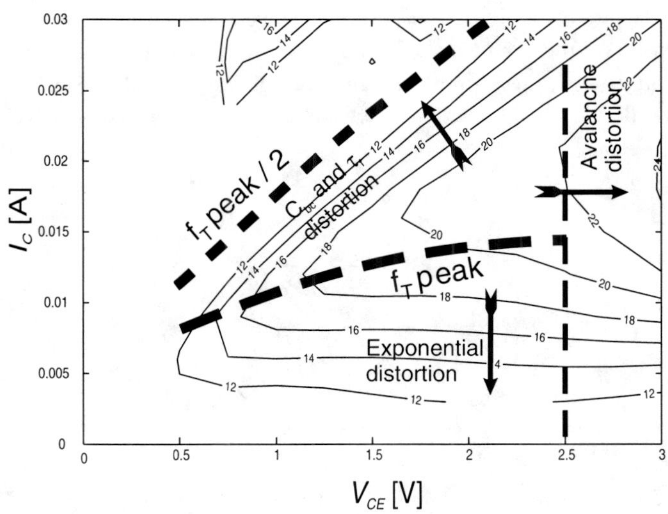

FIGURE 6.9 Constant OIP3-contours in dBm on the $I_C - V_{CE}$ plane for the SiGe reference device, simulated in the circuit of Figure 6.6. Indicated are the dominant distortion phenomena and the bias conditions for the peak-f_T and the points where the f_T has dropped to halve its peak value due to high current effects.

"weak" avalanche effects. In practical circuit implementations we avoid device operation under these bias conditions in order to guarantee a reliable operation.

Based on the previous observations, we define a simplified equivalent schematic in Figure 6.10 of a bipolar transistor, including the most dominant distortion sources. For the functional description of these sources and charges we refer to Ref. [14]. In the remainder of this chapter we will use simplifications of this model in order to evaluate BJT device linearity for practical circuit conditions.

Having the previous linearity considerations of the benchmark circuit in mind combined with this simple model; we are now able to make classifications for design strategies commonly found in "classical" RF circuit design. For this purpose we will distinguish between low and high current/power operation.

Low Current/Power Operation

From the above it is obvious that the exponential relationship between input voltage and currents and charges dictates low-current applications. For this reason, one is in practice focused on the exploration of the "linear" input-to-output current relation or the so-called current-mode type of design [15]. A simple practical way to achieve this is the use of high-Ohmic source and low-Ohmic load conditions of the transistor stages as will be discussed in Section 6.4. Although simple to use and very effective, this technique is not directly compatible with the universal desire of RF engineers to design for characteristic "50-Ohm" input- or output impedances. For these typically low-characteristic source impedances we find a significant deviation from the ideal current-driven situation, which is based on the use of infinite source impedance. Therefore, the device will be driven by input power rather than input current. Consequently, the device will also be voltage driven. This results again in the situation that one suffers from the exponential relation between input-voltage, current, and diffusion charge. This simple fact has triggered various design techniques for improving linearity, ranging from local feedback (e.g. inductive emitter degeneration [16]), overall feedback [11,17], to multi-tanh [18] or ultra-multi-tanh stages [19] in order to mask the exponential transfer function. Although effective for many applications, the general trend in RF circuitry is to increase operating frequency and reduce the DC power consumption and supply voltage, yielding significant limitations when implementing the desired linear RF circuit functions without compromising other specifications.

In respect of this, recently new circuit design techniques became available for the design of LNAs, mixers, and PAs that eliminate the classical trade-offs between DC-power consumption, gain, noise, and linearity [16,20–22], facilitating a high linearity performance at low current levels and high frequencies. These designs are based on the use of out-of-band matching techniques in order to fully exploit IM3 cancellation effects. In general, these circuit designs can operate at much lower DC powers for a given linearity requirement. Besides that, the amount of trade-offs can be reduced, which are required between

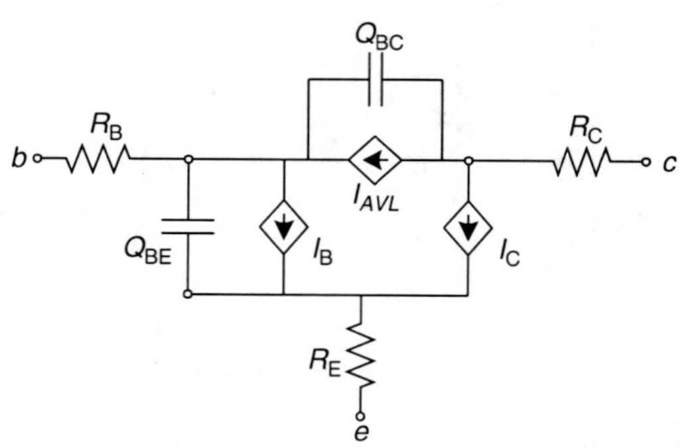

FIGURE 6.10 Simplified large signal equivalent circuit of a bipolar transistor with all dominant nonlinearities in forward active operation, R_B, I_B, I_C, I_{AVL}, Q_{BE}, and Q_{BC}.

the various specifications. Although very promising, the latter design technique is in general more difficult to implement, since the proper impedances must be provided not only at the fundamental, but also at the second-harmonic as well as the IF or base-band frequencies. We will describe these techniques in Section 6.4.

High Current/Power Operation

At higher DC current levels, for a properly scaled device in class-A operation, the influence of the exponential distortion can be neglected (see also (6.25) for the OIP3). Consequently, the nonlinear feedback capacitance C_{bc} and the dependency of the delay time τ_f on the output voltage are most troublesome for linearity, since both cause a nonlinear feedback action over the device. Due to the fact that the transfer function of a network is predominantly determined by its feedback elements, the transfer-function will also become nonlinear. Design techniques should therefore target the reduction of this undesired non-linear feedback by harmonic loading [23], unilateralization, or by modifying the transistor technology for the lightly doped epilayer [24]. Note that for hard driven devices quasisaturation effects come into play [8,14], which yield a strong increase of the total base charge and consequently also C_{bc} and τ_f. Besides the undesired nonlinear feedback through the base charge modulation, also weak avalanche effects play a role since they cause a nonlinear feedback action to the internal base node as well. Normally this operation region will be avoided in most designs. However, if one wants to push a PA stage to its maximum performance for a given technology this effect might be important. In these special cases, attention should be given to bias circuitry in order to handle the negative going base current. At the same time, one can use harmonic terminations at the output in order to shape the waveform in such a way that the peak excursion remains limited [23,25,26]. The use of out-of-band terminations at the input can be used to further boost the linearity and enhance self-biasing effects to improve the power compression behavior of the amplifier [27].

6.4 RF Design Techniques for Linearity

In the previous section various linearization techniques were mentioned, which will be discussed and studied here in more detail. In order to link our discussion to practical design examples, we will use the previously introduced 70 GHz SiGe BJT as reference device. This 0.5 μm $\times$ 20.3 μm BJT, which has its peak f_T of 70 GHz @ 10 mA, will be modeled by a simplified analytical Volterra series model to highlight the basic distortion mechanisms of a particular circuit design, as well as, by the full Mextram model to illustrate the validity of the analysis and its related assumptions.

In sequence, we will discuss the current-mode or translinear design approach, the use of feedback design techniques with a focus on inductive emitter degeneration, the applicability of multi-tanh-based circuits for RF design and conclude with an inventory of linearization techniques based on the utilization of out-of-band terminations.

Current-Mode Operation

One of the most effective ways to design highly linear circuitry at lower power levels is the use of current-mode design techniques [15], which are based on the "linear" current relationship between the collector and the base current. The basic assumption in current-mode design is that the input signal is delivered by a current source with an infinitely high drive impedance. In that case we operate the transistor in current-mode in contrast to our previous analysis in Section 6.2, where we considered ideal bipolar devices without memory effects under voltage-driven conditions.

In order to investigate the applicability of this design technique with respect to linearity at RF frequencies, this section studies the influence of the source impedance on the linearity of a bipolar device *with memory* (capacitances) and evaluates the results towards different design approaches. For the moment we will assume *constant* (resistive) terminations at the input- and output of the stage under consideration and study the influence of the source resistance R_S on the common-emitter stage and the common-base stage.

Analysis of a CE Stage in Current-Mode

To investigate the influence of the source resistance on linearity, we analyze the circuit of Figure 6.11 using a Volterra series analysis. The full analysis for the CE-stage is given in detail in the Appendix, however, in this analysis we have used the Miller approximation for the base–collector capacitance, since it is assumed that the collector is terminated with a low impedance. Besides that, taking the feedback of the base–collector capacitance into account, does not significantly improve the accuracy of our analysis, while the equation complexity would increase dramatically. Furthermore, the series feedback through the emitter resistance r_E is neglected for simplicity.

First of all, the overall linear transfer function relating the source voltage V_s to the output current I_c is given by

$$H_{1_c}(s) = \frac{I_c}{V_s} = g_m H_{1_b}(s) \tag{6.36}$$

in which the linear transfer function $H_{1_b}(s)$, relating the source voltage V_s to base–emitter voltage V_{be} is given by

$$H_{1_b}(s) = \frac{V_{be}}{V_s} = \frac{1}{1 + g_m R_S/\beta_f + sC_\pi R_s} = \frac{R_P}{R_S}\frac{1}{1 + sR_P C_\pi} \tag{6.37}$$

where $R_P = r_\pi // R_S$, $r_\pi = \beta_f/g_m$, $C_\pi = C_{jE} + \tau_f g_m$, and $s = j\omega$. In order to make a fair comparison, *we must use the output-current referred third-order intercept point* OIP3 rather than the IIP3, when driving the transistor with *a very low impedance* (voltage-mode) or *a very high impedance* (current-mode). This can be understood by considering the extreme case ($R_S \to \infty$) in Figure 6.11, where the high source impedance *prevents* any signal power going into the device, yielding a very high IIP3. Therefore, we focus on the OIP3, which is in this case the only viable means of comparison. Then, for the simplified circuit in Figure 6.11 we obtain

$$\mathrm{OIP3_{CE}} = \sqrt{\frac{4}{3}\frac{H_{1_c^3(s_2)}}{H_{3_c(s_2,\, s_2,\, -s_1)}}} \approx \sqrt{\frac{4}{3}\frac{g_m^3(s)}{|\varepsilon(\Delta s, 2s)| \cdot |A(s)|}} \tag{6.38}$$

in which

$$A(s) = \frac{R_P(1 + sR_S C_{jE})}{R_S(1 + sR_P C_\pi)} \tag{6.39}$$

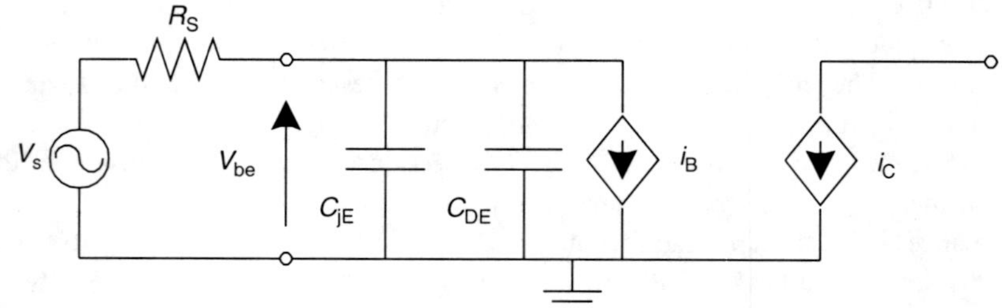

FIGURE 6.11 The simplified large-signal model of a CE-stage for studying the intermodulation distortion in current-mode.

and

$$\varepsilon(\Delta s, 2s) = \frac{g_m}{3V_T^2}\left[\frac{R_P}{2R_S}\frac{(1+2sR_S C_{jE})}{(1+2sR_P C_\pi)} + \frac{R_P}{R_S}\frac{(1+\Delta s R_S C_{jE})}{(1+\Delta s R_P C_\pi)} - 1\right] \tag{6.40}$$

When substituting $g_m = I_C/V_T$ and assuming $\Delta s = \pm j(\omega_2 - \omega_1) \approx 0$, we obtain

$$\mathrm{OIP3_{CE}} = \frac{2I_C}{\sqrt{\left|\frac{R_P}{R_S}\frac{(1+sR_S C_{jE})}{(1+sR_P C_\pi)}\right| \cdot \left|\left(\frac{R_P}{2R_S}\frac{(1+2sR_S C_{jE})}{(1+2sR_P C_\pi)} + \frac{R_P}{R_S} - 1\right)\right|}} \tag{6.41}$$

As we will see, the ratio of R_S and r_π plays a primary role, since $R_P/R_S = (1 + R_S/r_\pi)^{-1}$. This is most evident when we consider the low frequency limit or "DC-behavior" of (6.38) by setting $s = 0$, yielding

$$\mathrm{OIP3_{CE,DC}} = 2I_C\sqrt{2\left|\frac{(1+R_S/r_\pi)^2}{1-2R_S/r_\pi}\right|} \tag{6.42}$$

The denominator of (6.42) indicates that an IM3-cancellation effect exists for a specific base-current level, namely $I_B = V_T/2R_S$. This "low"-frequency IM3-cancellation phenomena was already reported by Reynolds [28] and results from the interaction of the source resistance with the nonlinear base–emitter junction. We will address IM3-cancellation in more detail in "Harmonic Matching Techniques" and here only focus on the current-mode design approach.

For this reason we will now compare as a function of frequency the current-mode situation where $R_S \gg r_\pi$ and the voltage-mode situation where $R_S \ll r_\pi$. In current-mode, setting $R_S \gg r_\pi$ in (6.41) yields approximately

$$\mathrm{OIP3_{CE,cm}} \approx 2I_C\sqrt{\frac{R_S}{r_\pi}\left|\frac{1+sr_\pi C_\pi}{1+sR_S C_{jE}}\right|} \tag{6.43}$$

which for low frequencies ($\mathrm{OIP3_{CE}} \approx 2I_C\sqrt{R_S/r_\pi}$) is a strong function of the ratio R_S/r_π. When plotting (6.43) as a function of frequency (Figure 6.12), the $\mathrm{OIP3}_{CE}$ starts decreasing at $1/2\pi R_S C_{jE}$ and levels off just beyond the frequency $\omega = \omega_T / \beta_f$ to $\mathrm{OIP3_{CE}} \approx 2I_C\sqrt{C_\pi/C_{jE}}$. When considering voltage-mode, setting $R_S \ll r_\pi$ in (6.41) yields

$$\mathrm{OIP3_{CE,vm}} \approx 2I_C\sqrt{2\left|\frac{(1+sR_S C_\pi)(1+2sR_S C_\pi)}{(1+sR_S C_{jE})(1+2sR_S C_{jE})}\right|} \tag{6.44}$$

which shows that $\mathrm{OIP3_{CE}} \approx \sqrt{8}I_C$ over a wide frequency range and slightly increases at high frequencies, since $C_\pi > C_{jE}$. To illustrate the previous equations, Figure 6.12 plots the calculated and simulated $\mathrm{OIP3_{CE}}$ versus frequency in current-mode ($R_S = 50k$) and in voltage mode ($R_S \approx 0$). This leads to the conclusion that a current-mode design with a CE-stage is beneficial when the driving impedance R_S is much higher compared to r_π and when we operate the circuit below ω_T/β_f. In practice this puts relatively high demands on the driving circuit. Also note that the gain versus frequency performance of this stage is strongly frequency dependent. To improve for these points, when aiming design for current-mode operation, people quite often prefer the use of a common-base (CB) configuration. For this reason, we will now also study the linearity of the CB-stage.

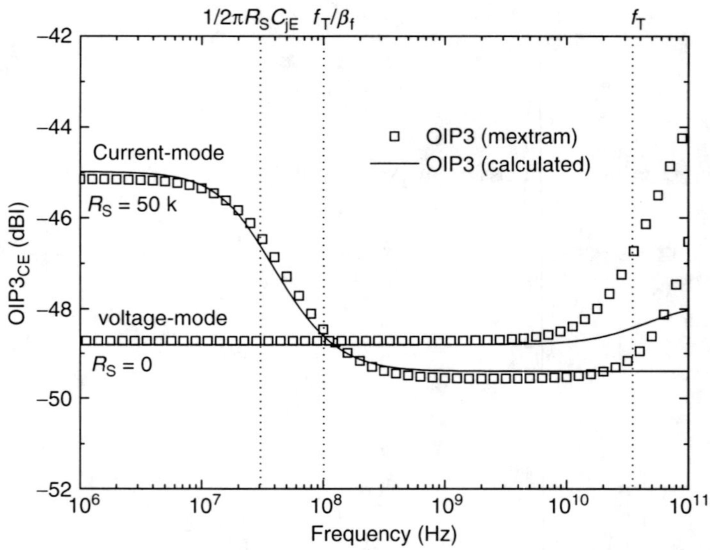

FIGURE 6.12 The OIP3 in dBI versus frequency of a CE-stage in current-mode and voltage mode.

Analysis of a CB-Stage in Current-Mode

The circuit of the CB-stage is given in Figure 6.13. Since $\beta \gg 1$ for typical SiGe devices we can neglect the base current nonlinearity with respect to the collector current nonlinearity. In this case, the expression for OIP3 is similar to the one in the previous example with the only difference that $R_P = r_e // R_S$ with $r_e = 1/g_m$, yielding

$$\mathrm{OIP3_{CB}} = \frac{2I_C}{\sqrt{\left|\frac{R_P}{R_S}\frac{(1+sR_SC_{jE})}{(1+sR_PC_\pi)}\right| \cdot \left|\left(\frac{R_P}{2R_S}\frac{(1+2sR_SC_{jE})}{(1+2sR_PC_\pi)}+\frac{R_P}{R_S}-1\right)\right|}} \tag{6.45}$$

This means that now the ratio of R_S and r_e plays a primary role. Again we compare the current-mode ($R_S \gg r_e$) to the voltage-mode situation ($R_S \ll r_e$), which for the current-mode ($R_S \gg r_e$) in (6.45), yields

$$\mathrm{OIP3_{CB,cm}} \approx 2I_C\sqrt{\frac{R_S}{r_e}\left|\frac{1+sr_eC_\pi}{1+sR_SC_{jE}}\right|} \tag{6.46}$$

In the low-frequency limit (6.46) becomes $\mathrm{OIP3_{DC}} \approx 2I_C\sqrt{R_S/r_e}$, which is a strong function of the ratio R_S/r_e. Moreover, the linearity improvement of a current-mode CB-stage extends up to ω_T, making current-mode design attractive also at high frequencies (Figure 6.14). In voltage-mode, setting $R_S \ll r_e$ in (6.45) yields

$$\mathrm{OIP3_{CB,vm}} \approx 2I_C\sqrt{2\left|\frac{(1+sR_SC_\pi)(1+2sR_SC_\pi)}{(1+sR_SC_{jE})(1+2sR_SC_{jE})}\right|} \tag{6.47}$$

which is equal to that of the CE-stage in voltage mode. Most importantly, these equations show again that the lowest distortion is obtained in current-mode when $R_S \gg r_e$. Figure 6.14 plots the $\mathrm{OIP3_{CB}}$ as function of frequency for a CB-stage in current-mode ($R_S = 5k$) and voltage-mode ($R_S \approx 0$). This figure clearly shows that in current-mode operation the linearity improvement of a CB-stage extends to almost the cutoff frequency and is a strong function of R_S and r_e, of which the latter is set by I_C.

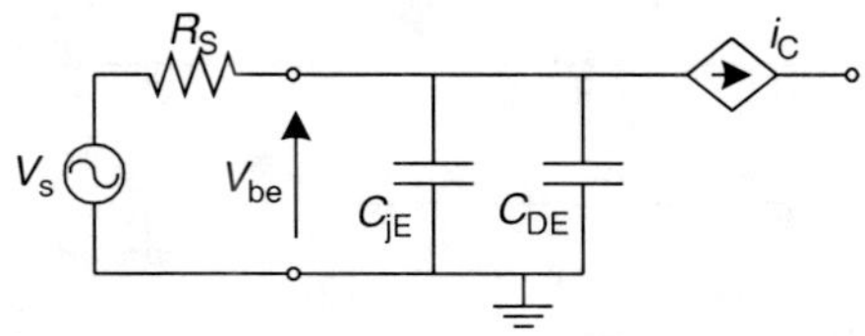

FIGURE 6.13 The simplified large-signal model of a CB-stage for studying the intermodulation distortion in current-mode.

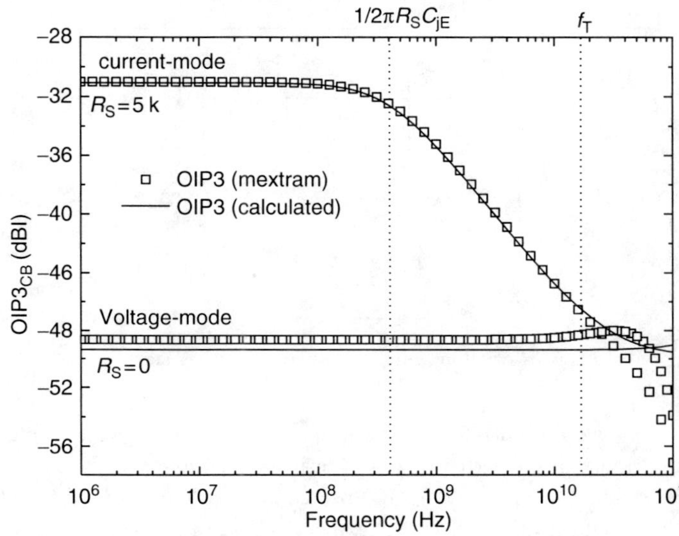

FIGURE 6.14 The OIP3 in dBI versus frequency of a CB-stage in current-mode and voltage mode.

From the previous analyses we can conclude that once again, when the source impedance is high compared to the input impedance of the transistor stage (current-mode) we can obtain a significant reduction of the IM3 compared to the voltage-driven situation. However, since the input impedance of a CB-stage at lower frequencies is approximately a factor β_f lower than for a CE-stage, current-mode operation is enforced for much lower values of the source resistance. This freedom can be utilized to obtain linear transfer functions, which are less frequency-dependent compared to the CE stage, yielding improved current-mode operation up to very high frequencies (Figure 6.14). A practical situation where we use the current mode for the CB-stage is for example in the Gilbert mixer where the switching core is driven by the high ohmic output of the driving stage (see Figure 6.15).

Note that, although a mixer in general is considered as a nonlinear component due to the up- or down-conversion of the RF signal by the switching/mixing action with the LO signal, the RF signal transfer of the lower stage to the switching core must be linear in order to obtain a proper "distortion-free" up- or down-frequency conversion. In practice, current-mode signal transfer of the driving stage to the switching core almost automatically arranges this property. Consequently, most difficulties with mixer nonlinearities are found in converting the input signal power to a current-mode signal by the driving stage. To deal with this problem, we will consider various approaches in "Negative Feedback Techniques" and "Shaping the Transfer Function."

Translinear Design Techniques

The translinear loop (TL) principle is also a form of current-mode, which can be applied in order to obtain the desired overall linear current-to-current relation [15]. It is based on intermediate transitions from current-mode to voltage-mode, and vice versa. In practice, when a closed loop of base–emitter junctions is formed, a TL is created, which can be configured to yield linear current-to-current transfer

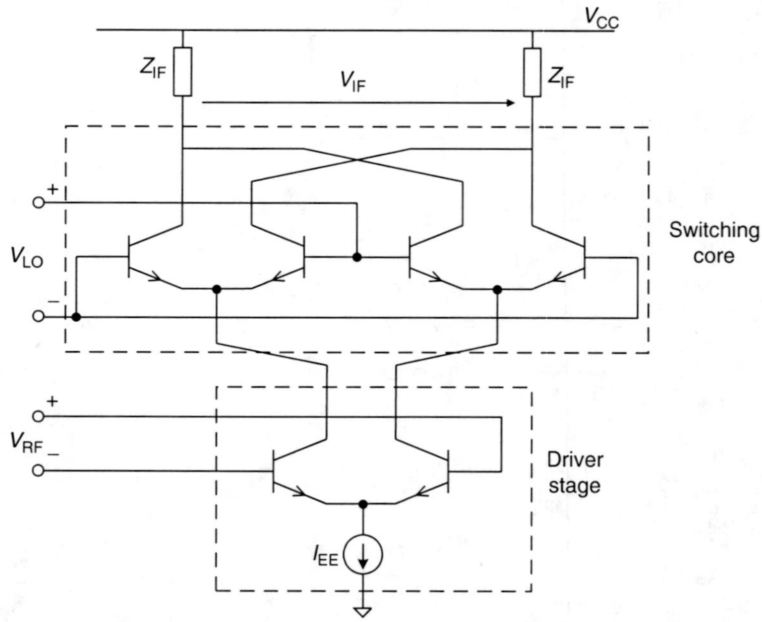

FIGURE 6.15 Gilbert cell mixer with switching core and differential driver stage.

functions. Figure 6.16 shows the current-mirror, which is one of the most simple examples of this principle. The input current I_i is converted to a nonlinear base–emitter voltage $V_{be} = V_T \ln(I_i/I_S)$ and translated again to a current $I_o = I_S \exp(V_{be}/V_T)$, which is linearly proportional to the original input current and scaled by the ratio of the active device area A_{e2}/A_{e1}. In this context, however, we will treat these techniques as an extension of the current-mode design approach. An extensive discussion of various translinear circuit design techniques that also addresses distortion generated by device mismatch and β-nonlinearity can be found in Ref. [15]. A practical example of a PA driver based on the TL principle can be found in Ref. [29], where the input drive signal is a current, which is converted to an output signal power. Another nice example that makes use of the TL principle is the Micromixer [30]. It uses a TL driver stage as shown in Figure 6.17, which can handle very large input current signals without introducing significant distortion. Although elegant, this design approach does not provide a solution for the transfer of the input signal power to a current representation. This is caused by the voltage-dependent input impedance of the micromixer. As a result, the overall linearity of the micromixer is dominated by this aspect.

Discussion

When applicable, current-mode design provides you with a direct and straightforward implementation of the desired linear circuit functions. Although efficient for many applications, one should be aware of the limitations of this technique. This becomes apparent when considering higher frequencies of operation, where the relative high base–emitter capacitance of a bipolar transistor limits the proper use of current-mode design for CE-stages ($\omega \approx \omega_T / \beta_f$). CB-stages are more tolerable at this point due to their much lower input impedance (r_e), which lowers the impact of the base–emitter capacitance at higher frequencies. Another advantage of current-mode design techniques is the ease of combining output signals of various circuit blocks by summing the output currents. An illustrative example of this technique is given by Aggarwal et al. [29] where the outputs are current-summed at all times.

As closing remarks on the current-mode design approach, we mention that although translinear design techniques are available [15], which relax the influence of β-nonlinearity; one should verify that when designing at very low currents or high currents levels, the collector-to-base current ratio is not significantly degraded by the nonideal base-current or high-current effects [14]. In addition, for very

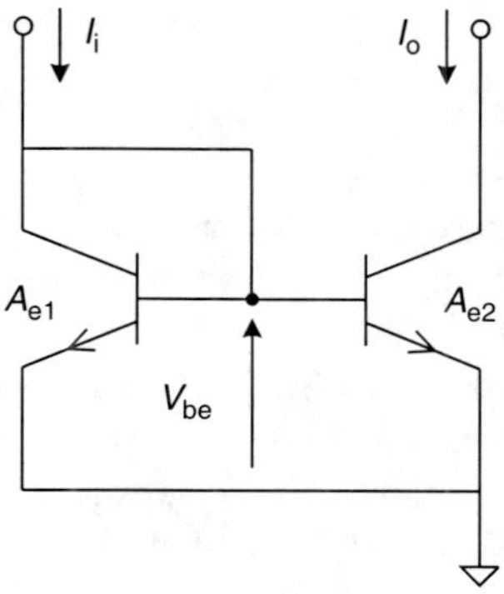

FIGURE 6.16 The current mirror as an example of a simple translinear circuit.

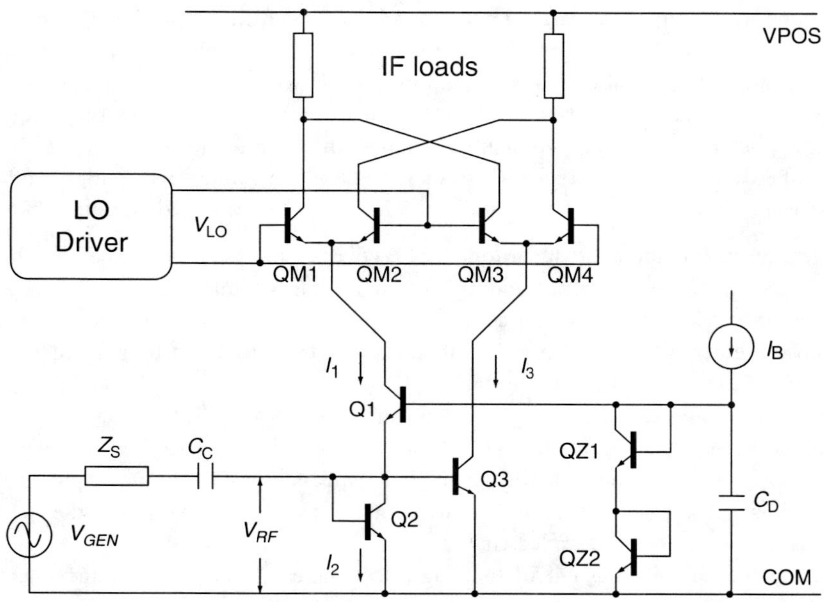

FIGURE 6.17 The micromixer uses a translinear circuit to implement a linear current-to-current transfer between the RF-input and the switching core. (From B. Gilbert. *IEEE J. Solid-State Circ.* 32(9), 1412–1423, 1997. With permission. Copyright 1997 IEEE.)

highly linear applications, attention should be paid to the β-nonlinearity caused by the reverse Early effect [14,31]. Although this is in general a weak nonlinearity that depends on the semiconductor technology, it may limit the linearity of current-mode designs to some extent. Finally, it must be mentioned that current-mode design techniques often lack noise optimization and do not provide a real solution for the nonlinear transfer of the input signal voltage (or power) to a current representation. This makes this design technique less favorable for a front-end LNA design operating at several GHzs. For this reason, the next section considers other design techniques, which overcome these problems.

Negative Feedback Techniques

Feedback design techniques are commonly used in electronic circuits to improve the circuit performance in terms of well-defined (wideband) transfer characteristics and optimum impedance matching conditions for both low return loss and low noise. In addition, the use of negative feedback makes the circuit less sensitive to process, temperature variations and here most importantly, it reduces the error in the transfer characteristics, making the overall circuit more linear.

Overall Feedback

For these reasons and as became apparent from the analysis of "Feedback Configurations," the use of negative feedback is at first sight an effective and attractive technique to improve the linearity of a system. Therefore it is not surprising that overall feedback based on multiple transistor stages is one of the favored design techniques at lower frequencies [11,32]. At high frequencies, however, overall feedback design is somewhat less common. Here conventional microwave techniques or the use of only local feedback stages are mostly favored. The reasons for this are the following: firstly, at higher frequencies the gain of the active devices drops and therefore the use of negative feedback is less attractive, since it lowers the gain even more. Secondly, RF applications require device operation closer to the f_T of the device, yielding a relative large phase shift per stage. Consequently, combining more stages within a feedback loop can easily yield an instable situation. Finally, wireless communication circuits are quite often more bandpass oriented than lowpass, which is traditionally the case for most feedback designs. Note that a bandpass characteristic makes the circuit less sensitive for jamming or blocking by an interfering signal outside the band of interest, while a lowpass characteristic would not suppress interfering signals at lower frequencies. Although these considerations put some restrictions to the blind application of feedback design techniques, the recent improvements in high-frequency performance of SiGe devices facilitate the use of these feedback techniques also at higher frequencies. In view of this, we provide here some design guidelines for linear negative feedback amplifiers.

- A large loop-gain reduces the distortion (see "Feedback Configurations") therefore the loop-gain must be maximized, preferably by adding more active stages, while assuming the overall gain to be fixed by the feedback network. In practice this approach will be limited by stability considerations that pose a maximum on the number of transistor stages that can be used (e.g., for practical circuits in the order of two to three stages [11,33].
- When using a cascaded chain of gain stages, the conclusion of "Cascaded Systems" apply and consequently, the stage with the highest nonlinearity must be placed at the beginning of the amplifying chain, while the stage with the highest gain and linearity should be placed at the end of the amplifying chain. Note that the requirements on the signal handling of the stages increases with their position in the cascaded chain.
- Increasing the DC-bias current of the stages will increase their transconductance and consequently their gain, yielding a higher overall loop gain. Furthermore, the relative current swing (see Equation 6.22) for the stages will be reduced. Both effects will yield an improved linearity at the cost of DC power consumption.
- The use of local feedback within the active loop should in general be avoided, since the linearity improvement of the local stage will vanish compared to the drop in linearity as a result of the lower overall loop gain [32].
- Since the resulting linearity is not only depending on the loop-gain itself (consider e.g., the loop-gain term $a_1 f$ in (6.33)–(6.35)), but also on the magnitude of the nonlinearity to be reduced (e.g., a_2 and a_3 in (6.33)–(6.35)), it is also important that these "loop errors" are reduced as well [34]. This can be accomplished by applying current-mode coupling of the gain stages within the loop, or when considering a voltage-current transfer (due to a nonideal current-driven stage at the beginning of the chain of cascade stages) the inclusion of a stage with a shaped (linearized) transconductance function like the multi-tanh doublet [18], which we discuss later in "Shaping the Transfer Function".

A practical example of the application of overall feedback in a RF mixer is shown in Figure 6.18, where overall feedback is applied in order to implement the LNA driver block of the switching core [17]. In this way the inherently nonlinear transfer of the "power" input signal to the output drive current for the switching core is linearized.

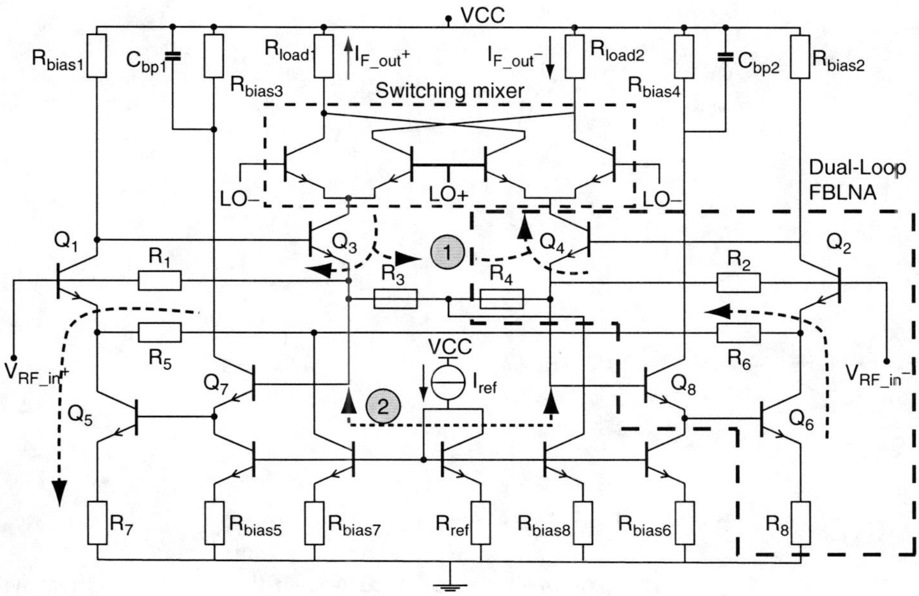

FIGURE 6.18 Dual-loop wide-band active-feedback LNA and switching mixer. (From Adiseno, M Ismail, and H Olsson. *IEEE J. Solid-State Circ.* 37(9), 1162–1168, 2002. Copyright 2002 IEEE.)

Local Feedback

As mentioned before, traditionally only local feedback is applied at "very" high frequencies to improve the linearity of a transistor stage. For this purpose, the most frequently used configuration is the emitter degenerated CE-stage. Depending on the type of emitter degeneration, resistive, capacitive, or inductive, the linearity of the stage can be improved to a great extent. This has been analyzed by Keng and Meyer [16], who discuss the results for various emitter degenerated CE and differential stages. In order to follow the discussion of Keng and Meyer, we consider the schematic of Figure 6.19. In the following discussion we assume again low-current operation, and consequently that the dominant source of nonlinearity is formed by the exponential distortion of the base–emitter junction, which is reflected in i_b, i_c, and the diffusion capacitance C_{DE}. The complete analysis is given in the Appendix, but in this study we neglect the influence of C_{bc}. This assumption simplifies the equations, without losing generality of the main aspects of series feedback. The source impedance Z_S and base resistance r_B are both absorbed in Z_b and the emitter resistance r_E and any externally applied emitter impedance Z_E are absorbed in Z_e. Consequently, when applying a two-tone excitation, assuming $s \approx s_1 \approx s_2$ and $\Delta_s = (s_1 - s_2) \ll s$, we can write the input-referred IM3 product as

$$\begin{aligned} \text{IM3} &= \frac{3}{4}|\varepsilon(\Delta s, 2s)| \cdot |H_{1_b}(s)|^3 \cdot |D(s)| \cdot \hat{V}_s^2 \\ &= \frac{1}{4}\frac{\hat{V}_s^2}{V_T^2}|H_{1_b}(s)|^3 \cdot |\boldsymbol{T}(s)| \cdot \left|\left(\frac{1}{2}H_{1_b}(2s) + \boldsymbol{T}(\mathbf{2s}) + H_{1_b}(\Delta s)\boldsymbol{T}(\Delta s) - 1\right)\right| \end{aligned} \tag{6.48}$$

in which the transfer from the source voltage to the internal base–emitter voltage is defined as

$$H_{1_b}(s) = \frac{V_{be}}{V_s} = \frac{1}{1 + Z(s)(g_\pi + sC_\pi) + g_m Z_e(s)} \tag{6.49}$$

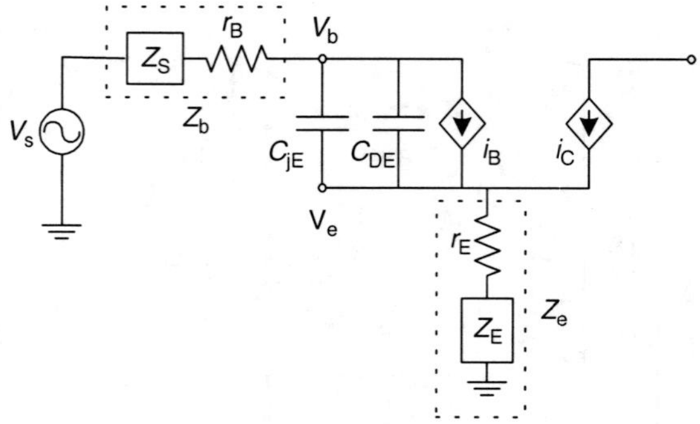

FIGURE 6.19 Large-signal model of a CE-stage for studying the effects of series feedback.

and

$$T(s) = 1 + sC_{jE}Z(s) \tag{6.50}$$

where $Z(s) = Z_b(s) + Z_e(s)$ is the sum of the series impedances in the base and emitter, the device parasitics, and the characteristic impedance of the driving source.

From (6.49) we observe that the use of a high emitter impedance Z_e as emitter degradation effectively lowers the voltage transfer of the source to the internal base–emitter junction by H_{1_b}, yielding directly to a reduced IM3 level as can be seen in (6.48). In addition to this basic feedback phenomenon, the IM3 level is also influenced by a factor $T(s)$ in (6.48). Here $T(s)$ represents the interaction of $Z(s)$ with C_{jE} at the fundamental frequency, while $T(\Delta s)$ and $T(2s)$ address these effects at the baseband and double frequencies. Note that these later frequencies are sometimes also referred to as out-of-band frequencies. The influence of the factors, $T(s)$, $T(\Delta s)$, and $T(2s)$ on the IM3 level have been the subject of various studies in literature. In Ref. [16] focus was placed on partial IM3-cancellation effects through $T(s)$, while the influence of $T(\Delta s)$ and $T(2s)$ was recognized but considered as a secondary effect. In the study of [16] three choices for $Z_e(s)$ where considered, namely:

- **Capacitive.** Choosing Z_e capacitive, yields a real positive value for $sC_{jE}Z_e(s)$, which adds up to the '1'-term of $T(s)$ in (6.50), consequently, the IM3 level increases by this factor.
- **Resistive.** Choosing $Z_e = R_e$ will result in a positive imaginary value of $sC_{jE}Z_e(s)$, which again yields an increase of $T(s)$, however less dramatically as in choosing Z_e capacitive.
- **Inductive.** Choosing Z_e inductive proves to be the most beneficial, since it results in a negative real value for $sC_{jE}Z_e(s)$, lowering the factor $T(s)$, and consequently the IM3 level.

Note that perfect IM3-cancellation through $T(s)$ requires that $Z(s)$ is purely inductive. In practical situations this will never be the case since $Z(s)$ is composed out of: Z_e, Z_b the device series resistances r_B, r_E, and the characteristic impedance of the driving source ($R_S = R_0$). As a result, only partial IM3-cancellation will occur based on the influence of the factor $T(s)$ alone.

Inspection of (6.48) also reveals that the IM3 level is influenced by $T(\Delta s)$ and $T(2s)$ in the most righthand factor of (6.48), which relate to the device-circuit interactions at the baseband and double frequency. Consequently, a specific choice of $Z(s)$-like inductive emitter degradation in combination with a proper choice of other circuit elements can also yield "partial" IM3-cancellation effects through these terms. The relevance of the cancellation effects via $T(\Delta s)$ and $T(2s)$ becomes much more apparent when considering the noise and power matching conditions for an inductively degenerated emitter stage (Figure 6.20). For this configuration it can be shown, that by proper selection of L_E, g_m, and the active device area A_e the optimum noise match at the base can be manipulated close to the conjugate value of the input impedance of the stage [35]. As a result, simultaneous noise and impedance matching for an inductively degenerated emitter stage can be easily implemented. By adding the remaining impedance matching network to the base (in its most simple form an inductor), an easy to implement, high-

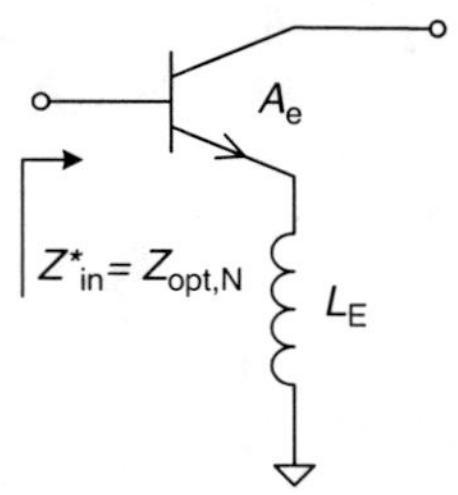

FIGURE 6.20 CE-stage with inductive emitter degeneration.

performance LNA stage is created, which for most applications offers a very acceptable noise, impedance match, and linearity performance. Therefore, this configuration is currently one of the most popular input stages in RF circuit design. It must be mentioned, however, that for this configuration the requirements for simultaneous noise and power match prove to be incompatible with the requirements for IM3-cancellation through the factor $T(s)$, since this would simply require a too high value of L_E, L_B, and/or A_e compared to the component values required for optimum noise and impedance match ($Z_S = Z^*_{in} = Z_{opt,\,N}$). In addition, $Z(s)$ always contains a real part associated with device parasitics and source resistance obstructing full IM3-cancellation. Therefore, the sometimes-remarkable high linearity results reported for this topology, must be explained by the "partial" IM3-cancellation effects through the terms $T(\Delta s)$ and $T(2s)$ [16,36]. A practical example of these phenomena can be found in Ref. [7], which gives a simulation based experiment of an inductively degenerated CE-stage, to study the best compromise of various design parameters with respect to: noise, impedance matching, gain, and linearity for a given SiGe technology. In Refs. [7,16], the distortion reduction through the baseband frequency and second harmonics is considered as a positive side effect, and not fully explored. Consequently, high linearity and low noise can be obtained, but at a relative high current level for a limited modulation bandwidth.

Cascode LNA Design

To illustrate the above we will now consider the linearity performance of an inductively degenerated cascode-LNA design in Figure 6.21 operating at 5 GHz using our reference transistor. The cascode-LNA has been optimized for simultaneous noise and impedance matching conditions by proper selection of L_E, g_m, and the active emitter area A_e. The base inductor L_B takes care for the remaining imaginary part of the input matching. In our design the input bias-decoupling capacitance ($C_D = 20$ pF) behaves like a short for the fundamental, while blocking the low-frequency component at the baseband (or envelope) frequency. This allows us to tweak the linearity using the elements in the base biasing circuitry, which basically defines $Z(\Delta s)$ seen by the bipolar device at its internal base–emitter junction. Note that the inductor in the bias path ($L_{CH} = 20$ nH) will represent a large impedance value at the fundamental frequency, eliminating any influence of the biasing circuitry on the noise level of the LNA. The simulation results of this circuit are given in Figure 6.22, which shows the minimum noise figure (F_{min}), the noise factor (F), power gain, and return loss in dB as function of collector current I_C. Evidently, simultaneous noise and impedance matching conditions have been achieved at a gain level of 15 dB and a bias current of $I_C = 5$ mA. When considering the linearity in terms of low- and high-band IIP3, for a small tone spacing ($\Delta f = 0.1$ MHz) we find a "partial" IM3-cancellation conditions around 6 mA as shown in Figure 6.23. Note that this distortion optimum slightly differs from the current for minimum noise level. When repeating this simulation using a larger tone spacing ($\Delta f = 100$ MHz), we observe that asymmetry in the IM3 distortion products causes a difference in the low- and high-band IIP3, while the peaking becomes less pronounced in the higher band. Note that the lowest IIP3 level indicates the usable linearity performance of the stage. Finally, when considering the IIP3 for the LNA using a different value of the resistor in the bias path ($R_B = 3k$), the distortion cancellation mechanism disappears resulting in a much lower linearity, while the noise performance and matching conditions remain unchanged. The latter experiment clearly indicates that indeed $Z(\Delta s)$ through the factor $T(\Delta s)$ has a significant impact on the linearity performance of this LNA configuration, which is in agreement with the results in Ref. [36].

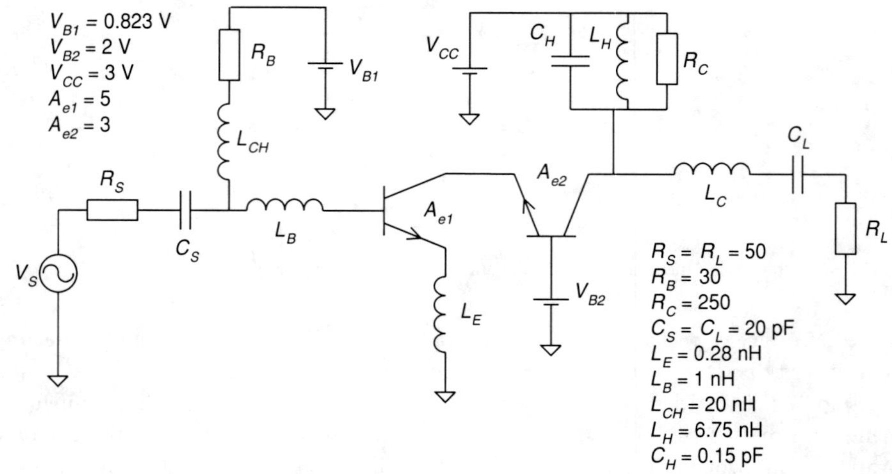

FIGURE 6.21 A 5-GHz cascode LNA with inductive emitter degeneration.

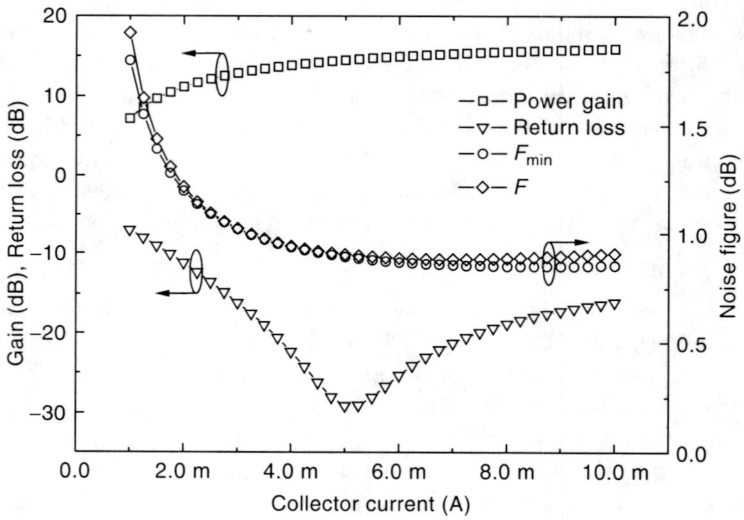

FIGURE 6.22 Simulated small-signal power gain, return-loss, and noise figure as function of collector current for the 5 GHz cascode LNA with inductive emitter degeneration.

When considering the differential pair transconductance stage [16] for their linearity, similar conclusions can be found as for the single-ended stage when comparing the linearity performance of capacitive, resistive, or inductive emitter degeneration, with again finding the best performances for the inductive case. However, it must be noted that the IM3 of a degenerated differential pair is at least twice as large as that of a degenerated CE-stage with the same bias current and degenerated transconductance* [16]. A nice example of a class-AB mixer with inductive degeneration is given in Figure 6.24.

Discussion

The ease of integration, the high linearity, combined with the inherent simultaneous noise and impedance matching properties, provided by the inductively (differential or single-ended) degenerated CE-stage, has led to its great popularity in wireless applications. In spite of this, there are also some

*Without degeneration the *IM*3 of the differential pair is twice as small as that of a CE-stage (see "Two-Tone Excitation").

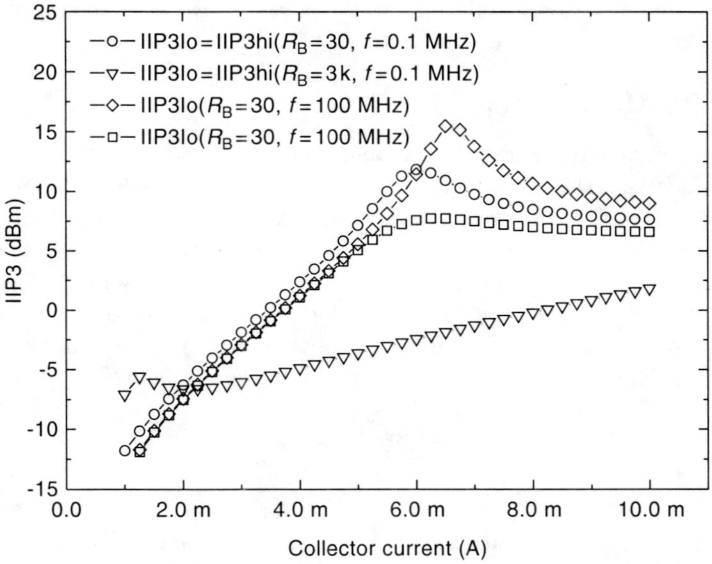

FIGURE 6.23 Simulated input referred intercept point (IIP3) as function of collector current for the 5 GHz cascode LNA with inductive emitter degeneration.

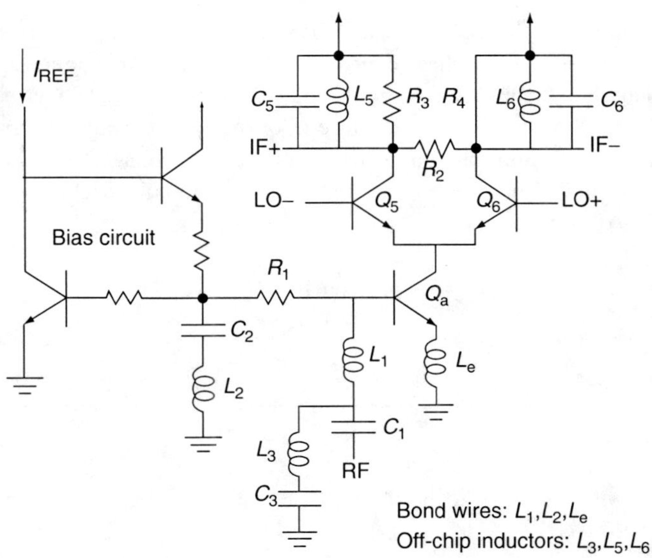

FIGURE 6.24 A class-AB mixer, which uses inductive emitter degeneration to achieve low noise and high linearity performance. (From Keng Leong Fong and RG Meyer. *IEEE J. Solid-State Circ.* 33(4), 548–555, 1998. Copyright 1998 IEEE.)

"minor" drawbacks. Namely, inductive degeneration means negative feedback, consequently the gain of the stage is reduced. As a result, sufficiently high gain (and linearity) is paid with a relative high DC current that can be disadvantageous for battery-operated applications. Finally, since noise matching, impedance matching, gain, and linearity, all have to be addressed through the same, limited number of design parameters (L_E, the emitter area A_e and g_m or I_C), suboptimum solutions are sometimes unavoidable. In order to overcome these limitations, we will develop in "Harmonic Matching Techniques" "out-of-band" IM3-cancellation techniques to their full extent. This new design technique offers more design freedom to the designer, which can be utilized to realize circuits operating at a very low DC power, while providing high gain and linearity over a large bandwidth, at the expense of a somewhat more difficult design procedure/implementation.

Shaping the Transfer Function

In the previous section negative feedback was promoted as design technique to improve the linearity of a signal transfer function. This worked out to be particularly useful when considering the strongly nonlinear voltage-to-current transfer of a bipolar device. While the basic principle of negative feedback is based on the reduction of the transfer error by increasing the return difference and loop-gain (Equation 6.35 in "Feedback Configurations"), the multi-tanh design technique [18], which we describe in this paragraph, is focused on the linearization of the nonlinear voltage–current transfer itself. To introduce these topologies let us first consider the input-voltage to output-current relation of an ideal differential pair stage (see Figure 6.2) without any memory effects, which can be written as [5]:

$$I_{\mathrm{OUT}} = I_{\mathrm{EE}} \tan\mathrm{h}\left(\frac{V_{\mathrm{IN}}}{2V_{\mathrm{T}}}\right) \tag{6.51}$$

When considering the small-signal transconductance of the differential pair we find

$$g_{\mathrm{mDP}}(V_{\mathrm{IN}}) = \frac{\mathrm{d}I_{\mathrm{OUT}}}{\mathrm{d}V_{\mathrm{IN}}} = \frac{I_{\mathrm{EE}}}{2V_{\mathrm{T}}} \sec\mathrm{h}^2 \frac{V_{\mathrm{IN}}}{2V_{\mathrm{T}}} \tag{6.52}$$

which is strongly dependent on the input voltage, as is visualized in Figure 6.25. It is not surprising that without any additional measures this voltage dependency of the transconductance significantly limits the input voltage signal handling capability, as is expressed by the input referred voltage IP3 we found earlier in Equation 6.18 of "Two-Tone Excitation." To improve for the voltage handling capability and make the transconductance more linear, multiple differential pairs with DC voltage offsets at their inputs can be combined [18] as shown in Figure 6.26. With all voltage offset and tail currents basically free, we obtain the following generalized expression:

$$I_{\mathrm{OUT}} = \sum_{j=1}^{N} I_j \tan\mathrm{h}\left(\frac{V_{\mathrm{IN}} + V_j}{2V_{\mathrm{T}}}\right) \tag{6.53}$$

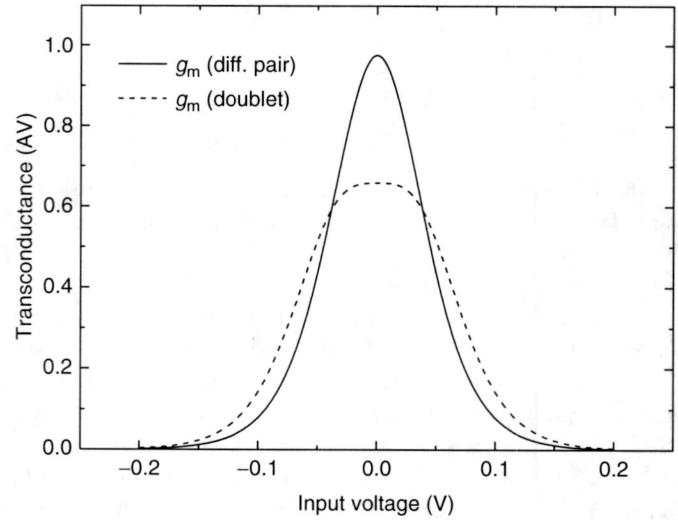

FIGURE 6.25 The normalized transconductance of a differential pair and a multi-tanh doublet.

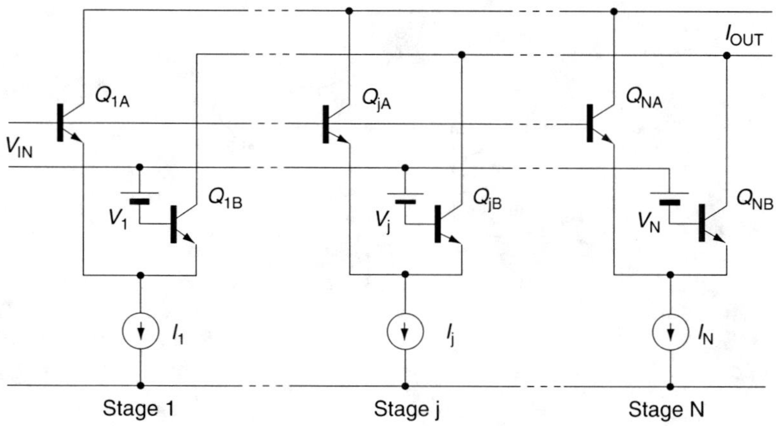

FIGURE 6.26 A generalized multi-tanh system composed out of several differential pair stages. (From B Gilbert. *IEEE J. Solid-State Circ.* 33(1), 2–17, 1998. Copyright 1998 IEEE.)

with its related transconductance as

$$g_{m_{MT}} = \sum_{j=1}^{N} \frac{I_j}{2V_T} \operatorname{sec h}^2\left(\frac{V_{IN} + V_j}{2V_T}\right) \tag{6.54}$$

Note that the voltage offsets can be introduced in various ways [18,19], however, in this chapter we will restrict ourselves to the most well-known implementation; the multi-tanh doublet ($N = 2$). The doublet uses equal tail currents and introduces the required voltage offsets by combining different active emitter areas of the transistors in a differential pair. The transconductance of the doublet is plotted as function of the input voltage in Figure 6.25 for an emitter area ratio of 3.75. This figure shows that the maximum transconductance of the doublet is reduced compared to the differential pair with the same total tail current. However, the transconductance as a whole has become much more constant over a larger input voltage range. The optimum active emitter area ratio for a multi-tanh doublet can be analyzed for a maximum flat behavior of g_m and is found to be close to 3.75 [18].

Although elegant in their basic operation, most multi-tanh implementations are based on low-frequency considerations, which ignore device parasitics like the base and emitter series resistances and junction capacitances. Consequently, when applying the multi-tanh concept at higher frequencies one should carefully verify if for a given application and operating frequency, the multi-tanh concept indeed offers advantages over other design techniques as the resistively or inductively degenerated differential pair. For this reason, we will compare now the performance of the multi-tanh doublet with a resistively and inductively degenerated differential pair at low and high frequencies. We perform this simulation experiment with our SiGe reference device for a voltage driven situation and monitor the fundamental and third-order IMD differential output currents as function of the input voltage amplitude. Figure 6.27 shows the circuits used for our comparison, which all operate with the same current budget ($I_{C_{tot}} = 1$ mA). The area ratio of the multi-tanh doublet is set to 3.75, while the current density of the transistors is kept well below the current of maximum f_T ($I_C = 10$ mA for $A_e = 10\ \mu m^2$). The emitter resistance or inductance for the differential pair is chosen in such a way that the resulting small-signal transconductance of the stage at the operating frequency of 5 GHz is equal to that of the multi-tanh doublet. In addition, the active emitter area of the inductively degenerated stage is scaled to the point where

$$\omega^2 C_{jE} L_E \approx 1 \tag{6.55}$$

which yields a distortion cancellation condition for the inductively degenerated differential pair due to the factor $T(s)$, which acts on the fundamental frequency (See Equation 6.50 for the single-ended case).

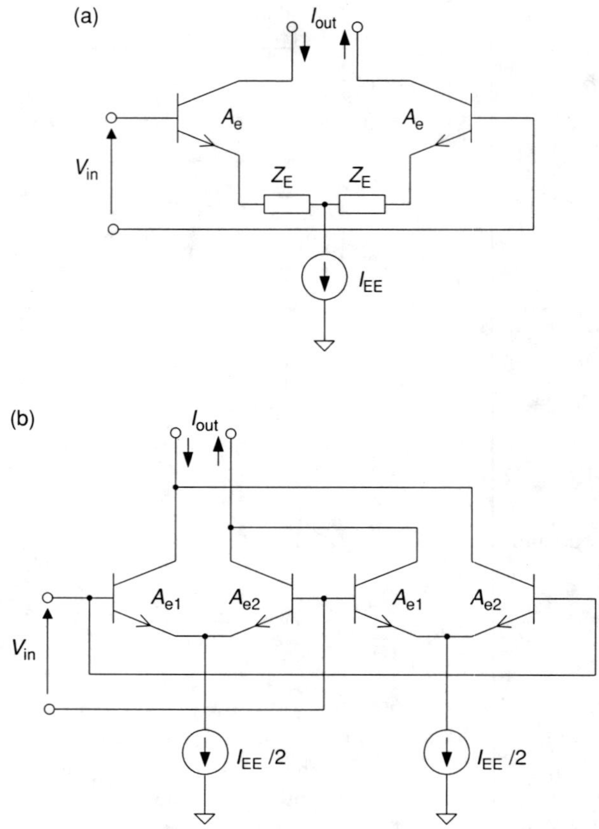

FIGURE 6.27 (a) The emitter degenerated differential pair stage and (b) the multi-tanh doublet.

The results of this simulation experiment are given in Figure 6.28, where the third-order IMD current is plotted as a function of the input voltage amplitude at 5 GHz; and in Figure 6.29, where the gain and the voltage IIP3 are plotted as function of frequency. As can be observed from these plots, by far the best results can be obtained at 5 GHz using the inductively degenerated differential pair. The best wide-band performance for the linearity is given by the multi-tanh doublet, while the resistively degenerated differential pair proves to be inferior. Clearly visible in Figure 6.29 is the partial distortion cancellation at 5 GHz for the inductively degenerated differential pair, illustrating the relevance of a careful dimensioning of the circuit components involved. Note that the series resistance of the device, r_B and r_E in combination with the source resistance and ohmic losses of the inductor, will determine how effective the IM3 cancellation is.

Harmonic Matching Techniques

Previously, we have examined several techniques based on the linearization of the overall transfer-characteristic, like feedback and the multi-tanh approach. Although effective, these techniques require in general a higher DC power consumption in order to compensate for the loss in gain. In RF circuit design for mobile applications, however, it is quite often desirable to keep the DC power consumption as low as possible to save the battery lifetime, while maintaining useful gain, noise, and linearity specifications. This section presents circuit design techniques that do not compromise the gain and power consumption requirements for increased linearity. These techniques make explicit use of the fact that third-order distortion products are generated in two ways: first, the third-order distortion terms generated directly

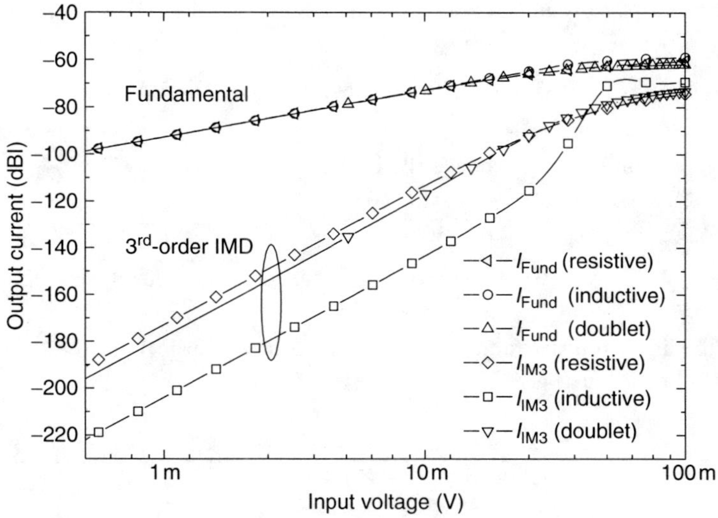

FIGURE 6.28 The fundamental and third-order intermodulation current in dBI as a function of input voltage for the emitter degenerated differential pair and multi-tanh doublet of Figure 6.27.

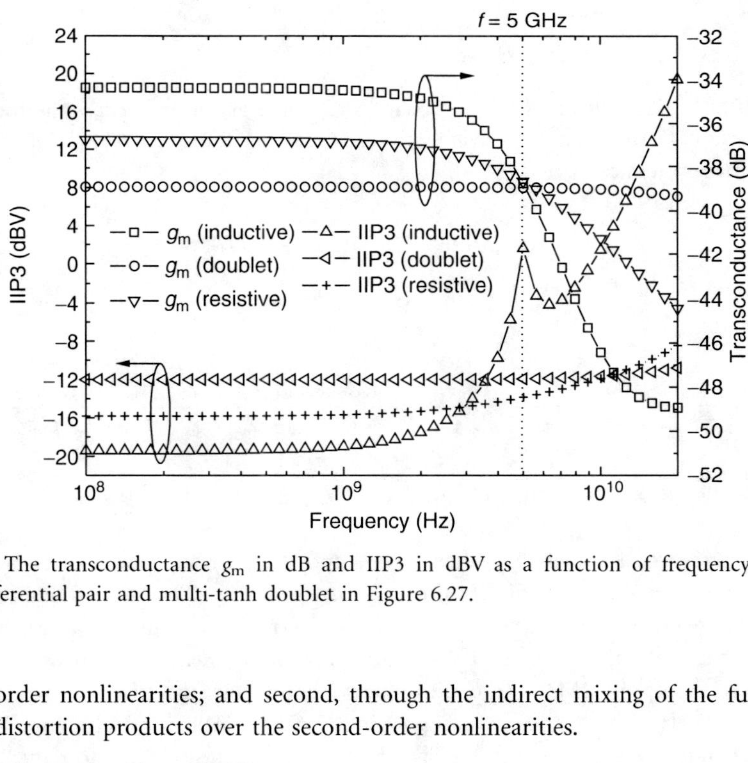

FIGURE 6.29 The transconductance g_m in dB and IIP3 in dBV as a function of frequency for the emitter degenerated differential pair and multi-tanh doublet in Figure 6.27.

via the third-order nonlinearities; and second, through the indirect mixing of the fundamental and second-order distortion products over the second-order nonlinearities.

Third-order Distortion Cancellation

Figure 6.37 shows the large-signal model of a common-emitter stage, which is used for this analysis. As in the previous discussions, we assume again the exponential current-nonlinearity to be the single dominant source of third-order distortion. However, in contrast to the previous analyses, the base–collector capacitance is taken into account as well, since the condition for distortion cancellation is sensitive to all circuit elements, especially at higher frequencies. The complete analysis for the CE-stage is given in the Appendix, and we will discuss here the most important results.

Again we assume $s \approx s_1 \approx s_2$ and $\Delta s = s_1 - s_2 \ll s$, for which the IM3 is expressed as

$$\mathrm{IM3} = \frac{3}{4} \cdot |H_{1_b}(s)|^3 \cdot |D(s)| \cdot |\varepsilon(\Delta s, 2s)| \cdot \hat{V}_s^2 \quad (6.56)$$

and consequently, the input referred voltage IP3 is given by

$$\mathrm{IIP3_v} = \sqrt{\frac{4}{3} \frac{|H_{1_c}(s_2)|}{|H_{3_c}(s_2, s_2, -s_1)|}} \approx \sqrt{\frac{4}{3} \frac{1}{|\varepsilon(\Delta s, 2s)| \cdot |H_{1_b}(s)|^3 \cdot |D(s)|}} \quad (6.57)$$

in which $H1_b$ is the linear transfer function relating the source voltage V_s to the internal base–emitter voltage V_{be}, which for $\beta \gg 1$ is given by

$$H_{1_b}(s) = \frac{1 + sC_{bc}Z_c}{1 + g_m(Z_b/\beta + Z_e) + s[C_\pi(Z_b + Z_e) + C_{bc}(Z_b + Z_c + g_m\bar{Z})] + s^2 C_\pi C_{bc}\bar{Z}} \quad (6.58)$$

where $\bar{Z} = Z_b Z_e + Z_b Z_c + Z_e Z_c$. The term $D(s)$ linearly relates the third-order nonlinear terms at the base–emitter junction to the collector node and is expressed as

$$D(s) = \frac{A(s)}{H_{1c}(s)} = \frac{1}{g_m} \frac{1 + s[C_{jE}(Z_b + Z_e) + C_{bc}Z_b] + s^2 C_{bc}\tau Z_b}{1 - sC_{bc}(r_e + Z_e) - s^2 C_\pi C_{bc} r_e Z_e} \quad (6.59)$$

Finally, the term $\varepsilon(\Delta s, 2s)$ indicates how the second-order interaction influences the total IM3 distortion. This important factor is most conveniently expressed by Equation 6.93 in the Appendix and repeated here for convenience:

$$\varepsilon(\Delta s, 2s) = \frac{g_{m,3}}{3}[2F(\Delta s) + F(2s)] \quad (6.60)$$

where

$$F(s) = \frac{1 - 2g_m(Z_b/\beta + Z_e) + s[(C_{jE} - 2\tau g_m)(Z_b + Z_e) + C_{bc}(Z_b + Z_c - 2g_m\bar{Z})]}{1 + g_m(Z_b/\beta + Z_e) + s[C_\pi(Z_b + Z_e) + C_{bc}(Z_b + Z_c + g_m\bar{Z})] + s^2 C_{bc} C_\pi \bar{Z}} + \frac{s^2 C_{bc}(C_{jE} - 2\tau g_m)\bar{Z}}{1 + g_m(Z_b/\beta + Z_e) + s[C_\pi(Z_b + Z_e) + C_{bc}(Z_b + Z_c + g_m\bar{Z})] + s^2 C_{bc} C_\pi \bar{Z}} \quad (6.61)$$

Inspection of Equation 6.61 shows that when Z_b, Z_e, and Z_c approach zero, $F(s)$ becomes 1 and the IM3 is completely determined by the direct third-order nonlinearity $g_{m,3}$ and the fundamental response. In practice this is never the case, due to the finite series resistances in the device and the rather unpractical circuit conditions. Therefore, if we exclude for the moment the influence of $\varepsilon(\Delta s, 2s)$, the only practical way to improve IM3 is through inductive emitter degeneration to increase the denominators of $H_{1_b}(s)$ and $F(s)$ as we already discussed in "Negative Feedback Techniques." However, this results in lower gain of the amplifying stage. Therefore, the most effective way to reduce the IM3 is by careful selection of the terminal impedances of the device to enforce cancellation of the direct third-order distortion by nullifying $F(\Delta s)$ and $F(2s)$. Since (6.60) is a function of both Δs and $2s$, it is rather difficult to find a general solution for these out-of-band impedances. However, several practical implementations have been reported [20–22,27,36,37], which we discuss here in more detail.

When using the out-of-band third-order distortion cancellation technique, it is evident from (6.56) and (6.57) that the term $\varepsilon(\Delta s, 2s)$ and consequently $F(\Delta s)$ and $F(2s)$ are our design equations. We will use a first-order approximation of (6.61) to introduce the most practical IM3-cancellation conditions for a CE-stage. In addition, we will assume for simplicity that $Z_c(\Delta s) = Z_c(2s) \approx 0$ and $\Delta s \approx 0$, yielding

$$F(\Delta s) \approx \frac{1 - 2g_m(Z_b/\beta + Z_e)}{1 + g_m(Z_b/\beta + Z_e)} \tag{6.62}$$

and

$$F(2s) \approx \frac{1 - 2g_m(Z_b/\beta + Z_e) + 2s[(C_{je} - 2\tau g_m)(Z_b + Z_e) + C_{bc}Z_b(1 - 2g_m Z_e)]}{1 + g_m(Z_b/\beta + Z_e) + 2s[(C_{je} + \tau g_m)(Z_b + Z_e) + C_{bc}Z_b(1 + g_m Z_e)]} \tag{6.63}$$

Obviously, $F(s)$ is a strong function of the out-of-band impedances Z_b and Z_e, the transconductance $g_m = Ic/Vt$ and the emitter area A_e. We can make F zero by solving the numerator. This gives a continuous set of solutions for the out-of-band impedances. Here we highlight the most relevant solutions with their reference to actual circuit implementations. We perform this exercise by substitution of specific out-of-band terminations at the base and emitter terminal, yielding a simplified expression for $\varepsilon(\Delta s, 2s)$. This expression can be used to find the optimum circuit conditions for IM3-cancellation. Doing so, we obtain the following general set of solutions:

1. Setting $Z_e(\Delta s) = Z_e(2s) \approx 0$ (fully grounded emitter) and $Z_b(\Delta s) = Z_b(2s) = \beta/2g_m$ [21], yields

$$\varepsilon(\Delta s, 2s) \approx \frac{2s(C_{jE} + C_{bc} - 2\tau g_m)r_\pi/3}{1 + 2s(C_{jE} + C_{bc} + \tau g_m)r_\pi/3} \tag{6.64}$$

2. Setting $Z_b(\Delta s) = Z_b(2s) \approx 0$ and $Z_e(\Delta s) = Z_e(2s) = 1/2g_m$ [22], yields

$$\varepsilon(\Delta s, 2s) \approx \frac{2s(C_{jE} - 2\tau g_m)r_e/3}{1 + 2s(C_{jE} + \tau g_m)r_e/3} \tag{6.65}$$

3. Any combination of $Z_e(\Delta s) = Z_e(2s) = R_e$ and $Z_b(\Delta s) = Z_b(2s) = R_b$, which fulfills

$$R_b/\beta + R_e = 1/2g_m \tag{6.66}$$

4. Setting $Z_e(\Delta s) = Z_b(\Delta s) \approx 0$ [37,38], yields

$$\varepsilon(\Delta s, 2s) \approx \frac{1 + 2s[C_{je}(Z_b + Z_e) + C_\mu Z_b]}{1 + g_m(Z_b/\beta + Z_e) + 2s[(C_{je} + \tau g_m)(Z_b + Z_e) + C_\mu Z_b(1 + g_m Z_e)]} \tag{6.67}$$

which needs a second-harmonic inductive termination at the base and/or emitter to make the numerator zero, similarly to the discussion of 6.4

5. Setting $Z_e(2s) = Z_b(2s) \approx 0$ [21], yields

$$\varepsilon(\Delta s, 2s) \approx \frac{1 - g_m(Z_b/\beta + Z_e)}{1 + g_m(Z_b/\beta + Z_e)} \tag{6.68}$$

which needs a resistive baseband termination at the base and/or emitter node to make the numerator zero.

One could argue that choosing any complex impedance at the base and/or emitter node basically yields the same result, however, a purely resistive termination at the baseband and double frequency results in a more broadband IM3-cancellation [21]. Furthermore, if the terminations, especially at the baseband, are not resistive; asymmetry can arise in the IM3 sidebands for large envelope frequencies [20]. Therefore, these cases will not be considered in this text.

To compare the two main circuit solutions (1 and 2), which are shown in Figure 6.30, we will consider the IIP3 as available source power in dBm for $Z_s(f_0) = Z_L(f_0) = 50$. We can calculate this quantity from the voltage referred IIP3 as

$$\mathrm{IIP3_P} = \frac{\mathrm{IIP3_V^2}}{8R_0} = \frac{1}{6R_0|H_{1_b}(s)|^3 \cdot |D(s)| \cdot |\varepsilon(\Delta s, 2s)|} \tag{6.69}$$

Low-Frequency IM3-Cancellation

To study the IM3-cancellation of our reference transistor, we first plot the low-frequency IIP3 as a function of collector current in Figure 6.31 to illustrate the bias dependency of the cancellation condition. The IIP3 is calculated using the simplified model of Figure 6.37 (optimized at $I_C = 1$ mA) and simulated with the full Mextram model. As a reference, the IIP3 is plotted for the case where we only have harmonic shorts at the input and the output of the device. For the reference case, cancellation effects occur around 5 mA, which are due to the intrinsic series resistances of the device (r_E and r_B). The other curve shows a cancellation at 1 mA, for a properly dimensioned emitter resistance ($R_E = 1/2g_m - r_E$). Note that the results are quite accurate around 1 mA, but slightly differ at higher currents approaching the peak-f_T value.

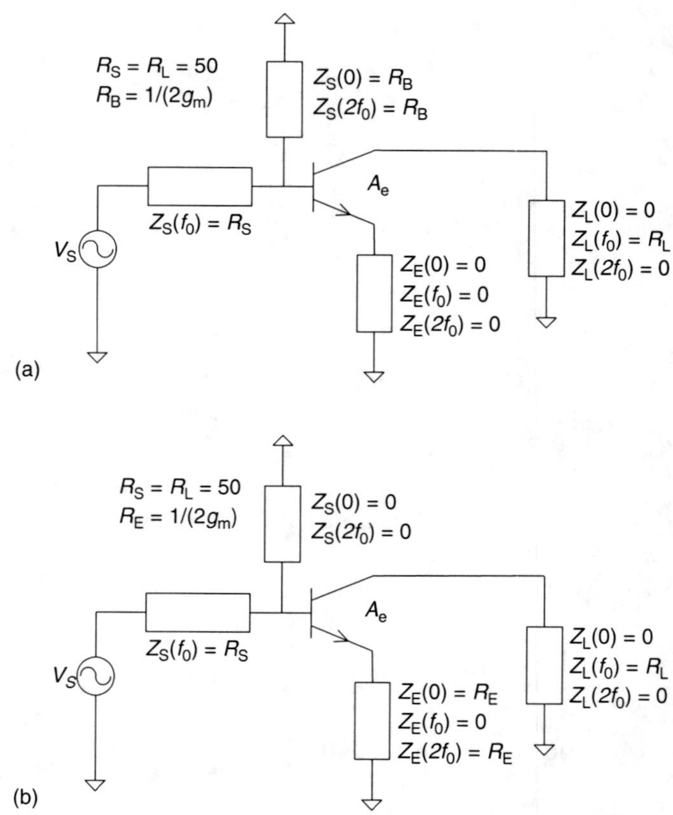

FIGURE 6.30 The AC equivalent circuit of a CE-stage with out-of-band IM3-cancellation terminations at the (a) base and (b) emitter.

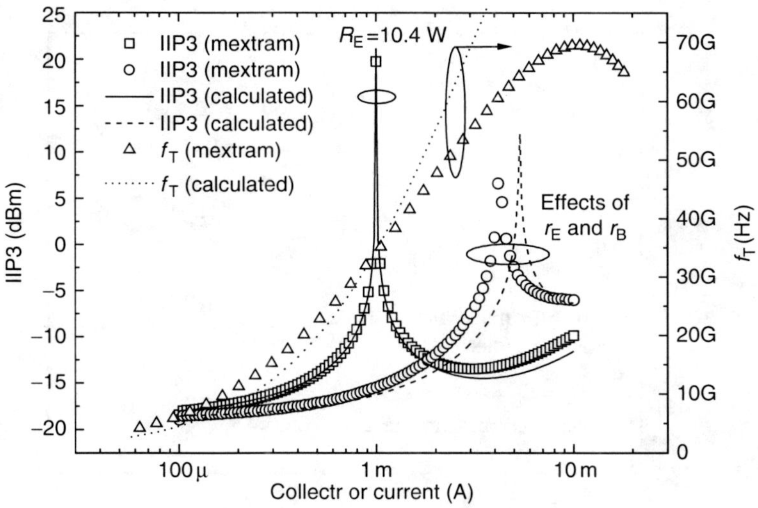

FIGURE 6.31 The cut-off frequency f_T and low-frequency IIP3 as a function of collector current with proper resistive out-of-band terminations in the base or emitter of a CE-stage and with these terminations set to zero.

To understand the differences between out-of-band tuning at the emitter and the base terminal, Figure 6.32 plots the IIP3 as a function of fundamental frequency for both cases. Apparently, applying the harmonic tuning in the emitter gives a linearity improvement over a much broader bandwidth even when no high-frequency IM3-cancellation is applied. This is because $\varepsilon(\Delta s, 2s)$ has a pole around $\frac{3}{2}\omega_T/\beta_f$ according to (6.64), when the device is properly terminated for IM3-cancellation at its base terminal. While according to (6.65) a pole around $\frac{3}{2}\omega_T$ exists, when the device is properly terminated for IM3-cancellation at its emitter terminal. Note that this latter pole is negligible, since it lies far beyond the frequency of operation of the device. Since the IIP3 is proportional to $\varepsilon(\Delta s, 2s)^{-1}$, these poles translate to zeros in the IIP3 versus frequency plot. The fact that proper IM3-termination at the emitter gives a more broadband improvement, can also be explained by considering the transistor as a common-base stage for the even-harmonic frequencies like we have discussed previously in section "Analysis of a CB-Stage in Current-Mode."

High-Frequency IM3-Cancellation

In the previous situations we observed that IM3-cancellation becomes less effective at higher frequencies. This linearity degradation can be explained by the increased influence of the total base charge ($C_{be} + C_{bc}$). We can compensate for this in various ways. Assuming that Z_c is truly zero, we can improve for the high-frequency linearity by choosing the optimum transconductance or by optimum scaling of the emitter area in (6.64) or (6.65) to satisfy

$$g_{m,opt} = \frac{C_{jE} + C_{bc}}{2\tau_f} \tag{6.70}$$

Note that R_b or R_e need adjustment if we choose to modify g_m to satisfy (6.70), rather than to change the active emitter area. Figure 6.33 shows that in case we terminate the emitter correctly for the low-frequency IM3-cancellation condition, fulfilling (6.70) also yields an effective IM3-cancellation at very high frequencies.

As became clear from our previous exercise, IM3-cancellation techniques at higher frequencies rely on the small details of the design, which at first sight, seem to be irrelevant for the basic circuit performance in terms of gain and noise. However, as shown above, it works out that these details can trigger major

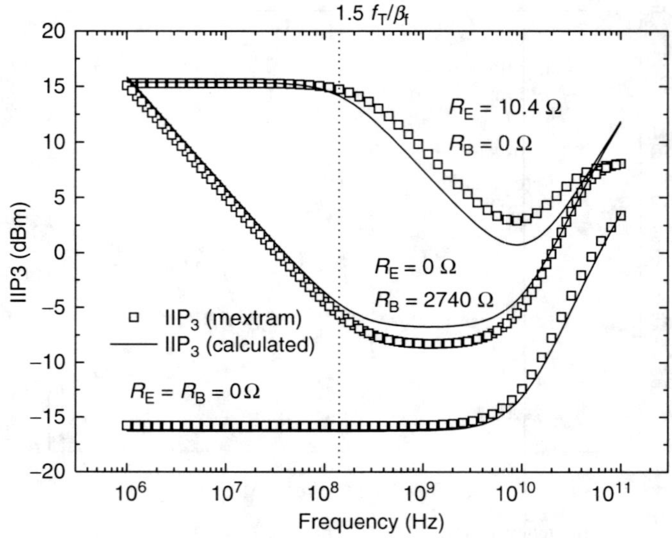

FIGURE 6.32 The IIP3 as a function of frequency for the CE-stages in Figure 6.30 with proper out-of-band terminations in the base and the emitter and with both these terminations set to zero.

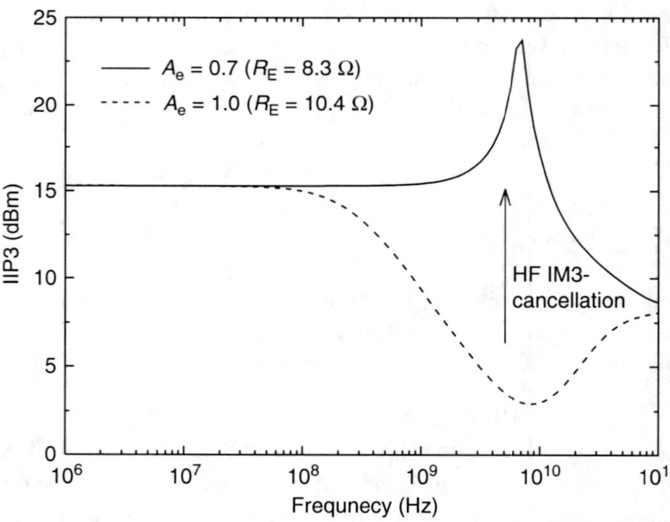

FIGURE 6.33 High-frequency IM3-cancellation for the CE-stage in Figure 6.30b with proper out-of-band terminations in the emitter.

improvements of the circuit linearity, even at high frequencies and low DC powers. For this reason, basic understanding of nonlinear distortion phenomena in bipolar circuits is mandatory for the reliable implementation of highly linear power-efficient circuit blocks. This understanding goes hand in hand with accurate models for the active, as well as the passive devices models. To illustrate the potential of the proposed design techniques we will now briefly discuss two practical implementations.

Practical Examples

From the previous analysis, it is obvious that terminating the base or emitter impedance correctly at the baseband and second-order frequency gives the best trade-off in gain, linearity, and DC power consumption. Two practical circuit examples, in which these techniques were used are given in Ref. [22]

for a down-conversion double-balanced active mixer and in Ref. [38] for an LNA where we simultaneously match for noise, impedance, and IIP3 at low DC-power consumption. Figure 6.34 shows the double-balanced active mixer design which is operated from a 2.7 V DC supply and consumes approximately 2.2 mA current at 2.1 GHz. The mixer is designed with proper out-of-band impedance at the common-mode node of the emitter terminal according to case 2 in the previous subsection. In Ref. [22], the linearity of the mixer with third-order cancellation is compared to a multi-tanh input stage that is not terminated with proper second-harmonic terminations. The IIP3 of the mixer is approximately 10 dB better than the linearity of the multi-tanh stage. This is in agreement with the results in "Shaping the Transfer Function" where we have seen similar linearity improvements, when comparing the transconductance linearity of the doublet to a inductively degenerated differential pair fulfilling the IM3-cancellation condition.

Figure 6.35 shows an example of a current-feedback LNA operating from a 1.5 V DC supply consuming 2.5 mA current at 900 MHz [38]. In this design the IM3-cancellation is implemented according to case 4, presented earlier. By applying a proper inductive second-harmonic source termination, an improvement of approximately 15 dB in terms of IIP3 has been obtained compared to the case without this second-harmonic termination, while gain, noise, impedance match, and bias were not compromised by the use of out-of-band matching techniques.

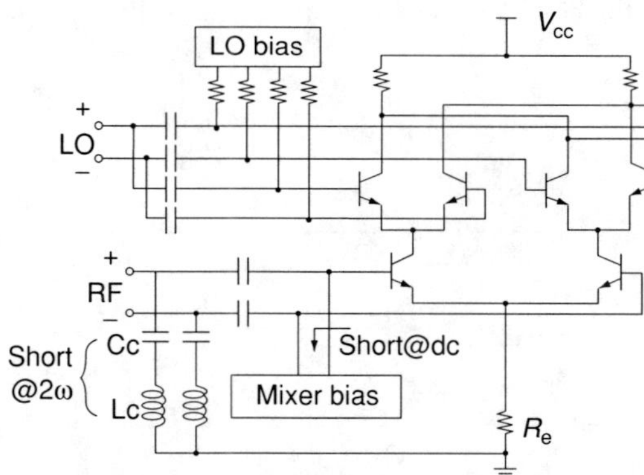

FIGURE 6.34 Down-conversion double-balanced active mixer with third-order cancellation impedance in the emitter and second-harmonic shorts in the base. (From Liwei Sheng and L.E. Larson. *IEEE Trans. Microwave Theory Techn.* 51(11), 2211–2220, 2003. Copyright 2003 IEEE.)

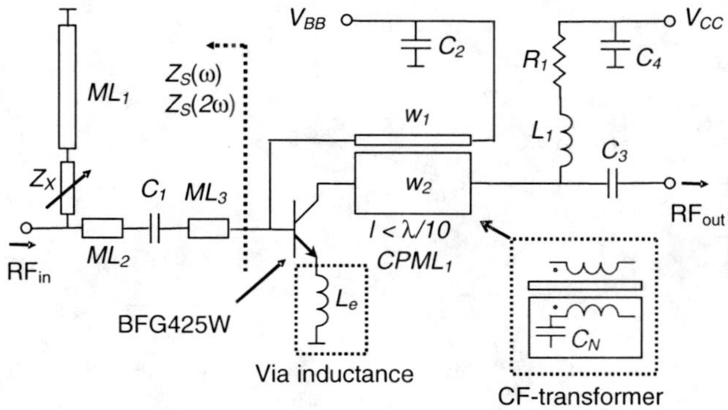

FIGURE 6.35 Low-noise amplifier with simultaneous noise, impedance, and IIP3 match. (From M.P. van der Heijden, L.C.N. de Vreede, F. van Straten, and J.N. Burghartz. Proceedings of the 2003 Bipolar/BiCMOS Circuits and Technology Meeting, pp. 45–48, September 2003. Copyright 2003 IEEE.)

Figure 6.36 shows an example of a InGaP HBT differential driver stage operating from a 3 V DC supply and consumes 30 mA current at 2.0 GHz [39]. In this design the IM3-cancellation is implemented according to case 1 in the previous subsection. By applying a proper resistive second-harmonic common-mode source termination, an improvement of approximately 20 dB in terms of IM3 has been obtained up to the 1 dB output power compression point, compared to the case without this second-harmonic termination.

Discussion

It is clear that the proper use of out-of-band terminations for improved linearity is more demanding for the designer than the previously discussed techniques. This is mainly caused by the fact that the device matching conditions not only have to be fulfilled for the fundamental, but also for the out-of-band frequencies. Although perfect IM3-cancellation conditions for the active device are sometimes difficult to implement, suboptimum solutions already provide sufficient linearity advantages (e.g., the cascode LNA in "Cascode LNA Design"). When considering differential topologies, the orthogonal behavior of the even- and odd-distortion components, facilitate a straightforward implementation of "even" out-of-band terminations through the use of the center tap [21,22], which acts as a virtual ground for the "odd" fundamental. Note that the larger number of design variables translates itself to more design freedom and therefore facilitates superior amplifier stages that make no compromise whatsoever in gain, noise, impedance matching, and linearity performance, while operating at lowest DC power consumption possible [37,38]. Consequently, the best values currently reported for the dynamic range figure-of-merit

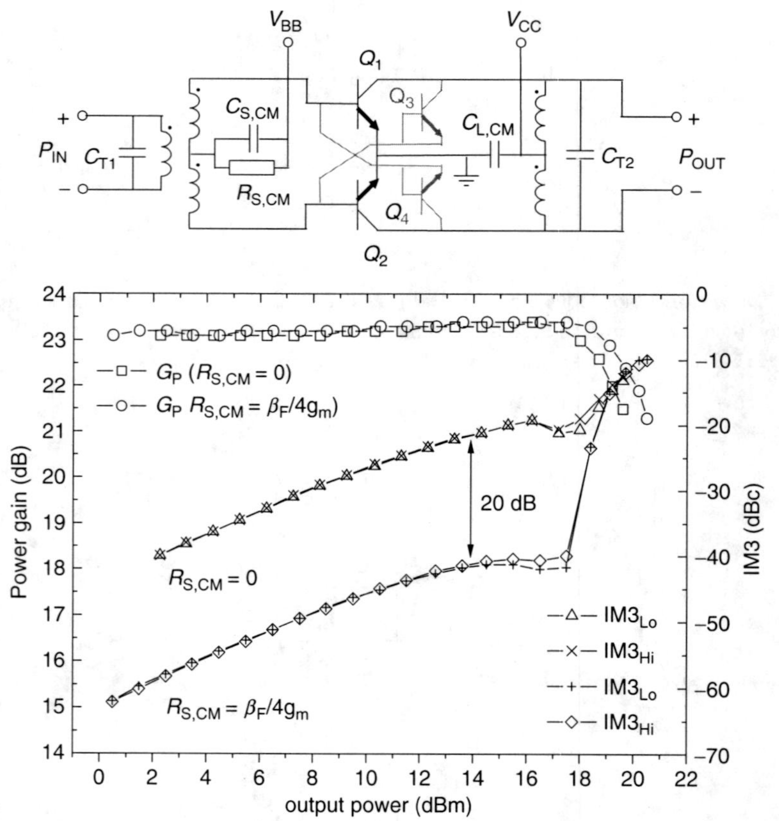

FIGURE 6.36 Measured power gain and IM3 versus output power of a 2 GHz InGaP HBT unilateral differential driver stage operating from $V_{CC} = 3$ V and $I_C = 30$ mA. (From MP van der Heijden, M Spirito, LCN de Vreede, F van Straten, and JN Burghartz. *IEEE MTT-S Digest* 1, 235–238, 2003. With permission.)

[38] are designs using out-of-band terminations. Based on this observation, we expect that future bipolar RF front-end designs will be implemented with some form of third-order distortion cancellation technique in the first stages of the receiving chain, followed by current-mode based designs for the subsequent stages.

Appendix

To investigate the influence of the externally applied impedances on the linearity of a common-emitter stage, we performed a Volterra series analysis as described in Refs. [3,4] using the large-signal equivalent circuit of Figure 6.37. All relevant parameters for our circuit analysis are obtained from the Mextram model of our reference transistor with a peak f_T of 70 GHz at 10 mA. The parameters below have been extracted at $I_C = 1$ mA ($f_T \approx 35$ GHz) and are given by:

- The base–emitter depletion capacitance $C_{jE} = 11$ fF
- The collector–base depletion capacitance $C_{bc} = 20$ fF
- The collector–substrate depletion capacitance $C_{CS} = 5.8$ fF
- The forward base–emitter transit time* $\tau_f = 0.8$ psec
- The forward current-gain $\beta_f = 350$
- The base resistance $r_B = 15\,\Omega$
- The emitter resistance $r_E = 2.35\,\Omega$
- The collector resistance $r_C = 15\,\Omega$

For the low-current operation of the device, we assume the exponential input-voltage-to-current relation to be the dominant nonlinearity. Consequently, we write for the collector current

$$i_C = I_S \exp\left(v_{BE}/V_T\right) \tag{6.71}$$

and approximate this function with a third-order power series

$$i_c = g_m v_{be} + g_{m2} v_{be}^2 + g_{m3} v_{be}^3 \tag{6.72}$$

in which the Taylor coefficients at the DC-collector current I_C are given by

$$g_m = \frac{I_C}{V_T}, \quad g_{m,2} = \frac{I_C}{2V_T^2}, \quad g_{m,3} = \frac{I_C}{6V_T^3} \tag{6.73}$$

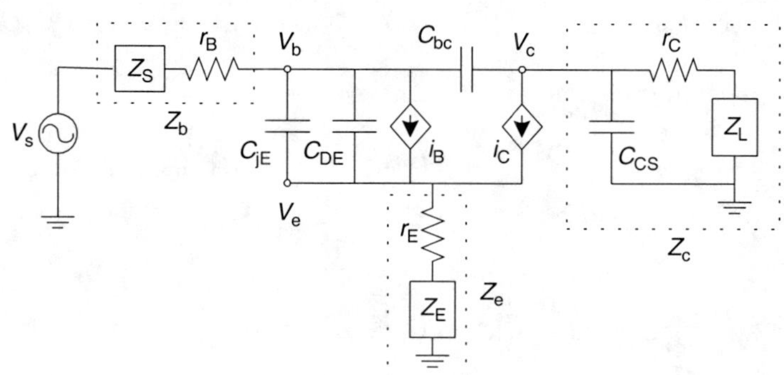

FIGURE 6.37 Large-signal model of a CE-stage with only exponential nonlinearities.

*For our distortion calculations we need the "pure" base–emitter transit time for the base–emitter diffusion capacitance and not the more common total delay time, which includes all delays.

The base current i_b and the base–emitter diffusion current are both linearly related to i_c via the current gain β_f and the transit time τ_f, which are both assumed to be constant.

$$i_b = \frac{i_c}{\beta_f} \tag{6.74}$$

$$i_{C_{DE}} = \tau_f \frac{d}{dt} i_c \tag{6.75}$$

Furthermore, the base–emitter depletion capacitance is considered to be constant as well, since its nonlinearity is small in comparison with the nonlinearity of the base–emitter diffusion capacitance C_{DE}.

The linear transfer function, which relates the source voltage V_s to the internal base–emitter voltage V_{be}, is given by

$$H_{1_b}(s) = \frac{V_{be}}{V_s} = \frac{1 + sC_{bc}Z_c}{1 + g_m(Z_b/\beta + Z_e) + s[C_\pi(Z_b + Z_e) + C_{bc}(Z_b + Z_c + g_m\bar{Z})] + s^2 C_\pi C_{bc}\bar{Z}} \tag{6.76}$$

in which $\bar{Z} = Z_b Z_e + Z_b Z_c + Z_e Z_c$ and $C_\pi = C_{jE} + C_{DE}$ with $C_{DE} = \tau_f g_m$. The overall linear transfer function relating the source voltage V_s to the internal collector voltage V_c is given by

$$H_{1_c}(s) = \frac{V_c}{V_s} = H_{bc}(s)H_{1_b}(s) \tag{6.77}$$

in which H_{bc} is the linear transfer function relating the base–emitter voltage to the collector voltage:

$$H_{bc}(s) = \frac{V_c}{V_{be}} = \frac{-Z_c g_m}{1 + sC_{bc}Z_c}[1 - sC_{bc}(r_e + Z_e) - s^2 C_\pi C_{bc} r_e Z_e] \tag{6.78}$$

Following Ref. [3], the second-order nonlinear transfer function can be expressed as

$$H_{2_c}(s_1, s_2) = g_{m,2} H_{1_b}(s_1) H_{1_b}(s_2) A(s_1 + s_2) \tag{6.79}$$

in which $A(s)$ relates the nonlinear terms at the base–emitter junction to the collector node and is given by

$$A(s) = \frac{-Z_c(1 + s[C_{jE}(Z_b + Z_e) + C_{bc}Z_b] + s^2 C_{bc}\tau Z_b)}{1 + g_m(Z_b/\beta_f + Z_e) + s[C_\pi(Z_b + Z_e) + C_{bc}(Z_b + Z_c + g_m\bar{Z})] + s^2 C_\pi C_{bc}\bar{Z}} \tag{6.80}$$

and the third-order nonlinear transfer function can be expressed as

$$H_{3_c}(s_1, s_2, s_3) = \varepsilon(s_1, s_2, s_3) H_{1_b}(s_1) H_{1_b}(s_2) H_{1_b}(s_3) A(s_1 + s_2 + s_3) \tag{6.81}$$

where

$$\varepsilon(s_1, s_2, s_3) = g_{m,3} - \frac{2}{3} g_{m,2}^2 [B(s_1 + s_2) + B(s_1 + s_3) + B(s_2 + s_3)] \tag{6.82}$$

in which

$$B(s) = \frac{Z_b/\beta_f + Z_e + s[\tau(Z_b + Z_e) + C_{bc}\bar{Z}] + s^2 C_{bc}\tau\bar{Z}}{1 + g_m(Z_b/\beta_f + Z_e) + s[C_\pi(Z_b + Z_e) + C_{bc}(Z_b + Z_c + g_m\bar{Z})] + s^2 C_\pi C_{bc}\bar{Z}} \tag{6.83}$$

For a two-tone excitation $s_1 = s_3 = j\omega_1$ and $s_2 = j\omega_2$. We calculate IM3 assuming a small tone-spacing to keep our equations compact, $s_1 \approx s_2 \approx s$ and $\Delta s = s_1 - s_2 \ll s$:

$$\mathrm{IM3_{LO}} = \frac{3}{4}\left|\frac{H_{3_c}(s_1, s_1, -s_2)}{H_{1_c}(2s_1 - s_2)}\right| \hat{V}_s^2 \approx \frac{3}{4}|\varepsilon(\Delta s, 2s)| \cdot |H_{1_b}(s)|^3 \cdot |D(s)| \cdot \hat{V}_s^2 \tag{6.84}$$

$$\mathrm{IM3_{HI}} = \frac{3}{4}\left|\frac{H_{3_c}(s_2, s_2, -s_1)}{H_{1_c}(2s_2 - s_1)}\right| \hat{V}_s^2 \approx \frac{3}{4}|\varepsilon(\Delta s, 2s)| \cdot |H_{1_b}(s)|^3 \cdot |D(s)| \cdot \hat{V}_s^2 \tag{6.85}$$

where

$$D(s) = \frac{A(s)}{H_{1_c}(s)} = \frac{1}{g_m}\frac{1 + s[C_{jE}(Z_b + Z_e) + C_{bc}Z_b] + s^2 C_{bc}\tau Z_b}{1 - sC_{bc}(r_e + Z_e) - s^2 C_\pi C_{bc} r_e Z_e} \tag{6.86}$$

and

$$\varepsilon(\Delta s, 2s) = g_{m,3} - \frac{2}{3}g_{m,2}^2[2B(\Delta s) + B(2s)] \tag{6.87}$$

Having the expression for IM3, we can calculate the amplitude of the input signal tones $\hat{V}_s$ at which IM3 = 1, which is the input referred third-order intercept point (IIP3):

$$\mathrm{IIP3} = \sqrt{\frac{4}{3}\frac{|H_{1_c}(s_2)|}{|H_{3_c}(s_2, s_2, -s_1)|}} \approx \sqrt{\frac{4}{3}\frac{1}{|\varepsilon(\Delta s, 2s)| \cdot |H_{1_b}(s)|^3 |D(s)|}} \tag{6.88}$$

The output referred third-order intercept point OIP3 can be calculated by multiplying the IIP3 by the overall linear transfer function $H_{1_c}(s)$ at the fundamental frequency:

$$\mathrm{OIP3} = \sqrt{\frac{4}{3}\frac{|H_{1_c}(s_2)|^3}{|H_{3_c}(s_2, s_2, -s_1)|}} \approx \sqrt{\frac{4}{3}\frac{|H_{bc}(s)|^3}{|\varepsilon(\Delta s, 2s)| \cdot |A(s)|}} \tag{6.89}$$

The factor $\varepsilon(\Delta s, 2s)$ determines how the out-of-band impedances at the baseband and double frequencies influence the total third-order intermodulation distortion. This factor can be expressed in various ways in order to give more insight in how certain circuit parameters influence its behavior. Therefore, we first substitute the nonlinear Taylor coefficients of (6.73) in (6.87), yielding:

$$\varepsilon(\Delta s, 2s) = \frac{g_m}{6V_T^2}[1 - 2g_m B(\Delta s) - g_m B(2s)] \tag{6.90}$$

which we can can rewrite in the following form:

$$\varepsilon(\Delta s, 2s) = \frac{g_m}{3V_T^2}\left[E(\Delta s) + \frac{1}{2}E(2s) - 1\right] \tag{6.91}$$

where

$$E(s) = \frac{1 + s[C_{jE}(Z_b + Z_e) + C_{bc}(Z_b + Z_c)] + s^2 C_{jE} C_{bc}\bar{Z}}{1 + g_m(Z_b/\beta_f + Z_e) + s[C_\pi(Z_b + Z_e) + C_{bc}(Z_b + Z_c + g_m\bar{Z})] + s^2 C_\pi C_{bc}\bar{Z}} \tag{6.92}$$

Equation 6.92 clearly shows the interaction of the total series impedance connected to the device with the depletion capacitances. This form will be extensively used in "Negative Feedback Techniques." Another way of writing $\varepsilon(\Delta s, 2s)$ is given by

$$\varepsilon(\Delta s, 2s) = \frac{g_m}{18V_T^2}[2F(\Delta s) + F(2s)] \tag{6.93}$$

where

$$F(s) = \frac{1 - 2g_m(Z_b/\beta + Z_e) + s[(C_{jE} - 2\tau g_m)(Z_b + Z_e) + C_{bc}(Z_b + Z_c - 2g_m\bar{Z})]}{1 + g_m(Z_b/\beta + Z_e) + s[C_\pi(Z_b + Z_e) + C_{bc}(Z_b + Z_c + g_m\bar{Z})] + s^2 C_{bc}C_\pi\bar{Z}} + \frac{s^2 C_{bc}(C_{jE} - 2\tau g_m)\bar{Z}}{1 + g_m(Z_b/\beta + Z_e) + s[C_\pi(Z_b + Z_e) + C_{bc}(Z_b + Z_c + g_m\bar{Z})] + s^2 C_{bc}C_\pi\bar{Z}} \tag{6.94}$$

This form will be used to describe the IM3 cancellation conditions in "Harmonic Matching Techniques" since it clearly shows the low- and high-frequency circuit conditions for having broadband IM3-cancellation.

Acknowledgments

We would like to thank the TU Delft for giving the opportunity to conduct research in this field for many years and Philips Research and Philips Semiconductors for their steady encouraging support over many years. In addition we would like to acknowledge Agilent Technologies for software support in terms of MDS, ADS, and ICCAP, as well as BSW and Maury Microwave for their cooperation on large signal measurements. We would like to personally thank Rik Jos and Peter Deixler for their support. Finally, out special thanks to H. C. de Graaff for his very solid and encouraging mentorship over the past few years.

References

1. JCJ Paasschens and WJ Kloosterman. The Mextram bipolar transistor model, level 504. Unclassified Report NL-UR 2000/811, Philips National Laboratory, available at http://www.semiconductors.philips.com/Philips_Models/bipolar/mextram/
2. B Razavi. *RF Microelectronics.* Upper Saddle River, NJ: Prentice-Hall, 1998.
3. P Wamback and W Sansen. *Distortion Analysis of Analog Integrated Circuits.* Dordrecht: Kluwer Academic Publishers, 1998.
4. SA Maas. *Nonlinear Microwave and RF Circuits.* 2nd edition. Norwood, MA: Artech House, 2003.
5. PR Gray and RG Meyer. *Analysis and Design of Analog Integrated Circuits.* 3rd edition. New York: Wiley, 1993.
6. ML Lantz. Systematic Design of Linear Feedback Amplifiers. PhD dissertation, Lund University, Lund, Sweden, 2002.
7. JD Cressler and G Niu. *Silicon–Germanium Heterojunction Bipolar Transistors.* Boston: Artech House, 2003.
8. LCN de Vreede, HC de Graaff, K Mouthaan, M de Kok, JL Tauritz, and RGF Baets. Advanced modelling of distortion effects in bipolar transistors using the Mextram model. Proceedings of the 1994 Bipolar/BiCMOS Circuits and Technology Meeting, pp. 48–51, October 1994.
9. LCN de Vreede, HC de Graaff, JL Tauritz, and RGF Baets. Extension of the collector charge description for compact bipolar epilayer models. *IEEE Trans. Elect. Dev.* 45, 277–285, 1998.
10. M Schroter, DR Pehlke, and T-Y Lee. Compact modeling of high-frequency distortion in silicon integrated bipolar transistors. *IEEE Trans. Elect. Dev.* 47(7), 1529–1535, 2000.
11. EH Nordholt. *Design of High-Performance Negative Feedback Amplifiers.* Elsevier, 1983.
12. LCN de Vreede, HC de Graaff, JA Willemen, W van Noort, HFF Jos, LE Larson, JW Slotboom, and JL Tauritz. Bipolar transistor epilayer design using the MAIDS mixed-level simulator. *IEEE J. Solid-State Circ.* 34(9), 1331–1338, 1999.

13. LCN de Vreede, HC de Graaff, and B Rejeai. The effect of non-saturated electron drift velocity on bipolar device linearity. In: Proceedings of the 2002 Bipolar/BiCMOS Circuits and Technology Meeting, pp. 88–91, September 2002.
14. HC de Graaff and FM Klaasen. *Compact Transistor Modeling for Circuit Design.* New York: Springer-Verlag, 1990.
15. C Toumazou, FJ Lidgey, and DG Haigh (Eds). *Analogue IC Design: the Current-Mode Approach.* IEE Circuits and Systems Series 2. London: Perigrinus, 1990.
16. Keng Leong Fong and RG Meyer. High-frequency nonlinearity analysis of common-emitter and differential-pair transconductance stages. *IEEE J. Solid-State Circ.* 33(4), 548–555, 1998.
17. Adiseno, M Ismail, and H Olsson. A wide-band RF front-end for multiband multistandard high-linearity low-IF wireless receivers. *IEEE J. Solid-State Circ.* 37(9), 1162–1168, 2002.
18. B Gilbert. The multi-tanh principle: a tutorial overview. *IEEE J. Solid-State Circ.* 33(1), 2–17, 1998.
19. K Kimura. The ultra-multi-tanh technique for bipolar linear transconductance amplifiers. *IEEE Trans. Circ. Systems I.* 44(4), 288–302, 1997.
20. V Aparin and C Persico. Effect of out-of-band terminations on intermodulation distortion in common-emitter circuits. *IEEE MTT-S Digest* 3, 977–980, 1999.
21. MP van der Heijden, HC de Graaff, and LCN de Vreede. A novel frequency-independent third-order intermodulation distortion cancellation technique for BJT amplifiers. *IEEE J. Solid-State Circ.* 37(9), 1176–1183, 2002.
22. Liwei Sheng and LE Larson. An Si–SiGe BiCMOS direct-conversion mixer with second-order and third-order nonlinearity cancellation for WCDMA applications. *IEEE Trans. Microwave Theory Techn.* 51(11), 2211–2220, 2003.
23. SC Cripps. *RF Power Amplifiers for Wireless Communications.* Norwood, MA: Artech House, 1999.
24. WD van Noort, LCN de Vreede, HFF Jos, LK Nanver, and JW Slotboom. Reduction of UHF power transistor distortion with a nonuniform collector doping profile. *IEEE J. Solid-State Circ.* 36(9), 1399–1406, 2001.
25. F van Rijs, R Dekker, HA Visser, HGA Huizing, D Hartskeerl, PHC Magnee, and R Dondero. Influence of output impedance on power added efficiency of Si-bipolar power transistors. *IEEE MTT-S Digest*, 1945–1948, 2000.
26. M Spirito, LCN de Vreede, LK Nanver, S Weber, and JN Burghartz. Power amplifier PAE and ruggedness optimization by second-harmonic control. *IEEE J. Solid-State Circ.* 38(9), 1575–1583, 2003.
27. MP van der Heijden, M Spirito, M Pelk, LCN de Vreede, and JN Burghartz. On the optimum biasing and input out-of-band terminations of linear and power efficient class-AB bipolar RF amplifiers. Proceedings of the 2004 Bipolar/BiCMOS Circuits and Technology Meeting, September 2004.
28. J Reynolds. Nonlinear distortions and their cancellation in transistors. *IEEE Trans. Elect. Dev.* 12, 595–599, 1965.
29. S Aggarwal, A Daanen, M Locher, AL Landesman, M Judson, F Garrigues, M Bracey, H Xu, O Charlon, and M Gao. A highly integrated dual-band triple-mode transmit IC for cellular CDMA2000 applications. *IEEE J. Solid-State Circ.* 38, 1561–1569, 2003.
30. B Gilbert. The MICROMIXER: a highly linear variant of the Gilbert mixer using a bisymmetric class-AB input stage. *IEEE J. Solid-State Circ.* 32(9), 1412–1423, 1997.
31. P Antognetti and G Massobrio. *Semiconductor Device Modeling with SPICE.* New York: McGraw-Hill, 1987.
32. CJM Verhoeven, A van Staveren, GLE Monna, MHL Kouwenhoven, and E Yildiz. *Structured Electronic Design.* Dordrecht: Kluwer Academic Publishers, 2003.
33. R van den Brink. Low-Noise Wideband Feedback Amplifiers. PhD dissertation, Delft University of Technology, Delft, The Netherlands, 1994.
34. EM Cherry and DE Hooper. *Amplifying Devices and Low-Pass Amplifier Design.* New York: John Wiley and Sons, 1968.

35. SP Voinigescu, MC Maliepaard, JL Showell, GE Babock, D Marchesan, M Schroter, P Schvan, and DL Harame. A scalable high-frequency noise model for bipolar transistors with application to optimal transistor sizing for low-noise amplifier design. *IEEE J. Solid-State Circ.* 32, 1430–1439, 1997.
36. Keng Leong Fong. High-frequency analysis of linearity improvement technique of common-emitter transconductance stage using a low-frequency-trap network. *IEEE J. Solid-State Circ.* 35(8), 1249–1252, 2000.
37. V Aparin, E Zeisel, and P Gazzerro. Highly Linear SiGe BiCMOS LNA and Mixer for Cellular CDMA/AMPS Applications. Proceedings of 2002 IEEE RFIC, pp. 129–132, June 2002.
38. MP van der Heijden, LCN de Vreede, F van Straten, and JN Burghartz. A high performance unilateral 900 MHz LNA with simultaneous noise, impedance, and IP3 Match. Proceedings of the 2003 Bipolar/BiCMOS Circuits and Technology Meeting, pp. 45–48, September 2003.
39. MP van der Heijden, M Spirito, LCN de Vreede, F van Straten, and JN Burghartz. A 2 GHz high-gain differential InGaP HBT driver amplifier matched for high IP3. *IEEE MTT-S Digest* 1, 235–238, 2003.

7

SiGe MMICs

7.1 Introduction 7-1
Motivation: SiGe versus III–V MMICs • SiGe MMICs: What They Can and Cannot Do
7.2 The Substrate Issue 7-3
7.3 SiGe MMICS Using Heterojunction Bipolar Transistors 7-5
Narrow-Band Circuits • Wide-Band Circuits
7.4 SiGe MMICS Using Heterostructure Field-Effect Transistors 7-16
7.5 Summary 7-17

Hermann Schumacher
University of Ulm

7.1 Introduction

Often overlooked amid the spectacular advances in digital VLSI, monolithic microwave integrated circuits (MMICs) are crucial building blocks of modern wireless communications systems. Combining passive and active components of microwave circuits onto a single semiconductor substrate not only dramatically reduces size and weight, and eliminates the harmful effects of wire interconnects, but also reduces the cost and effort associated with assembly and most notably the manual pruning still very common in microwave subsystem assembly.

In the context of this chapter, the definition of MMICs will be extended to include very wide-band components commonly used in the analog portion of fiber-optic communications systems, e.g., the wide-band low-noise amplifier following the photodetector. These circuits use circuit design techniques very similar to microwave circuits in the narrower sense, and may find use also in microwave applications requiring multi-octave performance, such as in instrumentation and surveillance applications.

The discussion of MMICs using Si–SiGe heterostructure transistors will start with a brief comparison with its counterparts fabricated on III–V semiconductor substrates. It is important to understand that SiGe MMICs, while a very interesting option in many cases, are not a "cure-all" technology. They perform best and are most convincing when designed with its inherent limitations in mind.

Probably, the first issue that comes to mind when comparing Si and GaAs as materials for microwave circuits is the significantly higher loss encountered in Si substrates. The substrate issue hence deserves special treatment, and we will see that it is addressed not so much through substrate modifications, but through appropriate circuit design which has the added advantage of significantly reducing the chip size required for certain functions.

While SiGe heterojunction bipolar transistors (HBTs) are at the time of this writing the only commercially available device technology and deserve most of the attention, SiGe heterostructure field-effect transistors (HFETs) are a topic of intense research whose suitability for MMICs has already been demonstrated experimentally. Therefore, a brief section will treat SiGe HFET circuits as well.

Motivation: SiGe versus III–V MMICs

Monolithic microwave integrated circuits have been realized in the past primarily using III–V semiconductor technologies, predominantly GaAs. GaAs-based MMICs still have performance advantages, which are mostly derived from the significantly higher low-field electron mobilities of GaAs and InGaAs, the higher saturation velocity of electrons at medium electric fields, and the higher breakdown fields. The issue of the semiinsulating GaAs substrate, versus mostly conducting substrates for Si, will be dealt with in Section 7.2.

The main motivation for trying to replace III–V MMICs with Si-based circuits is cost. If the expected market volume for a given microwave application is large, Si–SiGe MMICs can be significantly cheaper than their III–V competitors, predominantly due to the much larger wafer sizes and the industrially very mature manufacturing technology. For small quantities, the advantage is not so clear—especially for large wafer sizes (>6 in.) and BiCMOS technologies, mask costs can be an order of magnitude higher than for typical III–V MMIC processes. An often overlooked point is that testing and packaging tend to render the cost ratios between Si and GaAs ICs considerably less impressive for the packaged products.

SiGe MMICs: What They Can and Cannot Do

Comparing SiGe and GaAs MMICs, one realizes that another difference is in the circuit complexity, in the number of individual functions integrated on a single chip, and in the total chip size.

GaAs MMICs cannot deny their descent from hybrid microwave integrated circuits (MICs), which combine a few active devices with planar transmission line structures, which also implement the necessary reactances (e.g., open-ended stubs as shunt capacitors). In doing so, the designer can rely on a vast body of microstrip and coplanar models. Due to the high substrate resistivity available in GaAs (about $10^7\ \Omega$ cm), designs can be easily transferred from nonsemiconducting substrates (e.g., alumina) to the monolithically integrated circuit. The resulting ICs are rather large in size (5 to 15 mm^2) and sparsely populated by active devices.

The designer of silicon MMICs, including Si–SiGe, will often have to live with substrate resistivities ranging from 20 Ω cm to may be 1000 Ω cm. Long transmission line segments would result in intolerable loss in the substrate. The Si MMIC designer will hence resort to realizing reactances using lumped passive elements, wherever possible, predominantly metal–insulator–metal (MIM) capacitors and spiral inductors. The latter still receive considerable research attention (see e.g., Ref. [1]). Longer transmission line segments are avoided, leading to very compact designs with a smaller chip area (sometimes <1 mm^2) and a considerably higher density of active devices (capacitors and inductors will still dominate the area consumed, though).

Integration density and the high yield of mature silicon processes make MMICs with a high number of integrated functions possible. This is also true for Si–SiGe HBT technologies that differ from all-Si processes only in the base design. Figure 7.1, which shows a chip photograph of an eight-channel receiver array for 24 GHz monolithically integrated in a Si–SiGe BiCMOS technology [2], shall serve as an example.

In SiGe HFETs, the deviation from standard Si processing is larger, and this explains the lag in commercialization there.

Integration density is assisted by another often overlooked difference between GaAs and Si substrates: the thermal conductivity of Si is about three times higher than for GaAs (1.45 W K^{-1} cm^{-1} versus 0.46 W K^{-1} cm^{-1}), which makes the thermal management of densely placed active devices considerably easier and prevents the formation of "hot spots" on the IC.

As pointed out in Ref. [3] for InP–InGaAs–InP DHBTs, the power dissipated in the collector region of high-speed bipolar transistors increases with the square of the transistor operational bandwidth. For the investigated InP-based HBTs, significant temperature increases were observed in simulation if an InGaAs etch-stop layer of only 50 nm thickness was present. SiGe HBTs should fare much better in this respect.

Raw device speed is no longer an issue in Si-based MMIC design. Si–SiGe HBT technologies with transit frequencies up to 350 GHz have been demonstrated [4], rivaling state-of-the-art results reported in III–V technologies. The focus of attention shifts toward compact ICs combining many analog and

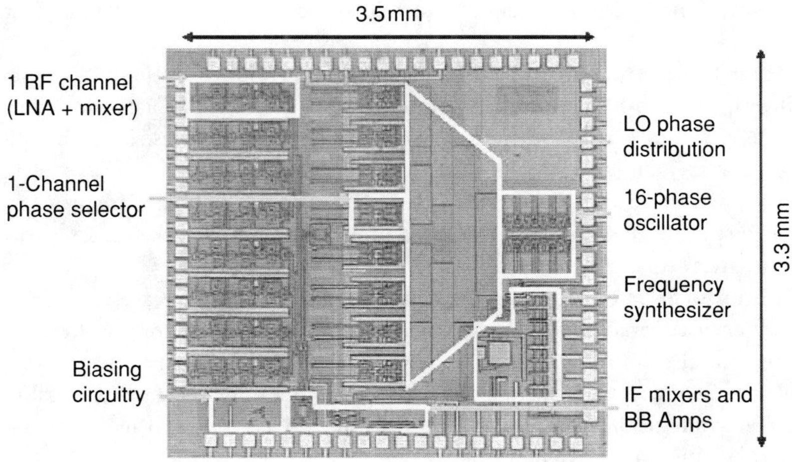

FIGURE 7.1 Chip photo of an eight-channel phased array receiver for 24 GHz in a SiGe BiCMOS technology, demonstrating the level of complexity possible in Si-based MMICs. (After H. Hashemi, X. Guan, and A Hajimiri. ISSCC 2004, San Francisco, CA, February 15–18, 2004. With permission.)

some digital (e.g., frequency dividers) onto a single chip. It shall not be overlooked though that III–V MMICs continue to have an edge in performance: the high low-field electron velocity translates into lower noise figures, while the higher average electron velocity at medium electric fields coupled with the higher breakdown fields lead to significantly higher breakdown voltages in III–V semiconductor devices, if we compare devices with the same f_T.

Excellent application opportunities for Si–SiGe MMICs are hence given if

- There is a significant market potential in terms of volume or
- Multi-functionality on a single chip provides clear benefits and
- Ultimate performance is not required

7.2 The Substrate Issue

In much of the early work on Si-based microwave and millimeter-wave integrated circuits, substrates with very high resistivity (e.g., $10^5\,\Omega$ cm) have been used [5]. These substrates have dielectric properties comparable to semi-insulating GaAs substrates (which have resistivities $>10^6\,\Omega$ cm), allowing the direct transfer of established MMIC design techniques from GaAs to Si substrates. These designs make heavy use of transmission line segments in the realization of reactances, phase adjustment, and impedance transformation and hence depend on low-loss substrates.

However, these high-resistivity substrates suffer a decrease in resistivity after high-temperature processing steps. They became only very recently commonly available in commercial bipolar and CMOS processes. Additionally, the SiO_2–Si interface on top of the substrate deserves very careful attention—frequently, parasitic surface inversion channels form in high-resistivity Si substrates, leading to strong fluctuations in the loss observed in transmission lines and spiral inductors. The common use of channel stopper implants defeats the purpose of having a high-resistivity substrate and has to be avoided here.

Hence, most Si MMICs reported today employ either low- or medium-resistivity Si substrates (1 to $50\,\Omega$ cm). The use of transmission lines is reduced in favor of concentrated reactances—predominantly spiral inductors and metal–insulator–metal (MIM) capacitors. As an added benefit, the resulting layout is necessarily very compact, leading to a further cost benefit over typical GaAs-based MMICs using transmission-line segments. On the other hand, extension of the concept to the upper microwave and the millimeter-wave range needs substantial effort in the characterization of passive structures and layout effects.

Where distributed structures have to be used on low- and medium-resistivity substrates, they can be modeled as either quasi-thin-film microstrip or lossy-conductor-backed coplanar waveguides [6]. Thin-film microstrip lines require vias to the substrate, while coplanar waveguides require additional space on chip for the ground metallization. In either case, the Si substrate serves as a lossy backplane. Strong dispersion (slow-wave effects) may exist at low frequencies, while at higher frequencies dispersion is tolerable (see Figure 7.2). Observed loss would be prohibitive, for example, quarter-wave impedance transformers, but are adequate for interconnects in typical dense MMICs.

A common feature of Si-based MMICs is the significant part of the chip real estate consumed by spiral inductors. An often-used small-signal equivalent circuit for spiral inductors on lossy Si substrates is shown in Figure 7.3a. The shunt elements at input and output model the capacitive coupling to the lossy substrate, while the frequency-dependent series resistance models the increase in metal resistance with increasing frequency due to the skin effect. More rigorous models may also include the effect of Eddy currents in the substrate (see e.g., Ref. [1]).

An interesting observation is that when moving into the millimeter-wave domain, practical spiral inductors can be modeled using simpler equivalent circuits (see Figure 7.3b). This is due to two separate effects:

- With increasing frequency, series resistance losses dominate due to the skin effect.
- As the frequency of operation increases, the necessary inductance in tuned circuits decreases proportionately. As a useful rule-of-thumb, the necessary impedance of a spiral inductor is in the same order of magnitude as the characteristic impedance of MMIC transmission lines, typically less than 90 Ω, which leads to a practical value of approximately 15 nH GHz for the product of inductance and frequency of operation. As an example, required inductor values at 24 GHz are a few tenths of a nano-Henry. With decreasing electrical value, the inductor geometry also shrinks, leading to a markedly decreased interaction with the lossy substrate.

Metal–insulator–metal capacitors are less critical in the frequency dependence of their impedance than spiral inductors, but may exhibit a parasitic series resistance L_S in their equivalent circuit, depending on the geometric shape (see Figure 7.4). The series resistance R_S can be substantial if the bottom electrode is not an Al interconnect level, but a silicide contact. In any case, the bottom electrode will have a nonnegligible capacitance to the lossy substrate.

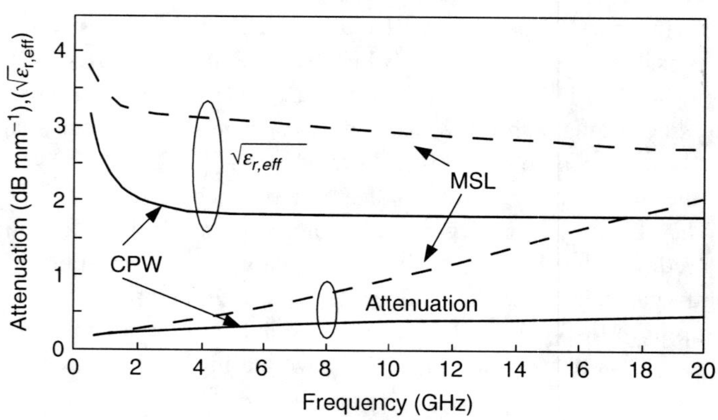

FIGURE 7.2 Measured attenuation and slowing factor for 4.2-μm wide microstrip (MSL—dashed curves) and coplanar transmission lines (CPW—solid curves) on 20 Ω cm silicon substrate. (After W. Dürr, U. Erben, A. Schüppen, H. Dietrich, and H. Schumacher. *IEEE Trans. Microwave Theory Tech.* 46, 712–715, 1998. With permission.)

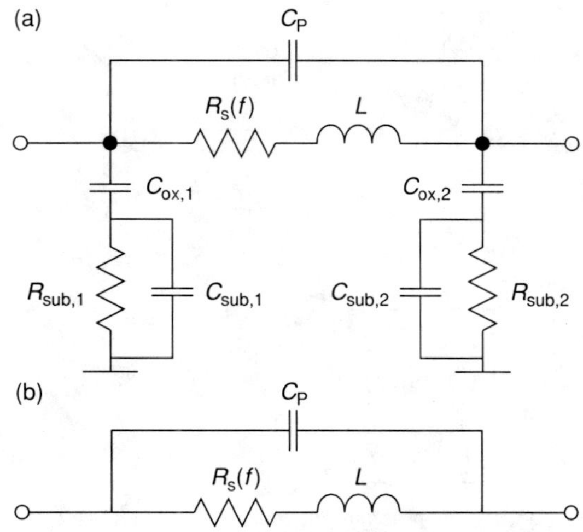

FIGURE 7.3 Typical equivalent circuit of a practical spiral inductor (a) at low-to-medium frequencies and (b) in the upper micro- and millimeter-wave range.

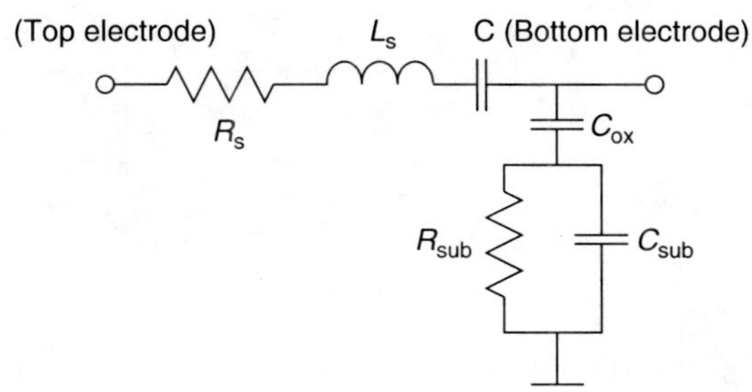

FIGURE 7.4 Equivalent circuit of a metal–insulator–metal (MIM) capacitor.

The substrate spreading resistance, which appears as R_{sub} in the equivalent circuits above, generates thermal noise, which is coupled into the circuit, and can substantially degrade performance, e.g., in the input circuitry of a low-noise amplifier. As Figure 7.5 shows, not only the input bond-pad, but also a DC blocking MIM capacitor is prone to ground-coupled noise.

Ground-coupled noise can be reduced or eliminated using a low-resistance ground shield, which can be the lowest-level interconnect metal or even the buried layer in a bipolar process. It provides a low-resistance path to ground for substrate noise. Figure 7.6 illustrates the concept for the case of a typical coplanar bond-pad arrangement, but it can be used also for the MIM capacitor in the above example.

In spiral inductors, patterned ground shields are often used which not only reduce substrate loss and ground-coupled noise, but where the shield pattern can strongly reduce Eddy currents in the substrate which otherwise would lower both the inductance and the quality factor [7].

7.3 SiGe MMICS Using Heterojunction Bipolar Transistors

As SiGe building blocks for wireless communications systems are covered elsewhere in this handbook, this chapter will cover some not-so-common circuits—for applications in the 10 to 30 GHz frequency range, and extremely wideband circuits for fiber-optic communications systems.

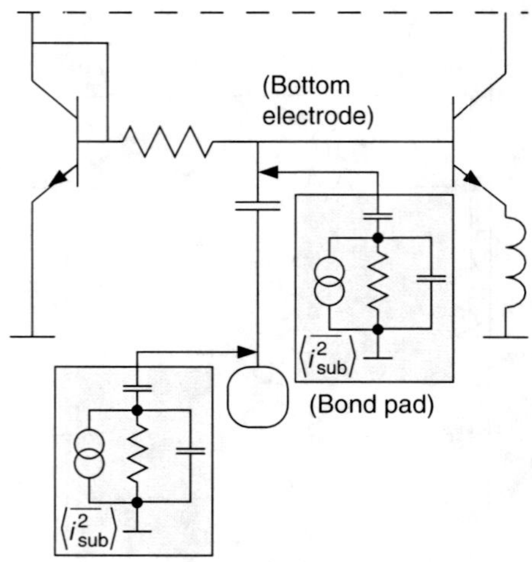

FIGURE 7.5 Input circuit of a typical low-noise amplifier, showing nodes prone to ground-coupled noise.

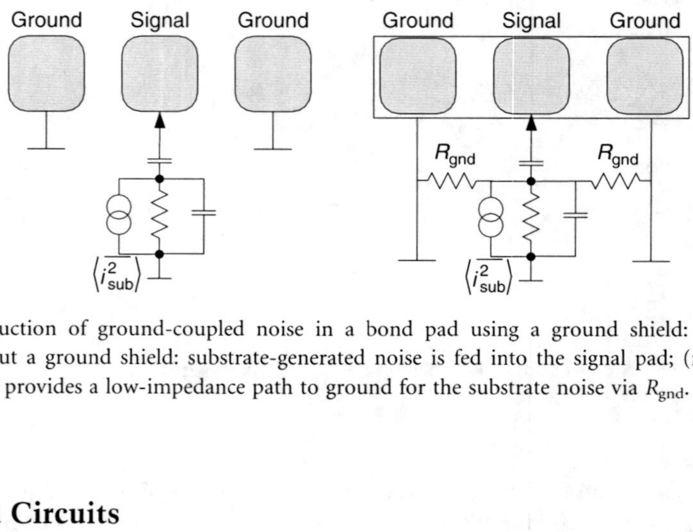

FIGURE 7.6 Reduction of ground-coupled noise in a bond pad using a ground shield: (left) the standard arrangement without a ground shield: substrate-generated noise is fed into the signal pad; (right) addition of a ground shield layer provides a low-impedance path to ground for the substrate noise via R_{gnd}.

Narrow-Band Circuits

In communication and sensor applications in the microwave range, e.g., for satellite communications systems in the Ku and Ka band, or sensors in the increasingly popular 24 GHz ISM band, local oscillator phase noise is often a critical parameter. It is sometimes advantageous to operate the oscillator at a lower frequency, where a higher resonator quality factor can be achieved, and then apply a frequency multiplier to reach the local oscillator frequency required. Frequency multiplication can be achieved in any nonlinear transfer characteristics, in FETs or bipolar transistors. However, unwanted spectral components need to be suppressed, which can be done more efficiently in dedicated frequency multiplier topologies than through filtering. Other critical parameters are required oscillator power, the conversion efficiency, and the necessary supply voltage.

The circuit shown in Figure 7.7a has been realized in an 80 GHz f_T SiGe HBT technology. Its key advantages are the low complexity, inherent suppression of fundamental frequency, and the ability to operate from a 1.2 V supply, provided that reactive loads are used. The circuit exploits the fact that the common emitter potential of a differential transistor pair fluctuates with twice the frequency of the driving waveform: the voltage is at its maximum if either $V_{in,1}$ or $V_{in,2}$ peak, and reaches its minimum when the two input voltages (which are 180° out of phase) balance. The potential fluctuation is then fed to a common-base transistor stage through a coupling capacitor, which transfers charge between the two

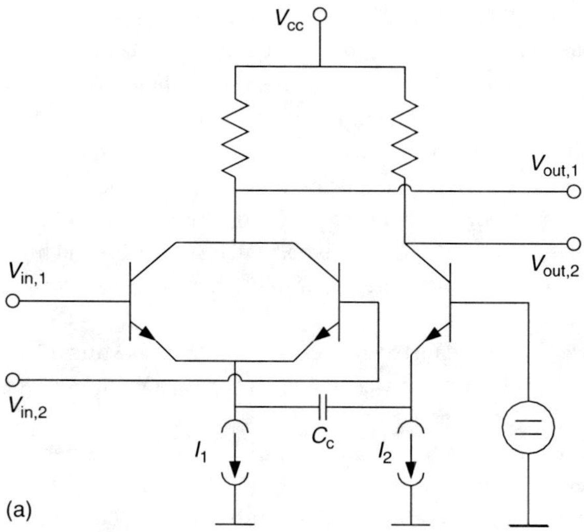

(a)

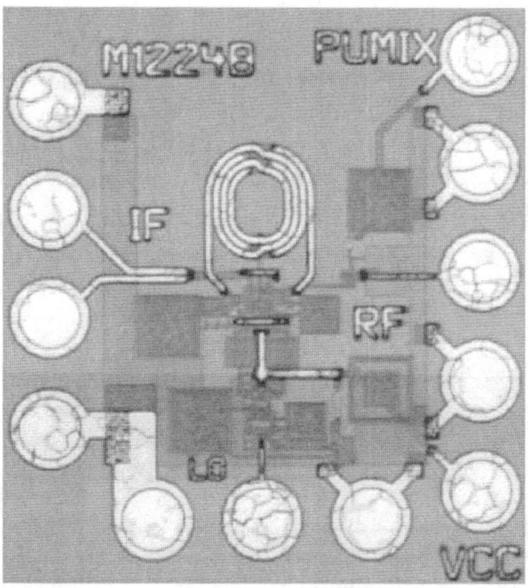

(b)

FIGURE 7.7 (a) Principle schematic of the frequency doubler cell; (b) chip photo of the realized Ku band subharmonically pumped downconverter. (After F. Gruson, H. Schumacher, and G. Bergmann. IEEE International Microwave Symposium 2004, Ft. Worth, TX, June 6–11, 2004, Paper TU2C-3. With permission.)

current sources I_1 and I_2, leading to the output signal at twice the input frequency. Resistors have been used as current sources in the practical implementation.

The doubler cell has been used in the design of a subharmonically pumped mixer cell [8], which is wideband, but has its optimum performance at 14 GHz (7 GHz LO input signal). The single-ended LO input signal is converted into a balanced signal in an active balun circuit, then applied to the frequency doubler cell, which drives a single-balanced Gilbert mixer. In the realization of the doubler stage, the collectors of the differential pair have been connected to a low-impedance node created by a second common-base transistor. The collector loads are inductive, realized as a single symmetrically tapped spiral inductor. The chip photo is shown in Figure 7.7b. The doubler has a conversion gain of 17 dB when converting from 7 to 14 GHz, while the mixer overall has a conversion gain of 8 dB for

$f_{RF} = 14$ GHz and $f_{IF} = 100$ MHz. The circuit consumes 48 mA from a 2.7 V supply. The suppression of the fundamental frequency has been determined at 5 GHz, where it was −43 dB.

Impressive results have been achieved also with frequency dividers using SiGe HBTs. Using a SiGe HBT technology with 200 GHz, a static frequency divider (divide by 32) with a maximum input frequency of 86.2 GHz, drawing 180 mA from a 5 V supply has been reported [9].

Even higher input frequencies are possible using dynamic frequency dividers. The concept is shown in Figure 7.8. Any spectral component at $\frac{1}{2}f_1$ at port 1 of the mixer is sufficient to produce $f_1 \pm \frac{1}{2}f_1$ at port 3. The active mixer has a low-pass characteristic, which suppresses the higher frequency component and passes the $\frac{1}{2}f_1$ spectral component, which is amplified and fed back to the mixer in a regenerative fashion.

The dynamic frequency divider reported in Ref. [9] operated beyond 110 GHz, the highest characterization frequency. Power consumption was 310 mW from a 5 V supply. The input sensitivity versus frequency is also shown in Figure 7.8.

Multifunctional ICs that combine most or all function, e.g., of a radio receiver on a single chip can be realized even in the upper microwave or millimeter-wave range. A fully integrated receiver IC using Si–SiGe HBTs on medium-resistivity silicon was already published in 2002 [10]. It contained all necessary circuits (preamplifier, VCO, mixer, and IF amplifier) of a 24-to-0.1 GHz downconverter, reducing the number of necessary external connections considerably. The IC used a 50 GHz f_T and f_{max} HBT process, which can be considered to be at the limits of its useful performance at 24 GHz.

Using a HBT technology with 75 GHz f_T and f_{max}, a downconverter IC for applications, e.g., in the 26 GHz LMDS band has been realized [11]. The layout is shown in Figure 7.9. Within a compact area of 700 μm × 170 μm, the IC consists of three preamplifier stages using LC interstage matching (the three left spiral inductors), a Gilbert-cell active mixer stage with the groundside current source replaced by spiral inductor, and an intermediate-frequency output buffer stage. The local oscillator is here not part

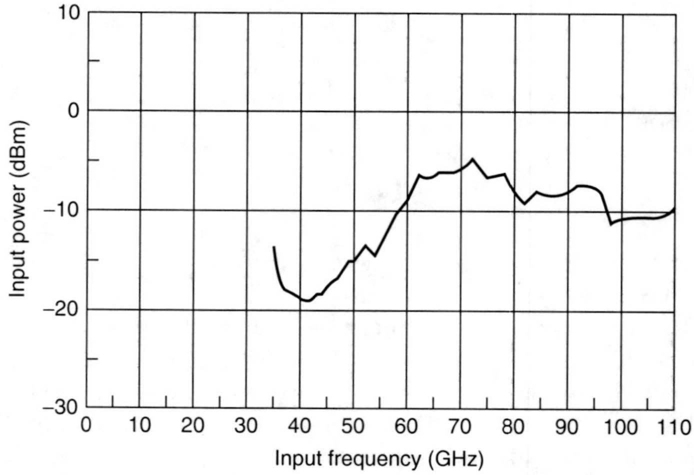

FIGURE 7.8 Concept of a dynamic frequency divider, and the result obtained with a SiGe HBT implementation. (From H. Knapp, M. Wurzer, T.F. Meister, K. Aufinger, J. Böck, S. Boguth, and H. Schäfer. International Microwave Symposium, Philadelphia, PA, June 8–13, 2003, Paper WE6C-5. With permission.)

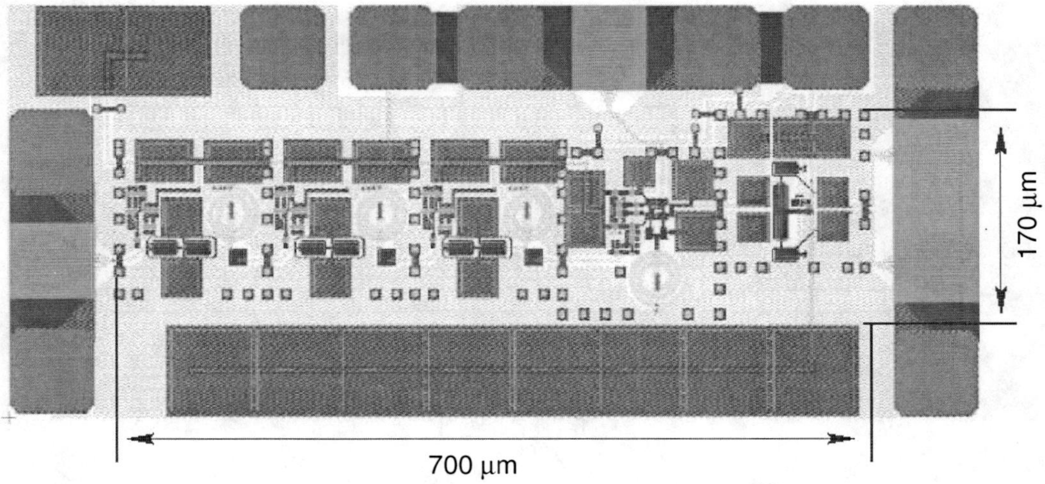

FIGURE 7.9 Layout of a 26 GHz SiGe HBT downconverter for LMDS applications. (After E. Sönmez, A. Trasser, P. Abele, K.-B. Schad, and H. Schumacher. Proc. European Microwave Conference (EUMC/EMW), Munich, Germany, October 6–10, 2003, pp. 399–402. With permission.)

of the chip, but is fed through the top coplanar port. The IF output buffer contains an RC low-pass filter with a 3-dB cutoff frequency of 1.5 GHz. The preamplifier stages are identical and consist of a cascode gain cell biased through a current mirror.

The circuit has a conversion gain of 24 dB with a differential IF load, and draws 120 mW from a 3 V supply.

Sensors operating in the 24 GHz ISM band and potentially at higher frequencies benefit from the low cost of SiGe HBT technologies as well. Doppler RADAR sensors can be realized quite simply and find use in many applications such as contactless speed measurement or motion detection.

In SiGe, Doppler sensors can be realized using inexpensive HBT technologies. In the following example, a commercial SiGe HBT technology with 50 GHz f_T and f_{max} has been used. The circuit uses only lumped elements, resulting in a chip area of 780 × 690 μm^2. It can be flip-chip mounted on a substrate carrying the base-band processing and the antennas.

In hybrid implementations of Doppler RADAR sensors, a single oscillator is used for receiver and transmitter. The RF power is split by means of a Wilkinson divider, which consumes considerable area and results in power loss. Here, a different concept has been pursued: two separate oscillators synchronized by a magnetic coupling loop (see Figure 7.10a). As an added benefit, the magnetic coupling allows a clear separation between the receive and transmit paths in the layout, which is shown in Figure 7.10b. The pitch of the flip-chip bond bumps is 200 μm^2.

The key component, the magnetically coupled inductors, is laid out in perfect symmetry. A two-turn coupling loop is formed in a metalization layer underneath. A lumped-element equivalent circuit has been used to simulate the coupled inductor four-port over the 0 to 50 GHz range with good accuracy.

The transmit amplifier feeds a differentially side-fed patch antenna; a single-ended side-fed receive antenna directly works into a Gilbert cell mixer. The IC has been incorporated in a very compact module including the antennas and simple base-band processing [12].

Wide-Band Circuits

Multi-octave bandwidth is commonly achieved through two different fundamental approaches:

- Through negative feedback. Single-stage wide-band amplifiers commonly use a parallel (output voltage to input current) feedback technique, often referred to as a transimpedance amplifier.

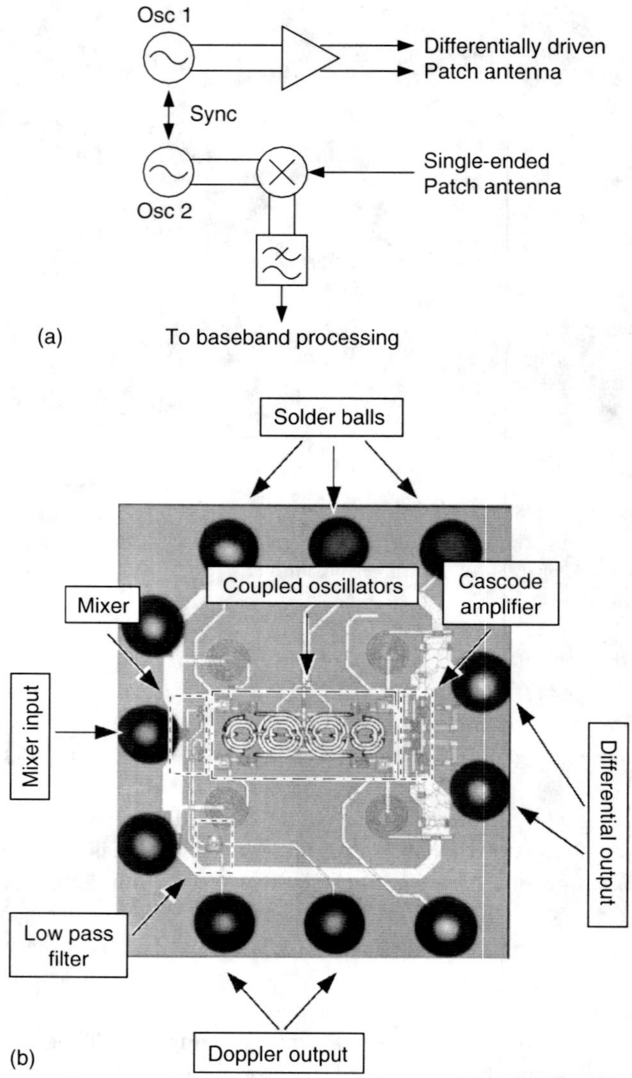

FIGURE 7.10 Block diagram (a) and chip photo (b) of a 24 GHz Doppler RADAR sensor using magnetically coupled synchronized oscillators.

Multistage wide-band amplifiers will use alternating parallel and series (output current to input voltage) feedback techniques—the Cherry–Hooper design principle.

- Through distributed amplifier concepts. Here, the input and output capacitances which form RC low-pass poles with the stage's source and load impedances are embedded into artificial transmission line structures formed either by lumped-element inductors or, more commonly, by high-characteristic-impedance transmission line segments. Figure 7.11 shows the concept for the case of field-effect transistors. Using this technique, the device input (C_{GS}) and output (C_{DS}) capacitances are completely absorbed over a wide frequency range in an artificial transmission line with characteristic impedance Z_L (up to the low-pass cutoff frequency of the cascaded transmission line segments), resulting simultaneously in a flat gain and excellent input and output matching conditions.

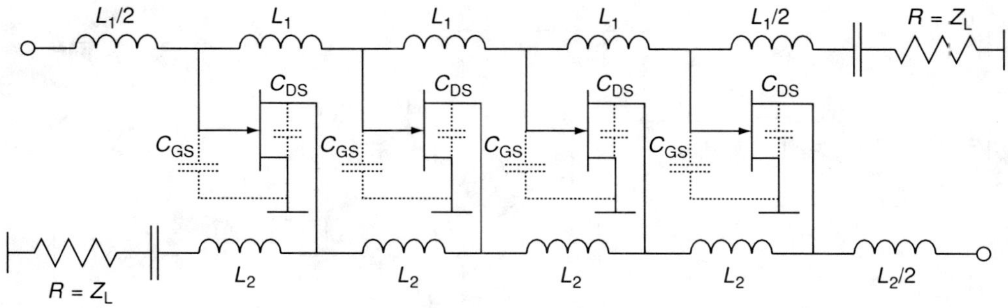

FIGURE 7.11 Generic concept of a distributed (or traveling-wave) amplifier.

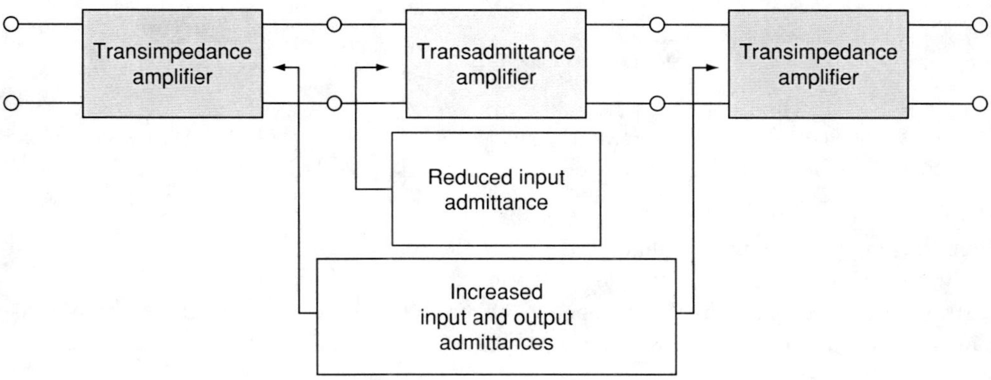

FIGURE 7.12 Generic concept of a Cherry–Hooper wideband amplifier.

Cherry–Hooper wideband amplifiers are commonly used in SiGe HBT amplifiers intended for high-bitrate fiber-optic communications systems. As indicated, they deliberately apply large interstage impedance mismatches and hence deviate significantly from common MMIC design techniques. The generic concept is shown in Figure 7.12. A wideband amplifier directly behind a photodiode will have a transimpedance stage as the first stage. A transimpedance stage employs parallel (output voltage to input current) feedback around an inverting amplifier stage, providing a low input impedance for the photodiode (which is essentially a current source). The parallel feedback reduces both the input and the output impedance. The next stage is then a transadmittance stage, using series (output current to input voltage) feedback, which decreases both the input and the output admittance. Most importantly, the input capacitance of the transadmittance stage is reduced by the series feedback, leading to a small characteristic time constant of the pole between the first and second stages. The pole between the second and third stages is reduced by the low input impedance of the third stage, which has a transimpedance topology again.

An excellent example of a Cherry–Hooper type amplifier is shown in Figure 7.13 [13]. The circuit has a fully differential topology to reduce problems with common-mode impedance in the ground and supply lines. The first stage is a transimpedance stage with feedback R_F. The photodiode is connected to one input, while the other is shorted to ground capacitively, but receives an offset-compensating control current. As a slight modification of the classic Cherry–Hooper design, the next block is a three-stage emitter follower, which provides the high input impedance of a transadmittance stage, but a low output impedance. The third stage is then a transadmittance stage with series-feedback resistors R_E and a cascode configuration. Using a SiGe HBT technology with 72 GHz f_T, the circuit proved capable of 40 Gbit/sec operation.

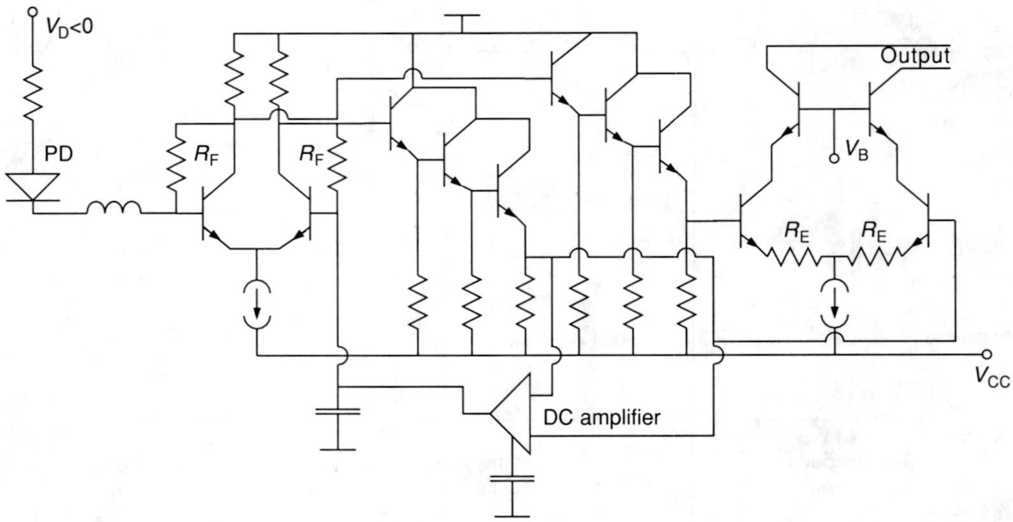

FIGURE 7.13 SiGe HBT limiting transimpedance amplifier for 40 Gbit/sec. (After J. Müllrich, T.F. Meister, M. Rest, W. Bogner, A. Schöpflin, and H.-M. Rein. *Electron. Lett.* 34, 452–453, 1998. With permission.)

Here, the amplifier combines a linear first stage (the transimpedance stage) with voltage-limiting action in the following stages, but this does not affect the general principle.

Distributed amplifiers are the circuit topology of choice in HFETs where the input admittance of a common-source stage is predominantly capacitive. An example using experimental SiGe HFETs will be shown in Section 7.4.

Compared to negative-feedback amplifiers, distributed amplifiers have the advantage that it is considerably easier to achieve a constant group velocity (a linearly varying phase) versus frequency. In digital communications systems, constant group velocity is a very critical parameter, at par with the more commonly considered power gain and input and output match. This inherent property makes distributed amplifiers interesting for implementation also in SiGe HBT technologies.

For bipolar transistors in common-emitter configuration, a real conductance $g_{be} = g_m/\beta$ appears at the input node in parallel (see Figure 7.14), and hence the input admittance is no longer purely imaginary, even at low frequencies:

$$y_{1,e} = \frac{g_m}{\beta} + j\omega C_{Te} \text{ or, as } f_T \approx \frac{g_m}{2\pi C_{Te}}:$$

$$y_{1,e} = \frac{g_m}{\beta}\left[1 + j\beta\frac{f}{f_T}\right]$$

where g_m is the transconductance, β the small-signal current gain, C_{te} the total base–emitter capacitance, and C_{bc} the base–collector space–charge capacitance. The real part would then seriously attenuate the wave propagating on the input transmission line.

The input impedance of the individual gain cell can be increased by using a common-collector (emitter follower) stage ahead of the common-emitter stage. In Figure 7.15, the combination of an emitter followed by a cascode stage is shown—the reason for the cascode cell will be discussed later.

Using the common-emitter input admittances

$$y_1 = \frac{g_{m1}}{\beta_1}\left[1 + j\beta_1\frac{f}{f_{T1}}\right] \text{ for the common-emitter transistor T1 and}$$

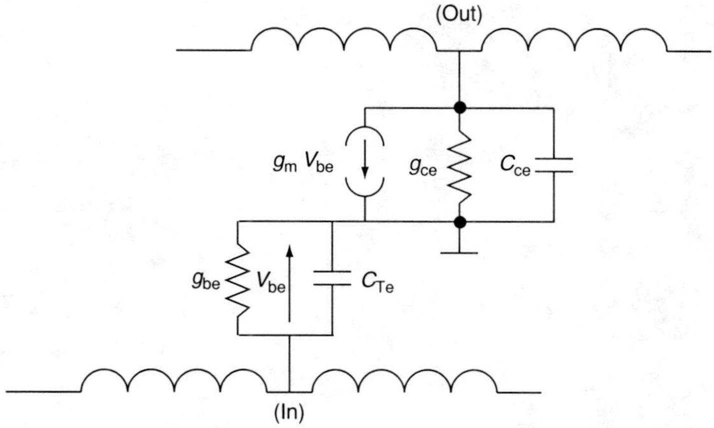

FIGURE 7.14 Simplified equivalent circuit of a common-emitter bipolar gain cell in a distributed amplifier.

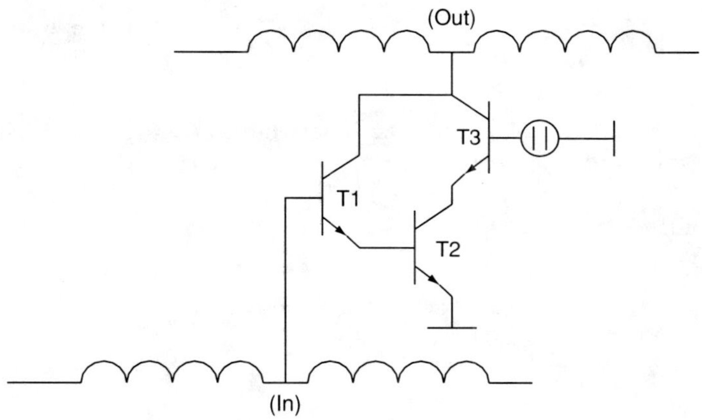

FIGURE 7.15 Simplified schematic of a bipolar gain cell in a distributed amplifier, using an emitter follower/cascode topology (biasing elements not shown).

$$y_2 = \frac{g_{m2}}{\beta_2}\left[1 + j\beta_2 \frac{f}{f_{T2}}\right] \quad \text{for the collector transistor T2}$$

the input admittance of cascaded common collector and common-emitter stages follows:

$$y_{in} = \frac{y_2}{1 + \frac{y_2}{y_1}(1+\beta_2)} \approx \frac{y_1}{\beta_2} \quad \text{if } \beta_2 \gg 1, \quad \frac{y_2}{y_1}(1+\beta_2) \gg 1$$

or

$$y_{in} = \frac{g_{m1}}{\beta_1 \beta_2}\left[1 + \beta_1 \frac{f}{f_{T1}}\right]$$

The cascode cell formed by transistors T2 and T3 provides a larger gain, reduces feedback between output and input, and reduces the attenuation on the output line. The latter point deserves some attention. Figure 7.16a shows the small-signal equivalent circuit for the cascode cell for the case $v_{be2} = 0$, i.e., for the base of T2 at a constant potential. The capacitances C_{ce2}, C_{ce3} in a bipolar transistor are

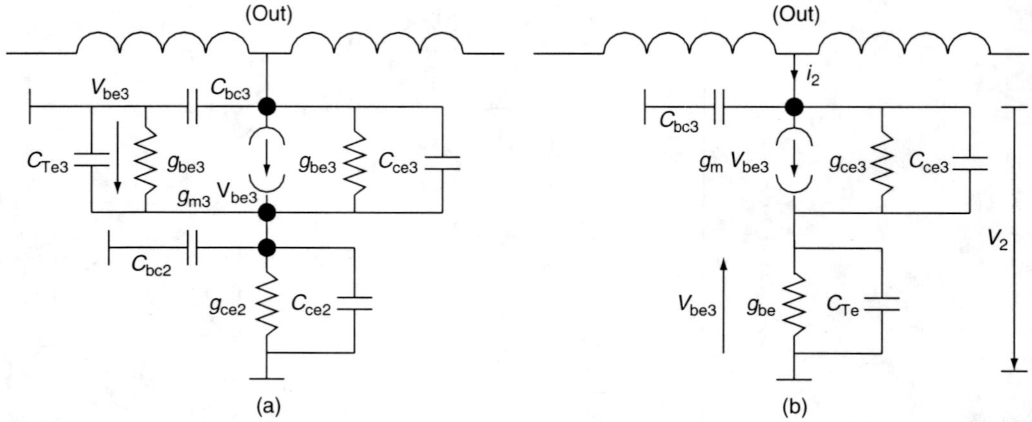

FIGURE 7.16 (a) Equivalent circuit of the cascode cell (T2 and T3 in Figure 7.15) for $v_{be2} = 0$. (b) Simplified equivalent circuit, assuming identical transistors and recognizing $I_{C2} = I_{C3}$, $g_{m2} = g_{m3} = g_m$, $C_{te2} = C_{Te3} = C_{Te}$, $g_{be2} = g_{be3} = g_{be}$.

parasitic. We can assume $C_{te} \gg C_{bc}, C_{ce}$ for both transistors, further $g_{be} = g_m/b = g_m$, $g_{be} \gg g_{ce} = I_C/V_A$, where V_A is the Early voltage. Using these assumptions, we arrive at the much simpler equivalent circuit in Figure 7.16b.

The output admittance is

$$Y_2 = \frac{(g_{ce} + j\omega C_{ce3})(g_{be} + j\omega C_{Te})}{g_{ce} + g_{be} + g_m + j\omega(C_{Te} + C_{ce3})} + j\omega C_{cb3}$$

Further simplification: $C_{Te} \gg C_{ce3}; g_{be} = \frac{I_C}{\beta V_T} \gg g_{ce} = \frac{I_C}{V_A}$:

$$Y_2 \approx \frac{g_{be}g_{ce} - \omega^2 C_{Te}C_{ce3} + j\omega(g_{be}C_{ce3} + g_{ce}C_{Te})}{g_m + j\omega C_{Te}^2} + j\omega C_{cb3}$$

We are particularly interested in the real part, which will attenuate the wave on the output line:

$$\mathcal{R}\{Y_2\} = \frac{\dfrac{g_{ce}}{\beta} - \dfrac{\omega^2}{\omega_T}\left(C_{ce3} - \dfrac{g_{ce}}{\omega_T}\right)}{1 + \left(\dfrac{\omega}{\omega_T}\right)^2}$$

using
$g_{be} = \dfrac{g_m}{\beta}; \dfrac{g_m}{C_{Te}} = \omega_T = 2\pi f_T$, the transit frequency.

We note that the resistive loading of the output line is reduced by the common-emitter current gain β (to g_{ce}/β). Provided that $C_{ce3} > g_{ce}\omega$ the output conductance may actually become negative at elevated frequencies. This can be used to further compensate loss on the output line, but may also lead to instability at high frequencies.

Losses on the input line can be compensated by introducing a series capacitor into the emitter of the common-collector transistor stage, which will result in a negative resistance at the base node [15].

Using gain cells consisting of a common-collector input stage, a differential amplifier gain cell and a cascode output cell, Wohlgemuth and coworkers realized a distributed amplifier capable of 80 Gbit/sec operation [15]. Figure 7.17 shows the block diagram. The design is fully differential, eliminating problems with common-node impedances when packaged. The transmission lines have been realized as ground-backed coplanar waveguides, using the top and bottom metallizations of the standard metallization stack in the SiGe HBT technology used, which has f_T and f_{max} in excess of 200 GHz.

The structure has been realized on a chip area of $1.3 \times 0.9\,mm^2$; it consumes 90 mA from a 5.5 V power supply. Characterization has been reported for single-ended performance only; the scattering parameters s_{11}, s_{22}, and s_{21} are shown in Figure 7.18.

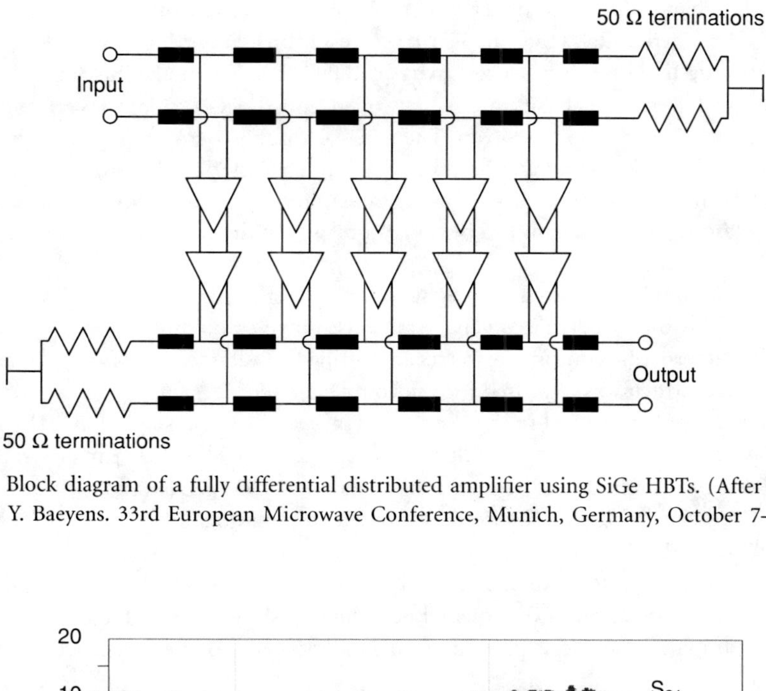

FIGURE 7.17 Block diagram of a fully differential distributed amplifier using SiGe HBTs. (After O. Wohlgemut, P. Paschke, and Y. Baeyens. 33rd European Microwave Conference, Munich, Germany, October 7–10, 2003. With permission.)

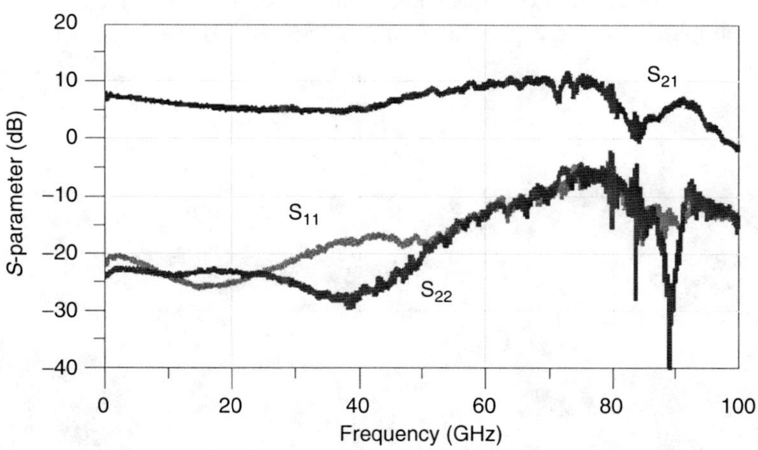

FIGURE 7.18 Experimental scattering parameters of the differential distributed amplifier shown in Figure 7.17. The measurement is single-ended. (After O. Wohlgemut, P. Paschke, and Y. Baeyens. 33rd European Microwave Conference, Munich, Germany, October 7–10, 2003. With permission.)

Calibration above 70 GHz was reported to be difficult with the GSSG wafer probes used, which may contribute to the artifacts observed above 80 GHz. The measurements demonstrate a 3-dB-bandwidth of 81 GHz. In differential mode, the gain-bandwidth product would be 350 GHz and be at par with the best values reported for III–V-based distributed amplifiers. The input and output match is good at least up to 60 GHz, above which it may be affected by the noted calibration problems.

7.4 SiGe MMICS Using Heterostructure Field-Effect Transistors

SiGe HFETs are an emerging technology, which uses strained Si–SiGe heterostructure channels to enhance low-field mobilities of electrons and holes. Both MOSFET-like structures (charge control through an MOS gate diode) and MODFET-like structures (charge control through a Schottky-gate diode) have been used.

A first ultrawide-band MMIC amplifier structure has been realized using an n-channel MODFET technology fully described in Ref. [16]. The FET layers reside on a fully relaxed $Si_{1-x}Ge_x$ buffer with $x =$ 40%. The Si substrate underneath has a resistivity of 1000 Ω cm. The n-channel forms in a 9-nm thick undoped Si channel layer, which sits between two Sb-doped SiGe supply layers and spacers. The Schottky gate sits on top of a 3.5 nm Si cap. The gate electrode is a Pt–Au T-gate with $L_G = 100$ nm defined by electron beam lithography. The transistors are characterized by a transconductance $g_m =$ 175 mS/mm, a transit frequency f_T of 52 GHz, and a maximum frequency of oscillation $f_{max} = 148$ GHz extracted from Mason's unilateral gain *U*. The transistors used in the circuit design had a total gate width of 100 μm.

The MODFETs were modeled using a large-signal equivalent circuit approach with globally continuous equations [17]. Models for the coplanar transmission lines on the virtual SiGe substrate were developed using test structures on similar wafers and fitting the parameter set for coplanar transmission lines in Agilent ADS, which was also used for the complete circuit design.

The distributed amplifier realized in this technology has six identical stages [18]. The gate transmission line has segments of 450 μm length with a center conductor width of 20 μm and signal-to-ground gaps of 60 μm, while the drain transmission line has segments of 530 μm length, with a center conductor width of 20 μm and gaps of 40 μm. Figure 7.19 shows a chip photograph.

The circuit draws 45.7 mA from a 2.3 V supply, the gate bias is held at $V_{GS} = -0.16$ V. At this bias point, the amplifier exhibits 5.5 ± 0.8 dB up to 32 GHz (see Figure 7.20). The comparison between modeled and experimental performance shows excellent agreement. The relatively low gain of the amplifier can be explained by the rather high source resistance of the FETs used ($R_S = 9\ \Omega$).

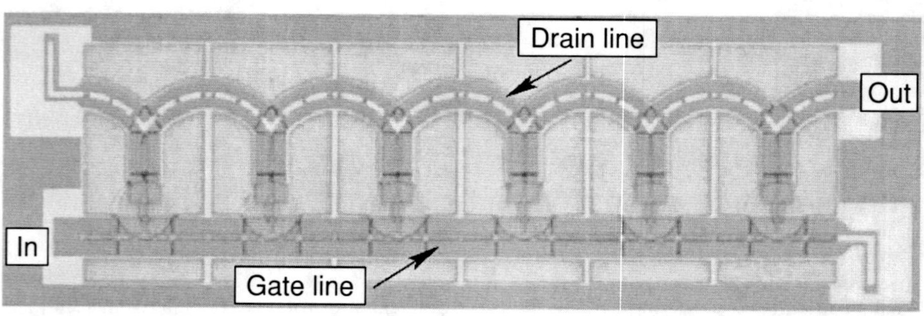

FIGURE 7.19 Chip photograph of a distributed amplifier using Si–SiGe n-channel MODFETs. (After P. Abele, M. Zeuner, I. Kallfass, J. Müller, H.L. Hiwilepo, D. Chrastina, H. Von Känel, U. König, and H. Schumacher. *Electron. Lett.* 39, 1448–1449, 2003. With permission.)

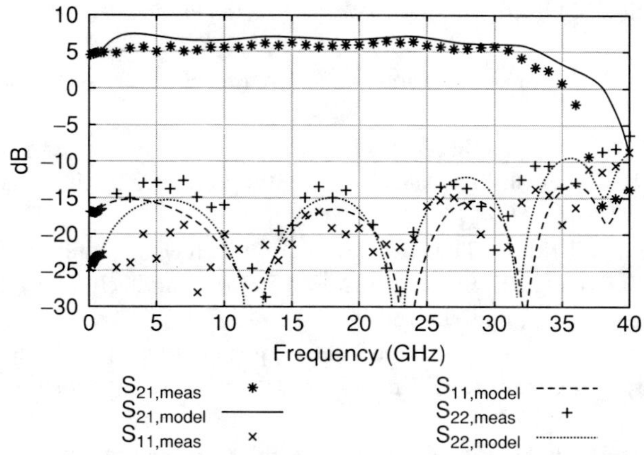

FIGURE 7.20 Measured and modeled scattering parameters for the Si/SiGe MODFET traveling-wave amplifier. (After I. Kallfass, T.J. Brazil, B. OhAnnaidh, P. Abele, Th. Hackbarth, M. Zeuner, U. König, and H. Schumacher. *Solid-State Electron.*, 48, 2004, 1433–1441. With permission.)

7.5 Summary

Monolithic microwave ICs using Si–SiGe HBTs and HFETs offer interesting design options for microwave, millimeter-wave, and wideband fiber-optic applications. The disadvantage of the lossy substrate, compared to GaAs, can be compensated by proper circuit design techniques. These silicon-based heterostructure ICs have their best application potential if

- A large market volume can be expected or
- High on-chip circuit complexities are needed and
- Low-noise and power specifications are moderate

Acknowledgments

The author gratefully acknowledges the contributions by and many fruitful discussions with his staff at the University of Ulm, particularly Peter Abele, Nabil Alomari, Frank Gruson, Martin Häfele, Christoph Schick, Ingmar Kallfass, Kai-Boris Schad, Ertugrul Sönmez, Andreas Trasser, and Shen Yan. The original work presented here would not have been possible without the continuous cooperation of Atmel Germany GmbH, Heilbronn, Germany for the HBT work, and the Daimler Chrysler Research Center, Ulm, Germany for the HFET work. Some of this research has been supported by the European Commission and the German Ministry of Education and Research.

References

1. A.C. Wilson, D. Melendy, P. Francis, K.Hwang, and A. Weisshaar. Comprehensive compact-modeling methodology for spiral inductors in silicon-based RFICs. *IEEE Trans. Microwave Theory Tech.* 52, 849–857, 2004.
2. H. Hashemi, X. Guan, and A. Hajimiri. A fully integrated 24-GHz 8-Path phased-array receiver in silicon. ISSCC 2004, San Francisco, CA, February 15–18, 2004.
3. I. Harrison, M. Dahlström, S. Krishnan, Z. Griffith, Y.M. Kim, and M.J.W. Rodwell. Thermal limitations of InP HBTs in 80- and 160 Gb Ics. *IEEE Trans. Electron Dev.* 51, 529–534, 2004.
4. J.-S. Rieh, B. Jagannathan, H. Chen, K.T. Schoenenberg, D. Angell, A. Chinthakindi, J. Florkey, F. Golan, D. Greenberg, S.-J. Jeng, M. Khater, F. Pagette, C. Schnabel, P. Smith, A. Stricker, K. Vaed, R. Volant, D. Ahlgren, G. Freeman, K. Stein, and S. Subbanna. SiGe HBTs with cut-off frequency of

350 GHz. International Electron Device Meeting, San Francisco, CA, December 8–11, 2002, pp. 771–774.

5. P. Russer. Si and SiGe millimeterwave integrated circuits. *IEEE Trans. Microwave Theory Tech.* 46, 590–603, 1998.
6. W. Dürr, U. Erben, A. Schüppen, H. Dietrich, and H. Schumacher. Investigation of microstrip and coplanar transmission lines on lossy silicon substrates without backside metalization. *IEEE Trans. Microwave Theory Tech.* 46, 712–715, 1998.
7. S.-M. Yim, T. Chen, and K.K.O. The effects of a ground shield on spiral inductors fabricated in a silicon bipolar technology. IEEE Bipolar/BiCMOS Circuits and Technology Meeting, Minneapolis, MN, September 24–26, 2000, pp. 157–160.
8. F. Gruson, H. Schumacher, and G. Bergmann. A frequency doubler with high conversion gain and good fundamental suppression. IEEE International Microwave Symposium 2004, Ft. Worth, TX, June 6–11, 2004, Paper TU2C-3.
9. H. Knapp, M. Wurzer, T.F. Meister, K. Aufinger, J. Böck, S. Boguth, and H. Schäfer. 86 GHz static and 110 GHz dynamic frequency dividers in SiGe HBT technology. International Microwave Symposium, Philadelphia, PA, June 8–13, 2003, Paper WE6C-5.
10. E. Sönmez, A. Trasser, K.-B. Schad, P. Abele, and H. Schumacher. A single-chip 24 GHz receiver front-end using a commercially available SiGe HBT foundry process. RFIC Conference, Seattle, USA, June 2–4, 2002, pp. 159–162.
11. E. Sönmez, A. Trasser, P. Abele, K.-B. Schad, and H. Schumacher. Integrated receiver components for low-cost 26 GHz LMDS applications using an 0.8 μm SiGe HBT technology. Proc. European Microwave Conference (EUMC/EMW), Munich, Germany, October 6–10, 2003, pp. 399–402.
12. P. Abele, A. Trasser, E. Sönmez, K.-B. Schad, A. Munding, and H. Schumacher. A compact low-cost doppler sensor using SiGe HBT technology and patch antennas for the ISM Band at 24 GHz. 34th European Microwave Conference, Amsterdam, The Netherlands. October 12–14, 2004.
13. J. Müllrich, T.F. Meister, M. Rest, W. Bogner, A. Schöpflin, and H.-M. Rein. 40 Gbit/s transimpedance amplifier in SiGe bipolar technology for the receiver in optical-fibre TDM links. *Electron. Lett.* 34, 452–453, 1998.
14. S. Deibele and J.B. Beyer. Attenuation compensation in distributed amplifier design. *IEEE Trans. Microwave Theory Tech.* 37, 1425–1433, 1989.
15. O. Wohlgemut, P. Paschke, and Y. Baeyens. SiGe broadband amplifiers with up to 80 GHz bandwidth for optical applications at 43 Gbit/s and beyond. 33rd European Microwave Conference, Munich, Germany, October 7–10, 2003.
16. T. Mack, T. Hackbarth, H.-J. Herzog, H. Von Känel, M. Kummer, J. Ramm, and R. Sauer. Si/SiGe FETs grown by MBE on LEPECVD grown virtual substrate. *Mater. Sci. Eng. B* 39, 368–372, 2002.
17. I. Kallfass, M. Zeuner, U. König, H. Schumacher, and T.J. Brazil. A DC to 40 GHz large signal model for n-channel SiGe HFET transistors including low-frequency dispersion. 32nd Europ. Microwave Conf. (EuMC), Milano, Italy, 2002.
18. P. Abele, M. Zeuner, I. Kallfass, J. Müller, H.L. Hiwilepo, D. Chrastina, H. Von Känel, U. König, and H. Schumacher. A 32 GHz MMIC distributed amplifier based n-channel SiGe MODFETs. *Electron. Lett.* 39, 1448–1449, 2003.
19. I. Kallfass, T.J. Brazil, B. OhAnnaidh, P. Abele, Th. Hackbarth, M. Zeuner, U. König, and H. Schumacher. Large-signal modelling including low-frequency dispersion of N-channel SiGe MODFETs and MMIC applications. *Solid-State Electronics.* 48, 1433–1441, 2004.

8

SiGe Millimeter-Wave ICs

8.1 Introduction 8-1
8.2 Silicon as a Substrate 8-1
8.3 Millimeter-Wave Generation 8-2
Oscillators • Coplanar Oscillators • Microstrip Oscillators • Multiplier
8.4 Amplifier 8-6
8.5 Mixer 8-6
8.6 MMICS with SiGe Diodes 8-7
Switches • Mott Diodes • IMPATT Diodes
8.7 Summary 8-7

Johann-Friedrich Luy
DaimlerChrysler Research

8.1 Introduction

The millimeter-wave range usually starts at 30 GHz and ends at 300 GHz entering the sub-millimeter-wave range. Below 100 GHz there are many possible applications of wireless communication systems, radars, and navigation systems, which are handicapped by the high costs of the front-end electronics. Beyond 100 GHz the applications are still explorative. A major breakthrough is expected for all these applications, if low-cost silicon technologies can serve the markets. Currently, the silicon–germanium HBT technology is the most promising approach to fulfill the requirements. The design, the state-of-the-art performance, and some technological aspects of analog silicon-based millimeter-wave integrated circuits using mostly HBT technologies are discussed in this chapter.

8.2 Silicon as a Substrate

Almost 20 years after the report of microstrip attenuation values below 1 dB/cm at 90 GHz on high-resistivity silicon (HRS) substrate [1], the question of the most suited transmission line on silicon substrates is still discussed. If HRS material is considered, microstrip or coplanar transmission lines may be the choice (Figure 8.1). There is a high technological effort for the fabrication of microstrip circuits with a thickness of 100 μm including via holes in silicon. Large area coplanar lines with similar attenuation values on HRS may be an alternative. Heat removal and packaging difficulties are the most relevant drawbacks of coplanar circuits. Using low-resistivity substrates (LRS), thin film microstrip lines might be attractive due to their attenuation independent on the substrate resistivity.

The question of the most suited transmission line cannot be answered without consideration of the circuit function. If extremely low attenuation values are required (application: high-Q circuits for low-noise MMICs), then microstrip circuits on 100-μm thick HRS substrate are superior with respect to the other transmission line concepts. Additionally, microstrip resonators can be easily coupled to a DR to further increase the Q-value [2]. Figure 8.2 shows the Q-value of different transmission lines calculated with field theoretical models [3]. The properties of coplanar lines depend on the line geometry.

FIGURE 8.1 Planar transmission lines: thin film microstrip, microstrip, and coplanar.

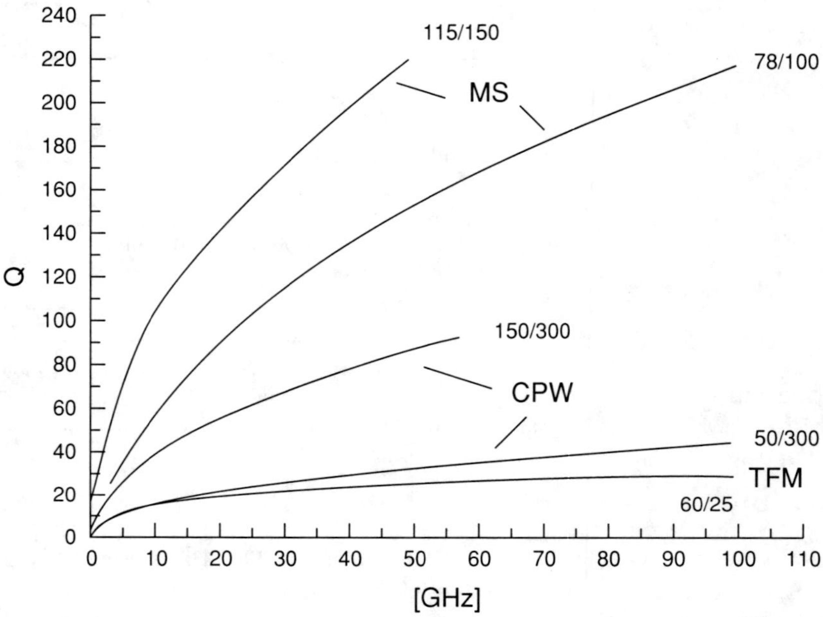

FIGURE 8.2 Quality factor of transmission lines as a function of the frequency. Numbers indicate width of line (μm)/substrate height (μm). CPW: ground-to-ground spacing.

The quality factor of coplanar transmission lines is usually below that of microstrip lines. Thin-film microstrip lines are suited for interconnects and frequencies below 10 GHz.

Silicon as a substrate material is advantageous due to a thermal conductivity three times higher than that of GaAs [4]. This is of great importance when active devices are to be integrated. The low thermal expansion coefficient of silicon may be a further advantage for a practical application of silicon-based millimeter-wave integrated circuits.

8.3 Millimeter-Wave Generation

Oscillators

Negative resistance circuits are widely used to realize low-phase noise oscillators. An active device is used to compensate for the attenuation in a resonator and to generate oscillations.

Coplanar Oscillators

Coplanar MMICs with SiGe HBTs may be realized on HRS substrates. The coplanar concept is advantageous with regard to the microstrip technology as it avoids backside processing and provides easy measurement access by on-wafer probing. One of the first publications on coplanar millimeter-wave oscillator SiGe MMICs reports on a 29 GHz oscillator [5]. A double mesa HBT process is used with an $f_{max} = 80$ GHz. Coplanar waveguides with miniaturized dimensions (50 μm ground-to-ground spacing) are used. The transistor is operated in common-base configuration, one emitter finger is connected to an open stub, which realizes the capacitive feedback of the reflection type oscillator. Bias is fed to the second emitter finger. A short stub is connected to the base to generate maximum negative resistance at the collector. At the collector side an output circuit is designed that matches the oscillator impedance to the 50 Ω load. A maximum output power of 1 dBm at 29.3 GHz is measured with a collector current of $I_C = 10$ mA. Modeling of the passive elements plays an important role for an accurate chip design. Missing the center frequency by more than 10% is an indication of larger differences between models and experiments. Improved modeling as well as improved technology properties lead to an optimized oscillator circuit at 47 GHz. The layout of this oscillator is shown in Figure 8.3. The chip size is 1.9×1.1 mm^2. The HBT is operated in a common-base configuration. DC bias is fed in symmetric lines to the six-finger HBT without air bridges. The inductive feedback from the base is caused by a coplanar line element. An output network matches the collector to 50 Ω. MIM capacitances in the emitter and collector DC bias lines block the RF signals. The ground-to-ground spacing of the coplanar lines is 50 μm. In order to optimize the oscillator circuit,

1. The Rollett factor is minimized.
2. The phase slope of the output matching network is maximized.

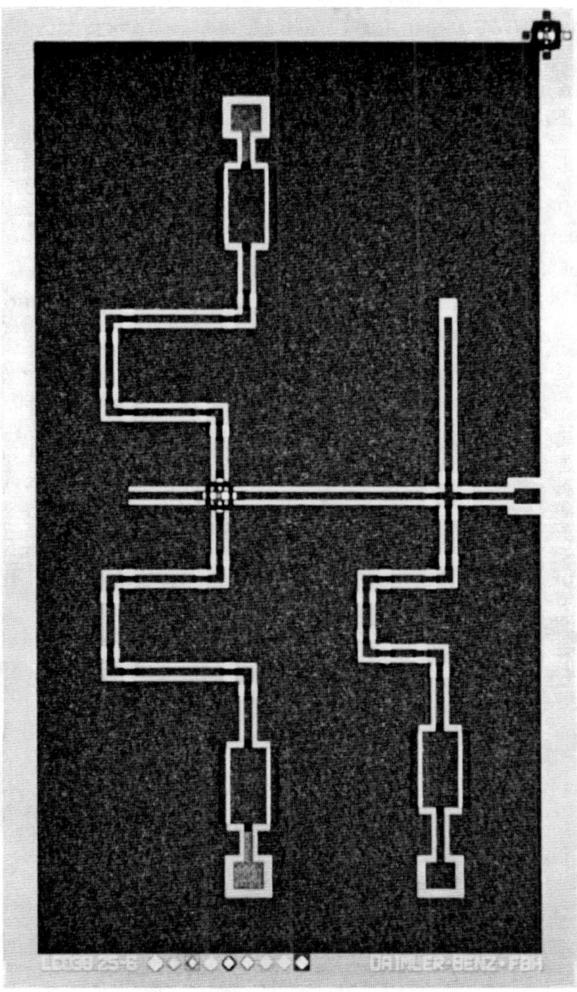

FIGURE 8.3 Layout of a coplanar 47 GHz oscillator.

From this SiGe MMIC oscillator an output power of 13 dBm at 47 GHz is measured at a collector current of $I_C = 32$ mA [6]. The corresponding conversion efficiency is 13.6%. The phase noise is below −91 dBc/Hz at 100 kHz (Figure 8.4). Similar performance data are obtained at 24 GHz. An output power of 8.7 dBm with a collector current of 27 mA, the SSB phase noise is −99 dBc/Hz at 100 kHz [7]. The coplanar LC resonator–oscillator design concept on HRS is suited to provide high Q-factors for excellent phase noise behavior.

Varactor diodes may be integrated in the SiGe HBT process using the base–collector junction as active layers of a varactor diode. A varactor capacitance can change effective lengths of lines, and therefore, tune oscillators. A tuning range of 5% is achieved at Ka band frequencies [7].

So far, coplanar oscillator designs on HRS using distributed elements are discussed. With lumped elements on LRS (20 Ω cm) miniaturized oscillators may be realized with good performance based on a commercial foundry process. The output power from a chip sized $0.75 \times 1.1\ \text{mm}^2$ reaches 1.2 dBm at 25 GHz with an SSB-phase noise of −90 dBc/Hz at 100 kHz [8]. In a fully differential circuit configuration, designed for 47 GHz operation, an output power of 5.6 dBm is obtained from the differential output employing five HBTs on a commercial process with $f_T \approx f_{max} \approx 75$ GHz. The phase noise is between −103 and −108 dBc/Hz at 100 kHz [9]. This design concept is further improved and applied to a process with $f_T \approx 150$ GHz and $f_{max} \approx 180$ GHz [10]. An output buffer comprising an emitter follower and an emitter-coupled stage as well a differential grounded-base stage followed by a network to match external 50 Ω output loads is added to the oscillator core (Figure 8.5). A total signal power of 14 dBm is obtained from the output ports at a center frequency for 77 GHz. The phase noise is −95 dBc/Hz at 1 MHz and −75 dBc/Hz at 100 kHz [11]. A 117 GHz LC oscillator is fabricated in a 200 GHz SiGe:C BiCMOS technology with 0.25 mm minimum feature size [12]. An output power of −12 dBm is achieved.

Microstrip Oscillators

In spite of the increased technological efforts (wafer thinning down to 100 μm, backside processing, via-hole technology) microstrip transmission (MST) line technology is investigated as well. MST enables high-Q resonators, lower attenuation, better RF packaging characteristics, and well-known design and simulation software. An MST technology uses 100 μm substrate thickness and circular via-holes with vertical sidewalls and a diameter of 80 μm, passive elements like resistors, MIM capacitors, and spiral inductors are realized, tested, and modeled [13].

This MST technology is combined with a Si–SiGe HBT technology, which uses for the microstrip technology differentially grown Si MBE layers with 30% Ge in the base layer, a double mesa fabrication

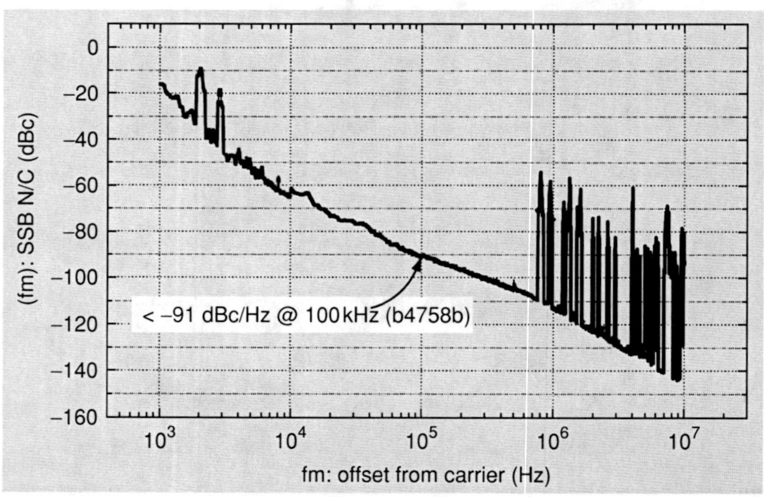

FIGURE 8.4 Single sideband phase noise of the 47 GHz oscillator.

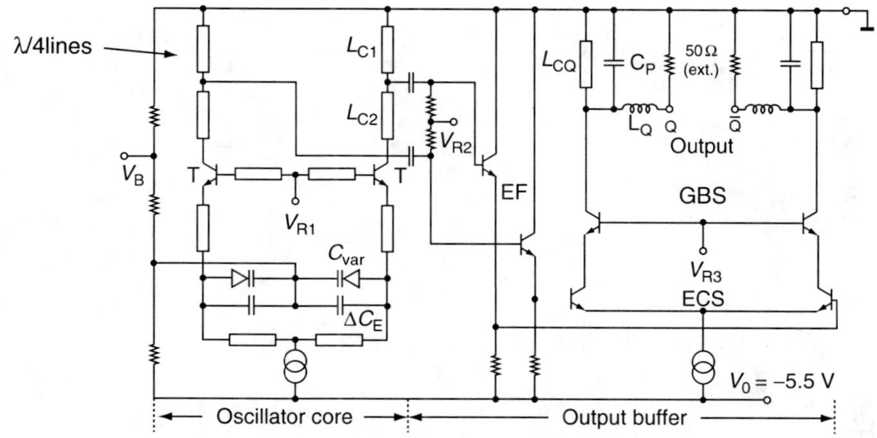

FIGURE 8.5 Circuit diagram of the 77 GHz VCO with output buffer.

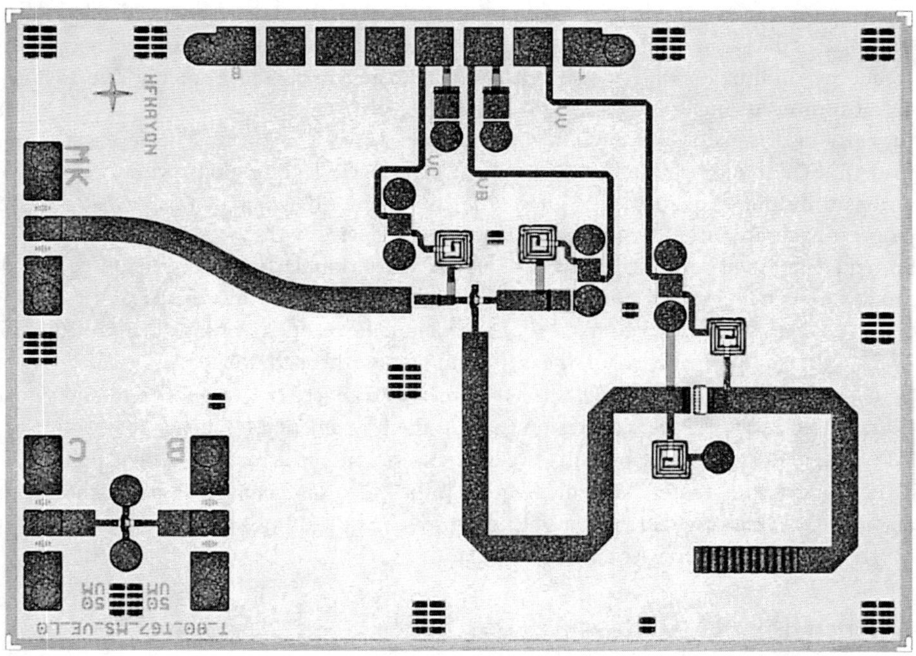

FIGURE 8.6 Chip photo of 27 GHz microstrip VCO. Chip size is 2.7 mm × 1.9 mm.

process with self-stopping emitter etching, self-aligned base contacts, trench-isolation etching, and low temperature-chemical vapor deposition (LT-CVD) deposited silicon nitride isolation. The f_T and f_{max} values of the differential grown and processed SiGe HBT are 40 and 60 GHz. The f_{max} value is quite low due to the not optimized differential processing. The base and collector layers are used for the varactor. A reflection-type oscillator design in common-base configuration is used [7]. The emitter is connected to a resonator line, which can be tuned by a varactor. The circuit design and chip photo are shown in Figure 8.6. From the output spectrum an SSB-phase noise value of better than −90 dBc/Hz at 100 kHz offset is found, which represents an extremely good value for a voltage-controlled oscillator. For a varactor tuning voltage from 1 to 7 V the output power remains constant. The total output power is 5 dBm at 27 GHz in this 400 MHz tuning range. The total frequency change is 800 MHz.

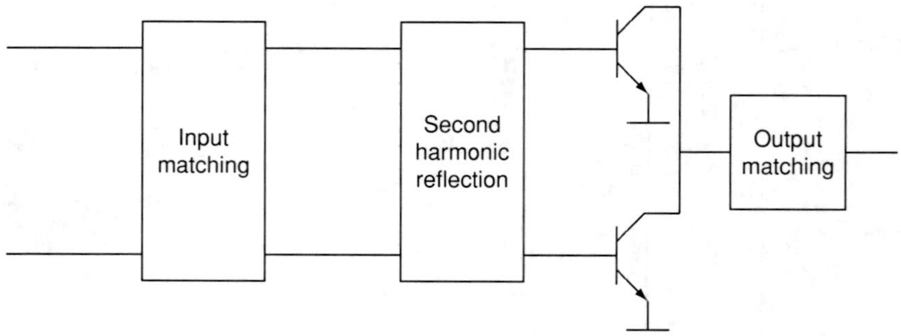

FIGURE 8.7 Scheme of a doubler in differential technology.

Multiplier

There are two important reasons to deal with multipliers: First, it is easier to generate oscillations with transistors at microwave frequencies and then to multiply the signal. Second, in most system applications frequency dividers are needed for stabilization purposes, which are rarely available at millimeter waves. High harmonics can be generated with nonlinear electronic devices as diodes or transistors. Active frequency multipliers with transistors have increasing importance mainly due to their small conversion loss or even conversion gain.

The majority of designs of active millimeter multipliers are based on FET devices as the nonlinear component. A first frequency multiplier based on a SiGe HBT technology with $f_{\max} \approx 67\,\text{GHz}$ is designed for an output frequency of 55 GHz. The conversion efficiency of this circuit is better than $-12\,\text{dB}$, operating at an output frequency near $f_{\max}$ of the device [14].

In the frequency range between 15 and 18 GHz an active doubler with conversion gain is demonstrated. In a differential circuit configuration the input signals are fed into the base of the transistors, which are biased near the class B region in order to generate the second harmonic components efficiently (Figure 8.7). Due to the feedback of the $2f_0$ components through the base–collector capacitor, a resonator is needed at the base of transistors T1 and T2 to reflect the second harmonic signals back into the doubler circuit for higher conversion gain. At the fundamental frequency the two input signals are out of phase, and the reflector presents two shunt capacitors, which are treated as part of the input matching network. At the second harmonic frequency, the $2f_0$ components are in phase and the reflector is equivalent to two series resonators. An output power of 5 to 6 dBm is obtained from 15 to 18 GHz, and the corresponding input power is 1.5 dBm [15].

8.4 Amplifier

A first HBT-based amplifier MMIC is demonstrated at Ka-band frequencies on high-resistivity substrate [16]. The design is based on the coplanar concept. Air bridges are used to suppress undesired slot-line modes. Spiral inductors serve as RF blocks. The passivation of the metal-free regions between the line conductors must be removed in order to prevent surface charges inducing parasitic conducting layers. Recent results using IBM's BiCMOS 7HP process prove that SiGe distributed amplifier MMICs for the frequency range from 0.1 to 50 GHz are competitive with III–V solutions [17]. The passband gain from 100 MHz to 50 GHz for this amplifier varies between 9 and 5 dB.

8.5 Mixer

One specific application area that has received much attention in recent years is the automotive area, especially as the FCC granted regulatory approval in the United States for ultra-wideband (UWB) emissions in 2002. For the use in direct-down conversion I/Q receivers in short-pulse radar systems a high instantaneous bandwidth is required. Over the 3 to 35 GHz frequency range for an IF frequency of 500 MHz and an LO drive of 0 dBm, a SiGe mixer MMIC displays a conversion gain of over 20 dB, including the contribution of the baseband output stage [18]. The design comprises an LO input stage, a double-balanced Gilbert cell for mixing and multiplication, and an IF output stage. The double sideband (DSB) noise figure of this mixer, which is designed and realized on an Atmel process using 1000 Ω cm substrate, is less than 8 dB at 24 GHz. The chip size is 250 μm × 280 μm.

In the Gilbert cell, the use of current sources should be avoided due to their deteriorating effect on the noise performance [19]. In an optimized design, the usual transistor current source is replaced by a spiral inductance of two turns. With a three-stage preamplifier in cascode configuration and a single-ended input, a Gilbert cell mixer with a single-ended LO-drive input and a differential-output IF buffer amplifier circuit using lumped elements (MIM capacitors and spiral inductors), a conversion gain of 40 dB for an IF frequency of 100 MHz can be achieved [20].

8.6 MMICS with SiGe Diodes

Switches

Pin diodes are used in switches as they can provide a very low on-resistance and low switching times. SPDT switches with lateral pin diodes for 77 GHz operation are demonstrated on HRS providing the I-region and p^+ and n^+ layers formed by implantation [21]. These circuits may be fabricated on a wafer before the silicon–germanium process sequence is started. Vertical pin diodes may be created using the collector layer of a HBT as "intrinsic" layer. Shunt diodes based on the IBM SiGe foundry process are demonstrated, which achieve an insertion loss of <0.5 dB and an isolation of >20 dB across the 1 to 20 GHz band [22]. A full transfer switch containing a five-port design with 10 pin diodes is demonstrated with a path loss of 1.4 to 2.1 dB across 7 to 11 GHz; the isolation over the same band is 41 to 75 dB. Utilizing this diode, several control circuits including a broadband (1 to 20 GHz) monolithic single-pole double throw switch, a five-port transfer switch, a six-bit phase shifter, and a five-bit attenuator are realized [23].

Mott Diodes

Schottky diodes with fully depleted active layer at zero bias are known as Mott diodes and especially useful in zero bias mixers. In order to adjust the barrier height of the Schottky contact delta-doped layers or SiGe layers may be inserted between the Schottky metal and the active layer [24]. A thin (8 nm) SiGe layer on a n-doped silicon layer increases the barrier height, whereas a SiGe layer on p-layers reduces the barrier height [25].

IMPATT Diodes

Impatt diodes are prone to be noisy. This is only partly true, as IMPATT diode design regarding the injection angle and the transit angle can improve the noise behavior significantly [26]. A much more severe drawback is the limited tunability and stability of the IMPATT diodes, which is simply a consequence of the two terminal device. Injection locking is one possible approach to stabilize an IMPATT oscillator [27]. A 73 GHz monolithic IMPATT oscillator can be subharmonically injection locked by a 24 GHz SiGe HBT oscillator [28]. Synchronization is achieved on the third harmonic. Changing the collector current of the HBT-MMIC results in a frequency modulation of the 73 GHz signal.

8.7 Summary

This chapter tries to give a necessarily selective overview on state-of-the-art SiGe MMIC results. The performance of SiGe MMICs has to be viewed under the consideration of competing approaches—especially CMOS. Recently, K-band low-noise amplifiers using a 0.18 μm CMOS technology with operating frequencies up to 24 GHz, a gain of more than 12 dB, and a noise figure of 5.6 dB have been reported [29]. At 26 GHz still 8.9 dB of gain and a noise figure of 6.9 dB are obtained. The current consumption is 30 mA from a 1.8 V power supply. This result represents the highest operation frequencies reported for LNAs in a standard CMOS process.

These results have to be compared with the latest SiGe device and circuit results. Cutoff frequencies are now approaching 400 GHz [30] and the circuit performance at W-band frequencies approaches or even exceeds [11] that of III–V MMICs. The most important advantage of SiGe MMIC concepts is probably the possible co-integration of low-noise, high-current building blocks (e.g., oscillators) with CMOS circuits in order to realize complete single-chip solutions—e.g., a phased array transceiver [31].

References

1. K.M. Strohm, J. Büchler, P. Russer, and E. Kasper. Silicon high resistivity substrate millimeter wave technology. IEEE-MTT-S, 1986, pp. 93–97.
2. U. Guettich. Low cost voltage controlled oscillators for X-band mobile communication purposes realised with Si BTs and SiGe HBTs. Topical Meeting on Silicon Monolithic Integrated Circuits in RF Systems, September 17–18, 1998, Michigan.
3. W. Heinrich. Coplanar waveguide silicon MMICs. Topical Meeting on Silicon Monolithic Integrated Circuits in RF Systems, September 17–18, 1998, Michigan.
4. J.-F. Luy, K.M. Strohm, J. Buechler, and P. Russer. Silicon monolithic millimeter-wave integrated circuits. *IEE Proceedings-H* 139(3), 1992, 209–216.
5. C. Rheinfelder, K. Strohm, F. Beißwanger, J. Gerdes, F.J. Schmückle, J.-F. Luy, and W. Heinrich. 26 GHz coplanar SiGeMMICs. IEEE-MTT-S, 1996, pp. 273–276.
6. C. Rheinfelder, K. Strohm, L. Metzger, H. Kibbel, J.-F. Luy, and W. Heinrich. 47 GHz SiGe MMIC oscillator. IEEE-MTT-S, 1999, pp. 5–8.
7. K.M. Strohm, C.N. Rheinfelder, J.-F. Luy, P. Nüchter, T. Hess, W. Heinrich, H. Kuhnert, M. Nadarassin, C. Warns, and W. Menzel. Coplanar and microstrip oscillators in SiGeSIMMWIC technology. IEEE-MTT-S, 2001, pp. 1563–1566.
8. H. Kuhnert, W. Heinrich, W. Schwerzel, and A. Schüppen. 25 GHz MMIC oscillator fabricated using commercial SiGe HBT process. *Electron. Lett.* 36, 2000, 218–220.
9. H. Li, H.-M. Rein, R. Kreienkamp, and W. Klein. 47 GHz VCO with low phase noise fabricated in a SiGe bipolar production technology. IEEE-MWCL 12(3), 2002, 79–81.
10. H. Li, H.-M. Rein, R.-E. Makon, and M. Schwerd. Wide-band VCOs in SiGe production technology operating up to about 70 GHz. IEEE-MWCL 13(10), 2003, 425–427.
11. H. Li, H.-M. Rein, and T. Suttorp. Design of W-band VCOs with high output power for potential application in 77 GHz automotive radar systems. GaAs IC Symposium, 2003, pp. 263–266.
12. W. Winkler, J. Borngräber, and B. Heinemann. A 117 GHz LC oscillator in SiGeC BiCMOS technology. International SiGe Technology and Device Meeting, 2004, pp. 71–72.
13. K.M. Strohm, P. Nuechter, C.N. Rheinfelder, and R. Guehl. Via hole technology for microstrip transmission lines and passive elements on high resistivity silicon. IEEE-MTT-S, 1999, pp. 581–584.
14. S.P.O. Bruce, A. Rydberg, M. Kim, F.J. Beißwanger, J.-F. Luy, H. Schumacher, U. Erben, M. Willander, and M. Karlsteen. Design and realization of a millimeter-wave Si/SiGe HBT frequency multiplier. IEEE-MTT 46(5), 1998, 695–700.
15. J.-J. Hung, T.M. Hancock, and G. Rebeiz. A high-efficiency miniaturized SiGe Ku-band balanced frequency doubler. IEEE-MTT-S, 2004.
16. K.M. Strohm, J.-F. Luy, F. Schäffler, H. Jorke, H. Kibbel, C. Rheinfelder, R. Doerner, J. Gerdes, F.J. Schmückle, and W. Heinrich. Coplanar Ka-band SiGe-MMIC amplifier. *Electron. Lett.* 31(16), 1995, 1353–1354.
17. J. Aguirre and C. Plett. 50 GHz SiGe HBT distributed amplifiers employing constant-k and m-derived filter section. IEEE-MTT 52(5), 2004, 1573–1578.

18. I. Gresham and A. Jenkins. A low-noise broadband SiGe mixer for 24 GHz ultra-wideband automotive applications. RAWCON 2003, pp. 361–364.
19. H. Schumacher, P. Abele, E. Soenmez, K. Schad, and T. Teppo. Low-cost Ku to Ka band MMICs using a commercially available Si/SiGe HBT process. Asia Pacific Microwave Conference, 2000.
20. E. Sönmez, A. Trasser, P. Abele, F. Gruson, K.-B. Schad, and H. Schumacher. 24 GHz high sensitivity downconverter using a commercial SiGe HBT MMIC foundry technology. Topical Meeting on Silicon Monolithic Integrated Circuits in RF Systems, 2003, pp. 68–71.
21. J.-F. Luy, K.M. Strohm, H.-E. Sasse, A. Schüppen, J. Büchler, M. Wollitzer, A. Gruhle, F. Schäffler, U. Güttich, and A. Klaaßen. Si/SiGe MMICs. IEEE-MTT 43(4), 1995, 705–714.
22. R. Tayrani, G. Sakamoto, P. Chan, R.v. Leeuwen, and T. Nguyen. X-band SiGe monolithic control circuits. IEEE-MTT-S 1998, 126–134.
23. R. Tayrani, M.A. Teshiba, G.M. Sakamoto, Q. Chaudry, R. Alidio, Y. Kang, I.S. Ahmad, T.C. Cisco, and M. Hauhe. Broad-band SiGe MMICs for phased-array radar applications. *IEEE J. Solid State Circuits* 38(9), 2003, 1462–1469.
24. K.M. Strohm, J.-F. Luy, T. Hackbarth, and S. Kosslowski. MOTT SiGe SIMMWICs. IEEE-MTT-S 1998, 1691–1694.
25. S. Chattopadyay, L.K. Bera, K. Maharatna, S. Chakrabarti, S. Dhar, S.K. Ray, and C.K. Maiti. Schottky diode characteristics of Ti on strained-Si. *Solid-State Electron.* 41, 1997, 1891–1893.
26. J.F. Luy and P. Russer, eds. *Silicon-Based Millimeter-Wave Devices.* Springer-Verlag, Berlin, 1994.
27. J.-F. Luy and P. Russer. SiGe SIMMWICs. IEEE-RFIC Symposium, 1997, pp. 105–108.
28. K.M. Strohm, F. Beisswanger, and J.-F. Luy. A 73 GHz SiGe SIMMWIC module. ESSDERC 97. Proceedings of the 27th European Solid State Conference, Stuttgart, September 1997, pp. 728–731.
29. K.-W. Yu, Y.-L. Lu, D.-C. Chang, V. Liang, and M.F. Chang. K-band low-noise amplifiers using 0.18 μm CMOS technology. IEEE-MWCL 14(3), 106–108.
30. A.D. Stricker, G. Freeman, M. Khater, and J.S. Rieh. Evaluating and designing the optimal 2D collector profile for a 300 GHz SiGe HBT. International SiGe Technology and Device Meeting, ISTDM 2004, pp. 97–98.
31. H. Hashemi, X. Guan, and A. Hajimiri. A fully integrated 24-GHz 8-path phased-array receiver in silicon. IEEE International Solid-State Circuits Conference, San Francisco, February 2004.

9

Wireless Building Blocks Using SiGe HBTs

9.1 Introduction 9-1
What is Different About Designing Circuits at RF?
9.2 Monolithic Components for RFIC Design 9-5
Passives • Bipolar versus CMOS Transistors
9.3 RF Amplifiers 9-14
Low-Noise Amplifiers
9.4 RF Mixers 9-22
Transformer Balun-Coupled Mixer • Mixer Comparison and Summary
9.5 Voltage-Controlled Oscillators 9-27
Resonant Tanks • Integrated Circuit Oscillators • VCO Comparison and Summary

John R. Long
Delft University of Technology

9.1 Introduction

Radio-frequency (RF) circuit design using silicon technologies has progressed over the past decade from small-scale building blocks to complete systems-on-a-chip enabling applications such as 3G wireless telephony [1]. The hundreds of active and passive components used in a handset in the mid-1990s were reduced to a single silicon chip and a handful of passives (filters, capacitors, etc.) by 2000 [2,3]. This was achieved through improvements in the technology, smaller passive components (small and lighter filter packages), and innovations in circuit design and radio architecture. It can be argued that the current success of wireless technology as a consumer product with mass-market appeal has been made possible by the cost, size, and performance advantages of integration on silicon. The performance improvements offered by silicon–germanium heterostructure bipolar transistors (SiGe HBTs) in BiCMOS technologies are enabling even further advances in wireless transceiver performance and integration. The first section of this chapter reviews the aspects of RF circuit design relevant to the design of wireless building blocks. Low-noise amplifier (LNA), mixer, and voltage-controlled oscillator (VCO) circuits are then examined in detail.

Radio-frequency integrated circuits (RFICs) are constructed using both active and passive devices fabricated directly on a semiconducting substrate. Chip-level integration is possible within the monolithic context, where low complexity circuit blocks (e.g., oscillator, mixer, amplifier) may be combined into a more complex functional block (e.g., transceiver front-end, as shown in Figure 9.1). A single-chip transceiver for GSM (global system for mobile communications) capable of operating in both 900 MHz and 1800/1900 MHz bands is a current example of this type of functional integration.

The gain-bandwidth product (i.e., f_T) of active devices fabricated in both silicon and III–V technologies and representing the current state-of-the-art in production are compared in Figure 9.2. It is evident

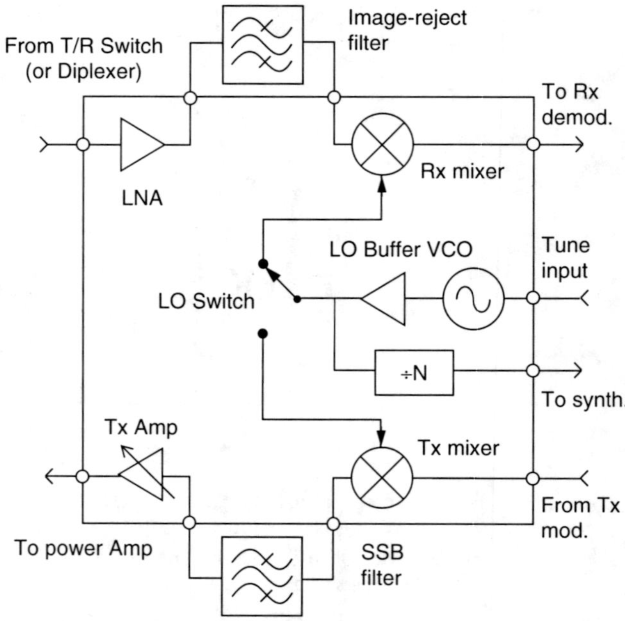

FIGURE 9.1 RFIC wireless front-end building blocks.

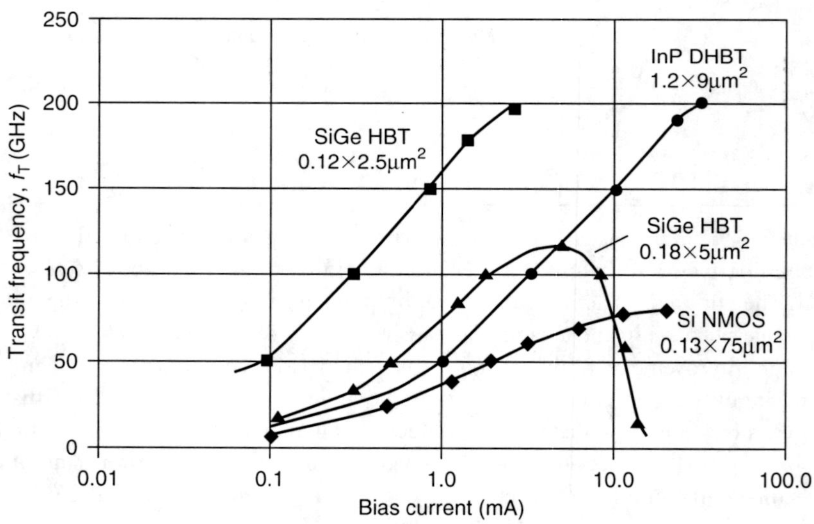

FIGURE 9.2 Gain-bandwidth product (f_T) versus bias current for transistors fabricated using various semiconductor technologies (year 2003).

that SiGe HBTs are now competitive with the latest devices fabricated in more exotic materials such as InP. Silicon NMOS devices with 0.13 μm length gates have peak f_Ts of approximately 70 GHz while the SiGe bipolar devices have an f_T of 200 GHz at the same lithography level ($0.12 \times 2.5\ \mu m^2$). The f_T for a $0.12 \times 2.5\ \mu m^2$ SiGe transistor peaks at about 2 mA, while typical Si-NMOS or InP devices require almost ten times more current to reach their f_T peak. Consequently, circuits can be constructed in silicon–germanium technology that consume less current (and therefore less power) than in other technologies. SiGe BiCMOS technology offers the added advantage that many more transistors—meaning more memory and more system functions—can be integrated onto a single chip. In many wireless circuits the use of SiGe devices lowers the number of parts needed to build a mobile phone handset which also lowers the cost, and improves the talk-time from a single charge of the batteries.

The impressive improvements in active device bandwidth that result from scaling are likely to continue. However, the performance of on-chip passive components and interconnections as well as the packaging technology for off-chip interconnections have not kept pace with the improvements in active device performance. This presents a bottleneck to the development of silicon as a true RF/MMIC platform, where both high-performance passive and active components are integrated together using the same technology.

What is Different About Designing Circuits at RF?

There are a number of aspects of circuit design at radio frequencies that make it different from circuit design for lower frequency applications. Some are related to frequency, that is, attempting to operate a transistor or circuit close to its frequency limit, and the difficulties of testing and measuring such a circuit or component in the GHz frequency range. Other differences arise from the way the circuit is used in a system, and the specifications and terminology used to describe RF and high-speed circuitry. In addition, there are many simulators designed specifically to solve RF design problems, so designers must be familiar with the various simulation techniques and design methodologies.

Analog RF circuit requirements demand much more than fast switching speeds between binary states and the capability to pack an enormous number of devices onto a single chip. For example, the analog interface between the RF communications channel and the baseband digital signal processor in a cellular telephone demands the ability to detect microvolt signal levels, while keeping the harmonic and intermodulation distortion produced by much stronger interferers within acceptable limits. Also, tetherless communications devices such as cellular telephones and wireless LAN interfaces must be light and portable, which makes factors such as size and battery lifetime important to the RF circuit designer.

High Frequency and High Speed Defined

Many linear circuits might not appear to involve "high frequencies" or "high speed," but these terms are related to the bandwidth limitations of the active devices used in the design. The frequency range over which bandwidth-limited instability occurs in most circuits lies between the -3 dB bandwidth, f_{bw}, and the unity gain frequency, $f_{unity\text{-}gain}$. Thus, we will define the "high-frequency" region to be between f_{bw} and $f_{unity\text{-}gain}$. For a bipolar transistor, this region lies between f_β and f_T. For a SiGe HBT with a current gain β of 100 at kHz frequencies and an f_T of 100 GHz, this implies that the high-frequency range lies above 1 GHz. "High-speed" implies a device or circuit with a frequency range extending into the high-frequency regime, where the maximum frequency of operation lies in the range from $0.1f_T$ to $0.5f_T$.

Transmission Line Effects

One of the benefits of the small physical dimensions inherent in integrated circuit technology is that propagation delays along interconnections are usually dominated by component parasitics, which makes them relatively easy to analyze. For example, a 1 GHz signal sees a phase shift of less than 3° across a 1 mm connection corresponding to a traveling time or "time-of-flight" of 8 psec. This is tiny compared to the delays caused by resistances and capacitances present in CMOS circuits. Because these conditions have prevailed since the early years of IC development, popular circuit analysis programs such as SPICE and Spectre® implicitly assume that all circuit components behave as lumped elements.

The "time-of-flight" over interconnect wiring is normally neglected but becomes significant at multi-GHz frequencies. This is partly due to the fact that the average wiring length in circuits is not shrinking as quickly as the transistor dimensions. The physical space required by resistors, capacitors and inductors, power supply wiring, and interconnect bondpads, as well as the need to physically separate circuits as a way of reducing electrical interference between blocks, keeps wiring lengths between 10 and 100 μm within circuit blocks and approaching 0.5 to 1 mm between blocks of circuitry.

The millimeter-wave frequency range (wavelengths of 10 mm or less in air) corresponds to a frequency of 12 GHz or more for signals traveling on a conventional silicon IC (effective permittivity of ~6). Above these frequencies, the wavelength encountered on-chip approaches the length of a typical interconnect wire and time-of-flight delays become important. Wires no longer behave as a simple lumped resistor–capacitor filter, but begin to look like transmission lines. This indicates that a change in the computer-aided design (CAD) tools used to analyze circuits is needed, and also that a shift toward the methods used by microwave circuit designers to synthesize higher frequency circuits will follow.

For off-chip interconnections, however, these transmission line effects occur at much lower frequencies. Any interconnection that has a physical length greater than one-tenth the wavelength of the highest frequency component in the signal must be treated as a transmission line, because the effects of time delay over the interconnection affect the circuit behavior. Aside from an RF transmitter and antenna, common examples of transmission lines are: the data bus between a microprocessor and RAM module in a personal computer, a hydroelectric generating plant and an electrical substation several hundred miles away, or the video connection between a DVD player and video monitor.

To the circuit designer, transmission line effects imply that impedance must also be closely controlled so that signal reflections and the ringing of pulse waveforms (or "signal integrity") do not impair performance. For example, an ideal voltage amplifier (infinite input impedance, zero output impedance) would not be suitable as an interface to an RF antenna as power could not be efficiently transferred to or received from the antenna. Also, since interconnect cable lengths in a typical measurement setup for an RF circuit are approximately 1 m or more (about 0.2λ in electrical length at 1 GHz), transmission lines and their behavior must be understood in order to test and characterize RF and high-speed circuits.

While transistor transit frequencies in production technologies move beyond 100 GHz, interconnect bandwidth and passive component parasitics are not scaling with the minimum transistor feature size (e.g., MOSFET gate length). As a result, improvements in RF circuit performance that could capitalize on advances from scaling continue to be constrained by the environment surrounding nanometer-sized active devices. This includes on-chip interconnects, imperfections in passive components that process the voltages and currents in an analog circuit, packaging parasitics (e.g., bondwires), as well as the printed circuit or other sub-assemblies. Despite these problems, there are components available at RF that are not realizable at lower frequencies as illustrated by the chip micrograph of Figure 9.3 [4]. In this circuit, a symmetric inductor (synthesized from on-chip transmission lines) improves input power matching and also minimizes the noise figure of a differential low-noise preamplifier (LNA). The transmission line interconnection between the bondpad and LNA form part of the matching network. Interstage signal coupling and bias isolation between the LNA and two balanced mixers are implemented using an on-chip three-filament transformer. Magnetic components enable sub-1 V operation with reduced current consumption and wider dynamic range (i.e., lower noise and distortion) at the cost of additional circuit area. Superior device matching and potential for large-scale integration on silicon

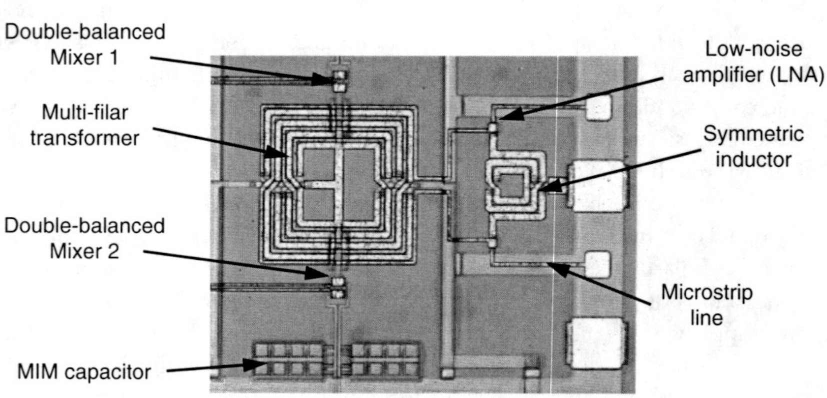

FIGURE 9.3 Passives on a 5 to 6 GHz band single-sideband receiver IC.

chips enable system architectures such as single-sideband converters (e.g., Figure 9.3) with baseband signal process and self-calibration schemes. Consequently, this saves on component costs and size while at the same time improving system reliability and manufacturability.

9.2 Monolithic Components for RFIC Design

Building blocks for wireless applications perform a variety of specialized tasks that have stringent requirements on sensitivity, distortion, bandwidth, spectral purity, and power output. Circuit performance can be optimized for a given task when a wide variety of on-chip components are available for designers to choose from. Many SiGe BiCMOS technologies offer processing options (e.g., thicker interconnect metal, linear capacitor, polysilicon resistor, etc.) that improve analog–RF performance beyond what is available from a high-volume technology designed for digital applications. The following sections review the attributes of passive and active components typically found in a SiGe BiCMOS technology from an RF circuit design perspective.

Passives

Monolithic circuit technologies for RF and microwave circuit fabrication usually consist of a mixture of passive lumped elements (i.e., R or C components with electrical dimensions less than approximately 0.1 times the wavelength) and passive distributed elements (e.g., components constructed from transmission line sections that are a electrically large, such as spiral inductors). However, it is difficult to realize a pure "lumped" element within the monolithic context, because the parasitics to ground or to the substrate affect the performance of any on-chip component. Lumped elements are relatively simple to model for circuit simulation, and these models are suitable for design up to approximately 20 GHz. At frequencies beyond 20 GHz, distributed element circuit models are the most useful for a manufacturable design.

Transmission Line Interconnections

Transmission lines interconnect circuit components, subsystems, and systems. At millimeter-wave frequencies (i.e., above $\sim$12 GHz on a silicon chip), all on-chip interconnections are considered transmission lines and impedance matching between any two-circuit nodes is necessary to preserve signal integrity and maximize efficiency. There are three basic conductor configurations for IC transmission lines: microstrip, coplanar waveguide, and coplanar stripline, where microstrip is the simplest and most popular configuration. The unique properties of transmission lines enable the implementation of monolithic inductors, balanced-to-unbalanced transformers (baluns), and other components that often outperform active circuit equivalents at microwave frequencies. However, existing microwave circuit designs cannot be ported directly to a silicon RF IC without initially considering the limitations and properties of interconnections, which are described in this section.

In modern silicon technologies, up to eight layers of metal are now used to reduce the "interconnect bottleneck" in digital VLSI circuits. Minimum dimensions are on the order of the metal thickness, which is currently about 0.5 μm for most metal layers, and 1 to 2 μm for top metal supply and ground busses. The shift to multilevel metallization schemes has also led to an increase in the thickness of the intermetal dielectric that separates the top metal layer and the semiconducting substrate (typically 5 to 8 μm). An additional benefit of thicker oxide is lower attenuation for microstrip transmission lines fabricated in production silicon processes, at least in the 1 to 5 GHz range of frequencies. However, the limitations imposed by the interconnecting metals and the conductive substrate continue to constrain the performance of circuits.

The semiconducting substrate degrades the performance of any transmission line fabricated on a silicon chip. "Skin-effect" and "slow-wave" modes of propagation may exist in addition to the quasi-transverse electromagnetic (quasi-TEM) mode seen in microstrip lines on insulating substrates [5]. SiGe-BiCMOS technologies developed for mixed-signal and RF applications employ medium resistivity substrates (usually between 10 and 20 Ω-cm) where the skin-effect mode is not present. Therefore, the slow-wave and quasi-TEM modes are of more interest for RF IC design.

The quality (or Q) factor is often used to compare passive components such as inductors and capacitors. The transmission line Q-factor is derived from a resonant tank built using a one-quarter of a wavelength long line at the resonant frequency, ω_0. The Q is then given by the ratio of bandwidth to ω_0, or $Q = \omega_{-3\,\text{dB}}/\omega_0$. Energy dissipated in the conductor metals, the substrate, or radiated to the surrounding environment reduces the Q, as illustrated in Figure 9.4a for three widths of microstrip transmission line (5, 10, and 20 μm). The relationship between Q-factor, strip width, and frequency follows from the attenuation of the microstrip line, with the wider lines having a larger Q at very low frequencies and the narrower lines having higher Qs above 3 to 4 GHz. It is interesting to note that the Q initially rises, dips, and then rises again as the frequency increases. This follows from the plateau in attenuation—as the attenuation per wavelength decreases at higher frequencies, Q-factor improves [6].

Since radiation is negligible for electrically short transmission lines (i.e., they are not antennas), the total Q-factor can be expressed as

$$Q_{\text{Total}} \approx Q_{\text{Conductor}} \parallel Q_{\text{Substrate}} \tag{9.1}$$

where $Q_{\text{Conductor}}$ is the Q-factor associated with the current loop (e.g., conductor metal and losses due to current induced in the substrate) and $Q_{\text{Substrate}}$ is the quality factor of the shunt parasitics (e.g., dissipation of the electric field in the silicon). As seen from Figure 9.4b, the total Q is dominated by

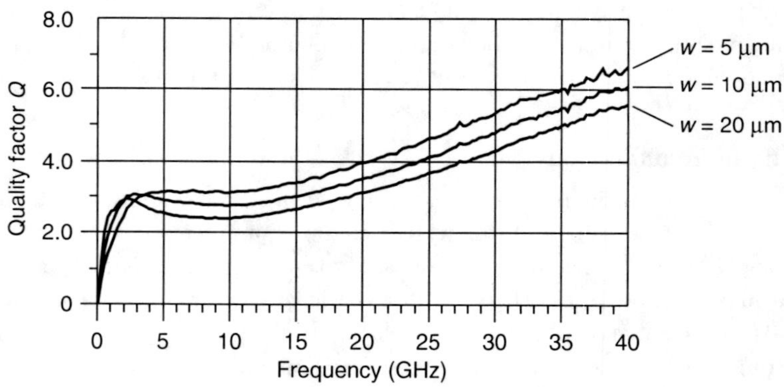

(a) Measured Q-Factor for 3 different widths of microstrip on silicon

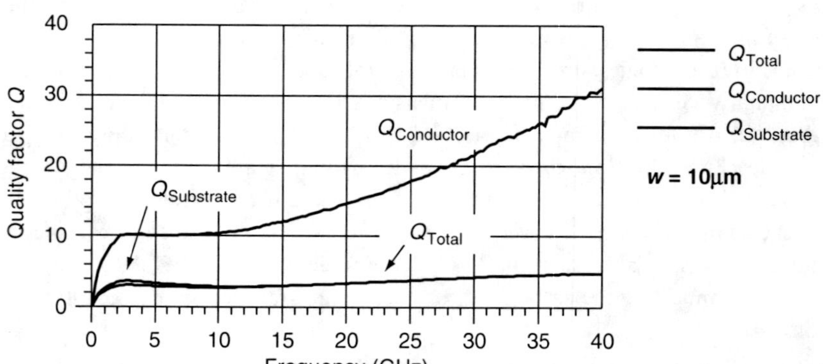

(b) Constituent components of the Q-Factor for w = 10μm (microstrip on silicon)

FIGURE 9.4 Measured Q-factor for microstrip lines on a 15 Ω-cm substrate (t_{oxide} = 5.8 μm, t_{silicon} = 200 μm, t_{metal} = 2.1 μm aluminum).

shunt parasitics. These results are typical of current silicon technologies. The greatest improvement in Q comes from reduction of losses to the substrate, and there is less to be gained from increasing metal thickness.

A relatively simple RLC model for microstrip lines on silicon is shown in Figure 9.5. It consists of a constant inductance in series with a frequency-dependent resistor that models skin effect in the conductors. Coupling to the substrate is via capacitors C_{ox}, with dissipation modeled by resistor r_{Si}. Capacitor C_{Si} models the transition between slow-wave and quasi-TEM modes. An L–R ladder network (lower part of Figure 9.5) models the skin effect for time-domain simulations [7]. At low frequencies, the total dissipation is the combined effect of the parallel network of resistors $r_1 - r_m$. As the frequency increases, series inductors $L_{sk1} - L_{skm}$ block current flow through the resistors and cause the desired frequency-dependent dissipation.

The measured attenuation constant for a 1 mm long microstrip line is compared to one- and two-section lumped element models extracted from measurements in Figure 9.6 [8]. The lumped element equivalent circuit is valid over the frequency range where the component length (in this case 1 mm) is less than approximately one-tenth of the signal wavelength. This corresponds to a maximum frequency of 13 to 14 GHz assuming quasi-TEM propagation with an effective permittivity of 5. As seen in

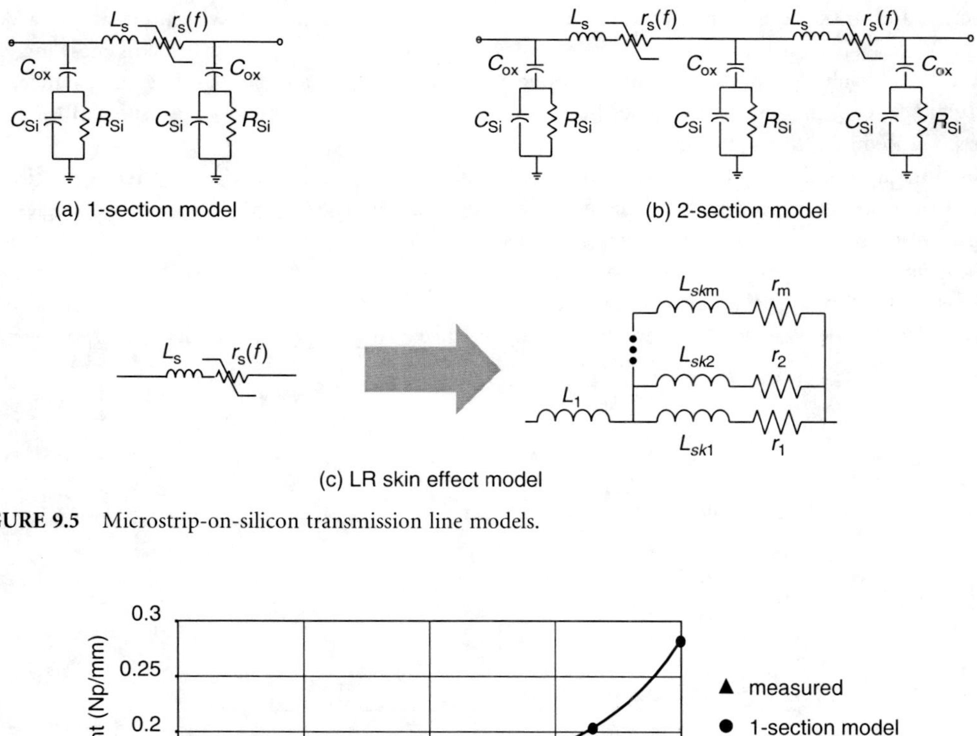

FIGURE 9.5 Microstrip-on-silicon transmission line models.

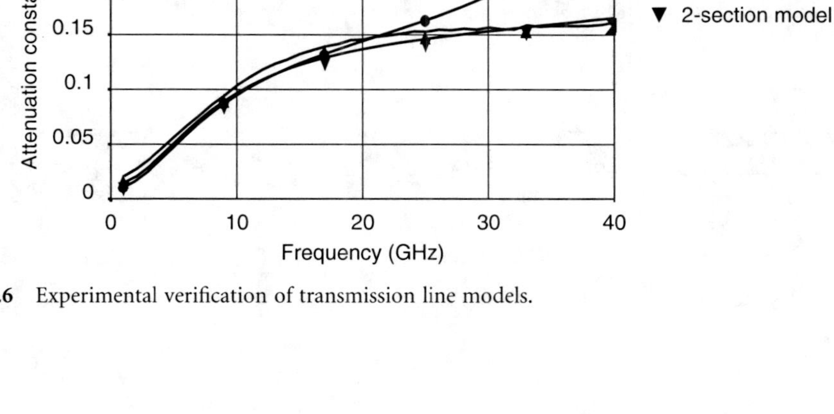

FIGURE 9.6 Experimental verification of transmission line models.

Figure 9.6, the one-section model begins to deviate significantly from the measured data around 15 GHz, while the two-section model remains reasonably accurate to almost 40 GHz. This validates both the lumped-element model and its accuracy for CAD purposes in both slow-wave and quasi-TEM modes for RF design up to mm-wave frequencies on silicon.

Capacitors and Resistors

Capacitors are used for ac coupling and decoupling as well as tuning of narrowband or resonant circuits. Larger capacitance values are realized using the metal–insulator–metal (MIM) structure or a combination of interdigitated metal fingers and multiple metal interconnect layers.

In a multi-metal layer technology, all but one of the metal layers can be used for the capacitor plates with the bottom layer as a substrate shield, which gives a specific capacitance of approximately 1 to 2 fF/μm^2 of chip area when five to six metal layers are used. A vertical plate configuration gives a high-density MIM capacitor where the plate separation is primarily defined photolithographically [9]. As a result it has less tolerance in processing than a horizontally stacked capacitor, where the insulator thickness can vary by 10% to 20%. A bottom metal shield blocks electric field leakage and associated losses to the underlying semiconductor. A similar layout scheme is used for bondpads (i.e., topmetal and shield layers only) connected to sensitive nodes, such as the input to a low-noise preamplifier.

The physical dimensions of MIM capacitors are normally small compared with a wavelength in the dielectric, and so the equivalent circuit of Figure 9.7 models the electrical behavior up to about 30 GHz. Resistance R_{plate} is typically two thirds of the total electrode resistance (note that the skin effect increases R over the dc value) and L_{plate} is the electrode inductance, C_{MIM} is plate-to-plate capacitance, and conductance R_{oxide} represents the losses in the dielectric film (often negligible). Parasitics of the shield are represented by a series RC circuit.

Resistors are required as loads and terminations and for the biasing of RF and microwave circuits. Polysilicon resistors are common in silicon IC technologies (see Figure 9.8). Often, only a single-sheet resistivity in the 10 to 30 Ω/sq. range (i.e., unsilicided gate polysilicon) is available, which limits applications to load resistances on the order of tens to a few hundred ohms. Polysilicon films in mixed-signal technologies are doped separately, allowing multiple sheet resistivities. Power handling is of the order of 100 $\mu W/\mu m^2$. The temperature dependence of doped polysilicon depends upon whether

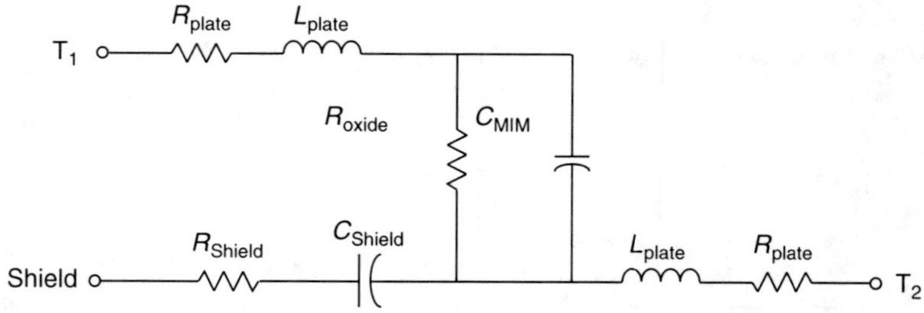

FIGURE 9.7 Lumped-element MIM capacitor model.

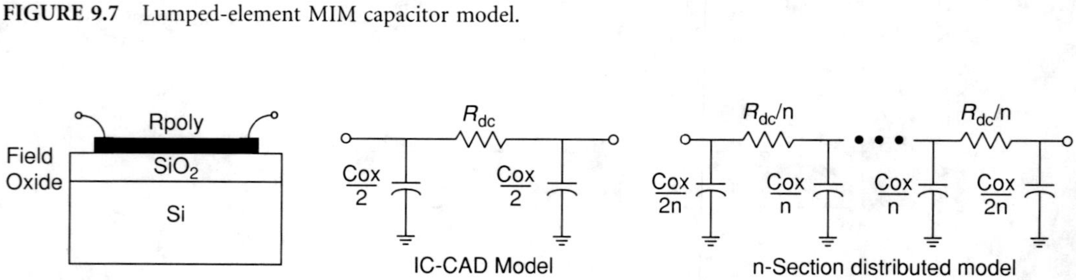

FIGURE 9.8 Polysilicon resistor structure and models.

p- or n-type dopant is used, with tolerances typically in the $\pm 15\%$ to 20% range. A metal film resistor, by contrast, has a typical sheet resistivity of 50 Ω/sq. with a tolerance of $\pm 5\%$.

The frequency limitation of resistors is normally not captured in analog IC design tools, and it is the designer's responsibility to adjust the resistor model by adding (RC) sections as required. The IC-CAD model for the resistor is a simple π-topology (Figure 9.8), which is valid to a frequency of approximately

$$f_{\text{max}} = 0.145/R_{\text{dc}}C_{\text{ox}} \tag{9.2}$$

R_{dc} is the dc resistance between the resistor's terminals and C_{ox} is the total parasitic capacitance to the substrate. Above this frequency, a larger number of π-sections are needed to model the distributed RC network formed by the polysilicon, oxide, and substrate. For example, at 2 GHz this implies 73 fF (maximum) parasitic capacitance for a single-section IC-CAD model of a 2 kΩ resistor. Above 2 GHz a lumped equivalent circuit can still be used, however multiple *R–C* sections in cascade (as shown in Figure 9.8) are required to accurately model the electrical behavior.

Inductors and Transformers

Planar inductors for monolithic circuits are useful for interstage matching and coupling, as resonant loads, and for biasing and bias circuit isolation in RF IC applications [10,11]. They can be realized in a number of configurations, all implemented (at minimum) using a single-layer metallization scheme and as on-chip transmission lines (i.e., a microstrip line). The total line length must be kept at a small fraction of a wavelength; otherwise the conductor cannot be treated as a lumped element. These components only approximate lumped inductors and in this sense they do not have a low-frequency equivalent for monolithic design compared to on-chip capacitors, which are also used in monolithic circuits for signal processing from audio to radio frequencies. They can also be used to implement components that are unique to RF circuits, such as transformers [7,12–14].

Resonant-tuned (LC) circuits offer many benefits to the designer of high-frequency circuits. Operation at a low supply voltage, simplified impedance matching between stages, and low dissipation for reduced circuit noise are just a few of the properties of LC circuits that can be exploited to achieve a higher level of performance. However, an on-chip inductance is required for the realization of LC networks for these purposes. At radio and microwave frequencies, a purely passive inductor is often preferable to synthesis of an inductive reactance with an active circuit. Passive components introduce less noise, consume less power, and have a wider bandwidth and linear operating range than their electronic equivalents, such as the gyrator. Prior to the mid-1990s, silicon integrated circuit (IC) technology was rarely used for analog applications in the radio and microwave range of frequencies, in part because transmission line structures performed poorly on a semi-conducting substrate. However, suitable performance can be realized when the limitations imposed by the technology are understood by designers and the components are accurately modeled and characterized.

For dimensions typically encountered in a commercial IC fabrication process (metal line-widths between 2.5 and 50 μm on a 350-μm thick substrate), the characteristic impedance of a microstrip line ranges from approximately 100 to 200 Ω. The substrate behavior depends upon both the resistivity and the frequency of the propagating wave. However, the substrate tends to behave as a lossy dielectric in modern silicon VLSI processes, where resistivities are typically in the 10 to 20 Ω-cm range (i.e., for "analog" or mixed-signal technologies) and operation is in the GHz range of frequencies. Low-ohmic substrates (i.e., resistivities below 0.1 Ω-cm) are often used for VLSI digital circuit fabrication to suppress latch-up when the highest packing density is required. Substrate current is easily induced immediately beneath a coil on a highly conductive substrate by the alternating magnetic field, which reduces the inductance and causes additional losses.

The operation of a passive transformer is based upon the mutual inductance between two or more conductors, or windings. The transformer is designed to couple alternating current from one winding to the other without a significant loss of power, and impedance levels between the windings are transformed in the process (i.e., the ratio of terminal voltage to current flow changes between windings). In addition, direct current flow is blocked by the transformer, allowing the windings to be biased at different potentials.

Interwinding microstrip spiral inductors to magnetically couple independent conductors is a logical extension of this concept, and results in a monolithic transformer as shown in Figure 9.9(a). Frlan and Rabjohn [7] demonstrated square spiral transformers on alumina and GaAs substrates, and developed circuit simulation tools based upon the extraction of a lumped element model for the transformer from physical and geometric parameters. This modeling technique was later extended to the analysis of planar structures on conductive substrates, such as silicon. In the recent literature, there are many examples of monolithic transformers fabricated in silicon IC technology for use in RF circuits, such as preamplifiers, oscillators, mixers, and power-amplifiers.

For the transformer, magnetic flux produced by current i_P flowing into the primary winding at terminal P, induces a current in the secondary winding that flows out of terminal S. This produces a positive voltage, v_S across a load connected between terminals S and $\bar{S}$. The main electrical parameters of interest to a circuit designer are the transformer turns ratio n, and the coefficient of magnetic coupling, k_m. The current and voltage transformations between windings in an ideal transformer are related to the turns ratio by the equation

$$n = \frac{v_S}{v_P} = \frac{i_P}{i_S} = \sqrt{\frac{L_S}{L_P}} \tag{9.3}$$

where the primary and secondary voltages (v_P, v_S) and currents (i_P, i_S) are defined in Figure 9.9(b), and L_P, L_S are the self-inductances of the primary and secondary windings, respectively. The strength of the magnetic coupling between windings is indicated by the k-factor,

$$k_m = \frac{M}{\sqrt{L_P L_S}} \tag{9.4}$$

where M is the mutual inductance between the primary and secondary windings. The self-inductance of a given winding is the inductance measured at the transformer terminals with all other windings open-circuited. If the magnetic coupling between windings is perfect (i.e., no leakage of the magnetic flux), k_m is unity, while uncoupled coils have a k-factor of zero. A practical transformer will have a k-factor somewhere between these two extremes. Since the materials used in the fabrication of an IC chip have magnetic properties similar to air, there is poor confinement of the magnetic flux in a monolithic

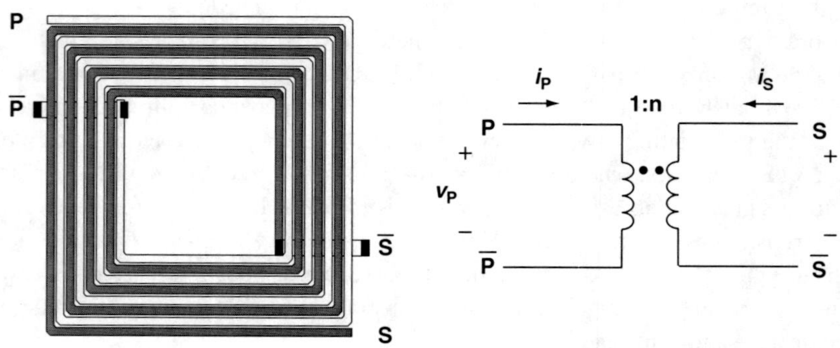

(a) Physical layout (b) Schematic symbol

FIGURE 9.9 A 1:1 Frlan-type planar transformer. (After JR Long. *IEEE Journal of Solid-State Circuits* 35:1368–1382, 2000. With permission.)

transformer and $M < \sqrt{L_P L_S}$. Thus, the *k*-factor is always substantially less than one for a monolithic transformer, however, coupling coefficients as high as 0.9 are realizable on-chip.

The phase of the voltage induced at the secondary of the transformer depends upon the choice of the reference terminal. For an ac signal source with the output and ground applied between terminals P and $\bar{P}$, there is minimal phase shift of the signal at the secondary if the load is connected to terminal S (with $\bar{S}$ grounded). This is the noninverting connection. In the inverting connection, terminal $\bar{S}$ is grounded and S is connected to the load so that the secondary output is antiphase to the signal applied to the primary. Aside from the phase shift between input and output ports, other aspects of the transformer's electrical behavior depend upon the choice of terminal configuration.

An example of a compact model for a transformer with four independently driven ports (i.e., P, $\bar{P}$, S, and $\bar{S}$) and turns ratio 1:*n* is shown in Figure 9.10. In many applications, the compact model can be further simplified because one (or more) of the ports is grounded. At the core of the model is an ideal linear transformer with magnetizing inductance, L_m and turns ratio 1:*n*. The path for magnetic flux between windings has the same permeability as free space for conventional IC technologies (unless, for example, a ferromagnetic layer is used in fabrication), and therefore, the magnetizing inductance is lossless and the magnetic path is highly linear. Note that this is an advantage in RF IC applications, where the dynamic range (and hence linearity) requirements are very demanding. Inductances L_P and L_S are placed in series with the primary and secondary windings of the linear transformer to account for imperfect coupling or leakage of the magnetic flux between the windings. Resistors r_P and r_S are placed in series with the leakage inductances representing ohmic losses in the windings, which are significant due to the relatively thin layers of metal available in an IC process. The interwinding capacitance is modeled by capacitors connected between primary and secondary, C_0 and C_x. The dominant capacitive parasitics between each winding and the underlying substrate (Z_{sh}) are represented by the series connection of capacitors C_{ox} and C_{Si}, and substrate loss is included through the addition of resistor r_{Si} in parallel with C_{Si}, as in the inductor models.

Multifilament transformers can also be constructed on-chip. These devices are used to implement power dividers or combiners and baluns. A balun is a device, which couples a balanced circuit to an unbalanced one. There are many structures used to implement baluns at RF and microwave frequencies, although a differential amplifier is the most commonly used circuit for unbalanced to balanced signal conversion on-chip. Microwave balun structures such as the Lange, rat-race, and branch line coupler

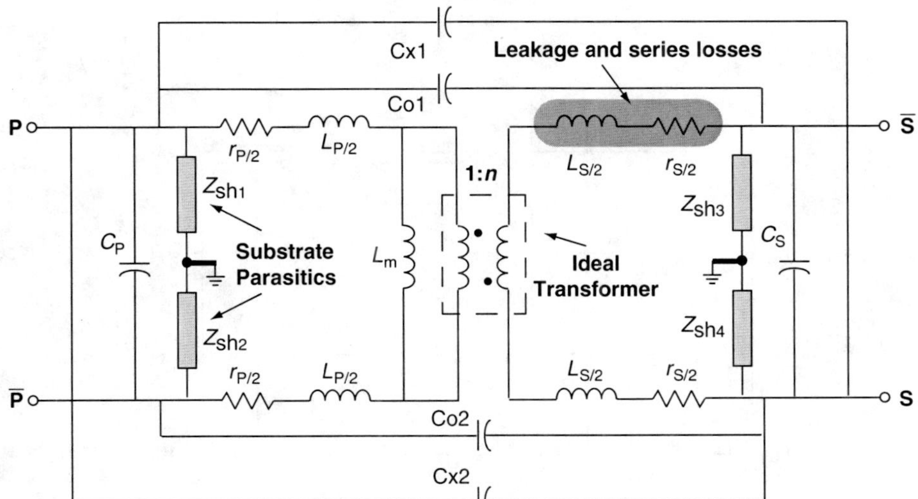

FIGURE 9.10 1:*n* transformer compact model. (After JR Long. *IEEE Journal of Solid-State Circuits* 35:1368–1382, 2000. With permission.)

require physical dimensions on the order of the signal wavelength and so these devices consume too much chip area when operating below approximately 15 GHz. The transformer shown in Figure 9.9 can also implement a balun by grounding one of the windings at the electrical center, or center tap. The electrical and physical centers of an asymmetric winding differ, which is a disadvantage of asymmetric layouts.

A square symmetric layout, first proposed by Rabjohn [7] and illustrated in Figure 9.11, solves this problem. This transformer consists of two groups of interwound microstrip lines that are divided along a line of symmetry running horizontally, as shown in the figure. The groups of lines are interconnected in a way, which brings all four terminals to the outside edge of the transformer layout, which is an advantage when connecting the transformer terminals to other circuitry. Also, the mid-point between the terminals on each winding, or the center tap, can be located precisely in the symmetric layout as indicated in Figure 9.11. The turns ratio for the example shown is 4:5 between primary and secondary.

The measured and simulated responses for this balun are compared in Figure 9.12. The experimental transformer is designed with OD = 325 μm, 8 μm linewidth, and 3 μm line spacing. The slight

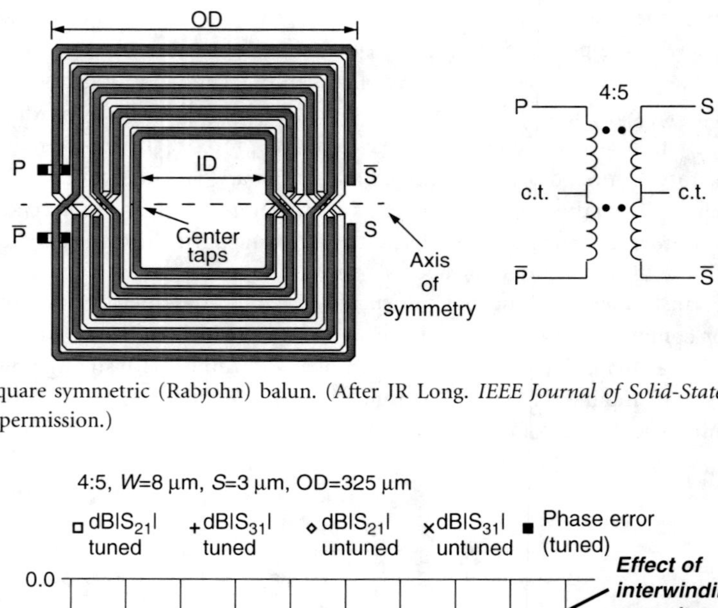

FIGURE 9.11 Square symmetric (Rabjohn) balun. (After JR Long. *IEEE Journal of Solid-State Circuits* 35:1368–1382, 2000. With permission.)

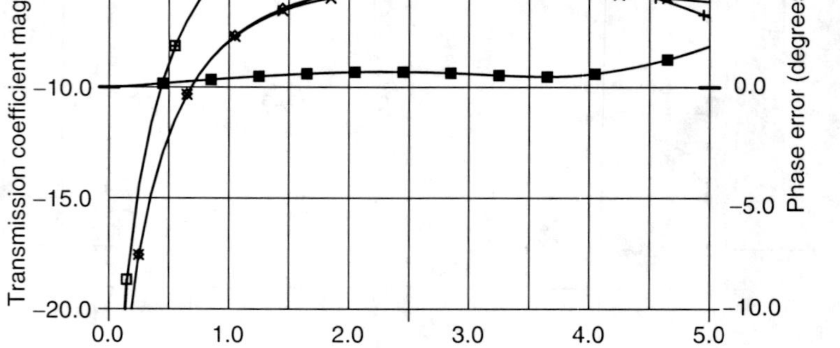

FIGURE 9.12 Frequency response of a monolithic transformer balun. (After JR Long. *IEEE Journal of Solid-State Circuits* 35:1368–1382, 2000. With permission.)

difference in magnitude response at the inverting and noninverting secondary ports is clearly seen from the measurements. This is due to the effect of interwinding capacitance. The effect is not reduced by adding tuning capacitance in shunt with the transformer ports. Capacitors connected at the input and output ports (425 fF across the primary and 1.7 pF across each secondary winding) tune the balun to match the (50 Ω) source to the secondary load, and the tuned response is also plotted in Figure 9.12. The measured transmission loss is reduced from over 5 dB to very close to the ideal (3 dB) by tuning. The phase error between secondary ports of the tuned balun is also shown in the figure, where the phase error is the deviation from a 180° phase difference between ports. This error is on the order of 1° in the desired passband for the balun (2 to 3 GHz).

Bipolar versus CMOS Transistors

Relevant performance attributes of current silicon MOS and bipolar (BJT) devices are compared in Table 9.1. The ratio of bias current to transconductance (I/g_m) is an indication of the gain that can be realized at a given supply current. This ratio for a MOSFET depends on the effective gate bias voltage ($V_{GS}-V_T$), which is on the order of 300 mV, compared to the thermal voltage (kT/q, approximately 25 mV at 27°C) for the bipolar. The impedance levels in RF and high-speed circuits are usually kept low because of bandwidth restrictions, so transconductance largely defines the active gain at RF (note that voltage or current gain can also arise from impedance matching; this is a "passive" gain), while the product of the transconductance and output resistance ($g_m * r_o$) sets the maximum active gain at low frequency. Gain also depends on parasitics, and the ratio of input to ground to the Miller capacitance (C_{in}/C_{μ}) in the common-source (or common-emitter) configuration indicates the relative importance of the device parasitic capacitances. Bipolar devices have a clear advantage in both gain and bandwidth. In addition, the input impedance at RF is relatively low compared to a MOSFET. MOS devices can realize voltage gain at the input if a matching network is used to match the transistor input impedance to an RF source (e.g., a 50 Ω generator). However, low-loss matching elements (with a high *Q*-factor) are needed. This gives a narrowband frequency response that is susceptible to tolerances and usually requires trimming in manufacture. A bipolar transistor, on the other hand, has a relatively low input impedance making the transistor easier to match to the typical RF source off-chip using a simple, low-*Q* matching network.

Flicker noise is important at frequencies up to the flicker corner frequency ($1/f$ corner), where thermal and flicker noise levels in a device are equal. Aside from baseband circuitry, $1/f$ noise affects oscillator phase noise and mixer noise figure in homodyne receivers with intermediate frequency stages that operate below the flicker corner frequency. Therefore, it is desirable that $1/f$ corner frequency should be kept as low as possible.

Breakdown voltage is important in circuits where either high voltages or high power outputs are required. For MOS devices, the breakdown depends mainly on the gate oxide, which will fail when the applied voltage exceeds the breakdown strength. This is 1.8 V for 0.18 μm gate length devices, but downscaling of the gate length in future generations of MOS devices will lower the breakdown voltage even further (e.g., 1.2 V for 0.13 μm technologies). In bipolar technology, breakdown voltage is defined by the collector–emitter breakdown voltage when there is a high impedance path for current flow from the base terminal to ground. The collector–base breakdown voltage (which is two to three times higher than V_{CEO}) is more relevant to RF circuits, since excess electrons created by impact ionization in the base

TABLE 9.1 Silicon MOSFET and BJT Comparison

Parameter	MOS	BJT
I/g_m	$V_{eff}/2$	V_T
C_{in}/C_{μ}	3	10
$1/f$ corner	1 MHz	1 kHz
r_o at 5 mA bias	1 kΩ	10 kΩ
Breakdown voltage	1.8 V	2–5 V

region have a relatively low impedance path out of the transistor's base terminal. The Johnson limit, which is the product of breakdown voltage (V_{CEO}) and f_T for the BJT is approximately constant, and is approaching 300 GHz-V for current SiGe devices.

Other considerations for high-frequency are CAD model accuracy and the accuracy of the CAD design kits that are available to the designer. Deep sub-micron MOS models (e.g., the industry-standard, BSIM3) are not scalable or accurate enough for RF analog design without extensive modification. Newer generations of MOS models (e.g., BSIM4 or Philips' MOS Model 13 and above) offer substantial improvements in small- and large-signal simulation accuracy. Bipolar models have improved considerably over the past decade. The latest bipolar models (VBIC, MEXTRAM, and HiCuM) are fully scalable and have demonstrated excellent accuracy in RF and high-speed applications.

9.3 RF Amplifiers

Radio-frequency amplifiers are used in both broadband and narrowband applications. Broadband applications in the 1990s were dominated by high-speed data communication using optical fiber. However, the demand for wireless functionality and greater data throughput have opened up new opportunities for wireless circuits operating in bands from below 500 MHz to above 60 GHz. These systems can be considered "broadband," as their operating bandwidth is greater than 10% of the mid-band frequency. For example, 1 GHz of bandwidth centered at 10 GHz is a broadband signal according to this definition. SiGe bipolar and BiCMOS are enabling technologies for these new wireless applications, which will develop over the coming decade. However, SiGe technology has already had a considerable impact on the evolution of cellular telephony and third-generation (3G) mobile telephone hardware. LNAs for receivers are described in this section.

Low-Noise Amplifiers

The low-noise preamplifier (LNA) suppresses noise introduced by the mixer and subsequent stages in the receive chain. The other design objectives for the preamplifier in a radio receiver are: adequate gain, low intermodulation distortion, and terminal impedances, which match the system impedance (50 Ω is usually assumed). The maximum input signal range for the receiver should be determined by the mixer, so the linearity of the LNA must be greater than the linearity of the mixer (i.e., when mixer linearity is referred to the receiver input). The demand for greater linearity drives up power consumption as more quiescent current is needed to increase linearity for a class A preamplifier. In addition, interfaces at the input and output of the integrated circuit amplifier are, in general, off-chip. Therefore, transmission line effects and impedance matching must be considered in the design.

Amplifiers or mixers intended for use in the RF and microwave frequency range are characterized using performance measures that are often very different from those used for low-frequency circuits. The power gain and terminal impedances over the amplifier passband are usually specified by the four S-parameters for a two-port network. In addition to these small-signal characteristics, the dynamic range of the amplifier must also be specified. The spurious-free dynamic range (SFDR) is defined by the difference between the signal overload point and the minimum discernible or acceptable signal level at the amplifier input. The minimum input signal level, or sensitivity, can be determined from the amplifier noise figure. The upper limit of the dynamic range is usually set by the distortion (e.g., intermodulation distortion) or gain compression characteristics of the circuit.

Third-order intermodulation distortion (i.e., frequencies $2f_1-f_2$ and $2f_2-f_1$ produced by closely spaced input tones at f_1 and f_2) falls within the desired intermediate frequency (IF) channel bandwidth, and could interfere with signal reception in a heterodyne (or low-IF) radio. The second-order intermodulation distortion products (i.e., frequencies f_1-f_2 and f_2-f_1 produced by closely spaced input tones at f_1 and f_2) have a similar effect in homodyne receivers. The intermodulation distortion is characterized by the intercept specification (e.g., third-order intercept, IP_3 or second-order intercept, IP_2). The intercept point is measured by applying two signal frequencies at the mixer RF input (the two-tone test) and observing the third-order distortion products at the output. The third-order intercept point referred to the input (IIP_3) is related to the output intercept (OIP_3) by the amplifier gain.

Two possible topologies for a common-emitter preamplifier are shown in Figure 9.13. The common-emitter (CE) configuration is the best compromise between noise figure, gain and terminal impedances at frequencies approaching the device f_T. Parasitic lead inductance from a bond-wire and package pin connected in series with the common (i.e., emitter) terminal adds degeneration to the LNA and lowers the gain. The gain is also affected by the finite output conductance of the transistor, the load impedance, and feedback via the Miller capacitance. A cascode preamplifier (Figure 9.13b) uses the low impedance of a common-base amplifier to reduce the gain of the input common-emitter stage and suppress the Miller effect. There are other methods of achieving this, however, the cascode is a relatively simple circuit which increases the noise figure by only a small amount (typically 0.5 dB). With a 2.7 V supply, the two transistors in cascode can be biased with sufficient headroom if a resonant tank is used for the output load.

Modern SiGe bipolar devices in production have f_Ts exceeding 100 GHz, which is well above the operating frequencies for most commercial wireless products (e.g., mobile telephones or wireless computer networking equipment). Therefore, bandwidth can be traded-off for reduced power consumption in the LNA. In addition, the collector–base feedback capacitance (which causes the Miller effect) in a typical common-emitter preamplifier is low enough that a single device (i.e., not a cascode) can be used to realize a gain of 15 to 20 dB at a bias current of a few milli-Amperes. When on-chip resonant circuits are used as loads, the transistor can be biased from a supply as low as V_{BE} (approximately 0.9 V), thereby realizing a true low-voltage and low-power LNA with excellent RF performance.

The common-mode rejection inherent in differential and symmetric circuit topologies helps improve the isolation between RF blocks in integrated radio circuits, which is one of the main reasons why the higher power consumption in these circuits is tolerated. The differential pair (Figure 9.14a) has (ideally) the same noise figure and gain as the single-ended equivalent, but doubles the power consumption compared to a single transistor LNA. As the RF input power is now split across two transistor base–emitter junctions, the distortion produced by the amplifier for a given input power level is lower than for a single transistor amplifier. Also, an explicit ground connection is not required as there is a virtual ground at the common terminal in a differential amplifier driven by a balanced (i.e., differential) signal, so ground path parasitics (e.g., bondwire inductance and package inductance, Figure 9.14b) do not affect the amplifier's gain.

When distortion is considered, the linearity of the common-emitter amplifier alone (i.e., without feedback) is poor, with the input-referred third-order intercept (IIP_3) typically in the -12 to -20 dBm range. Intermodulation and harmonic distortions can be avoided by increasing the transistor bias

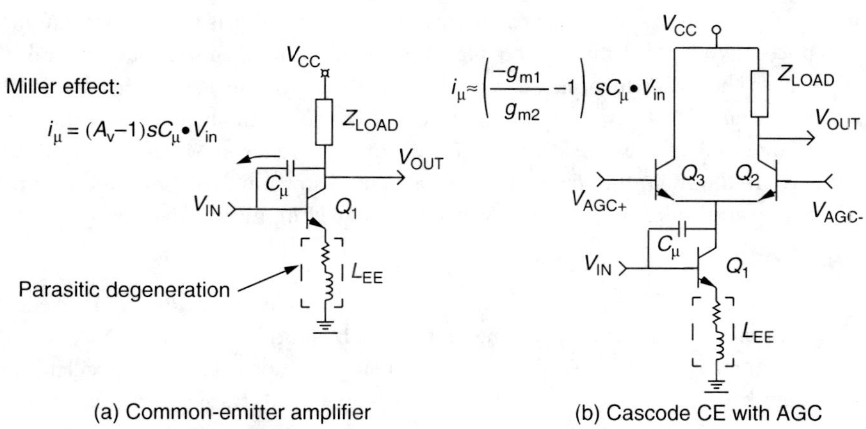

FIGURE 9.13 Simple common-emitter and cascode preamplifier topologies.

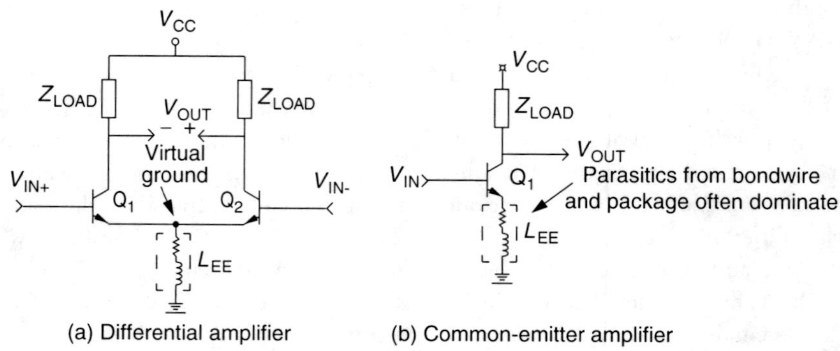

FIGURE 9.14 Differential and common-emitter preamplifiers.

current and the power supply voltage. However, a large bias current and a high supply voltage (which would be chosen to achieve high gain and good linearity) cannot be used in a low-power or low-voltage amplifier design. The amplitude of the base–emitter voltage drives the "active" or device-related distortion produced by the transistor. Negative feedback is often applied to an amplifier to reduce the base–emitter voltage for a given output voltage swing, thereby improving the linearity at the expense of lower gain. Hence, feedback can be an alternative to increasing the bias current for a low-power design, since it is well known that negative feedback can improve the linear input signal range of an amplifier, even though the active device itself might be operating in a relatively non-linear manner.

The impedances presented at the input and output ports of an RF amplifier at frequencies other than the desired operating frequency (i.e., frequencies other than the fundamental) also have a profound effect on the intermodulation (IM) distortion in common-emitter amplifiers [15–17]. The amplitude and phase of unwanted sum and difference frequencies caused by device nonlinearities contribute to third-order intermodulation distortion by mixing with the RF signal at the input. One source of distortion is the difference frequency between two in-band RF input tones. This relatively low frequency signal can be attenuated by keeping the impedance of the bias path at the amplifier input as small as possible at low frequencies. However, it has been shown that the optimum value of the impedance seen at low frequencies is not zero, but rather finite and complex [16]. Also, the second harmonic of the RF signal is fed from the output back to input via the collector–base capacitance, where it mixes with the fundamental tones thereby generating third-order IM distortion. Attenuating the second harmonic at the RF input will reduce this source of distortion. Neutralizing the feedback from output back to input will also diminish this source of inter-modulation distortion. Circuitry (often passive), which eliminates the low frequency and second harmonic signal at the input, are called out-of-band terminations.

Fifth-order intermodulation also has a greater effect on signal fidelity as third-order intermodulation distortion is reduced. While these impedances placed at the input to terminate out-of-band harmonics demonstrably reduce the IM distortion, experimental measurements also show that the improvement decreases with input amplitude and operating frequency. As the fundamental frequency approaches the bandwidth limitations of the transistor, the IM distortion increases. Moreover, the IM distortion improvement is realized only in the small-signal regime and distortion increases rapidly with RF input amplitude (power levels greater than about −30 dBm for a 50 Ω input impedance).

Cascode LNAs

Cascode preamplifier designs are popular because they offer wide bandwidth and excellent reverse isolation, which simplifies impedance matching and minimizes local oscillator (LO) leakage from the mixer back to the antenna via the LNA. The tendency of single-ended cascodes to oscillate is lower for a differential design because the base terminals of the cascode devices are at a virtual ground for the differential-mode signal. However, the series connection of two devices does require a larger supply voltage for proper biasing. A single-ended equivalent circuit for a differential cascode preamplifier biased from a 2.7 V supply is shown in Figure 9.15. The collector inductor has virtually no dc voltage drop and allows the output node to swing to $2V_{CC}$. Saturation of Q_1 must be avoided so V_{CE1} is set to 1 V (about 100 mV greater than V_{BE}). The maximum voltage swing across Q_2 is therefore 1.5 V-pk if 0.2 V is allowed for V_{CEsat} of Q_2. The bias current is typically chosen to meet linearity requirements from consideration of the clipping level at the output.

Insight into the optimization of microwave transistor noise figure comes from an analysis by Fukui, who derived an analytical solution for the transistor noise figure based on the hybrid-π model for the BJT. Fukui derived complete algebraic expressions for the four noise parameters in terms of the hybrid-π equivalent circuit parameters, including the effects of packaging parasitics. These expressions contain a large number of terms, and such detail is already captured in a modern CAD tool for use by RF designers today. However, the results of the analysis show that the minimum noise figure at high frequency for the device can be estimated using the following formulation for the minimum noise factor, F_{min},

$$F_{min\,|high\ frequency} \approx 1 + h\left(1 + \sqrt{1 + \frac{2}{h}}\right) \tag{9.5}$$

where

$$h = g_m \cdot r'_{bb}\left(\frac{f}{f_T}\right)^2 \tag{9.6}$$

Here, it is assumed that the dc and low-frequency current gains (β_{dc} and β_o) of the transistor are large. These current gains depend upon a number of factors, such as the technology (e.g., base doping), the collector current and the ambient temperature. The dc and ac current gains for SiGe bipolar transistors are both approximately 100 at low-medium collector current densities, so this condition is normally fulfilled. There is an additional restriction; the dc current gain should satisfy the following condition:

$$\beta_{dc} > 10\left(\frac{f_T}{f}\right)^2 \tag{9.7}$$

Assuming an operating margin (i.e., f_T/f) greater than three and a dc current gain of 100, this restriction will also be satisfied. Fukui's estimate for the minimum noise factor at high frequency (Equation 9.5 and Equation 9.6) indicates that the collector bias current, which determines the transconductance (g_m) and the transit frequency (f_T) of the device, must be selected in conjunction with the device size in order to minimize the transistor noise figure. As the frequency increases, the factor "h" also increases,

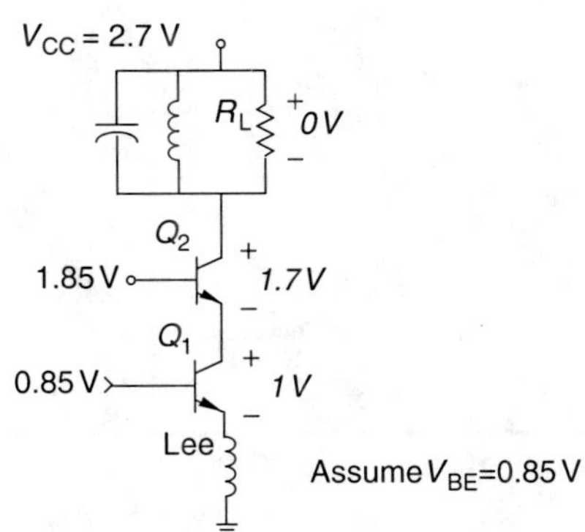

FIGURE 9.15 Single-ended equivalent circuit for a differential bipolar preamplifier showing bias voltages.

which results in an increase in the minimum noise factor as predicted by Equation 9.5 for a typical bipolar device. Thus, the dominant term in the noise factor for a common-emitter amplifier at high frequencies is

$$\text{Noise factor} = \alpha g_{\text{m}} \cdot r'_{\text{bb}} \left(\frac{f}{f_{\text{T}}} \right)^2 \tag{9.8}$$

This prediction is consistent with the dominant noise sources being collector current shot noise (since collector current density determines f_T) and the thermal noise generated by r'_{bb}. At a given frequency, temperature and bias current, only r'_{bb} and f_T in Equation 9.8 vary with emitter area. Thus, the device emitter area can be increased in order to reduce r'_{bb}, until the collector current shot noise causes an unacceptable degradation of the noise figure because of a reduction in the transistor f_T.

Example plots of the minimum noise figure for three emitter areas are plotted as a function of base current density in Figure 9.16. The flat portion of the noise figure curve (i.e., below 0.3 mA/μm) corresponds to the region where the noise figure is dominated by the extrinsic base resistance of the transistor. As the emitter area of the BJT increases, the base resistance decreases, however, the collector current also increases for a given current density and so the minimum noise figure remains almost constant (as predicted by Equation 9.5 and Equation 9.6). At large collector currents the f_T of the transistor begins to decrease, causing a rise in the minimum noise figure. Note that the minimum noise figure is almost independent of the emitter area selected, and that there is a broad range of bias points over which the noise figure of a typical common emitter amplifier can be minimized.

The noise figure is also frequency dependent. As the operating frequency begins to approach the device f_T, the gain of the transistor begins to decrease and hence the collector current shot noise begins to dominate. For a constant collector current, the transistor transit frequency is lowest for larger area devices, and therefore the noise figure of a large area transistor will increase more rapidly with frequency than for smaller transistor designs.

Since the relative contribution of the transistor noise sources to the overall signal-to-noise ratio depends upon the source impedance seen at the transistor input terminal, there is an optimum source impedance which results in the lowest noise figure. In general, this optimum noise match is not equal to the conjugate of the transistor input impedance required for maximum power transfer. An inductor placed in series with the emitter lead of the BJT modifies the optimum noise match, and under certain conditions the minimum noise figure and the maximum power transfer at the input can be achieved simultaneously, making this approach very attractive [18]. However, the amount of feedback which can

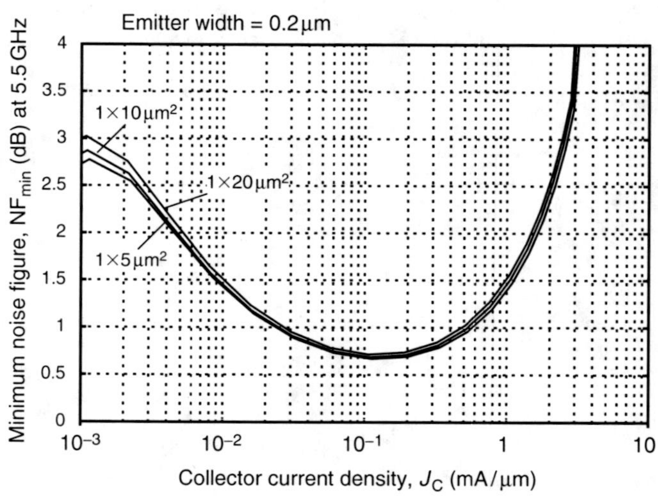

FIGURE 9.16 NF_{min} versus current density for various emitter lengths at $V_{CE} = 1.5\,V$, $f = 5.5\,GHz$.

be applied to the preamplifier is limited by the constraints of gain and power consumption, and in most cases only a small amount of feedback can be applied [19].

Examples of impedance matching networks are shown in Figure 9.17. The input is matched by adding inductance in series with the transistor input. At the output, a tapped capacitor impedance transformer matches a low impedance (e.g., 50 Ω) load to the collector impedance.

A possible physical layout for the tapped capacitor is also shown in Figure 9.17, where metal–insulator–metal capacitors are implemented using the multiple layers of metal in a VLSI technology. Note that the bottom plate of capacitor C_B shields the tapped capacitor from the conductive substrate.

Transformer Feedback Amplifiers

As the supply voltage of digital circuitry shrinks with technology scaling, RF circuit topologies require sub-1 V operation. This is because integration of analog–RF and digital circuitry on the same die is desirable from both cost and packaging considerations. In addition, as operating frequencies increase, amplifier designers can no longer neglect the effects of the collector–base feedback capacitance C_{bc}, on performance. Feedback via C_{bc} is reduced using a cascode configuration, however, a two-transistor stack is not optimal for operation at the lowest possible supply voltage. The transformer feedback amplifier employs reactive negative feedback through an on-chip transformer to neutralize C_{bc}, while also allowing a collector bias voltage equal to the supply voltage (i.e., $V_{CE} = V_{CC}$). As a result, gain and dynamic range are not compromised when only a single active device is used.

Circuit techniques that mitigate the effect of C_{bc} are usually grouped into two categories: unilateralization and neutralization. Unilateralization decreases reverse signal flow and thus coupling between output and input ports of an amplifier. Neutralization cancels signal flow through C_{bc} by adding signal paths around the amplifier that cancel signal flow via C_{bc}. This technique increases the forward gain and reverse isolation for a given power consumption but does not necessarily reduce the effect of C_{bc} on the input capacitance.

An alternative approach to neutralization uses a feedback transformer, which introduces magnetic coupling between collector and emitter inductors of a common-emitter transistor amplifier as shown in Figure 9.18 [20,21]. Feeding back a portion of the output signal via the transformer can effectively cancel the feedback from output to input through the Miller capacitance (C_{bc}) and neutralize the amplifier. This increases the amplifier gain for a given bias current and improves the isolation between output and input. The circuit parameters that define this condition are

$$\frac{n}{k} \approx \frac{C_{be}}{C_{bc}} \tag{9.9}$$

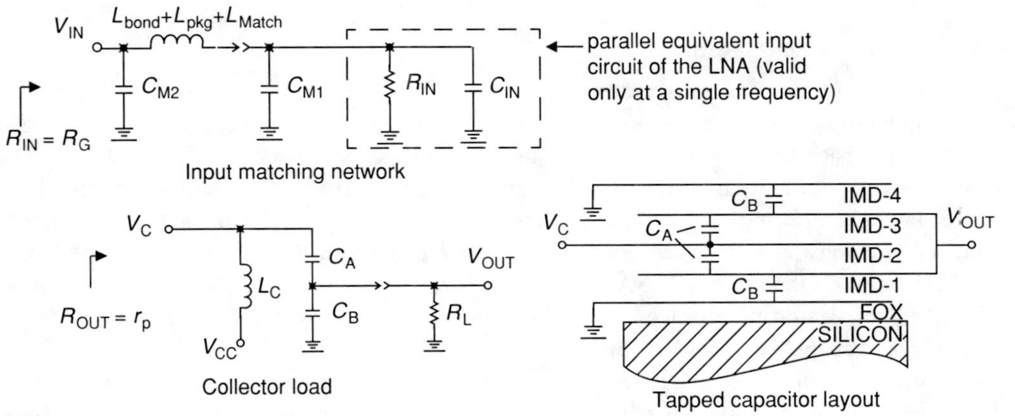

FIGURE 9.17 LNA impedance matching (single-ended equivalent networks).

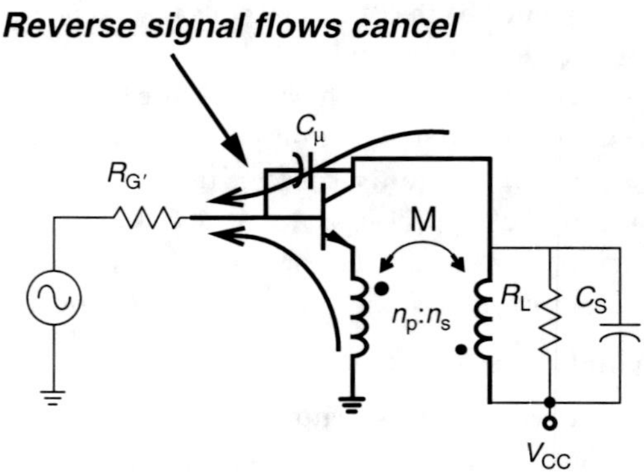

FIGURE 9.18 A transformer feedback amplifier.

using nonideal magnetic coupling in the transformer. Therefore, neutralization is achieved when the effective transformer turns ratio n/k is set equal to capacitance ratio C_{be}/C_{bc}. Note that there is no frequency dependence in the neutralization condition, implying that transformer-feedback can be used as a wide-bandwidth neutralization technique restricted only by the bandwidth of the transformer. For a given LNA design, the transformer turns ratio (n) is often constrained by linearity, gain, and noise specifications. In these cases, the coupling coefficient (k) is the extra degree of freedom that can be adjusted to achieve amplifier neutralization. This can be accomplished by adjusting the spacing between the transformer primary and secondary windings.

An example of a transformer feedback amplifier fabricated in a 0.5 μm SiGe bipolar technology is shown in Figure 9.19 [22]. The 2.4 GHz LNA draws 2.5 mA from a 0.9 V supply. The step-up ratio between primary and secondary of the transformer is realized by sectioning one winding (e.g., the primary) into a number of single turns rather than one continuous winding. These single-turn windings are then connected in parallel to form the step-up ratio between primary and secondary of the transformer. The 1:4 step-up design shown in Figure 9.19 consists of eight turns of 10-μm wide topmetal with a 3 μm conductor spacing, and measures 350 μm on each side. The step-up transformer is an almost ideal feedback element for an RF amplifier, and can be used as a narrowband alternative to a broadband, resistive network. The impedance match for the LNA input is off-chip so that either 50 Ω or minimum noise figure matching between the source and the amplifier input could be selected. When matched for optimal noise performance, the measured preamplifier noise figure is 0.95 dB at a gain of 10.5 dB (biased at 2.5 mA from 0.9 V). When a 50 Ω impedance match is used, the gain rises to 11 dB at a noise figure of 1.75 dB. At the same bias point, the third-order intercept point (IIP_3) of the preamplifier is −4.5 dBm for both matching situations. This combined performance (i.e., noise figure, gain, power consumption, and IIP_3) is excellent, regardless of the technology.

LNA Comparison and Summary

A performance summary of some recently reported LNAs fabricated in SiGe are compared to representative examples from other technologies in Table 9.2. The first entry in the table is an early example of a transformer-feedback amplifier as described in the previous section. This design is a demonstration of very low current consumption from a sub-1 V supply that can realize a noise figure of 0.95 dB. The gain is limited by the transformer step-up ratio and the circuit was not optimized for IM3 performance. The second entry demonstrates the benefits of eliminating low frequency at second-order harmonic feedback at the LNA input on the IP3 of an LNA [23].

The amplifier demonstrates an excellent third-order intercept point (IIP_3) of +8 dBm for PCS telephony using a single 0.5 μm SiGe BJT (i.e., no cascode). The low-frequency signal path at the RF input is via an external choke to an on-chip dc bias circuit. The third entry in Table 9.2 is a 0.25 μm SiGe bipolar LNA that consumes a comparable amount of DC power to the single-ended LNA from Ref. [23], but includes automatic gain control (AGC) in a fully differential circuit [24]. Two recently reported LNAs

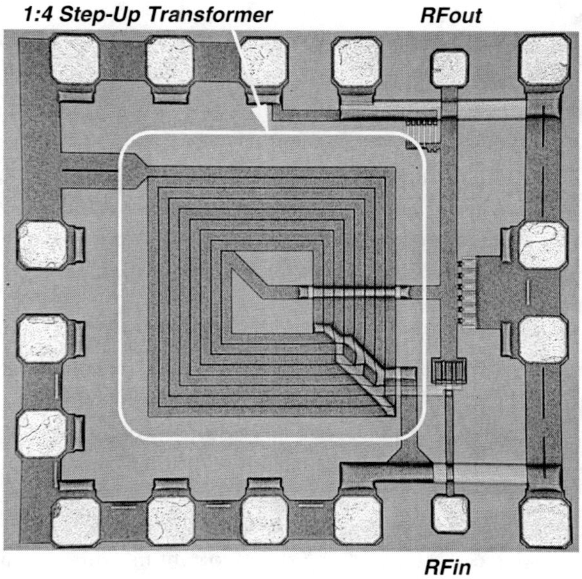

FIGURE 9.19 A 2.4 GHz transformer feedback amplifier in 0.5 μm SiGe bipolar technology. (After JR Long, M Copeland, S Kovacic, D Malhi, and D Harame. Proceedings of the International Solid-State Circuits Conference, 1996, pp. 82–83. With permission.)

TABLE 9.2 LNA Comparison

Frequency in GHz	Gain in dB	NF in dB	IIP_3 in dBm	V_{CC} in Volts	P_D in mW	Notes	Technology
2.4	10.5	0.95	−4	0.9	2.25	Transformer feedback	0.5 μm SiGe bipolar [22]
1.96	15.4	1.6	8.3	3	23	c-e with harmonic termination	0.5 μm SiGe-BiCMOS [23]
2	14.5	1.6	5	3	30	Differential with AGC	0.25 μm SiGe-BiCMOS [24]
17.2	17.5	2.4	−0.5	2.2	10.6	Differential cascode	0.18 μm SiGe-BiCMOS [25]
61.5	17	4.2	−8.5	1.8	10.8	c-b/c-b cascode	0.13 μm SiGe-BiCMOS [26]
2.14	15	1	7.3	3	25.5	LNA + bypass sw.	0.5 μm PHEMT [27]
1.57	16.5	1.3	−5	1.5	9	CMOS cascode	0.25 μm CMOS [28]

Sources: JR Long, M Copeland, S Kovacic, D Malhi, and D Harame. Proceedings of the International Solid-State Circuits Conference, 1996, pp. 82–83. V Aparin, P Gazzerro, J Shou, B Sun, S Szabo, E Zeisel, T Segoria, S Ciccarelli, C Persico, C Narathong, and R Sridhara. IEEE International Solid-State Circuits Conference, 2002, pp. 234–235. M Tanabe, T Fukuda, and K Nishii. IEEE Radio Frequency Integrated Circuits Symposium, 2002, pp. 209–212. SY Yue, D Ma, JR Long, and DL Harame. IEEE International Solid-State Circuits Conference, 2004, pp. 388–399. S Reynolds, B Floyd, U Pfeiffer, and T Zwick. IEEE International Solid-State Circuits Conference, 2004, pp. 442–443. S Kumar, M Vice, H Morkner, and W Lam. Proceedings of the IEEE International Microwave Symposium, 2003, pp. 1577–1580. P Leroux, M Steyaert, V Vassilev, and H Maes. Proceeding of ESSCIRC, 2002, pp. 335–338. With permission.

designed for millimeter-wave frequencies follow Refs. [25,26], which demonstrate the high operating frequencies enabled by the latest deep submicron SiGe-BiCMOS technologies. The final two entries in the table list recently reported data for LNAs designed around a III–V PHEMT [27] and an n-type MOSFET from a 0.25 μm CMOS technology [28]. In comparison, the SiGe BJT offers comparable RF performance (i.e., noise figure and linearity) to the PHEMT LNA and it allows greater integration of RF and baseband functions in a lower cost BiCMOS technology. Also, the SiGe LNAs enable operation at higher frequencies or with lower power consumption than state-of-the-art CMOS alternatives.

9.4 RF Mixers

The RF mixer is one of the most important and, from a theoretical standpoint, poorly understood components in radio engineering. The nonlinear nature of the mixing process makes analysis of some important performance aspects of the mixer, such as the signal-to-noise ratio, a formidable task. Consequently, discrete mixer circuits have primarily evolved through experimental refinement, rather than a design evolution based upon a solid theoretical foundation and engineering principles. The factors that influence the noise figure of a single diode mixer, for example, have been known since the early days of radio. However, a consistent theoretical analysis of single-diode mixer noise was not published until 1978.

The mixer in a radio receiver is an analog multiplier, which downconverts the received signal from the RF band to an IF, or directly to the baseband for demodulation and detection (the mixer in a transmitter up converts a signal from baseband or IF to RF). The effectiveness of the mixer as a frequency converter is measured by the degradation in signal-to-noise ratio and distortion of the input signal. The impedances at the mixer ports must be well defined and controlled, because in a typical radio implementation, the mixer interfaces with off-chip signals at all three I/O ports. Isolation of large signals (such as the LO) from the smaller signals at the other mixer ports, such as the (potentially) tiny received signal at the receive mixer RF input, must also be considered. This is particularly important in a direct conversion receiver design.

The range of the input signal from the minimum detectable level to the overload point defines the dynamic range. These limitations define the performance capabilities of the receiver to a large extent, although noise introduced by the mixer in a receiver is normally suppressed by a low-noise preamplifier. Aside from these two main performance specifications, other important mixer characteristics are: the port-to-port signal isolation, the port impedances as a function of frequency, and the mixer conversion gain.

An ideal downconverting mixer in a radio receiver produces an IF output signal only at the sum and difference frequencies between the RF and the LO input signals. However, in a practical circuit there are a number of spurious components that are generated by the mixing process, some of which may be suppressed through the use of balanced circuitry. Also, the circuit complexity and the minimum noise figure, which can be realized for a given topology, are related. An elaborate circuit topology that places more active and passive components and their associated noise sources in the RF signal path will usually have a higher noise figure.

The most widely used mixer in silicon technology is Gilbert's four-quadrant multiplier (shown in Figure 9.20). The RF input is converted to in-phase and anti-phase signal currents by the differential transistor pair, Q_1 and Q_2, and fed to the upper four transistors, Q_3 to Q_6 (often referred to as the switching quad), which are driven by the LO signal. Differential to single-ended conversion is normally required at the mixer output because most passive filters are single-ended, although balanced filters are available. The outputs of the mixer can be buffered to a low impedance load (e.g., 50 Ω) by a single-ended output buffer circuit, or the differential signal can be combined by a passive power combining circuit to drive the following stages. The emitter degeneration resistor, R_{ee}, improves the linear range of the differential input amplifier and it is connected so that no dc bias current flows through the resistor. This is done to ensure that the voltage drop across all components in the bias current path between the power supply and ground is as large as possible. Even so, the absolute minimum supply voltage (V_{CC}) for this circuit is approximately 2.7 V.

In order to minimize the noise contributed by the upper four transistors, the LO input is normally driven with a large amplitude signal. This minimizes the period of time when all four transistors in the quad are in the active region and contribute noise to the IF output. Ideally, the large LO drive signal (i.e., $v_{LO} > V_T$) switches the collector current in each transistor quickly between αI_{EE} and zero. When this assumption is made, the differential output current can be modeled as the input signal current at the collectors of Q_1 and Q_2 multiplied by a switching function which alternates between $+\alpha I_{EE}$ and $-\alpha I_{EE}$ at the LO frequency (ω_{LO}).

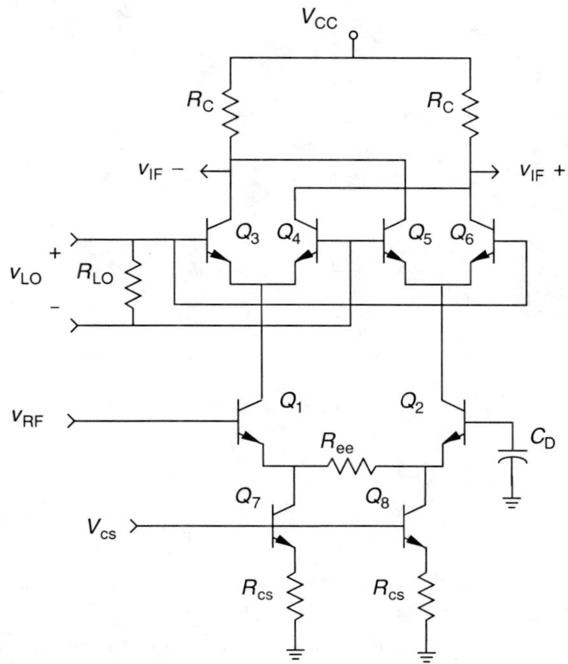

FIGURE 9.20 An IC mixer based upon the Gilbert multiplier.

Many of the limitations of the Gilbert multiplier as an RF mixer arise from design constraints imposed by the input differential pair. The linear range of the mixer, for example, is determined by the linearity of the input pair. This is approximately equal to the input voltage required to force all the bias current through one of the two transistors, either Q_1 or Q_2. It can be shown that this voltage is in the order of the bias current–emitter degeneration resistance product, and hence the linear range of the mixer can be improved by increasing the degeneration at the emitter of the input pair. This could be accomplished by either increasing the total bias current from Q_7 and Q_8, or by increasing R_{ee}. However, a large bias current increases both the power dissipation and the noise from transistor shot noise sources, and an increase in the emitter degeneration resistance also degrades the signal-to-noise ratio at the IF output. Thus, using emitter degeneration to reduce distortion will increase the noise figure of the Gilbert multiplier and as a result, realizing a dynamic range suitable for radio receiver applications has proven difficult using this circuit.

The linearity of the input pair (which acts as a preamplifier for the mixing quad) can also be improved by removing the low frequency and second harmonic of the RF signal using passive circuits in a manner similar to the technique described in the "Low Noise Amplifier" section.

Another disadvantage of the Gilbert multiplier as a high-frequency mixer is the relatively high input impedance of the differential input pair of transistors at RF. When the mixer RF input is driven from another circuit on the same chip, impedance matching is not an issue. When driven by an off-chip source, an impedance match is necessary at the mixer RF input, and off-chip components are required for the best performance (i.e., low dissipation and low tolerance). This increases the receiver parts count and makes an integrated solution less attractive in terms of cost and performance when compared to a discrete implementation, where components can be selected for optimum performance and lowest cost.

A Class-AB input stage [29,30] of Figure 9.21 has been developed to lower the distortion produced by an IC mixer for large-signal inputs, since it is the linearity of the mixer that normally defines the overall linearity in a properly designed RF receiver. This consists of a common-base amplifier Q_1 and current mirror Q_2/Q_3. Small-signal RF input signals are buffered to one pair of mixing transistors in the Gilbert

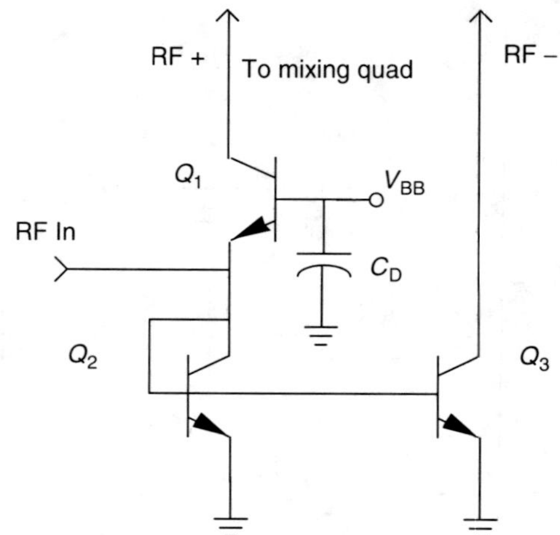

FIGURE 9.21 Class-AB mixer input stage.

quad by the common-base stage, while the mirror inverts the input signal and couples it to the second pair of mixing devices in the quad. For large input signal amplitudes, the current mirror–common-base combination operates in Class B mode, where the signal current in the common-base or current mirror stages increases in response to negative and positive excursions of the input signal, respectively. Diode connected transistor Q_2 is required to bias transistor Q_3, but because it is in the RF signal path it degrades the signal-to-noise ratio and noise figure of the mixer.

Transformer Balun-Coupled Mixer

The transformer-coupled mixer also offers an improvement in performance when compared to the Gilbert multiplier [20]. The input differential pair of the Gilbert multiplier acts like an "active" balun circuit, because it splits the RF input signal into two components, one in-phase and the other anti-phase to the RF input signal. An alternate method of deriving differential signal currents uses a monolithic transformer balun to split the RF input signal into differential components, which are then fed to the switching quad of transistors as shown in Figure 9.22. This circuit configuration retains many of the benefits of the Gilbert multiplier, such as the balanced topology and compatibility with silicon monolithic circuit technology, while allowing the designer the flexibility to design a mixer that has a wider dynamic range and operates with lower power consumption.

A balanced circuit is required to obtain good port-to-port isolation and rejection of spurious signals. The switching quad from the Gilbert multiplier is well suited to monolithic integration. Transformer coupling of the RF input signal retains the advantages of a doubly balanced topology while improving upon the performance of the active circuit used to generate differential RF signal currents in Gilbert's multiplier. The RF input signal to the mixer (refer to Figure 9.22) is split into in-phase and anti-phase components by balun T_1. These signals are then fed to the cross-coupled switching quad of transistors, Q_1 to Q_4. Bias current is fed from current source Q_5 to the switching quad through the center-tap in the balun secondary. The signal current is chopped by the transistor quad at the LO rate in order to downconvert the input signal from RF to the desired IF. Package and bondwire parasitics have a relatively small influence on the IF port matching and a good impedance match can be achieved through the proper selection of the collector load resistance, R_C.

Approximately, ideal transformer behavior can be assumed as the balun primary and secondary windings are resonant tuned. The LO inputs of the switching quad are driven with a large amplitude signal, and therefore, two of the transistors in the quad are biased in the active region, and the other transistors in the quad are cutoff for a large portion of each cycle (e.g., when LO+ is much larger than

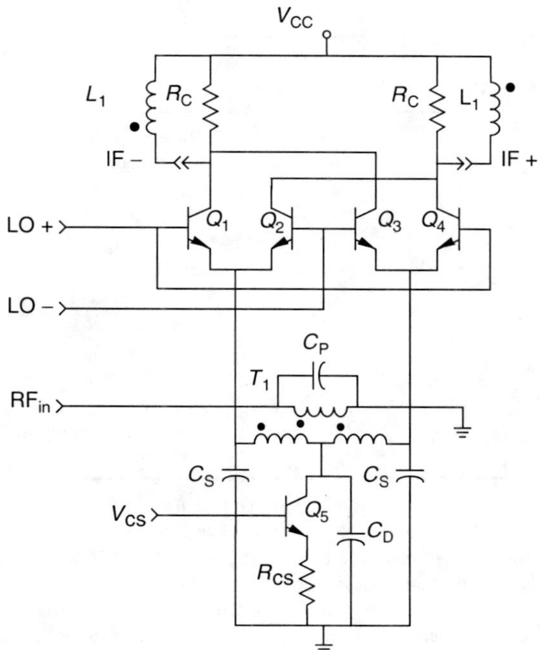

FIGURE 9.22 Balun-coupled mixer.

LO−, Q_1 and Q_4 are "on" and Q_2 and Q_3 are cutoff in Figure 9.22). The transistors biased in the active mode operate in the common-base configuration and amplify each phase of the received signal to the intermediate frequency output. The transformed source resistance (r_{SRC}) degenerates the common-base amplifier and extends its linear range of operation. The transformer has been used advantageously here to match the source impedance to the mixer and improve the mixer linearity without causing a significant increase in the overall noise figure. This occurs because no additional dissipation has been added to the circuit other than the losses in the transformer windings, which are relatively small. Linearization of conventional IC mixers, such as the Gilbert-type balanced demodulator, will require the addition of degeneration resistance which causes a large degradation in the mixer noise figure when high linearity is desired.

The noise introduced by the mixing process is difficult to determine analytically, but can be simulated. There is some degradation caused by the dominant transistor noise sources of the common-base amplifier, which are collector current shot noise and thermally generated noise from the extrinsic base resistance. Operating the mixer at a low bias current reduces the shot noise contributed by each active device. However, the switching speed of the transistors in the quad is also important when attempting to realize a lower noise figure. When the LO inputs are close to the same potential, very little of the signal at the mixer RF input appears at the IF output, because of the balanced circuit connection. All four transistors in the quad are forward biased in this condition and contribute noise to the IF output. Thus, the signal-to-noise ratio at the IF output is very low during the switching interval, and fast switching of the transistor quad is needed to reduce this portion of each LO cycle. Switching speed is not the only consideration because of the trade-off between emitter area and the extrinsic base resistance of the bipolar transistor. Careful selection of the emitter area for the transistors in the switching quad is therefore required in order to achieve a good mixer noise figure. A compromise is needed between a small transistor that can switch quickly between states and a larger transistor with less thermally generated noise from the transistor extrinsic base resistance, r'_{bb}

TABLE 9.3 Mixer Comparison

RF Input in GHz	Gain in dB	SSB NF in dB	IIP_3 in dBm	V_{CC} in Volts	P_D in mW	Notes	Technology
2.4	8	9.5	4	1.2	6	Balun-coupled	0.5 μm SiGe bipolar [22]
1.96	13.4	7.1	4.7	3	63.3	Single-balanced mixer with harmonic term, including LO buffer	0.5 μm SiGe-BiCMOS [23]
0.9	2	12	9	3	12	Class-AB input	0.4 μm Si-bipolar [31]
17.1	6.0	18	3.6	2.2	19.4	Double-balanced image-reject	0.18 μm SiGe-BiCMOS [25]
62	9	13	—	2.7	19.2	Single-balanced	0.13 μm SiGe-BiCMOS [26]
2.14	12	7	0	3	24	Gilbert mixer including LO buffer	0.5 μm PHEMT [27]
1.9	6.5	8.5	−3	3	39.3	Gilbert mixer including LO buffer	0.8 μm CMOS [32]

Sources: JR Long, M Copeland, S Kovacic, D Malhi, and D Harame. Proceedings of the International Solid-State Circuits Conference, 1996, pp. 82–83. V Aparin, P Gazzerro, J Shou, B Sun, S Szabo, E Zeisel, T Segoria, S Ciccarelli, C Persico, C Narathong, and R Sridhara. IEEE International Solid-State Circuits Conference, 2002, pp. 234–235. SY Yue, D Ma, JR Long, and DL Harame. IEEE International Solid-State Circuits Conference, 2004, pp. 388–399. S Reynolds, B Floyd, U Pfeiffer, and T Zwick. IEEE International Solid-State Circuits Conference, 2004, pp. 442–443. S Kumar, M Vice, H Morkner, and W Lam. Proceedings of the IEEE International Microwave Symposium, 2003, pp. 1577–1580. J Durec. *IEEE Journal of Solid-State Circuits* 33:1352–1372, 1998. PJ Sullivan, BA Xavier, and WH Ku. *IEEE Journal of Solid-State Circuits* 32:1151–1155, 1997. With permission.

Mixer Comparison and Summary

A summary of recently reported mixers fabricated in SiGe are compared to representative examples from other technologies in Table 9.3. The first entry in the table is a transformer balun-coupled double-balanced mixer as described in the previous section. This design demonstrates low voltage operation and low current consumption (5 mA from a 1.2 V supply). The supply voltage could be pushed below 1 V if the bias current source (Q_5 from Figure 9.22) were replaced by a current limiting resistor. The 4:5 balun step-up ratio allows a 50 Ω match to the RF input. The second entry in Table 9.3 demonstrates the improvement in IP3 that can be realized by eliminating low-frequency energy and second-order harmonic feedback at the input to a single-balanced multiplier [23]. The mixer demonstrates a high input third-order intercept point (IIP_3) with both high conversion gain (13.4 dB) and low noise figure (50 Ω SSB NF of 7.1 dB) in a 0.5 μm SiGe-BiCMOS technology. The third entry in Table 9.3 is a 0.4 μm silicon bipolar mixer with a Class-AB input stage [31]. This circuit offers high linearity at low dc bias current (+9 dBm IIP3 at 4 mA), but at the cost of noise figure. It is likely that a SiGe implementation would be capable of similar performance but with greater operating bandwidth. The fourth entry in the table is a full image-reject mixer (i.e., two double-balanced mixers driven by quadrature LOs) operating at 17.1 GHz. It demonstrates the high operating bandwidth offered by deep submicron SiGe-BiCMOS processes and low power consumption. The following entry is also a millimeter-wave mixer, but operating in the 60 GHz band proposed for IEEE 802.16 wireless communication. This single-balanced circuit was designed in a 0.13 μm BiCMOS technology. The final two entries in the table list data for mixers designed in a 0.5 μm GaAs PHEMT [27] and a 0.8 μm CMOS technologies [32]. SiGe BJT mixers offer comparable RF performance to their GaAs counterparts, but in a lower cost technology with large-scale integration possibilities. SiGe mixers are also competitive with CMOS designs, but offer lower current consumption or greater operating bandwidth at a given technology node.

9.5 Voltage-Controlled Oscillators

Oscillators are indispensable to electronic systems. They provide carrier signals for modulation in analog communications, and in a digital system, bit streams must be sampled or synchronized with some reference timing that originates from a stable, locally generated source.

Determination of the oscillation amplitude and frequency for weakly nonlinear oscillators has been considered by many researchers; van der Pol was probably the first to publish work in this area and many others have followed. The classical approaches to this problem start from a basic assumption about the nonlinear transfer characteristic within the active element and assume some filtering in the loop. The basic problem in oscillator design, however, is not the solution of the oscillation condition itself but characterization of the nonlinearities within the loop. Nonlinear CAD tools have been developed, which can predict the oscillation amplitude and the random fluctuations in the output phase (i.e., phase noise). This has given researchers the ability to identify the sources of phase noise degradation in the most popular circuit topologies and to develop modifications that improve performance.

The RF power spectrum is a common method of characterizing oscillator performance in the frequency domain. A plot of rms power versus frequency, such as seen on a spectrum analyzer, is a measure of phase noise when the noise introduced by amplitude modulation of the signal measured is insignificant. The spectrum is symmetric about the carrier frequency. Examining one side of this spectrum, the ratio of the single-sideband (SSB) to the carrier power spectral density at a frequency offset of f_d Hz away from the carrier, or $L(f_d)$, is used as a measure of the phase noise. The importance of the oscillator phase noise in a wireless transceiver application is illustrated in Figure 9.23. The received RF signal strength in the desired channel and the maximum signal strengths in the adjacent channels are shown in part (a) of the figure. The oscillator spectral characteristics are also shown, where the phase noise density of the oscillator must be well below the peak at a frequency offset equal to the channel spacing, f_{ch}. The down converted spectrum is illustrated as two parts in Figure 9.23b. One component is the "desired" block of signals in the lower part of the diagram, which are down converted by an (idealized) sinusoidal component at the nominal LO frequency, f_{LO}. The second or "interfering" block of signals arise from the down conversion of the RF signal spectrum by a sinusoidal component of the oscillator spectrum that is offset from the main carrier lobe by the channel spacing. The amplitude of the interference at the IF is directly proportional to the phase noise of the LO, and as a consequence, the level of adjacent channel interference increases as the phase noise of the LO increases. On the transmit side, the phase noise introduced by the LO used to up-convert a signal for transmission is a potential source of interference in the system.

The channel spacing in many cellular telephone systems is on the order of kHz, and consequently the demands on the spectral purity of the oscillator are much greater than in system such as cordless telephones where the channel spacing is on the order of 1 MHz. For example, in the American IS-54 system, the oscillator phase noise spectral density requirement is approximately −115 dBc/Hz at a 60 kHz offset from the carrier (the channel bandwidth in IS-54 is 30 kHz).

In an oscillator, the phase noise produced by the gain element (i.e., an amplifier) is passed through a frequency selective circuit or resonator (with quality factor Q), that acts as a bandpass filter to reject undesired noise components and feeds the desired frequency component from the amplifier output back to the input. For fluctuations in the output frequency measured with respect to the carrier frequency, the transfer function of the tuned circuit can be described by a lowpass filter response, $H(f_d)$, where

$$H(f_d) = \frac{1}{1 + \dfrac{j2Qf_d}{f_0}} \tag{9.10}$$

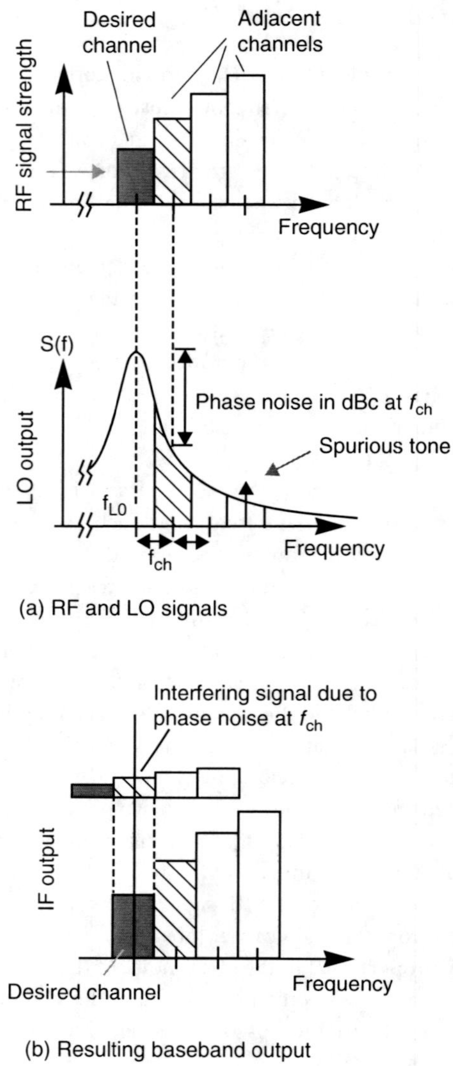

FIGURE 9.23 Effect of oscillator phase noise on down converted signal in a receiver.

$f_0/2Q$ is the half-bandwidth of the bandpass filter with quality factor Q, and $f_d = f - f_0$, which is the frequency deviation with respect to the carrier at f_0

Equation 9.10 describes the transfer function of the feedback network. The amplitude of the spectral components in the amplifier output that are produced by phase fluctuations at the input (i.e., the phase modulated sidebands) will pass unattenuated up to the half-bandwidth of the tuned circuit. The phase response of the resonator is given by

$$\theta(f_d) = -a\tan\left(\frac{2Qf_d}{f_0}\right) = \frac{-2Qf_d}{f_0} \quad \text{for small } f_d \tag{9.11}$$

Equation 9.11 shows that if the frequency deviation is small, the phase of the resonator output is linearly related to the frequency deviation, f_d. If the deviation is large, then the output phase approaches a constant value, which is independent of the frequency deviation.

The open-loop phase noise spectrum, $\Delta\phi_{in}(f_d)$, is produced by the effect of circuit noise sources on the amplifier input signal. The equivalent low-pass transfer function for the resonator ($H(f_d)$ from

Equation 9.10) is used to model the feedback network effects on the oscillator phase noise. Most of the oscillator noise close to the carrier in a high-quality oscillator is phase or FM noise, because the limiting mechanisms (e.g., an AGC circuit to control the amplifier gain) tend to eliminate AM noise. The closed-loop response of the phase feedback loop is given by

$$\Delta\phi_{\text{out}}(f_{\text{d}}) = \left[1 + \frac{jf_0}{2Qf_{\text{d}}}\right](\Delta\phi_{\text{in}}(f_{\text{d}})) \tag{9.12}$$

Equation 9.12 describes the phase modulation due to noise (or phase noise) at the amplifier output (output frequency f_{o}). The phase noise at the input to the amplifier ($\Delta\phi_{\text{in}}$) is enhanced by positive feedback because for small frequency deviations (i.e., small f_{d}) the output phase noise will be greater than the input phase noise. Or, as described by Equation 9.11, feedback through the resonator forces a phase change at the amplifier input (or resonator output) to be linearly related within the resonator bandwidth. Outside of this bandwidth, the feedback through the resonator approaches zero as f_{d} increases, and the output phase noise ($\Delta\phi_{\text{out}}$) is the same as the input phase noise ($\Delta\phi_{\text{in}}$) because the amplifier gain is unity.

A simplified model of oscillator phase noise called Leeson's model [33] predicts the spectral characteristics of a feedback oscillator in terms of known circuit parameters (the equivalent input noise of the simplify, the tuned circuit *Q*-factor, etc.) when a high *Q* tank is used. Leeson's model predicts that the single-sideband phase noise spectral density at frequency offset f_{d} ($L(f_{\text{d}})$) is given by the following equation:

$$L(f_\Delta) = \frac{1}{2}\left[1 + \left(\frac{f_{\text{o}}}{2f_\Delta}\right)^2\left(\frac{1}{Q_{\text{loaded}}}\right)^2\right]\left(1 + \frac{f_{\text{c}}}{f_\Delta}\right)\frac{FkTB}{P_{\text{sig}}} \tag{9.13}$$

where Q_{loaded} is the loaded *Q*-factor of the resonator. The first term (in square brackets) shows the strong effect of resonator *Q*-factor and the offset from the carrier frequency on the phase noise spectrum, the second term is due to the flicker noise sources in active devices (given by flicker corner frequency, f_{c}) and the last term is the ratio of thermal noise (gain block noise factor *F*, Boltzmann's constant *k*, noise bandwidth *B*, and temperature *T*) to output signal power (rms power P_{sig}). This equation predicts the four major causes of oscillator noise: upconverted $1/f$ noise or flicker FM noise, the thermal FM noise, the flicker phase noise, and the thermal noise floor.

Equation 9.13 has been confirmed by experimental measurements on high-quality microwave signal sources. Leeson's model predicts that oscillator phase noise is minimized by:

1. Minimizing the bandwidth of the feedback network (i.e., maximize loaded resonator *Q*)
2. Avoiding hard nonlinearity and saturation of the BJT. Nonlinearity in the oscillator increases the amplitude of noise components at frequencies close to the carrier, thereby increasing the phase modulation of the input signal ($\Delta\phi_{\text{in}}$)
3. The signal voltage relative to the equivalent noise voltage should be made as high as possible to reduce the phase modulation caused by additive noise sources
4. Choose active devices that have low flicker ($1/f$) noise

Highly integrated oscillators suffer from low-quality passive components compared to passive (off-chip) filtering technologies such as surface and bulk acoustic wave resonators. However, methods of predicting IC oscillator phase noise both quantitatively and qualitatively have been developed [34–38], and the important parameters that influence phase noise in the oscillator's output identified. These results confirm the guidelines 1 to 3 listed in the preceding paragraph derived heuristically by Leeson. However, it has also been shown that the effect of low-frequency noise sources such as flicker noise can be reduced substantially through the use of a fully symmetric circuit topology, such as the differential Colpitt's oscillator [39].

Resonant Tanks

Much of the work on oscillator development in monolithic circuit technologies has centered around the development of high-quality resonant tanks. This is understandable given the importance of the resonator Q on oscillator performance as indicated by Equation 9.13. The resonant tank in a VCO consists of a monolithic inductor in parallel with a capacitor that is electronically tunable (a varactor). The design and performance of these two components are outlined in the following sections.

Inductors

The peak Q-factor of an inductor fabricated on silicon has improved from approximately 5 in the early 1990s to 15 to 20 for an inductor in the 5 to 10 nH range. This progress is due to thicker metal films (e.g., 2 to 3 μm thick top metal) [40], differential drive [41], and substrate shielding [42]. Smaller inductance values (e.g., 1 to 3 nH) show inherently higher peak-Q, but realize their best performance above the frequency where Q peaks for a larger inductance. This implies a compromise between the operating frequency and performance factors related to tank Q, such as phase noise suppression in a VCO. Since the parallel equivalent impedance of a tank circuit at resonance is proportional to its inductance, multi-kΩ tank impedances are possible as the inductance L_p increases. Specifications like gain and power consumption are proportional to the tank impedance, so a large inductance is preferred from a low-power RF design perspective (other parameters being equal).

In a differential circuit implementation such as a typical monolithic VCO, a pair of spiral inductors are used in the physical layout (as shown in Figure 9.24a). Although the overall circuit may be differential, the excitation of each inductor is "single-ended." That is, one terminal of the spiral is

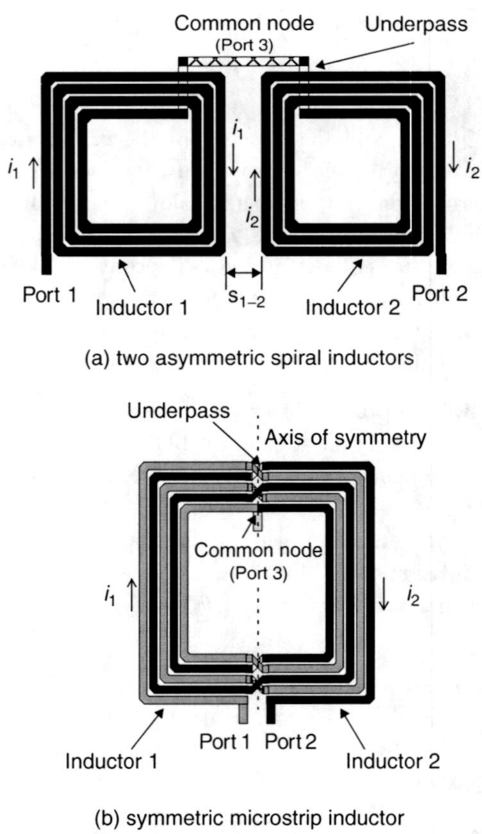

FIGURE 9.24 Microstrip inductor physical layouts for differential drive. (After M Danesh and JR Long. *IEEE Transactions on Microwave Theory and Techniques* 50:332–341, 2002. With permission.)

excited by an ac source while the other is connected to a common reference point (e.g., the supply voltage or ground). Note that signal currents associated with ports 1 and 2 (i.e., i_1 and i_2) flow in opposite directions, and hence some physical separation (s_{1-2}) is required to limit the negative mutual magnetic coupling between the two inductors.

The fully symmetric spiral inductor of Figure 9.24b is designed for differential excitation (i.e., voltages and currents at the terminals are 180° out of phase). When driven differentially, the voltages on adjacent conducting strips are anti-phase, however, current flows in the same direction along each adjacent conductor shown in Figure 9.24b (i.e., signal currents i_1 and i_2 flow in the same direction on any side). This reinforces the magnetic field produced by the parallel groups of conductors and increases the overall inductance per unit area.

The symmetric microstrip inductor is realized by joining groups of coupled microstrips from one side of an axis of symmetry to the other using a number of cross-over and cross-under connections. This style of winding was first applied to monolithic transformers for coupling both primary and secondary coils by Rabjohn [7]. One advantage of a fully symmetric layout is that the two separate spirals are replaced by a single coil, which has both electrical and geometric symmetry. This symmetry is important when locating the common node (a convenient bias point for active circuits), which separates the spiral into two inductances that have identical substrate parasitics at ports 1 and 2. As stated previously, a pair of asymmetric inductors must be spaced far enough apart to limit unwanted coupling (both magnetic and electric) between the inductor pair, which is not an issue for symmetric inductors. This is one of the reasons why a reduction in chip area results for the symmetric inductor. Also, the symmetric inductor is well suited for connection to active devices as the input terminals are on the same side of the structure.

Q-factor improvement resulting from differential drive can be estimated from the lumped-element equivalent circuit shown in Figure 9.25a. This equivalent circuit accurately models the electrical behavior of the inductor up to the first resonance frequency.

For differential excitation, the signal is applied between the terminals and the differential input impedance Z_d is the parallel combination of $2Z_P$ and Z_L. The substrate parasitics present higher equivalent shunt impedance in the differential case, and therefore Z_d approaches the value of Z_L over a wider range of frequencies than Z_{se}. At lower frequencies the input impedance in either the shunt or the differential connections is approximately the same, but as the frequency increases, substrate parasitics C_P and R_P come into play. For differential excitation, these parasitics have higher impedance at a given frequency than in the single-ended connection. This reduces the real part and increases the reactive component of the input impedance. Therefore, the inductor Q is improved when driven differentially, and the self-resonant frequency (or usable bandwidth of the inductor) increases due to the reduction in the effective parasitic capacitance from $C_P + C_O$ to $C_P/2 + C_O$.

From these simplified models, the ratio of differential to single-ended Q-factors is

$$\frac{Q_d}{Q_{se}} = \frac{2R_P \parallel R_L}{R_P \parallel R_L} \tag{9.14}$$

where and $R_L = r(1 + Q_L^2)$ for $Q_L = \omega L/r$. At low frequencies, $R_P \gg R_L$ and the two Q-factors are approximately the same. At lower frequencies, Q_L dominates in both cases and the Q-factor increases for increasing frequency. At higher frequencies, R_L is increasing (as $Q_L \propto f$) and R_P decreasing, so the differential Q-factor becomes larger than the single-ended Q. Eventually, R_P dominates the inductor dissipation and the Q-factor decreases with increasing frequency. The peak-Q occurs at a higher frequency when driven differentially due to the reduced effect of substrate parasitics in the differential case.

This analysis predicts Q improvement from differential excitation and that (ideally) the Q-factor can be doubled in the differential connection (i.e., when $R_P = R_L$ in Equation 9.14) with no modifications in IC technology or processing. The symmetric layout is also useful to preserve the balance desired in the differential implementations most often used on RF ICs.

A comparison between Q-factors from experimental measurements, three-dimensional electromagnetic simulation, and a lumped-element (SPICE) inductor model are shown in Figure 9.26. Good

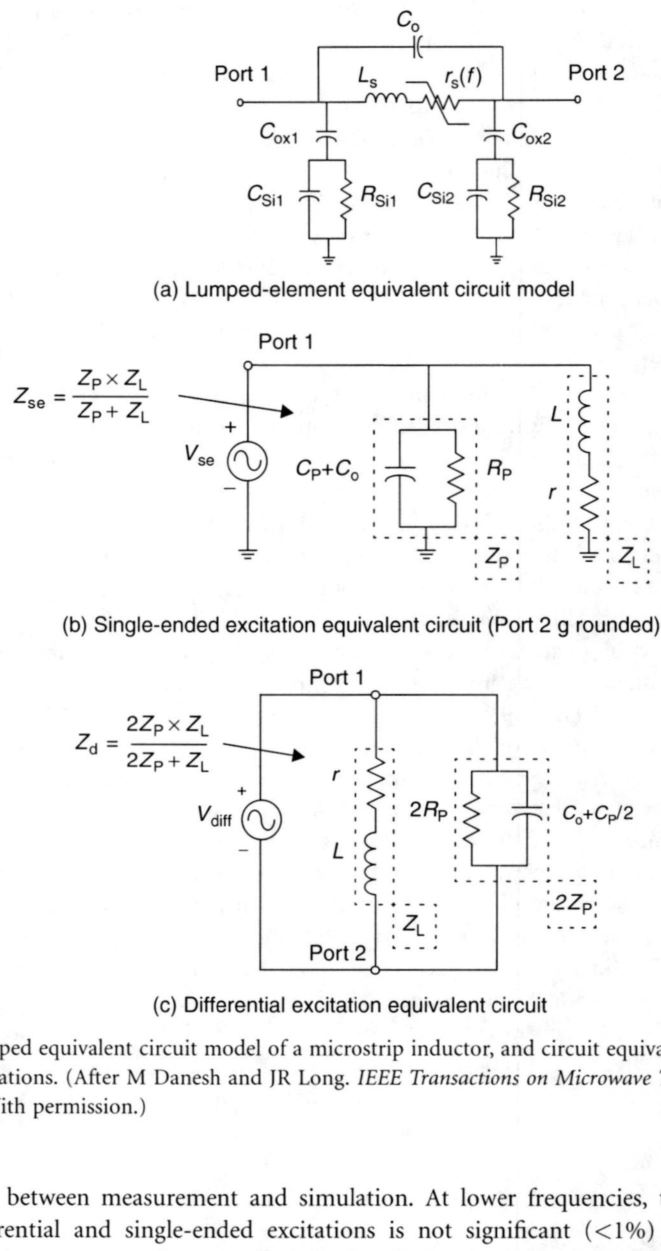

FIGURE 9.25 Lumped equivalent circuit model of a microstrip inductor, and circuit equivalents for single-ended and differential excitations. (After M Danesh and JR Long. *IEEE Transactions on Microwave Theory and Techniques* 50:332–341, 2002. With permission.)

agreement is seen between measurement and simulation. At lower frequencies, the difference in Q between the differential and single-ended excitations is not significant (<1%) because the shunt capacitive parasitic components do not affect the low-frequency input impedance. Hence, the two cases can be represented by a series L–r model. However, as the frequency increases, the difference between the input impedances becomes substantial and the difference between Q-factors in the differential and single-ended cases illustrates this point. The peak in the Q-factor is a result of the shunt parasitics as previously described. Lower parasitics for differential excitation result in a higher peak Q-factor and broadening of the Q peak compared with the single-ended (conventional) connection.

The Q of inductors in production VLSI technologies is primarily limited by the silicon substrate, which has a typical resistivity of 10 to 15 Ω-cm for mixed-signal and RF IC work. Substrate dissipation decreases when a high-resistivity substrate is used, or when a patterned shield connected to the on-chip ground is placed between the inductor and the substrate. A simplified illustration of this is shown in Figure 9.27a. The shield is designed to block current flow by magnetic induction but allows capacitive current to flow. The shield fingers also block the electric field from entering the underlying

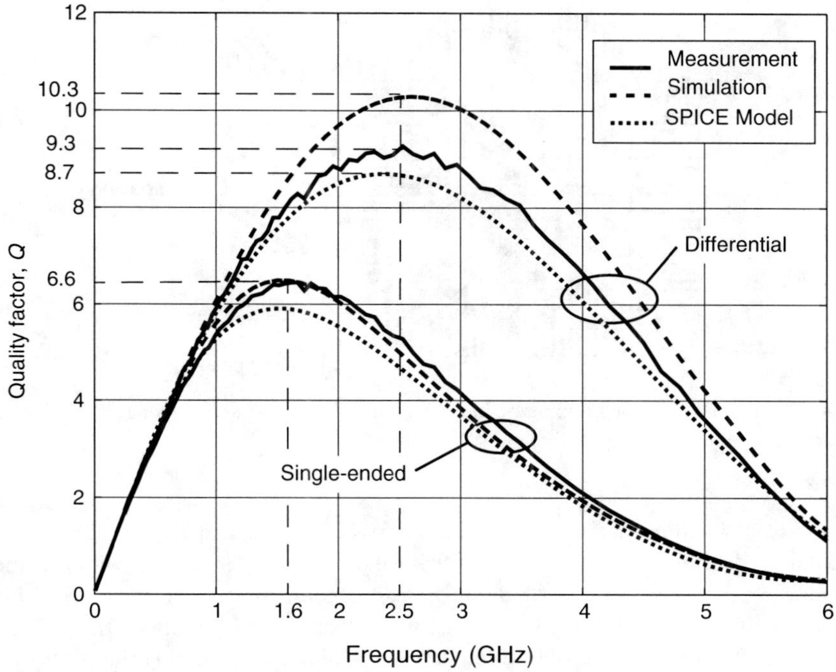

FIGURE 9.26 *Q*-factors for single-ended and differential excitations. (After M Danesh and JR Long. *IEEE Transactions on Microwave Theory and Techniques* 50:332–341, 2002. With permission.)

substrate. However, they add capacitance that reduces the inductor's self-resonant frequency. Gaps between the fingers block currents induced in the shield (limiting them to small eddy currents in each finger) but not in the ground line, which must be carefully designed. Any induced current decreases the total magnetic field, the inductance, and the *Q*.

Differential shielding does not require an explicit ground connection, leading to a simpler implementation. Figure 9.27b shows the differential shield implementation. A balanced transmission line is capacitively coupled to orthogonal metal strips placed beneath it. Due to the balanced excitation, a virtual ground on the strips shields the entire transmission line.

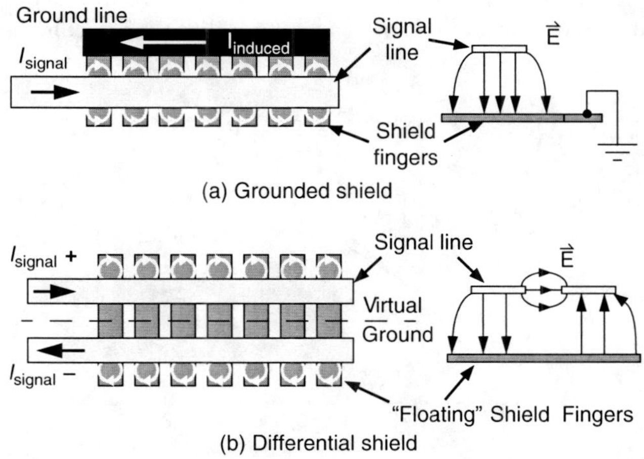

FIGURE 9.27 Conductor shielding. (After D Cheung and JR Long. IEEE Custom Integrated Circuits Conference, 2003, pp. 95–98. With permission.)

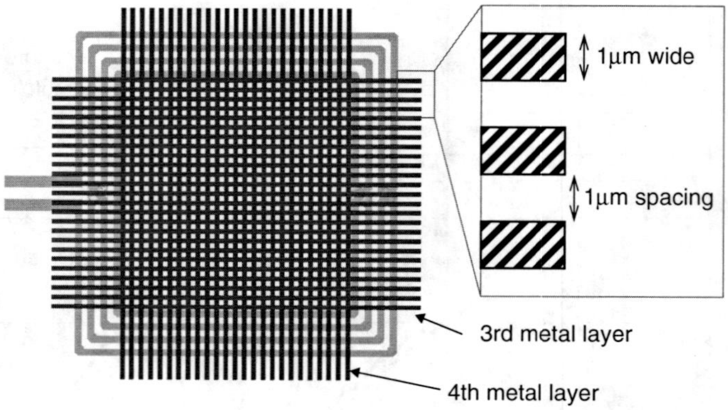

FIGURE 9.28 Differentially shielded symmetric inductor. (After D Cheung and JR Long. IEEE Custom Integrated Circuits Conference, 2003, pp. 95–98. With permission.)

An example of a shielding pattern for the inductors is illustrated in Figure 9.28 [42]. This mesh pattern is comprised of horizontal and vertical strips that span the length and width of the inductor. A voltage induced on a strip along one side of the coil is compensated for by an equal but opposite voltage induced at the other end of the strip, when the inductor is driven differentially. The net electric potential induced on the strips is zero. Induced current is inhibited by placing the strips orthogonal to the inductor winding.

The Q-factor for differential shielding is compared to an unshielded design in Figure 9.29. The measured low-frequency inductance is about 7.4 nH for both inductors. The parasitics of a differential shield lower the self-resonant frequency of the inductor by less than 3%. There is less than 2% difference in the inductance between shielded and unshielded inductors, which indicates that current induced in the shielding strips from the coil's magnetic field is very small. In other words, the addition of a differential shield does not diminish the useful frequency range or the inductance value. Figure 9.29 shows that differential shielding improves the Q-factor by 35%.

Varactors

The varactor is a voltage-variable capacitor (controlled by the dc bias voltage) that is used to vary the resonant frequency of an LC tank and the operating frequency of the VCO. It is a passive device which can be modeled using a series R–C equivalent circuit (e.g., lumped capacitor C_s, in series with resistor r_s) at a given frequency, resulting in a figure-of-merit Q-factor, $Q_v = 1/\omega C_s r_s$. Ideally r_s is zero, but losses of the ac current path through the varactor result in a finite Q. Unlike the inductor, the varactor Q is largest

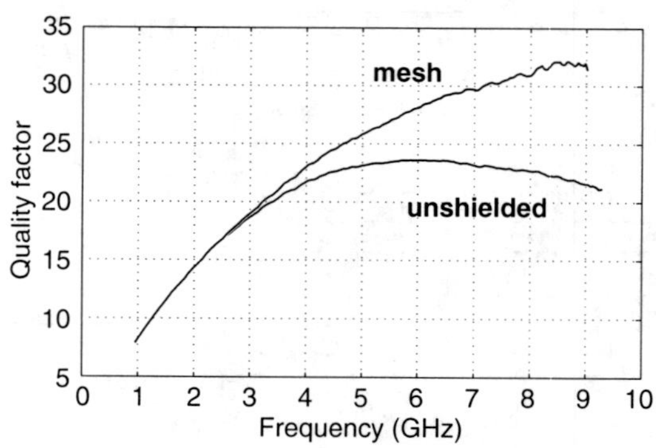

FIGURE 9.29 Measured inductor Q-factor on a 10 Ω-cm substrate. (After D Cheung and JR Long. IEEE Custom Integrated Circuits Conference, 2003, pp. 95–98. With permission.)

at low frequencies and decreases with increasing frequency. Typically, *Q*s on the order of 50 to 100 can be realized on-chip for a varactor capacitance of 1 pF at 1 GHz.

The bias voltage across the varactor changes the capacitance (C_V) seen across the tank according to

$$C_V = C_{j0}/(1 - V_D/V_j)^{mj} \tag{9.15}$$

where C_{j0} is the zero-bias capacitance and V_D is the applied voltage. A simple bias-dependent capacitor can be implemented using the depletion capacitance of a reverse biased p–n junction. The depletion region width of a reverse-biased diode varies inversely proportional to the reverse bias. If one side of the junction is lightly doped and the other highly doped, the depletion region is confined to the lightly doped layer, which can improve the linearity of the capacitance–voltage characteristic. Back-to-back diodes (see Figure 9.30a) are required to ensure that one diode is always reverse biased during the complete cycle of the oscillation across the tank in a differential circuit. Diode varactors can be implemented using the source–drain diffusions and well implant for an MOS device (either p+/n or n+/p junctions) of the collector–base junction of a bipolar transistor (n+−p junction) in a BiCMOS technology. The *Q* of these varactors is limited by ohmic losses in the semiconductor (i.e., depletion region and the bulk). Decreases in capacitance from zero to full reverse bias of about 1.5:1 for a 2 to 3 V change in bias voltage are typical for diodes fabricated in modern VLSI silicon technologies (see Figure 9.31 for an example). This can be improved to 3:1 if the p–n junction doping profile can be made hyper-abrupt using a dedicated implant step to form the varactor junction, although this adds an additional mask step in fabrication. In addition to a larger capacitance change with voltage, hyper-abrupt p–n junction varactors also can have higher *Q*-factors when properly designed.

MOS capacitors (i.e., gate-to-channel capacitance in the transition from depletion to accumulation, or between depletion and strong inversion modes) are also used to implement varactors on-chip. The capacitance–voltage (*C–V*) characteristics of the accumulation (A-MOS, Figure 9.30c) and inversion (I-MOS, Figure 9.30b) mode varactors are shown in Figure 9.31. When the channel underlying the gate oxide in an MOS transistor is in inversion (or accumulation) mode, the capacitance from gate-source–drain is maximum. As the carrier density decreases due to a change in bias voltage, the capacitance decreases, although the transition is much more abrupt than for a typical p–n junction varactor. The shape of the *C–V* curve is important as the nonlinearity of the capacitance with applied voltage does

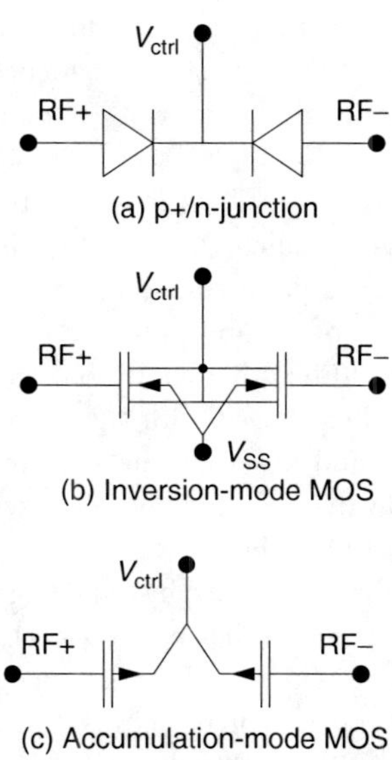

FIGURE 9.30 Varactor diodes for use in differential VCOs.

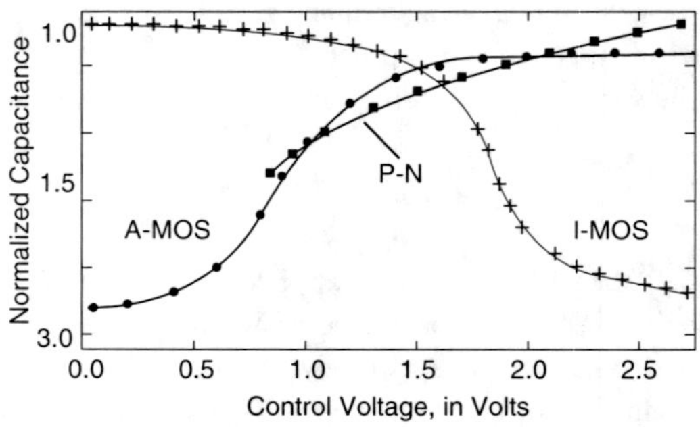

FIGURE 9.31 Normalized capacitance versus bias voltage for MOS and p–n junction varactors. (After P Andreani and S Mattisson. *IEEE Journal of Solid-State Circuits* 35:905–910, 2000. With permission.)

affect the phase noise of a VCO [44]. *Q*-factors for the MOS varactors are typically a factor of 2 higher than for a p–n junction varactor, and the capacitance variation between bias extremes is about 3:1. Technology scaling will improve this capacitance ratio as gate oxide thicknesses continue to decrease with newer generations. One advantage of most SiGe-BiCMOS technologies for RF design is the wealth of device options available. For varactors, there are usually three types of devices (one or more p–n junction, and also MOS varactors) available.

Differential VCO tuning is desirable for a VCO in order to reject common-mode noise generated by other wireless blocks at the tuning control input. This can improve phase noise and make the design more robust. By matching the *C–V* characteristics of inversion-mode PMOS and NMOS devices, a varactor which (to the first-order) responds to a differential tuning voltage can be implemented [45].

Integrated Circuit Oscillators

The full integration of a VCO with spectral purity sufficient to meet cellular telephony specification such as IS-95 has proven elusive. The greatest technical challenge preventing full integration is the quality of the resonator that can be implemented on-chip, due to the limited *Q*-factor of monolithic inductors and varactor diodes. However, there are a number of other constraints, which make integration a challenge. The recent trend to low supply voltages limits the available dynamic range and lowers the maximum signal-to-noise that is possible on-chip. Also, in any manufacturable design, the tuning range of the VCO must be sufficient to accommodate the entire range of component tolerances caused by process and temperature variations. However, as the tuning range is broadened, noise on the frequency control line of the VCO begins to cause a significant modulation of the oscillation frequency, resulting in higher phase noise from the VCO.

There are numerous circuit topologies for oscillators, such as the Colpitts, Clapp, Hartley, Pierce, and push–pull. The most widely used design for silicon RFICs is the differential Colpitts oscillator, consisting of a cross-coupled differential amplifier (a negative-R cell) and resonant load as shown in Figure 9.32. The differential topology offers rejection of common-mode noise, which improves the immunity of the circuit to potential interference generated by other circuit blocks on the same chip. Bipolar transistors have low $1/f$ noise and higher transconductance (g_m) for a given bias current compared to MOS devices, which can result in lower phase noise and lower current consumption for the VCO. The higher gain available in a single stage is also useful when designing power-efficient oscillator buffers to drive a mixer stage or buffer signals off-chip (Q_3 and Q_4 in Figure 9.32).

The resonant tank formed by the inductor shunted by opposing varactor tuning diodes and capacitors is shown in Figure 9.32 (left). When the $|-R|$ of the negative-R cell is less than the total shunt dissipation of the resonator, the circuit will oscillate. The negative resistance seen at the tank must be sufficient to overcome losses in the tank and other parts of the circuit. A simple low-frequency analysis of the negative resistance cell formed by the positive feedback around transistors Q_1 and Q_2 gives the following equation for resistance seen between the collectors of the differential pair:

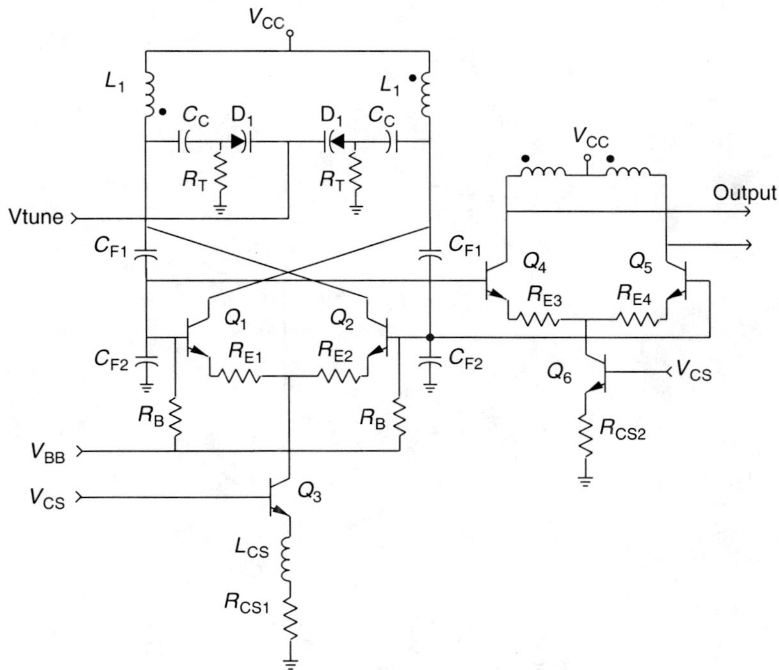

FIGURE 9.32 Negative *R* cell-based VCO and output buffer.

$$r_{\text{diff}} = \frac{-2(1 + g_{\text{m}} R_{\text{E}(1,2)})}{g_{\text{m}}} \tag{9.16}$$

where g_{m} is the transconductance of each transistor in the differential pair and R_{E} is the emitter degeneration resistance (i.e., $R_{\text{E}} = R_{\text{E1}} = R_{\text{E2}}$). The transconductance depends upon the selection of bias current.

Capacitive feedback using capacitors C_{F} (similar to the capacitive impedance transformer in a single-ended Colpitt's oscillator) closes the positive feedback loop. This improves the range of possible bias points for the $-R$ cell (using bias voltage V_{BB} via large-valued isolation resistors R_{B}) while eliminating the power consumed by emitter followers in a positive feedback network. It also ensures a large swing across the tank without driving transistors Q_1 and Q_2 into heavy saturation. Degeneration of current source Q_3 using a low-Q inductor ($Q < 3$) reduces phase noise caused by upconversion of noise from the bias circuitry.

The back-to-back tuning diodes (D_1) are referenced to ground and capacitively coupled (via C_{C}) to symmetric inductor L_2. This reduces frequency modulation of the VCO caused by changes in the supply voltage (i.e., supply pulling). Tuning voltage V_{tune} is isolated from the RF path by large resistors (R_{T}) as the bias current required by the tuning diodes is small. The processing tolerances inherent in varactors and on-chip inductors cause a shift in the resonant frequency away from the design value in manufacture. The tuning range of the resonant tank must be wide enough to accommodate this variation, which can be as high as 20%. Additional capacitors (switch selectable using digitally control switches) are typically used to adjust the oscillation frequency of a VCO in a power-up calibration sequence.

VCO Comparison and Summary

Numerous VCO designs have been described in conference and journal publications. A brief summary comparing silicon (Si) and SiGe bipolar with Si-CMOS VCOs is given in Table 9.4. The phase noise

TABLE 9.4 VCO Comparison

Maximum Oscillator Frequency in GHz	Phase Noise, in dBc/Hz (at $f\Delta$)	Adjusted Phase Noise (dBc)	Tuning Range (%)	RF P_{out} in dBm	P_D in mW (at VCC)	Notes	Technology
3.4	−129 at 1 MHz	199.6	25	—	22.5 (2.5 V)	Differential Colpitts	0.5 μm SiGe-bipolar [46]
1.2	−106 at 100 kHz	188	—	−28	5.0 (2.7 V)	Differential Colpitts with AGC	25GHz Si-bipolar [47]
1.56	−126 at 1 MHz	190	10	−10	29.7 (2.7 V)	Differential Hartley	25GHz Si-bipolar [48]
11	−87 at 100 kHz	188	5	−11.8	24.0 (3.0 V)	Single-ended Clapp	0.5 μm SiGe bipolar [49]
47	−108.5 at 1 MHz	201.9	8	5.6	280 (5.5 V)	Common-collector diff. Colpitts	70 GHz fT SiGe-bipolar [50]
2	−134 at 3 MHz	191	20	—	2.0 (2.0 V)	Differential Colpitts, bondwire inductor	0.5 μm CMOS [51]
1.8	−142 at 3 MHz	198	27.5	—	32.4 (1.8 V)	Differential Colpitts	0.25 μm CMOS [52]

Sources: J-M Mourant, J Imbornone, and T Tewksbury. IEEE Radio Frequency Integrated Circuits Symposium, 2000, pp. 65–68. MA Margarit, JL Tham, RG Meyer, and MJ Deen. *IEEE Journal of Solid-State Circuits* 34:761–771, 1999. M Zannoth, J Fenk, A Springer, and R Weigel. IEEE Radio Frequency Integrated Circuits Symposium, 1999, pp. 117–120. M Soyuer, JN Burghartz, HA Ainspan, KA Jenkins, P Xiao, AR Shahani, MS Dolan, and DL Harame. *IEEE Journal of Solid-State Circuits* 32:1451–1454, 1997. H Li, H-M Rein, R Kreienkamp, and W Klein. *IEEE Microwave and Wireless Components Letters* 12:79–81, 2002. F Svelto, S Deantoni, and R Castello. IEEE Custom Integrated Circuits Conference, 2000, pp. 577–580. B DeMuer, N Itoh, M Borremans, and M Steyaert. IEEE 2000 Custom Integrated Circuits Conference, 25-5, 2000, pp. 585–588. With permission.

reported by the authors for each oscillator from the literature has been adjusted to account for different center and offset frequencies using the following equation

$$\text{Adjusted phase noise} = 10\log\left[(f_o)^2/(f_\Delta)^2\right] - L(f_\Delta)\text{in dBc/Hz} \tag{9.17}$$

which gives a positive number (in units of dBc/Hz).

Many figures-of-merit (FOMs) for ranking oscillator performance have been proposed in the literature to account for power consumption, oscillator tuning sensitivity, and other factors in addition to phase noise. Rather than weighting the phase noise with one of more of these other factors, data for center frequency, tuning range, and power consumption are listed separately in Table 9.4.

The first three entries are differential Colpitts VCOs fabricated in SiGe [46] (entry 1) and silicon bipolar (entries 2 and 3) technologies [47,48]. The first entry illustrates the advantages of SiGe-BiCMOS technology: a wide tuning range, higher operating frequency, and lower phase noise than its silicon counterparts. The manufacturability of the SiGe VCO is extended by using MOS capacitors and MOSFET switches to either add or remove capacitance in parallel with the tank in order to trim the operating frequency.

The second entry in Table 9.4 is remarkable because it uses an AGC loop to regulate the amplitude of the oscillator's output rather than rely on limiting of the signal amplitude due to the finite supply voltage. In theory, this will result in lower phase noise from the oscillator. However, this is not apparent from comparison of the AGC oscillator with the 3.4 GHz SiGe design in Table 9.4. One reason is the tank Q-factor, which has a strong influence on the phase noise. Q-factor is inversely proportional to top metal thickness, which is 4 μm in the SiGe technology compared to 1 to 2 μm for the 25 GHz bipolar technology used to fabricate the oscillator with AGC. The third design uses coupled coils in a Hartley configuration to feedback the signal from the tank to the input of the differential pair, thereby replacing feedback capacitors C_F in Figure 9.32. This does not appear to offer any performance advantage or disadvantage from the reported results.

The following two entries in Table 9.4 (4 and 5) illustrate the higher frequency capabilities of SiGe technologies [49,50]. The 11 GHz VCO trades off reduced phase-noise performance for similar power consumption to the 3.4 GHz design, while the 47 GHz VCO attains similar phase noise performance at the expense of greater power consumption.

The last two entries in the table are results from two recently reported CMOS differential Colpitts designs. The 2 GHz VCO [51] uses a bondwire inductor, which gives a much greater Q-factor than can be achieved with an on-chip inductor and allows the designer the freedom to lower power consumption while preserving phase-noise performance. The 1.8 GHz CMOS design [52] demonstrates low phase noise for a fully monolithic oscillator, but requires a comparatively larger bias current. Both CMOS designs require a buffer in order to minimize loading of the oscillator by the load (such as a mixer) in an RF transceiver application.

Acknowledgments

The author would like to thank David Harame of IBM Microelectronics and Robert Hadaway of Northern Telecom Electronics for fabrication access.

References

1. J Sevenhans, F Op't Eynde, and P Reusens. The silicon radio decade. *IEEE Transactions on Microwave Theory and Techniques* 50:235–244, 2002.
2. AA Abidi, PR Gray, and RG Meyer. *Integrated Circuits for Wireless Communications.* Piscataway, NJ: IEEE Press, 1999.
3. AA Abidi. Low-power radio-frequency ICs for portable communications. *Proceedings of the IEEE* 83:544–569, 1995.

4. JP Maligeorgos and JR Long. A 5.1–5.8 GHz image-reject receiver with wide dynamic range. *IEEE Journal of Solid-State Circuits* 35:1917–1926, 2000.
5. H Hasegawa, M Furukawa, and H Yanai. Properties of microstriplines on Si–SiO_2 system. *IEEE Transactions on Microwave Theory and Techniques* 19:869–881, 1971.
6. JR Long. Passive components for silicon RF and MMIC design. *IEICE Transactions on Electronics of Solid-State Circuits* E86-C-6:1022–1031, 2003.
7. GG Rabjohn. Monolithic microwave transformers. M.Eng. thesis, Carleton University, 1991.
8. J Zhang, MK Jackson, and JR Long. Millimeter-wave characteristics of SiGe heterojunction bipolar transistors and monolithic interconnects in silicon technologies. Proceedings of the International Solid-State Circuits Conference, 2001, pp. 274–275.
9. R Aparicio and A Hajimiri. Capacity limits and matching properties of intermetal capacitors. *IEEE Journal of Solid-State Circuits* 37:384–393, 2000.
10. NM Nguyen and RG Meyer. Si IC-compatible inductors and LC passive filters. *IEEE Journal of Solid-State Circuits* 25:1028–1031, 1990.
11. JR Long and MA Copeland. The modeling, characterization and design of monolithic inductors for silicon RF ICs. *IEEE Journal of Solid-State Circuits* 32:357–369, 1997.
12. JR Long and MA Copeland. Modeling of monolithic inductors and transformers for silicon RFIC design. Proceedings of the IEEE MTT-S International Symposium on Technologies for Wireless Applications, 1995, pp. 129–134.
13. AM Niknejad and RG Meyer. Analysis, design, and optimization of spiral inductors and transformers for Si RF ICs. *IEEE Journal of Solid-State Circuits* 33:1470–1481, 1998.
14. JR Long. Monolithic transformers for silicon RF IC design. *IEEE Journal of Solid-State Circuits* 35:1368–1382, 2000.
15. SA Maas. *Nonlinear Microwave and RF Circuits.* Boston, MA: Artech House, 1997.
16. V Aparin and C Persico. Effect of out-of-band terminations on intermodulation distortion in common-emitter circuits. Proceedings of the IEEE International Microwave Symposium, 1999, pp. 977–980.
17. KL Fong. High-frequency analysis and linearity improvement technique of common-emitter transconductance stage using a low-frequency-trap network. *IEEE Journal of Solid-State Circuits* 35:1249–1252, 2000.
18. SP Voinigescu, MC Maliepaard, JL Showel, GE Babcock, D Marchesan, M Schroeter, P Schvan, and DL Harame. A scalable high-frequency noise model for bipolar transistors with application to optimal transistor sizing for low-noise amplifier design. *IEEE Journal of Solid-State Circuits* 32:1430–1439, 1997.
19. JR Long. A low-voltage 5.5–5.8 GHz image-reject downconverter RF IC. *IEEE Journal of Solid-State Circuits* 35:1320–1328, 2000.
20. JR Long and MA Copeland. A 1.9 GHz low-voltage silicon bipolar RF front-end for wireless personal communications systems. *IEEE Journal of Solid-State Circuits* 30:1438–1448, 1995.
21. D Cassan and JR Long. A 1 V transformer-feedback low noise amplifier for 5–6 GHz WLAN in 0.18 μm CMOS. *IEEE Journal of Solid-State Circuits* 38:427–435 2003.
22. JR Long, M Copeland, S Kovacic, D Malhi, and D Harame. RF analog and digital circuits in SiGe technology. Proceedings of the International Solid-State Circuits Conference, 1996, pp. 82–83.
23. V Aparin, P Gazzerro, J Shou, B Sun, S Szabo, E Zeisel, T Segoria, S Ciccarelli, C Persico, C Narathong, and R Sridhara. A highly-integrated tri-band/quad-mode SiGe BiCMOS RF-to-baseband receiver for wireless CDMA/WCDMA/AMPS applications with GPS capability. IEEE International Solid-State Circuits Conference, 2002, pp. 234–235.
24. M Tanabe, T Fukuda, and K Nishii. Step gain amplifier with impedance unchanged in gain control. IEEE Radio Frequency Integrated Circuits Symposium, 2002, pp. 209–212.
25. SY Yue, D Ma, JR Long, and DL Harame. A 17.1 to 17.3 GHz image-reject down-converter with phase-tunable LO using 3× subharmonic injection locking. IEEE International Solid-State Circuits Conference, 2004, pp. 388–399.

26. S Reynolds, B Floyd, U Pfeiffer, and T Zwick. 60 GHz transceiver circuits in SiGe bipolar technololgy. IEEE International Solid-State Circuits Conference, 2004, pp. 442–443.
27. S Kumar, M Vice, H Morkner, and W Lam. Enhancement mode GaAs PHEMT LNA with linearity control (IP3) and phased matched mitigated bypass switch and differential active mixer. Proceedings of the IEEE International Microwave Symposium, 2003, pp. 1577–1580.
28. P Leroux, M Steyaert, V Vassilev, and H Maes. A 1.3 dB NF CMOS LNA for GPS and 3 kV HBM ESD-protection. Proceeding of ESSCIRC, 2002, pp. 335–338.
29. WE Main and JC Durec. A linear class AB single-ended to differential transconverter suitable for RF circuits. Proceedings of the IEEE International Microwave Symposium, 1996, pp. 1071–1074.
30. B Gilbert. The micromixer: a highly linear variant of the Gilbert mixer using a bisymmetric class-AB input stage. *IEEE Journal of Solid-State Circuits* 32:1412–1423, 1997.
31. J Durec. An integrated silicon bipolar receiver subsystem for 900-MHz ISM band applications. *IEEE Journal of Solid-State Circuits* 33:1352–1372, 1998.
32. PJ Sullivan, BA Xavier, and WH Ku. Low voltage performance of a microwave CMOS Gilbert cell mixer. *IEEE Journal of Solid-State Circuits* 32:1151–1155, 1997.
33. DB Leeson. A simple model of feedback oscillator noise spectrum. *Proceedings of the IEEE* 54:329–330, 1966.
34. A Hajimiri and TH Lee. A general theory of phase noise in electrical oscillators. *IEEE Journal of Solid-State Circuits* 33:179–194, 1998.
35. A Demir, A Mehrotra, and J Roychowdhury. Phase noise in oscillators: a unifying theory and numerical methods for characterization. *IEEE Transactions on Circuits and Systems I* 47:655–674, 2000.
36. A Demir. Phase noise and timing jitter in oscillators with colored-noise sources. *IEEE Transactions on Circuits and Systems I* 49:1782–1791, 2002.
37. P Vanassche, G Gielen, and W Sansen. On the difference between two widely publicized methods for analyzing oscillator phase behavior. IEEE ICCAD, 2002, pp. 229–233.
38. Y Ou, N Barton, R Fetche, N Seshan, T Fiez, UK Moon, and K Mayaram. Phase noise simulation and estimation methods: A comparative study. *IEEE Transactions on Circuits and Systems I* 49:635–638, 2002.
39. A Hajimiri and TH Lee. Design issues in CMOS differential LC oscillators. *IEEE Journal of Solid-State Circuits* 34:717–724, 1999.
40. JN Burghartz, DC Edelstein, M Souyer, HA Ainspan, and KA Jenkins. RF circuit aspects of spiral inductors on silicon. *IEEE Journal of Solid-State Circuits* 33:2028–2034, 1998.
41. M Danesh and JR Long. Differentially-driven symmetric spiral inductors. *IEEE Transactions on Microwave Theory and Techniques* 50:332–341, 2002.
42. D Cheung and JR Long. Differentially shielded monolithic inductors. IEEE Custom Integrated Circuits Conference, 2003, pp. 95–98.
43. P Andreani and S Mattisson. On the use of MOS varactors in RF VCOs. *IEEE Journal of Solid-State Circuits* 35:905–910, 2000.
44. JWM Rogers, JA Macedo, and C Plett. The effect of varactor non-linearity on the phase noise of completely integrated VCOs. *IEEE Journal of Solid-State Circuits* 35:1360–1367, 2000.
45. NHW Fong, J-O Plouchart, N Zamdmer, L Duixian, LF Wagner, C Plett, and NG Tarr. A 1-V 3.8—5.7-GHz wide-band VCO with differentially tuned accumulation MOS varactors for common-mode noise rejection in CMOS SOI technology. *IEEE Transactions on Microwave Theory and Techniques* 51:1952–1959, 2003.
46. J-M Mourant, J Imbornone, and T Tewksbury. A low phase noise monolithic VCO in SiGe BiCMOS. IEEE Radio Frequency Integrated Circuits Symposium, 2000, pp. 65–68.
47. MA Margarit, JL Tham, RG Meyer, and MJ Deen. A low-noise, low-power VCO with automatic amplitude control for wireless applications. *IEEE Journal of Solid-State Circuits* 34:761–771, 1999.

48. M Zannoth, J Fenk, A Springer, and R Weigel. A single-chip Si-bipolar 1.6 GHz VCO with integrated bias network. IEEE Radio Frequency Integrated Circuits Symposium, 1999, pp. 117–120.
49. M Soyuer, JN Burghartz, HA Ainspan, KA Jenkins, P Xiao, AR Shahani, MS Dolan, and DL Harame. An 11-GHz 3-V SiGe voltage controlled oscillator with integrated resonator. *IEEE Journal of Solid-State Circuits* 32:1451–1454, 1997.
50. H Li, H-M Rein, R Kreienkamp, and W Klein. 47 GHz VCO with low phase noise fabricated in a SiGe bipolar production technology. *IEEE Microwave and Wireless Components Letters* 12:79–81, 2002.
51. F Svelto, S Deantoni, and R Castello. A 1 mA, −120.5 dbc/Hz at 600 kHz from 1.9 GHz fully tuneable LC CMOS VCO. IEEE Custom Integrated Circuits Conference, 2000, pp. 577–580.
52. B DeMuer, N Itoh, M Borremans, and M Steyaert. A 1.8 GHz highly-tunable low-phase-noise CMOS VCO. IEEE 2000 Custom Integrated Circuits Conference, 25-5, 2000, pp. 585–588.

10

Direct Conversion Architectures for SiGe Radios

10.1 Overview of Direct Conversion Architectures **10**-2
10.2 Direct Conversion System Aspects and IC Abstraction **10**-3
10.3 Choice of Differential Architecture **10**-3
10.4 Transmitter and Receiver Considerations **10**-6
Nonidealities in the Signal Paths • Considerations in the Transmitter Signal Path • Considerations in the Receiver Signal Path • Signal Generation Path for Transmitter and Receiver
10.5 Impacts of System Performance **10**-9
Impact of System Performance on Device Technology • Impact of System Performance on Circuits
10.6 SiGe Direct Conversion Radio: A Practical Example **10**-11
System Considerations • Circuit Design • Low-Noise Amplifier (LNA) • Downconversion Mixers • Downconversion Mixers: Practical Example • Signal Generation Path
10.7 The Integration: Interfaces and Layout **10**-19
10.8 Characterization of Receiver Front End **10**-21
10.9 Fabricated Hardware and Results **10**-21
10.10 Compensation and Corrections **10**-23
10.11 Future of Direct Conversion Radios **10**-24

Sudipto Chakraborty and Joy Laskar
Georgia Institute of Technology

In this chapter, we describe direct conversion architectures for radio applications and the impact of silicon germanium technology on such architectures. First, we describe the radio architecture and illustrate various challenges in the development of such architecture from an integration standpoint. Later, we demonstrate the development of two variants of fully monolithic radios in a second-generation silicon germanium BiCMOS technology. We also cover various practical aspects of these radios, with an appreciative viewpoint of both the simplicity as well as the complexity of such developments in a silicon-based environment. Afterwards, we present case studies of direct conversion radios for IEEE802.11a standard. While this standard is a representative of high data rate wireless system, similar approaches can be taken for other wireless standards as well.

10.1 Overview of Direct Conversion Architectures

Since its inception in the early 20th century, wireless engineering has come a long way. Most of the basic principles of the sophisticated radio architectures and their abstractions, as we see it today, were developed using vacuum tubes around 1930 [1]. Starting with the basic foundation laid down by Maxwell (1883), and with subsequent inventions in wave propagation and wireless telegraphy by Hertz, Bose, Marconi, and others, wireless technology was born around 1900 in a very primitive form. Demonstration of the superheterodyne receiver by Armstrong dates back to as early as 1924. Armstrong's superheterodyne receiver underwent considerable refinement during the 1920s and 1930s. This was the time when radio pioneers considered the use of homodyne architectures for single vacuum tube receivers. Following the invention of the transistor in 1947, and many more refinements in the semiconductor platform, the concept of integrated circuit gradually changed focus from small scale to medium and large scale, leading to the vision of "system-on-chip (SOC)" or "fully monolithic implementation" to support extremely diverse applications.

A radio is a complex mixture of highly diverse functionalities, and the challenges to realize a miniaturized high-performance integrated radio are quite significant. The challenges exist in all levels of communication systems, integrated circuits, and device technologies. Traditional superheterodyne architectures result in very high performance, while requiring high-quality factor components for radio functionality. The success of direct conversion architectures for practical applications was quite delayed due to its performance limitations, which require careful trade-off among system, circuit, and integration perspectives. Vance at ITT was the first to apply direct conversion for pager applications by means of a single-chip receiver [2]. During the early 1980s, direct conversion receivers were developed at Motorola as a possible way to implement compact radios. The major obstacle for the success of the direct conversion architectures is the presence of spectral component around DC for various physical layer standards, requiring careful system-level considerations.

The evolution of radio architectures, and their silicon abstractions has primarily focused on the frequency planning, and reducing the number of analog processing blocks (e.g., filters, amplifiers). This decision is also based on a number of unwanted external signals (e.g., blockers, cross-talk among various wireless terminals), as well as internal real estate (utilization of single oscillator versus multiple oscillators and PLLs). The benefits of superheterodyne architecture could not be fully realized without judicious frequency planning and utilization of high-quality filters. However, to minimize the real estate for radio abstractions in silicon, the architectures have continuously scaled from high intermediate frequency (IF) to medium IF to very low IF, leading to zero IF. In this section, we will limit our scope to system-level challenges with implementation details for a zero IF (or direct conversion) radio.

What is so different and attractive about a direct conversion radio? Why is every radio abstraction targeted to be direct conversion architecture in the first place irrespective of the application, given the fact that it is well known as "difficult-to-implement"? The answer to the first question is quite straightforward. A superheterodyne architecture requires more than one stage of frequency shifting (upconversion or downconversion, and use of two or more IF frequencies). This usually leads to the choice of multiple oscillators on chip, and filtering requirements for image rejection purposes. Direct conversion eliminates these considerations, and simplifies the task of a system designer in terms of a judicious choice of IF frequencies such as to achieve a desired radio performance in terms of blockers (modulated or unmodulated). The elimination of several IF stages also helps realize a compact die area for such implementations. The answer to the second question follows from the previous argument; proven the fact that it is lowest area approach, we always lean toward understanding the fundamental limitations of an integrated direct conversion radio abstraction as well as envisioning silicon germanium as a technology platform to implement such systems.

10.2 Direct Conversion System Aspects and IC Abstraction

Any communication system design is a combination of performance parameters of different building blocks, the big envelope of which includes gain, linearity, and noise. These three performance aspects are also accompanied by the power dissipation and the cost of implementation for each of the building blocks, and should be optimized by an intuitive, rather clever approach using a combination of techniques, such as frequency planning, various interfacing impedances, etc. To appreciate the classification of building blocks in this manner, we consider a high-frequency low-noise amplifier and a high-frequency oscillator. Noise contributed by each of these blocks is different not only from a circuit perspective, but also from fundamentals of device noise characteristics. The noise contribution from the amplifier is also dependent on the interfacing impedance at the input. The traditional 50 Ω interfacing impedance standard was established as a compromise between signal handling power and attenuation for coaxial cables since the World Ware II times, and has been used ever since. One could reasonably deviate from that for an integrated circuit perspective. Thus, the noise, gain, and linearity parameters can vary significantly depending on the interfacing impedance, and our definition and calculation based on 50 Ω interfaces need to be refined to consider such possibilities. Communication building blocks are usually separated from each other in terms of their functionalities (such as frequency conversion and amplification etc.). Hence, depending on the architecture, the choices and the distribution of the system parameters in terms of gain, linearity, and noise should be performed by system designer as prior to the silicon implementation. Often times the traditional 50-Ω-based interface does not prove optimum from power consumption and real estate perspectives. Also, given a fixed amount of power dissipation, one variable usually trades off with another in any particular circuit abstraction (e.g., for a fixed gain, the noise figure of a circuit might exhibit a trade-off with the linearity performance with a given power consumption).

In this section, we would classify the unwanted system variables in terms of external and internal effects. An interference signal (in-band or out of band, modulated or unmodulated) can be categorized as an external factor as it is picked up from the environment, and a leakage from the VCO to the other parts of the integrated system could be considered as an internal factor. Both categories impact the system in different ways, such as consuming more current in the circuit block to enhance linearity, or careful isolation and frequency planning techniques in the front-end to mitigate SNR degradation of the message spectrum. Like other radio architectures, direct conversion could also be visualized as a combination of two fundamental functionalities: down or up conversion and demodulation or modulation. In this section, we will focus primarily on the front-end implementations considering down or up conversion and their impact on the demodulation or modulation functionality. We will also see that some of the system impairments affect a superheterodyne architecture in a different manner compared to direct conversion architecture.

With this introduction, let us explore the various system-level challenges in a direct conversion radio architecture as shown in Figure 10.1. The critical considerations for such architectures are also illustrated in Table 10.1. The scheme is essentially an I/Q modulator in the transmitter path, followed by a single sideband mixing technique and amplification, depending on the necessary output power level and coverage. In the receiver, differential signals are generated at the amplifier or after the amplifier using a balun, with an I/Q signal planning for down conversion to baseband. In the following sections, we describe the key elements to the success of direct conversion architectures.

10.3 Choice of Differential Architecture

In both the transmitter and the receiver, LO leakage and linearity requirements pose restrictions on the system design. The architecture must conform to a fully differential topology to cancel any common mode noise components, an extremely critical consideration for high-frequency circuits and systems. These advantages trade-off with additional power and area requirements of such implementations to

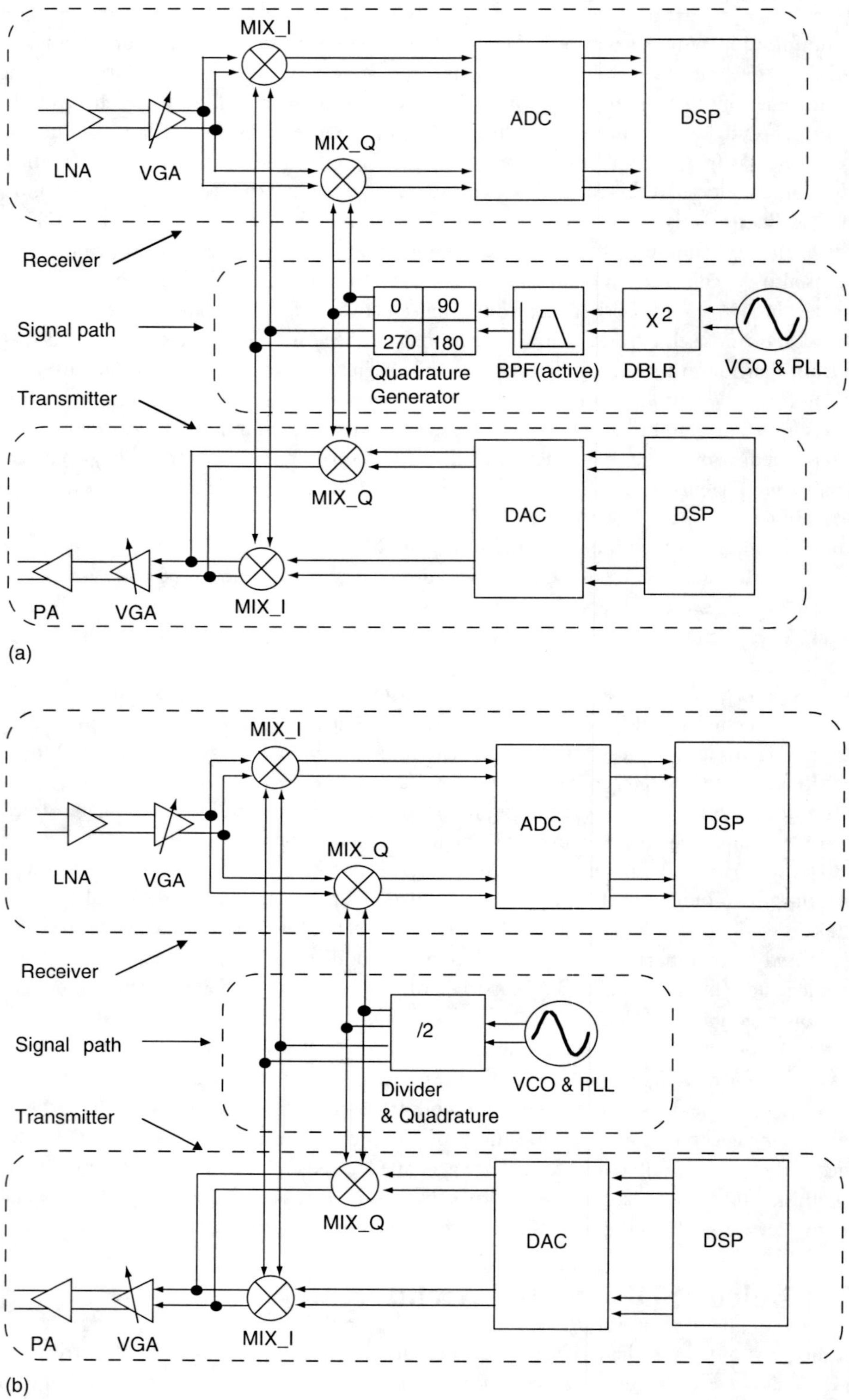

FIGURE 10.1 Two different variants of direct conversion radios: (a) 1/2X architecture and (b) 2X architecture.

TABLE 10.1 Impact of Building Blocks in Direct Conversion Radios

Building Blocks	Performance Metrics	Critical Considerations
VCO	Output voltage swing	Lower phase noise needed for closely spaced channels
	Usually differential topology	$1/f$ noise impacts the SNR of the message spectrum
	Tuning range	
	Phase noise	
	Area	
Quadrature generators	*Frequency dividers*	Frequency dividers
	I/Q accuracy	I/Q amplitude imbalance implies LO feedthrough in transmitter, phase imbalance implies sideband suppression
	Output voltage swing	Clock duty cycle error leads to the amplitude and phase imbalance
	Clock waveform characteristics	
	Division ratio: 2X/4X, etc.	
	Noise degradation	
	Polyphase networks	
	Signal loss	
	Bandwidth	
	Accuracy (I/Q imbalance)	
Mixers	Gain (Tx and Rx)	$1/f$ noise from the switch pairs in case of downconversion mixer
	Linearity (Tx and Rx)	Transconductor core for broadband and small narrowband linearity characteristics
	Noise figure (mostly Rx)	LO to RF isolation impacts the DC offset performance (dynamic)
	Power consumption (Tx and Rx)	Mismatch results in DC offset (static)
	Sideband suppression (mostly Tx)	IIP2 causes SNR degradation due to blocker (and cross modulation)
	LO to RF isolation	
	Mismatch	
	IIP2 (mostly receiver)	
Variable gain amplifier (VGA)	Gain	Different topologies (current steering, load switching, current switching) impact the performances in different ways
	Linearity	Care must be taken not to saturate these amplifiers High dynamic range (~80 to 90 dB) for cellular specification is a challenge
	Noise figure	
	Gain step	
	Dynamic range	
Low noise amplifier	Gain	Similar considerations for other receiver architectures
	Linearity	
	Noise figure	
	Stability	
	Power consumption	
	Single ended/differential	

their single-ended counterparts. However, often, due to the symmetrical nature of differential topology, it helps eliminate any common-mode parasitic components (such as bondwire inductance, or parasitic node capacitance), which may lead to stability considerations in single-ended counterparts at high frequencies. Differential topology also enhances the dynamic range, and suggests the possibility of neutralization techniques using parasitic components at opposite phases [3]. In modern integrated radios, the realization of a fully differential architecture has been a necessary practice. Even if there is an absolute necessity to be single ended (a case where single-ended antennas need to be integrated with

LNA or PA directly), it is desirable that we adopt to the differential topology in the signal path as early as possible.

10.4 Transmitter and Receiver Considerations

Nonidealities in the Signal Paths

As described earlier, tremendous space compaction in the integrated direct conversion radios takes place due to the elimination of IF processing blocks (filters and amplifiers) and multiple oscillators in various parts of the transmitter and receiver chains. However, from a practical standpoint, there exists nonidealities in signal paths mainly due to the two reasons: (a) finite reverse isolation of RF building blocks and (b) imbalance in the signal paths in terms of amplitude and phase, leading to the SNR degradations.

The first nonideality can be attributed to two effects. The first is the finite reverse isolation capability of the transistor (e.g., the reverse isolation is not infinite), which reflects to the finite reverse isolation of the whole circuit block. The second in this family is the finite isolation capability of low-resistivity semiconductor substrate (e.g., silicon). The substrate can be represented by a mesh of resistive and capacitive components in most part. In an integrated circuit, a signal can also propagate in the pad ring, causing unwanted disturbances. The imbalance in the signal path can be attributed to the presence of unequal traces in the signal path, and the accuracy with which the balance is maintained between different signal paths. A classic example of this effect would be the susceptibility of unequal load or presence of any signal dependent (nonlinear) behavior of components in the signal path.

Considerations in the Transmitter Signal Path

In case of direct conversion radios, any nonideality in the transmitter signal processing chain would result in an in-band SNR degradation around the LO carrier frequency. In the transmitter signal path, any imbalance in the I/Q signal paths would give rise to LO leakage. The amplitude imbalance usually leads to LO feed through, which creates in band SNR degradation, while phase imbalance is responsible for the sideband suppression. The transmitter also needs to accommodate for the dynamic range, which can be quite high for certain applications (e.g., cellular, ~80 to 90 dB dynamic range). It is quite important that the upconverted signal out of the I/Q modulators be at a level much higher than the input referred noise floor of the variable gain amplifier that is used to create such high dynamic range. Another important consideration for the transmitter is the error vector magnitude (EVM) requirements, which denotes the rotation of the constellation, and needs to be fairly low for a dense constellation (e.g., 64QAM, etc.). Apart from all these considerations, the transmitter must meet the FCC requirements in terms of spectral mask constraint. Noise is usually not a very critical consideration in the transmitter, and depending on the crest factor of the modulated signal judicious decision must be made in determining the amplifier operation (linear or nonlinear). However, it is quite challenging to accommodate the high dynamic range requirement in the VGA.

Considerations in the Receiver Signal Path

Analogous to the issues present the transmitter, nonidealities in the direct conversion receiver lead to impairments around the zero frequency (or DC). SNR degradation happens in this case due to DC offset and $1/f$ noise from the devices. DC offset can be divided into two categories: (a) static DC offset, which results from the mismatches in the circuit components, such as transistors, resistors, etc., (b) dynamic DC offset, which is essentially caused by the finite reverse isolation present in a down conversion mixer, caused by the self-mixing, or any other mechanism (such as substrate leakage), leading to the self-mixing phenomenon. A classic example of dynamic DC offset is the DC offset variation effect in the receiver when a wireless terminal experiences different signal reflections from surrounding objects in the environment, and the degradations vary with time. Static DC offsets can be tuned out by some form of a calibration mechanism performed when the IC is powered up, but the dynamic DC offset needs time-dependent averaging and cancellation, which is usually performed at baseband. Due to this effect, in any successful prototype demonstrating direct conversion architecture, the LO frequency is different from the RF frequency to minimize the number of nodes operating at the RF frequency (hence leading to self-mixing). The order of LO frequency determines the frequency scheme. If the LO frequency is *n* times the RF frequency, the architecture is referred to as "*nX*" architecture.

Along with the DC offset phenomenon, $1/f$ noise resulting from the devices is also one cause of impairment in case of receivers. Any device technology, which is fast enough to accommodate the RF carrier frequencies, while exhibiting a low $1/f$ corner, is definitely the choice in this case. SNR degradation due to $1/f$ noise is a resultant of two factors, the first one is from the phase noise behavior of the VCO (where the device flicker noise essentially gets upconverted to the carrier frequency), and the second one is from the baseband amplifiers. At the baseband processing blocks, it is prohibitive to utilize capacitors due to area restraint, and filter out the $1/f$ components. As with the case of any building blocks, the input referred noise is minimized by maximizing the gain, to support the incoming signal with the required bandwidth, the gain should increase with a considerable increment in bias current, thus the baseband blocks in a direct conversion tend to consume more current, and it is very difficult to attribute direct conversion architecture as the lowest power radio architecture. Another way to assess this aspect is that the direct conversion receiver usually provides lower gain in the receiver frontend, thus requiring more baseband gain (while supporting the entire message spectrum bandwidth) to fall within the dynamic range of the preceeding signal-processing blocks. Direct conversion radio architecture is a low real estate implementation, but its success is dependent on the area of the DC offset compensation circuits, which is often a function of the physical layer communication standard under consideration.

Apart from the fundamental analogies between direct conversion transmitter and receiver, one fundamental difference exists between the two in terms of second-order intermodulation parameter, IIP2. In direct conversion receivers, the second-order intermodulation becomes quite important to accommodate the effect of the blockers present with a reasonable SNR degradation. Intuitively, two blockers present at frequencies F1, and F2, where |F1–F2| is within the message bandwidth would lead to SNR degradation. The situation would be worse if there is a spreading effect in the process (i.e., any or both the blockers are modulated sources with some finite bandwidth around them), as the modulation can get transferred to the desired signal (also referred to as "cross modulation"). This is a major concern for many cellular-based systems. Figure 10.2 shows the various degradation aspects of direct conversion receivers.

Dynamic range of a receiver is determined by the noise floor and the maximum detectable signal level. The noise floor is essentially determined by the thermal noise floor along with the bandwidth, while the maximum allowable signal consideration is dependent on the linearity of the frontend. This is the case when transmitter and receiver do not operate simultaneously (time division duplex system, TDD). A fundamental deviation from this consideration occurs when transmitter and receiver operate simultaneously on closely spaced frequency bands (frequency division duplex, FDD). Such is the case with WCDMA systems and even using a diplexer system to perform frequency selective filtering, the finite amount of transmitter power leaks into the receiver and raises the noise floor. Thus, high-sensitivity FDD cellular phones usually pose most stringent system specification on practical implementations.

Signal Generation Path for Transmitter and Receiver

With the illustration of the signal paths of transmitter and receivers, let us consider the various aspects of the signal-generation path and impacts on the performance of the transceiver subsystem. The major considerations in this regard would be on the amplitude and phase imbalances of the various signal paths, and the frequency planning. Various circuit blocks generate harmonics (and not subharmonics) due to standard nonlinearity effects (such as odd-order frequency components in a hard-driven differential pair) or mismatch effects (a mismatched differentia pair would give rise to second- or even-order components). Let us consider the integrated direct conversion transceiver architecture shown in Figure 10.1. There have been two different approaches to design the signal generation path:

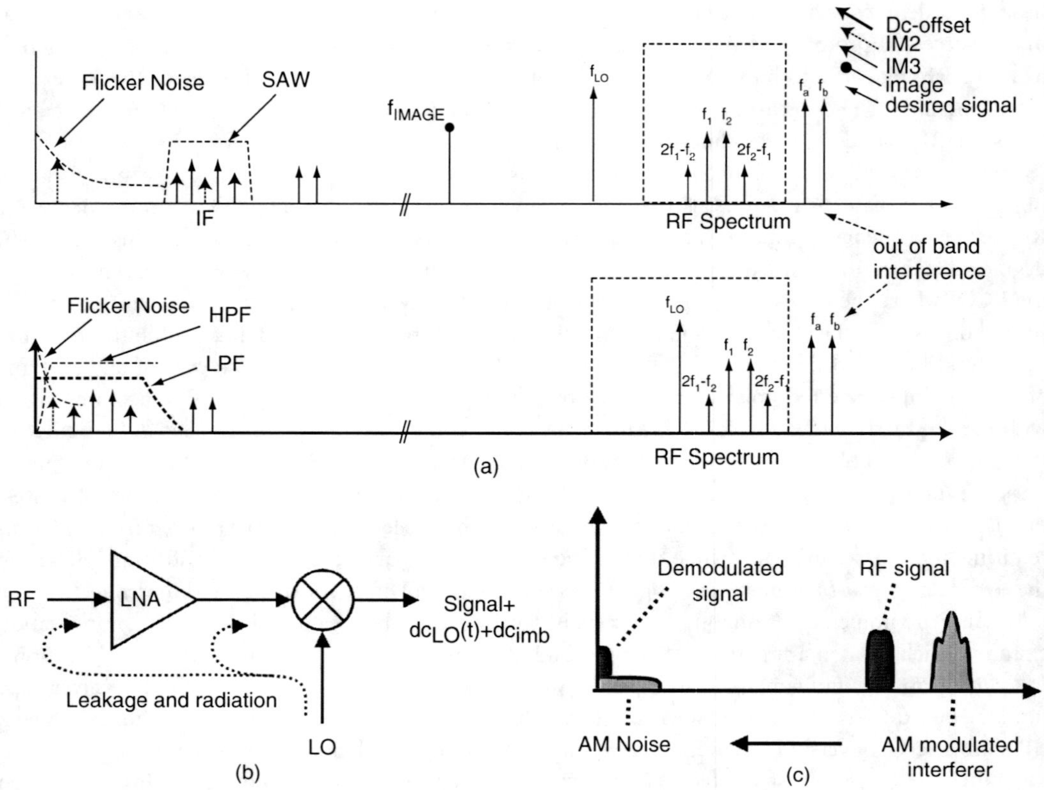

FIGURE 10.2 Illustrations of direct conversion frequency scheme: (a) comparison with superheterodyne scheme (the upper half of the down converted spectrum is described for illustration of $1/f$ noise and second-order intermodulation), (b) causes of DC offset, and (c) effect of second-order nonlinearity, and SNR degradation due to cross modulation.

(a) generate a lower frequency than the RF frequency (usually a subharmonic tone), and multiply by a frequency multiplier with the generation of necessary quadrature phases from a phase shifter, (b) generate a higher frequency than the RF frequency (usually a harmonic tone) and divide down by frequency dividers. The frequency dividers utilize static frequency division topology using D-type flip-flops, and a divide by 2 or a divide by 4 topology is fairly common. The choice of one architecture over the other is performed by system consideration in terms of I/Q accuracy, bandwidth of accurate quadrature generation, requirements of buffers, the power dissipation for the VCO operation at desired frequency, and the VCO pulling effects caused by the nonlinearity of other parts of the integrated transceiver. One example of approach (a) consisting of differential VCO, frequency multiplier, harmonic filter, and polyphase filter is shown in Figure 10.1a, with the summary shown in Table 10.2. The benefit of such approaches in terms of using a low-frequency VCO is offset by the losses and parasitics incurred in polyphase filters and the power consumption of the buffers for signal boosting. Polyphase filters are quite susceptible to the various injected harmonics of the signal, and an appropriate filtering must be performed. Use of a passive inductor for superior phase noise leads to bigger real estate. Also, such schemes prove immune to the pulling problems, as there is no scope of a subharmonic tone generation (and hence leakage associated) from the high-power amplifiers at high frequency. However, one drawback of this approach is the accuracy of quadrature generated from the polyphase filters over relatively narrow bandwidth, susceptibility to parasitics, and asymmetric phase shifts due to unequal loading effects. In comparison, the architecture option in (b) results in fairly accurate quadrature signals, while consuming more power in the differential VCO core (due to its high-frequency operation), as well as small real estate (due to miniaturization of inductor dimension at high frequency). However, this

TABLE 10.2 Comparison of the Different Architectures in the Signal Path for Direct Conversion Radios

Building Blocks	Submultiple Frequency Planning (1/2X)	Multiple Frequency Planning (2X)
VCO	Consumes less power Occupies more area (with passive inductor) Not susceptible to pulling effects	Consumes more power Occupies less area (with passive inductor) Susceptible to pulling effects
Frequency conversion blocks	*Frequency multiplier*	*Frequency divider*
	Accuracy is consideration Generates harmonics, requiring on-chip filtering Requires phase shifting in the input signals (usually 90°) to avoid any DC perturbation at the output	Usually quite accurate Usually static divider configuration using master/slave D flip-flops Usually no filtering needed
		Susceptible to duty cycle errors, clock skews, requiring cancellation circuitry
Polyphase filters	Generates quadrature Provides signal loss, leading to additional consumption in buffers Susceptible to parasitics Susceptible to imbalances in the signal path, and harmonics of the input tone Susceptible to unequal loading in the various signal paths	Not applicable

approach is susceptible to VCO pulling effects resulting from high-power PAs (VCO frequency is divided by 2 using dividers, and then second-order nonlinearity resulting from high-power PA results in the same frequency as the VCO and interacts with it). Some of these effects can be mitigated by using a fully differential scheme (generation of lower quantities of 2X frequencies due to mismatch), or moving to a 4X frequency scheme, leading to even lower VCO pulling and real estate, while consuming more power in the VCO core. Another important consideration is the susceptibility of such approaches to duty cycle errors in the differential signals resulting from the VCO. Transitioning of frequency divider at both the edges of signal waveforms would be quite susceptible to duty cycle errors, whereas a frequency divider operating at the same edges of the incoming VCO signal would be immune to such effects, but be susceptible to clock jitter. All the above effects are further complicated by the process and temperature variation effect in the semiconductor processes.

10.5 Impacts of System Performance

Impact of System Performance on Device Technology

It might seem at this point that availing the advantages of various semiconductor process technologies and integrating them in a system in package is the most optimum approach, as in that case we could avail the semi-insulating property of different substrates to reduce leakage and to realize high-quality factor passive components in the RF domain. While the option remains attractive from a fundamental standpoint, in most of the cases it is frustrated by the complexity in terms of interfacing impedances and packaging considerations. Silicon, as a platform is cheaper compared to its counterparts, and hence judicious circuit–system partitioning should be performed for the success of direct conversion radios. Hence, one should continuously look for silicon platforms, which support high cutoff frequency devices, and superior passives, while maintaining a reasonable cost.

Given the above considerations, we revert to the fundamental requirement of direct conversion architectures. Any high-frequency block requires high f_T, f_{max}, low-noise and high-breakdown capabilities. SiGe BiCMOS technology offers the flexibility to utilize bipolar and MOS transistors along with the high-quality factor passives. MOS transistors usually consume more current for the same transconductance in a given area, and exhibit more parasitic capacitances, thus lowering the speed. As a surface device, the noise characteristics of MOS transistors are usually an order of magnitude worse compared to that of the bipolar transistors. Broadband noise is important for all receivers, in terms of the noise floor, but $1/f$ noise is the most degrading factor in case of direct conversion architecture, indicating the superiority of bipolar devices in this regard. Also, the choice of bipolar transistors proves optimum as a low input impedance stage as part of a low-noise amplifier. Bipolars also prove superior as part of the LO buffers, as they exhibit very low input parasitic capacitance, eliminating the need of inductive loading. However, MOS proves advantageous compared to bipolars in various aspects such as (a) linear differential transconductor cells and (b) switching elements (MOS performs switching without a voltage shift from input to the output). Both devices are fundamentally capable of realizing direct conversion radio abstraction, but the high $1/f$ corner of MOS might prohibit its use in many sophisticated wireless applications. Another drawback of utilizing MOS transistors is the presence of body bias, which leads to significant nonlinearity effects at large swing operation. Inductors perform a major role to achieve various functionalities, and the benefit of passive inductors in terms of higher dynamic range over their active counterparts almost always come with the compromise in die area, and electromagnetic interactions with similar elements depending on the direction of current flow.

Apart from the standard technology (actives: MOS, bipolar, passives: resistor, capacitor, inductor, etc.), it sometimes proves quite effective to avail a Schottky barrier diode. Schottky barrier diodes are potentially the highest frequency nonlinear devices available in a given technology module, and can be effectively used as mixing devices. Mixing techniques using Schottky barrier diodes, as well as several interesting harmonic effects, are quite attractive for direct conversion architectures [4].

RF system performances are sometimes impacted by the mismatches in the devices, which essentially result from the inaccuracy caused by the variation in the minimum fabrication dimension (length in case of deep submicron CMOS, emitter width in case of bipolars). Hence, the choice of device dimension might trade-off with matching characteristics in some cases.

For highly sensitive building blocks at RF frequencies, various isolation structures are beneficial from a signal injection perspective, and the success of a mixed-signal system is heavily dependent on such features.

Impact of System Performance on Circuits

The circuits in direct conversion system must also provide a high level of reverse isolation. For the case of a LNA, a cascode topology provides about 30 dB reverse isolation, which is quite attractive in terms of direct conversion architecture. It might seem to the reader that an increasing number of such stages might give all the reverse isolation we would need to have DC offset under control, if we were to enjoy the freedom of consuming more current in the front end. While this proves to be a viable option, the actual implementation may be frustrated by the substrate leakage and presence of feedback loops, resulting in potential stability problems. The front-end circuits in their generic form consist of one or more of building blocks such as a transconductor core (V-I), a current switching core, an impedance network, etc. The signal levels and frequencies could define the functionality of the same circuit as a mixer, frequency multiplier, cascode amplifier, and a variable gain amplifier. Bipolar devices enjoy the benefit of the fact that the transconductance is only dependent on the bias currents, thus not requiring any scaling (hence lower parasitics) in geometry, leading to lower parasitics in most cases (except low noise requirements), implying the potential elimination of inductors in the LO buffer to minimize die area. Degeneration can be used to enhance linearity, and the output impedance from the current source should be high to realize higher IIP2 from the mixer cores, and high common mode rejection in case of differential amplifier topology. Input offset voltage and input referred noise of an amplifier could be reduced by increasing the gain of the amplifier. Offset voltage should be paid careful attention at low frequencies and near DC, while noise is important at RF frequencies.

Irrespective of various efforts and reports of research, it has been observed that fundamental developments in the area of circuit topology and functionality have been rather slow. Due to tremendous developments in the area of different communication technologies, and the continued maturity of semiconductor platforms and integration requirements, RF circuits have been looked into from different perspectives, each of which is quite different from the other. The beauty of RF circuits lies in their interrelations with other peripheral blocks, approaches from analog and microwave domains, and presence of more number of performance variables than the number of circuit components. While some of these are true for analog circuits, one key distinction enjoyed by RF circuits is the presence of distributed nature of circuit components, which are most appropriately captured in microwave domain. This vision, along with the fundamental considerations of device noise and physical behavior at high frequencies, presence of uncertainties make RF circuits ever interesting to work with.

10.6 SiGe Direct Conversion Radio: A Practical Example

With all the illustrations provided above, we now describe the design, implementation, and characterization of two SiGe BiCMOS radio front ends. For the sake of simplicity, and to illustrate the design complexity better, we focus mostly on the front end in this section. This implementation is done in 5 to 6 GHz band, with a target specification of IEEE802.11a high data rate wireless standard.

System Considerations

Let us start our approach by understanding the IEEE802.11a standards to design an integrated receiver. We focus on two approaches. In the first approach, we study the design and implementation of the receiver system based on a 50 Ω interfacing impedance, and in the second approach, we illustrate the development of the receiver radios in an integrated fashion (i.e., a non-50-Ω-based approach).

Operating in the 5 to 6 GHz frequency range, IEEE802.11a utilizes orthogonal frequency division multiplexing (OFDM) to achieve high data rate, while providing robustness to multipath interference effects (intersymbol interference). This standard is capable of supporting high data rates up to 54 Mbps and more. However, due to the high crest factor (also referred to as peak to average ratio) of the OFDM-modulated signal waveform, the RF front end blocks require much higher linearity than that for standard quadrature phase shift keying (QPSK) or FM receivers. However, success of OFDM scheme lies in the use of efficient signal-processing algorithms and simplicity in baseband processing. The 5 to 6 GHz band is divided into two parts. The lower U-NII band contains eight channels within 200 MHz (5.15 to 5.35 GHz) and the upper U-NII band contains four channels within 100 MHz (5.725 to 5.825 GHz) [5]. Figure 10.3 shows the frequency allocation of such systems.

An interesting observation can be made from Figure 10.3. Each channel consists of 52 OFDM subcarriers, arranged symmetrically around the zero carrier, each of which spans over 312.5 kHz. We can utilize the zeroth carrier as a null (no transmission at this subcarrier), leading to the implementation of "direct conversion friendly" system architecture due to the "DC-free" modulation scheme, with the compromise on the overall throughput by a small amount. At the receiver, a simple high pass filtering can be used to eliminate the unwanted degradation components (DC-offset and $1/f$ noise) without much SNR degradation for the message spectrum.

With these considerations, two simple architectures can be defined as shown in Figure 10.4. The first one is a traditional approach based on 50 Ω interfacing impedance, thus simplifying the system-level calculations to derive the specifications of building blocks. The second approach is based on more integration, eliminating the "50 Ω" concept for the interfacing circuits. Placement of the filter before the LNA provides a better system design for rejecting out of band interferers and facilitates comparison of both the system architectures. However, the placement of the filter in-between the LNA and mixer helps to lower the noise figure of the entire receiver, and reject out of band blockers. The dashed portions indicate the front-end circuit blocks that we will discuss. A differential configuration in the front-end mixers minimizes the effect of common mode noise and provides high IIP2 needed for direct conversion

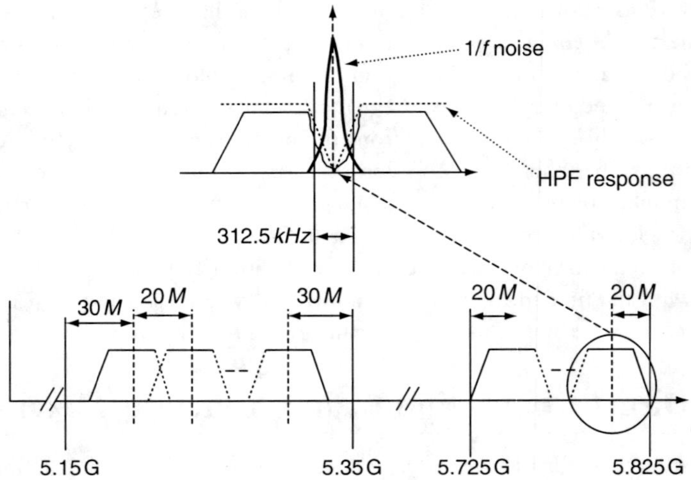

FIGURE 10.3 Illustration of frequency planning for IEEE802.11a, and the consideration of direct conversion receivers.

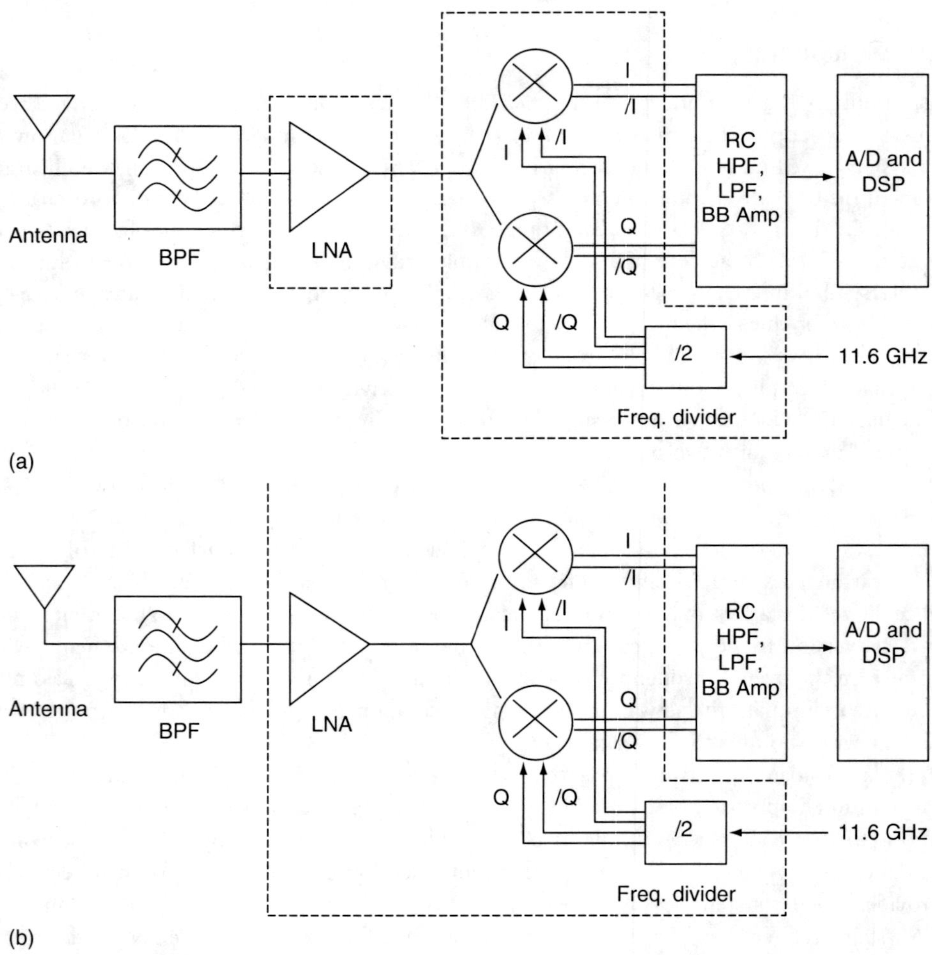

FIGURE 10.4 Two direct conversion receiver architectures under consideration: (a) 50-Ω-based interface and (b) non-50-Ω-based integrated approach.

receivers. A 2X frequency scheme has been adopted in this design for the ease of design for the frequency divider and the relatively high accuracy of the quadrature signals in the signal generation path. Thus, the frequency divider accepts differential signals at 11.6 GHz, while providing quadrature phases at 5.8 GHz at its output, to be interfaced with the downconverting mixers. The system calculations include the gain distribution, noise, I/Q imbalance, linearity parameters (IIP2, IIP3, P1dB). A detailed description of these calculations is provided in Ref. [6].

Circuit Design

The design platform under consideration is a second-generation SiGe BiCMOS technology consisting of five metal layers with an additional thick metal layer for high Q inductor realization. The substrate resistivity is 10 to 20 Ω cm and bipolar transistors with emitter widths of 0.32, 0.44, and 0.8 μm are available. The process utilizes high-performance graded bandgap base NPN with f_T and f_{MAX} of 47 and 60 GHz, respectively, for standard NPN HBTs, and f_T of 27 GHz for high breakdown NPN HBTs. It uses breakdown voltages of 3 and 5 V, respectively, for standard and high breakdown voltage devices. The high f_T transistors use a pedestal structure, which prevents spreading of the currents in different directions as opposed to the conventional BJT structure, thus improving the f_T. Gate oxide thickness of 5 and 7 nm used for MOSFETs (normal and 3.3 V, respectively). The NMOS and PMOS transistors have gate lengths of 0.24 μm. The top metal (called analog metal, AM) is 4-μm thick, and used for inductors and low loss interconnects. Isolation mechanisms include deep trench (DT) and shallow trench (ST).

While selecting the technology modules for circuit design, several trade-offs exist. The minimum width devices exhibit lowest possible noise contribution due to small base spreading resistance, and poor device matching properties due to susceptibility to process variations. However, lowest width devices exhibit more peak f_T compared to the other geometry variants. Choice of passive components (such as resistors and capacitors) is dependent on their parasitic component, temperature coefficient, and tolerance limits. These considerations result in an appropriate choice of components from a semiconductor platform, while building circuits.

Low-Noise Amplifier (LNA)

The LNA is realized in a cascode topology with a switched gain configuration as shown in Figure 10.5. As can be shown from the figure, Q2 transforms the input voltage to current, Q1 accepts input current

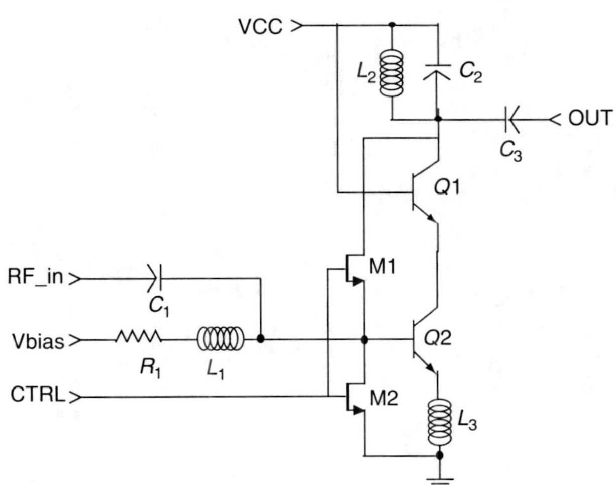

FIGURE 10.5 Circuit schematic for the cascode switch gain LNA.

and provides output current, and the load network generates the voltage at the output. While the functionality of the load network is quite linear, the major component of nonlinearity is contributed by Q2, which is often improved by using degeneration, as other analog feedback techniques prove detrimental in terms of parasitics and stability at high frequency. A cascode topology is often used for high-frequency low-noise amplifier (LNA) circuit due to several advantages: (a) isolation of input and output, leading to separate matching considerations at both ports, (b) reduction of Miller multiplication effect, (c) high reverse isolation from output to the input. While the first two are important in case of any radio architectures, the reverse isolation aspect is critical to the success of direct conversion radios. Use of inductive degeneration helps realize a 50 Ω impedance at the input in a narrowband fashion. The high transconductance and superior noise performance are the key performance parameters for the choice of bipolar transistors over MOS. Due to the higher transconductance, the inductor dimension gets reduced, leading to space compaction on die. Other benefits of using bipolars in the main signal path are the presence of very low parasitic capacitances, which reduces signal attenuation and noise contribution from the cascode transistor (Q1), and at the same time favoring the presence of low impedance, such as 50 Ω (the optimum noise impedance of bipolar is usually lower compared to that of MOS due to the presence of significant current noise spectral density at the base). The input and output matching is implemented using fully monolithic inductor implementation. The input transistor Q2 transforms the input voltage into current by the transconductance property.

Gain switching functionality has been realized by NMOS transistor pair, M1 and M2. When the control voltage is "LOW" or 0 V, both M1 and M2 are turned off, and the LNA functionality is governed by the bipolar transistors Q1 and Q2. This is referred to as the "high gain" mode operation of the LNA. The low gain mode occurs when the control voltage is at a "HIGH" state, typically raised to the supply voltage of the circuit. In this case, both the transistors, M1 and M2 are turned on. M2 provides a bypassing path to the input RF signal, and M1 provides a short between input and output terminals. This ensures a very little current flow to the base of the transistor, Q2, hence reduces the gain significantly.

This particular gain switching approach disturbs the input match by a significant amount in the low gain mode. The gain step is about 20 to 25 dB while incorporating this technique. However, if it is desirable that the input match and the RF characteristics (such as broadband noise and linearity) remain constant, the approaches shown in Figure 10.6 prove to be better alternatives than are shown in Figure 10.5. Figure 10.6a provides a scheme where current steering approach has been adopted, while Figure 10.6b provides a scheme where input and output match is kept constant while using the middle stage as a gain control stage (or bypassing stage). In practical receivers, the control voltage is determined by a

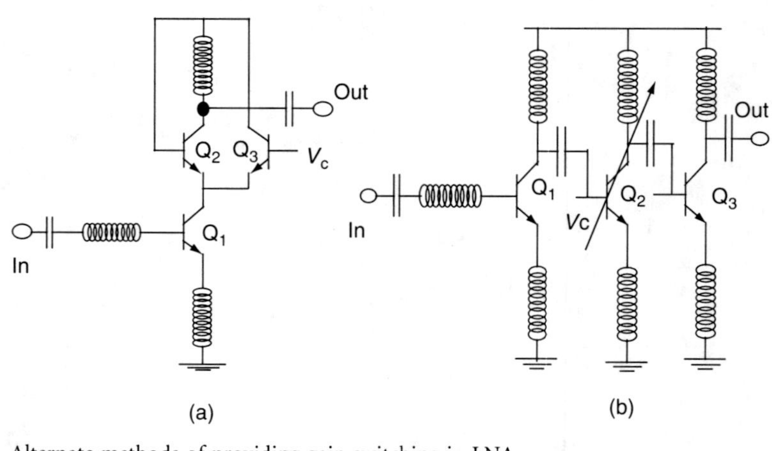

FIGURE 10.6 Alternate methods of providing gain switching in LNA.

digital control from the baseband DSP, after the signal strength has been detected by the receiver signal strength indicator (RSSI). A PTAT current source is desirable to obtain a constant transconductance ($g_m = I_c / V_t$) over the temperature range.

The input matching of the LNA is kept the same for both the architectures, while the output network is combined with the downconversion mixers.

Downconversion Mixers

The basis of the mixer architectures is the presence of a transconductor cell, followed by the switch core and the load network. The transconductor core converts the input voltage to current, and hence the dominating source of nonlinearity in the mixer. The switch network switches the current generated by the transconductor core, by the application of an auxiliary waveform called "local oscillator." The load network is used to perform current to voltage conversion, and works in a fairly linear fashion (exceptions to this are the use of a voltage-dependent capacitance in an RC low pass filter as the load network of a high voltage gain mixer, where the high output voltage swing can change the capacitance value, and hence the cutoff frequency). This configuration is the basis of a Gilbertcell mixer as shown in Figure 10.7 in terms of a single-ended interface, where a current source is also employed to provide a fixed current (which is essential to mixer's conversion gain, noise, and linearity performances). In this section, we first demonstrate the key functional elements of Gilbertcell mixers, followed by a detailed explanation of the specific implementation details. Many considerations apply to many similar families of differential circuits.

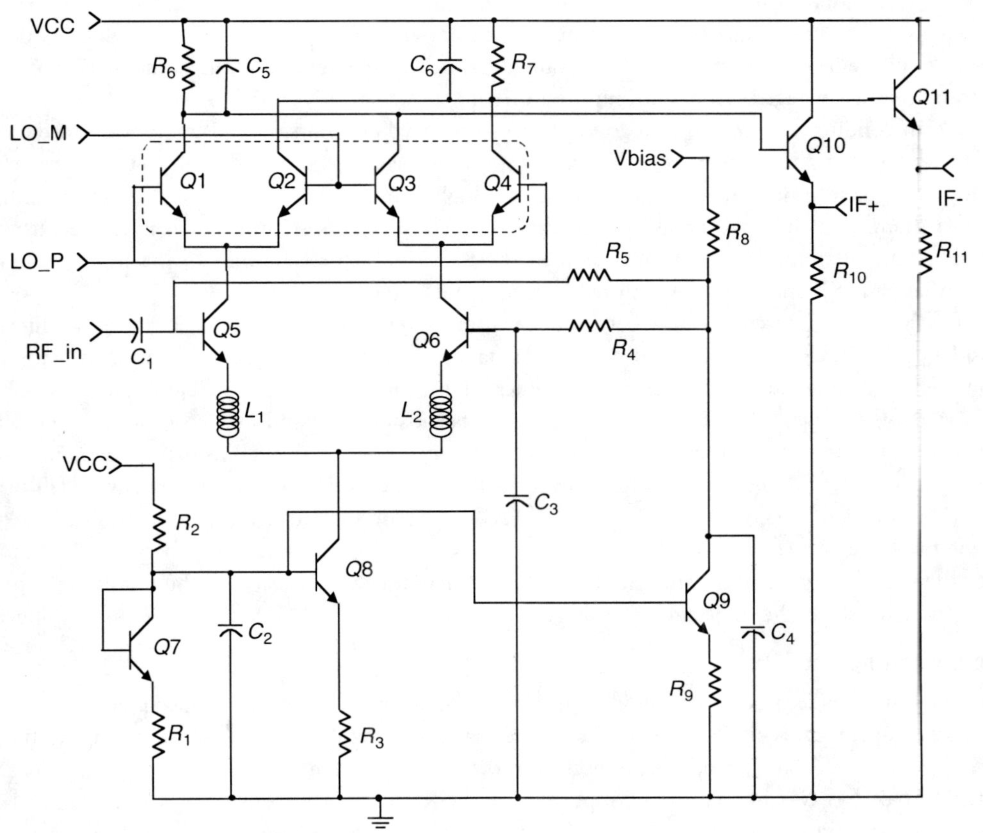

FIGURE 10.7 Circuit schematic of single-ended Gilbertcell mixer.

The Current Source

Let us pay close attention to the specific implementations of such mixers at high frequencies. The current source should provide high output impedance (for high rejection of common mode and second-order effects), low parasitic capacitance at the output node (not to create a pole at the high frequencies, which essentially is transformed into a zero at the output), low voltage swing (to reduce headroom, and minimum common mode voltage to accommodate high signal swing). Even order effects become significant when the transconductor core is driven asymmetrically (not a fully balanced fashion). Bipolar transistors prove to be superior compared to MOS in this regard, as the headroom needed is much less ($V_{ce,sat}$) in case of a bipolar transistor, while providing high output impedance, and much lower parasitic capacitance at the output node. In an MOS (rather NMOS) implementation, however, the common practice is using wide swing cascode current source, which exhibits significant parasitics at high frequency, as the headroom is an inverse square root function of the transistor geometry [7]. Placement of the current source in the middle of the differential pair would also cancel the current source noise in a correlated fashion at the differential outputs.

Choice of Transconductor

From a differential transconductor's perspective, a bipolar transconductor provides lower linearity (due to tanh transfer characteristics, which is limited till 78 mV of differential swing), compared to an MOS differential pair square law characteristics, limits set by the overdrive voltage. However, use of an MOS differential pair as a high impedance stage is frustrated by the presence of large input device capacitance. To achieve similar transconductance performance with a given bias current, MOS usually consumes more area compared to a bipolar, leading to higher parasitics. Thus, use of bipolars at current densities slightly lower than that to achieve peak (to accommodate variations) F_t is usually preferred for higher transconductance, and a degeneration inductor can be utilized to provide higher linearity, at the expense of reduced transconductance due to the presence of negative feedback. Inductors do not consume voltage headroom and do not generate noise. In a downconversion mixer, the thermal noise of the transconductor stage becomes an important design criterion.

The transconductor proves to be a good voltage-controlled current source if the input impedance of the stage is much higher compared to the driving source impedance (so that the full voltage drops across the input stage), and the output impedance is fairly high (which is basic property of a current source). We also desire that the node parasitics remain extremely small for any unwanted noise perturbation and signal attenuation. The basic operation of the transconductor is similar to the input stage of a differential pair [8], and would be omitted here to avoid duplicity.

A fundamental difference exists between the transmit and receive mixers in terms of linearity considerations. In case of transmitter transconductance core, the linearity consideration is a combination of broadband harmonic effects (harmonics of the input tones fall in band, such as 3X, 5X harmonics of the input differential tone at 10 MHz fall inside the band and corrupt the signal) and narrowband phenomenon (intermodulations of the input tones can also fall in band), while in case of receiver it is a narrowband phenomenon as the harmonics of the input tone almost always fall out of band (harmonics of 2.4 GHz tone falls at 7.2G, 12G etc., assuming a differential operation). However, the intermodulations (IM2 and IM3) usually fall in-band in case of a receiver, and cause degradation. Since the harmonic generation and intermodulation phenomena are caused by the nonlinearity of the transconductor stage, the transconductor stage at the transmitter would usually require higher linearity.

The Switching Pair

The switching pair is usually a combination of differential pairs, driven in proper phases. While the principle of operation is similar to that of a differential pair, careful attention should be paid to the use of choice of devices and important aspects for the RF performances, such as noise and linearity. Switching pair has two inputs: (a) an input current at RF frequency from a high output impedance source (e.g., the transconductor), (b) a switching waveform (called "local oscillator") in the voltage domain characterized by the amplitude and the frequency of operation. The switching core usually presents fairly low input impedance to the input RF current source (assuming the lead impedance is much smaller compared to the output impedance of the transistors, which is usually the case), which is desirable for a current amplifier interface. The input impedance of the switching pair is mostly dominated by parasitics, which requires inductors in the LO buffers for cancellation purposes. Hence, low parasitic and high-speed devices, such as bipolar transistors, are ideal candidates for the switching core. Due to the lower parasitic capacitance presented at various nodes, the bandwidth limitation in a bipolar implementation is also quite negligible. In the switching pair, the exponential characteristics of a bipolar differential pair become attractive, as a fairly low signal swing is needed to completely steer the current from one device to another in the differential pair, leading to low current consumption in the LO buffers and possible elimination of an inductive load, leading to further miniaturization.

It is important at this stage that we identify the sources of noise from a switching pair. The most dominating source of noise from the switching pair is the flicker noise, of the transistors, which is transferred to the output node without any frequency translation, at the zero crossing of the switching instants of the LO waveform. The noise contribution reduces with increasing the slope of the LO waveform, presence of lower parasitic capacitance at the emitter (or source), and inherent low flicker noise of the devices. Since the switching occurs in the current domain (and not in the voltage domain), there is no fundamental drawback such as voltage shift of bipolar transistors from emitter to collector (which hinders their use for a sample and hold switch) for using bipolars in a current mode switch core. Hence, bipolar transistors are the preferred choice for switch core devices. The high output impedance of bipolars also helps realize a summing node of the downconversion mixer core.

One important consideration in the switching core is the mismatch of the devices. A fully differential operation is conducive to high-performance direct conversion receivers in terms of perfect LO cancellation, and sideband suppression effects. However, in practical implementation, mismatches are always present, and difficult to completely eliminate even with the best preventive (such as common centroid layout) and predictive (such as extensive statistical simulations based on device mismatches) methods. Hence, the choice of devices usually is a judicious selection to satisfy the flicker noise performance, matching, input loading, and the optimum voltage swing for the switches.

The Load Network

The load network usually utilizes highly linear passive elements (such that the cutoff frequency does not vary with voltage swing across them). The commonly used components include resistors, capacitances, and inductances. The key aspects of these load networks include: (a) headroom consumed, (b) matching between the components to minimize DC offset (static), (c) noise contribution, and (d) bandwidth. Often times these networks describe a filtering action to eliminate the out of band high-frequency components. The signal swing can get higher at the transmitter and the receiver, leading to significant nonlinearity and shift in the filter poles when the voltage dependence of such components is considerable (which is usually the case with high-density capacitors, used in a low pass filter at the receiver output). The passive elements in the load network can also bring significant parasitic capacitance (distributed capacitance from a resistor or bottom plate parasitic of a capacitor), leading to bandwidth limitation and gain degradation for the circuit. To achieve optimum performance of the transmit and receive mixers, the switching core might require a different bias current compared to the transconductor core. A bleeder network (auxiliary current source and sink) can be used to achieve this functionality.

Buffers

A buffer circuitry serves the following purposes: (a) isolation between different functional blocks, (b) transforming impedances through interfaces, (c) boosting or attenuating voltages and currents, and (d) interfacing with the test equipment for controlling the output impedance of the test chip. Two common examples of buffer topologies are: (a) differential pair, class A stages (frequently used in the signal paths to control common mode voltages) and boosting gain and (b) emitter (or source) follower, which is used to isolate sensitive blocks, and realize a low output impedance.

Buffers are usually single-ended circuits, and consume a certain bias current for its functionality. The capacitance at the output node (emitter or source) should be observed for stability considerations. The emitter follower also exhibits nonlinearity effects at large signal swings, which is dependent on the quiescent current of the stage. Bipolar stages would be preferred here, due to higher transconductance and hence lower output impedance ($1/g_m$), lower parasitic capacitance, and absence of linearity problems due to body effects.

The Bias Network

Various types of bias circuits have been discussed in Refs. [1,7]. These include constant current, constant g_m, PTAT, constant V_{dsat}, etc. Bias circuits utilize large noise bypassing capacitors to reduce noise injection to sensitive RF circuits. It is also important that the output impedances of the bias circuit are carefully adjusted (high or low) depending on the application under consideration. Often times the bias circuits include several loops inside them and the loop gains must be carefully adjusted to prevent any oscillations at high frequency to any small part of the RF signal leaking to the bias network. Usually bias network uses many MOS transistors for current scaling purposes (as they do not exhibit a base current, eliminating the need of base current compensation techniques), and few bipolar devices to utilize their exponential characteristics.

Downconversion Mixers: Practical Example

Let us focus on the practical implementations of two different variants of Gilbertcell mixers. The first one is a micromixer as shown in Figure 10.8, originally proposed by Gilbert [9], using a class AB-type input stage using quasi-symmetric bipolar transistors (Q5 and Q10). The transistors Q5, Q6, Q9, and Q10 form a translinear loop, and the product of collector currents of Q5 and Q10 is equal to that from Q6 and Q9. The switching core is similar to a traditional Gilbertcell mixer. The input impedance of the circuit is formed by the parallel combination of the impedances seen in the two paths formed by: (a) the combined series impedance of L3, diode connected Q10, and R3, and (b) the combined series impedance of R1, L2, and emitter resistance of Q5 at high frequencies, it is quite important to maintain the balance between the two paths.

Micromixers exhibit several interesting properties. The first one is the large signal operation of the class—AB stage. Each of the transistors, Q5 and Q10 exhibit exponential nonlinearity, but the difference of the currents (which is significant for the switching operation) remains linear [9]. Thus, even if the individual stages exhibit nonlinearity, the differential operation remains linear, leading to very high linearity of such mixers at low bias currents. Another interesting property is their low input impedance due to high transconductance of bipolar transistors. The level of input impedance also varies with the input signal, leading to gain expansion at higher input signal swing as illustrated in Ref. [9]. The imbalance in the quasi-symmetric stages is adjusted by series combination of the passive elements. However, the diode-connected transistor at the input contributes significant noise, and thus micromixers remain attractive for a 50 Ω impedance standard, and high linearity, while contributing higher noise.

Transistors Q12 and Q13 form emitter-coupled buffers to isolate the mixer from the testing equipment. Capacitors C2 and C7 have been used as noise bypassing capacitors to minimize noise contributions from the bias circuit. Transistors Q7 and Q8, along with pull-down resistors, are used as offset nulling elements to cancel static DC-offset effects.

The second variant is a high impedance inductively degenerated Gilbertcell mixer (shown in Figure 10.8) used for a non-50-Ω-based architecture, and the similar descriptions would apply. The filter networks have been adjusted to accommodate the spectrum of interest, and the bipolar switching pair also requires low LO swing for the mixing operation. The operating frequency of the degeneration inductors should be less than: (a) self-resonating frequency (SRF) of the inductors (L1 and L2) and (b) resonating frequency of the network formed by the degeneration inductor (L1) and the base to emitter capacitance of the bipolar transistor (Q5) at the input stage.

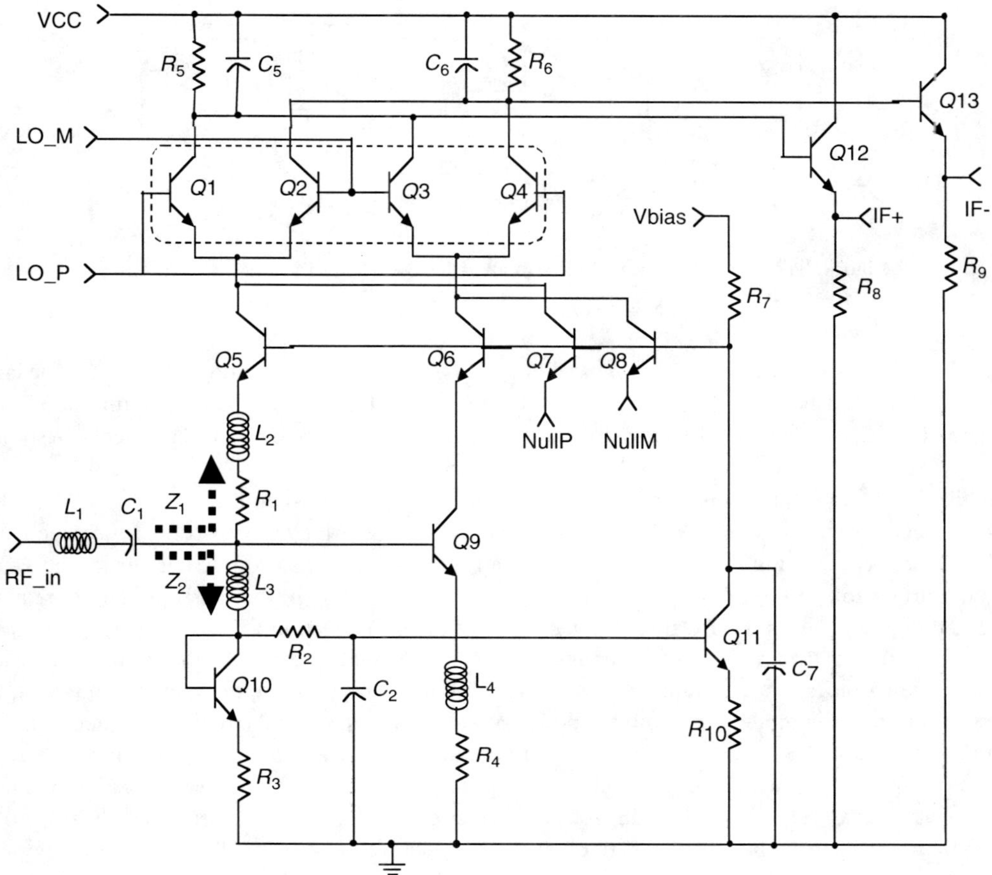

FIGURE 10.8 Circuit schematic of Gilbert-micromixer.

Signal Generation Path

In the signal generation path, 2X frequency scheme has been adopted due to the considerations illustrated earlier. Two emitter coupled D flip-flop circuits have been utilized to realize the master–slave configuration needed for a static frequency divider, which operate at different clock edges.

The scheme is shown in Figure 10.9. It starts with generation of differential signals from a single–ended input at 11.6 GHz, followed by the conversion to suitable voltage levels by appropriate buffers. The D flip-flop core generates quadrature signals, which in turn is amplified by the buffers to provide necessary LO swing to the mixer-switching core. The accuracy of such implementation is strongly dependent on the balance in the differential signal path, and any imbalance at the input of the divider network leads to the imbalance of the outputs in quadrature. Quasi-symmetric micromixer stage can be used here to achieve the desired matching for a 50 Ω interface.

10.7 The Integration: Interfaces and Layout

Having illustrated the details about the operation of the building blocks, we now focus on the integration of these circuits. From a circuit interface point of view, the frequency divider, along with the buffers, should be interfaced with the LO terminal of the mixers and the chain should be optimized for the desired signal swing at the LO terminals needed for switching. A resistive load in the LO buffer might suffice in this regard due to the lower parasitic capacitances of the bipolar transistors. In the signal

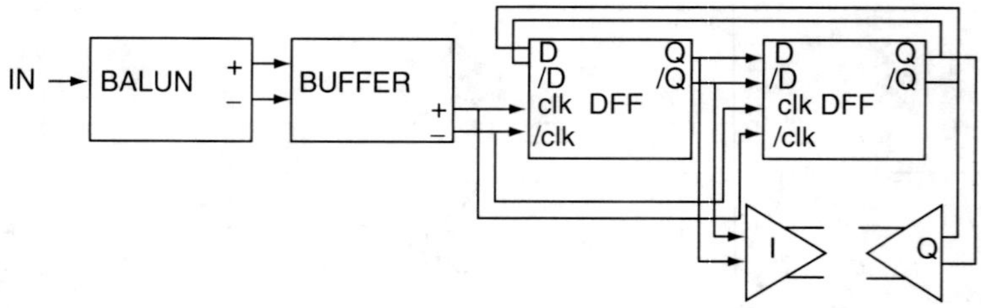

FIGURE 10.9 Block diagram of frequency divider (input at 11.6 GHz).

path, the voltage gain peak must be observed at the LNA-mixer interface, and the output impedance of the LNA should be combined with the input impedance of the mixer for area optimization. The integrated blocks should be designed for gain, noise, time-domain behavior, frequency response, linearity, and robustness to process and temperature variations, etc.

While the circuit design considerations are somewhat straightforward and intuitive to explain, there exist tremendous challenges in the layout of the integrated blocks, which usually come from experience, and mistakes performed in the early test chips. For any high-frequency silicon design, the layout plays a very important role in the circuit performance due to the lossy nature of the substrate. There are many generic design guidelines toward mixed signal layout, but in this section we will consider only those, which are pertinent toward the chosen application. Since the circuit integrates both sensitive analog and digital building blocks, care should be taken to separate the two power supplies, thus minimizing the effect of any common mode noise from the power supplies. Also, the digital circuitry (frequency divider in this case) should be surrounded by deep trench and substrate contacts to minimize the noise generated by digital switching on the other analog circuit blocks of the receiver. Substrate contacts (also referred to as "taps") also help in isolating one circuit block from another and fully absorb the ground current in the substrate. Input trace to the LNA should be kept small to minimize any noise figure degradation. Usually, the top metal is used for interconnection and routing due to lower parasitic capacitances. Separation and orientation of inductors are important factors to minimize the flux coupling effect.

Shielding of the input signals from the substrate is also very important. This is performed by having the input signals in topmost metal layer with the ground as stacked up M1 and M2 running together surrounding the chip and underneath the pads.

Layout symmetry is extremely critical in receiver front ends, and any asymmetry in this regard destroys the fully differential nature of the front end, leading to unwanted components. I/Q imbalance associated with the mixers is very sensitive to the layout symmetry. Hence, in the layout, equal trace length must be respected from the RF input to the mixers. This is even more critical in the D flip-flop layout in terms of the clock distribution network. If the transistors are not identical and the layout traces are not equal, it results in deviation of the duty cycle from 50%, directly affecting the system performance.

Sometimes shielding of the layout traces becomes very important. If the metal line connecting the base goes underneath the metal line connecting collector and vice versa, it gives rise to additional Miller capacitance, which might degrade the high-frequency performance. Hence, another metal layer can be used as shielding the two traces. DC bias lines may lead to unwanted inductive effect, which may affect the RF performance.

Large DC blocking capacitors should be provided between the power supply and ground to minimize the effect of noise from the supply voltages (supply voltage variation) as well as to prevent any kind of ringing effect from DC supplies. The building blocks for both the receiver architectures were laid out for on wafer probing technique, and all the inputs at high frequency were assumed to be single ended to minimize the number of pads, leading to area compaction.

10.8 Characterization of Receiver Front End

Any description of design and layout of integrated circuits is incomplete without its characterization and hardware debugging. A flow chart for the complete receiver testing is demonstrated in Figure 10.10. The measurement setups were performed exactly the way the front end was simulated. The micromixer and integrated receiver were characterized using *S*-parameters (by network analyzer), conversion gain (by spectrum analyzer), linearity (P1dB, IIP2, and IIP3) (by spectrum analyzer), I/Q imbalance (by oscilloscope), 50 Ω input match (by network analyzer), noise figure (by noise figure meter), and DC-offset (by multimeter). The developed circuits utilized wafer probing techniques for characterization purposes. During characterization, several subtle issues need to be addressed, such as:

(a) The power up sequence should be properly respected.
(b) The interfaces (such as differential to single-ended converter or transformer) used for noise characterization must utilize components that do not provide additional noise of their own.
(c) The interfaces (differential to single-ended buffer) used for gain and linearity characterization should utilize components with very high linearity and high input impedance, so that they do not introduce nonlinearity and do not load the signal path.
(d) Both time and frequency domain characterization should be performed to conclude about the functionality of the test-chip.
(e) During the IM2 test, it must be ensured that no component from IM3 tree affects the IM2 component. (This happens when the receiver is tested with 5.8003, and 5.8004 GHz tones in the presence of 11.6 GHz tone, so that the IM2 = 10 MHz coincides with IM5 tone.)
(f) For any transceiver characterization, the basic test should include characterization with discrete tones to eliminate the necessity for highly sophisticated modulated signal generation equipments.

10.9 Fabricated Hardware and Results

Following the above discussions, let us concentrate on the measurement results and learning from them. Figure 10.11 and Figure 10.12 show the microphotographs of the fabricated hardwares, with their performance descriptions provided in Table 10.3 and Table 10.4, respectively. From a direct conversion perspective, we would expect an extremely low value of DC offset, high value of IIP2, and lower value of

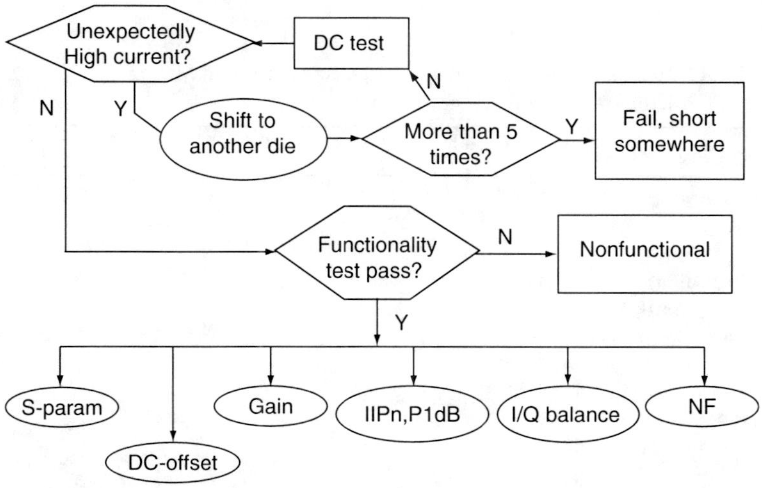

FIGURE 10.10 A complete flow chart for mixer or receiver testing.

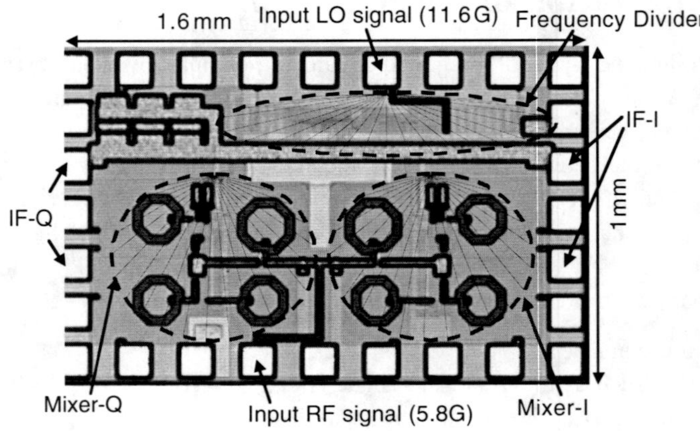

FIGURE 10.11 Die photograph of fully monolithic micromixer IC.

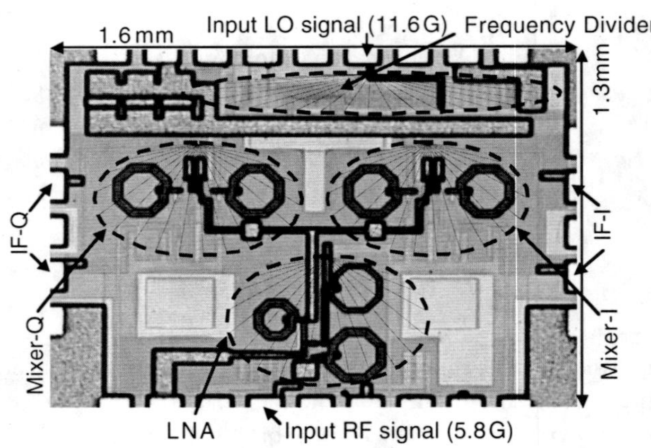

FIGURE 10.12 Die photograph of fully monolithic receiver IC.

TABLE 10.3 Performance Summary of Direct Conversion Micromixer IC (Two Micromixers with Frequency Divider)

Performance	Micromixer
Gain (dB)	9.23
NF (dB)	19.5
ICP (dBm)	−3
IIP3 (dBm)	+6
IIP2 (dBm)	+32
Amplitude imbalance (dB)	0.25
Phase imbalance (deg)	2.4
Power consumption (mW)	32
Input match (dB)	16
LO power (dBm)	−12
LO to RF leakage (dB)	58
Dynamic DC offset (mV)	<1.6

TABLE 10.4 Performance Summary of the Integrated Direct Conversion Receiver (LNA, I/Q Mixers, Frequency Divider)

Performance	Integrated Receiver
Gain (dB)	20.2
NF (dB)	7.1
ICP (dBm)	−15.5
IIP3 (dBm)	−3
IIP2 (dBm)	+31
Amplitude imbalance (dB)	0.9
Phase imbalance (deg)	5
Power consumption (mW)	64
Input match (dB)	15
LO power (dBm)	−10
LO to RF leakage (dB)	78
Dynamic DC offset (mV)	<1

I/Q imbalance parameters. While the first one is satisfied due to careful layout and reverse isolation techniques, the IIP2 is affected by the choice of a single-ended stage of the transconductors for simplicity of interfaces. Imbalances are a function of several aspects (assuming perfect layout symmetry): (a) inherent circuit imbalance due to unequal parasitic components at high frequency, (b) imbalances of single to differential conversion stage, (c) generation of additional imbalance from circuits in the signal path with imbalanced inputs. Another cause of imbalance comes from unequal grounding effects in the I/Q paths, which can cause severe degradation as noted in Table 10.4.

10.10 Compensation and Corrections

The above description of direct conversion radios addresses only its open loop operation. Although such description is quite simplistic to appreciate the fundamental operations from the circuits and device perspectives, it is far from complete from a system viewpoint. A major part of any transceiver system (and more so in case of direct conversion radio) is embedded in its baseband controls and signal-processing functionalities, offset cancellation, and tuning circuitry for robustness of operation. Any transceiver continuously communicates with its baseband controller during its operation, sometimes in an adaptive way. This is essential for DC offset cancellation, and any offset corrections. Fundamentally, the resultant effect of all mismatches leads to unwanted offset voltage at the baseband, which is stored on a capacitor, and by controlling the polarity of the capacitors, the offset is subtracted from the incoming signal. There could be other examples, such as comparison of circuit performances with a reference design to detect the process corner and compensate the circuit elements according to the process variations. Even if the principles are quite simple, such methods enhance the robustness of any successful prototypes. Although compensation schemes could be fundamentally feasible at RF [10], robust implementations and parasitic considerations at high frequency hinder their applicability in successful prototypes.

While static offset corrections are easier to perform (at the expense of more area on chip), and usually require a single time effort (usually at power up), dynamic corrections are extremely difficult to perform. For example, to compensate for dynamic DC offsets (whose levels can vary over several orders of magnitude depending on the surrounding environment), one can employ averaging techniques, or other second-order techniques, depending on the allowable time window in the communication protocol. It is inefficient to provide a filtering function, as it would degrade the SNR by considerable amount, consume area (large capacitors would be needed), and result in a long settling time.

10.11 Future of Direct Conversion Radios

Direct conversion architectures are viewed as potential candidates for the success of small form factor wireless transceivers. The implementation of direct conversion radios is a combination of architecture selection, communication technology, and semiconductor platform. Due to the presence of tremendous diversity in each of these selections, the direct conversion radios enjoy quite large variation in terms of their actual implementations. Future generation applications also require low voltage, low power implementations. An intuitive approach from circuit topologies indicates that to realize low power in the integrated radios, more functionality should be combined in a given bias current, leading to headroom problems if the blocks are stacked within the supply rails. Scaling of processes usually increases the fabrication cost (due to the need for sophisticated lithography techniques), but helps reduce the power consumption of the RF processing blocks by a significant amount. In case of CMOS scaling, the subthreshold region gets pushed further in high-frequency regime, thus potentially leading to low power operation. These approaches should be judiciously used to achieve a reasonable noise performance acceptable to wireless standards. To achieve high linearity in the front-end building blocks, number of voltage to current converter stages (V-I stages, e.g., transconductor) should be minimized, as this is a potential reason for nonlinearity in the front end.

An interesting consideration is the use of inductive components. Use of inductor not only increases the overall chip dimension, but also results in unwanted coupling with other inductors present in the radio architecture, leading to careful floor plan considerations. Quality factor of inductor should be carefully adjusted to accommodate narrowband and broadband communication systems. Active inductor eliminates these issues, while posing significant limits to the dynamic range. Hence, for superior performance, inductors should be judiciously used and designed for minimizing lateral area (footprint), and simultaneously realizing the necessary quality factor for the specific operations.

It should also be observed at this point that transceiver circuits pose many challenges to a circuit designer. Designing in a "CMOS-only" platform eliminates the possible use of any bipolar transistors except in the bias networks where substrate BJTs are used. At deep submicron regime, the gate leakage becomes quite significant, requiring similar techniques to base current compensation for bipolar transistors. Use of "bipolar-only" platform eliminates many considerations required for transistor scaling, etc. However, use of a high-speed platform such as "SiGe BiCMOS" leads to a much more variety of circuit topologies and the judicious use of MOS and bipolar transistors.

References

1. W.R. Maclaurin. *Invention and Innovation in the Radio Industry.* New York: The Macmillan Company, 1949.
2. I.A.W. Vance. Fully integrated radio paging receiver. *Proceedings of IEE* 129(1), 1982, 2–6.
3. T.H. Lee. *The Design of CMOS Radio-Frequency Integrated Circuits.* Cambridge University Press, 1998.
4. M. Shimozawa, T. Katsura, N. Suematsu, K. Itoh, Y. Isota, and O. Ishida. A passive-type even harmonic quadrature mixer using simple filter configuration for direct conversion receiver. IEEE International Microwave Symposium, June 2000, pp. 517–520.
5. IEEE Draft Supplement to IEEE Std 802.11, Wireless LAN Medium Access Control (MAC) and Physical Layer (PHY) Specifications: High-Speed Physical Layer in the 5 GHz Band, September 1999.
6. J. Laskar, B. Matinpour, and S. Chakraborty. *Modern Receiver Front-Ends: Systems, Circuits and Integration.* John Wiley Publisher, 2004.
7. P.E. Allen and D. Holdberg. *CMOS Analog Circuit Design*, 2nd ed. Oxford University Press, 2002.
8. P.R. Gray, P.J. Hurst, S.H. Lewis, and R.G. Meyer. *Analysis and Design of Analog Integrated Circuits*, 4th ed. John Wiley Publisher, 2001.

9. B. Gilbert. The MICROMIXER: a highly linear variant of Gilbert mixer using a bisymmetric class-AB input stage. *IEEE J. Solid State Circuits* 32(9), 1997, 1412–1423.
10. B. Matinpour, S. Chakraborty, and J. Laskar. Novel DC-offset cancellation techniques for even-harmonic direct conversion receiver. *IEEE Trans. Microwave Theory Tech.* 48(12), 2000, 2554–2559.

11

RF MEMS Techniques in Si/SiGe

11.1 Introduction 11-1
11.2 RF MEMS Switches 11-2
11.3 Application Circuits 11-6
Reconfigurable Low-Loss Resonators and Bandpass Filters • Reconfigurable Bandstop Filter • Reconfigurable Matching Networks (Tuners)

John Papapolymerou
Georgia Institute of Technology

11.1 Introduction

During the last four decades microwave and millimeter-wave systems have been used in a wide variety of commercial, medical, and military applications due to their advantages over other bands of the electromagnetic spectrum (unlike infrared and optical wavelengths they have the ability to permeate fog, dust, and smoke) and the lack of lower frequencies for new services. Some of these applications include satellite-to-earth communication systems, plane and missile navigation, early warning radars, earth remote sensing, short-haul line-of-sight transmission links for personal communication links (PCNs), wireless cable, wireless radio local area networks (LANs), mobile broadband systems, automotive anticollision radars, and hyperthermia for cancer treatment [1]. Until the early 1980s most of the microwave/mm-wave components were built with waveguide technology (rectangular or cylindrical) that offers low-loss and high quality factors at the price of increased size, weight, fabrication cost, and complexity [2]. In addition, the bandwidth of those systems was limited by the bandwidth of the employed waveguides.

With the maturity of integrated circuit techniques and solid-state device technology the electromagnetics research community started focusing its attention on monolithic microwave/mm-wave integrated circuits (MMICs) that dominate today's communication systems. MMICs offer the advantages of small size and weight, low cost, increased reproducibility and reliability, broadband operation, and easy integration of active devices [2]. All of the previously mentioned advantages are essential for today's cost-driven commercial and military microwave industry that seeks to achieve high-performance flexible systems with maximum density and speed. The requirements for high density, high speed, and high operating frequencies along with the advancements in materials science technology, semiconductor processing techniques, and solid-state devices have provided the stimulus for system-on-a-chip solutions. With these solutions, both active and passive components are monolithically integrated on a semiconductor substrate (Si/SiGe) for miniature lightweight and highly reliable microwave systems [3]. However, design and fabrication of entire microwave/mm-wave systems on a single chip pose new problems and challenges. Limitations in the performance of the different components relative to the waveguide ones, co-existence with other circuits (e.g., digital), power consumption and handling, signal switching, and packaging are a few significant samples. To address these issues, researchers in the last 10 years have implemented solutions based on micromachining and microelectromechanical systems (MEMS) techniques, widely used in the sensor and transducer areas since the 1960s [4].

Micromachining techniques generally refer to the selective removal of the silicon substrate material with a variety of methods (chemical, plasma, laser) to create new and interesting structures. The term also encloses the techniques where one builds relatively thick (more than 10 μm) structures on top of the substrate (LIGA, SU-8). MEMS generally relates to the development of electromechanical structures on the surface of the silicon substrate. All MEMS devices provide for a certain amount of mechanical flexibility in the physical construction of the circuit, in addition to the traditional electronic functions. The combination of micromachined and MEMS structures with microwave/mm-wave systems not only creates new interesting challenges, but also provides the opportunity for using revolutionary ideas and concepts to address next generation system requirements. The latter are dictated by the demand to create small chips that are able to sense, think, act, and communicate with the outside world. Monolithic integration, therefore, of Si (or SiGe) circuits, advanced MEMS devices, micromachined analog components, and digital CMOS-based processing circuits into one wafer is the next leap beyond the current state-of-the-art used in the system-on-a-chip approach [3].

In this chapter we will review briefly the fundamentals of RF MEMS switches and provide some application circuits on silicon substrates. Realizing that there exists a variety of MEMS structures that have been developed on silicon substrates, we will focus only on RF switches due to the extensive growth and research interest in their technology, as well as their dominant role in signal routing and distribution. Taking into account the extraordinary growth of the MEMS field up to this point and its market potential for the next 5–10 years, it is anticipated that MEMS technology will enable a new class of microwave/mm-wave components and subsystems and revolutionize the way people think about integrated systems on a chip.

11.2 RF MEMS Switches

RF MEMS switches can be developed by using *surface* micromachining, where thin-film layers (usually polysilicon, various metals, SiO_2, Si_3N_4) are deposited on the semiconductor surface (usually Si) and machined in a way that they can be released from the substrate and thus be able to move and flex in order to provide the desired function. In some cases, the free-standing structures are anchored at one or more positions, and in other cases, they are free to rotate around a pin joint [5]. The electromotive force used to "move" or rotate the above structures on the wafer surface is typically electrostatic attraction, although magnetic, thermal, and piezoelectric actuation mechanisms have been reported [6,7].

RF MEMS switch technology has emerged during the last 10 years as an alternative to traditional solid-state switching technology. RF MEMS switches typically consist of a moving metallic membrane that can "make" or "break" contact when it is actuated by a DC control voltage. The main advantages of RF MEMS switches are low insertion loss up to W-band (~0.3 dB), low intermodulation distortion (IIP3 > 65 dBm), negligible power consumption (~ a few μW), high isolation up to W-band, cutoff frequencies between 20 and 80 THz and low fabrication cost [8]. Switching times are in the range of 1–300 μs, while power handling capabilities are on the order of a few watts [8]. Several companies and universities have developed a wide variety of ohmic and capacitive switches on Si, GaAs, and quartz substrates with excellent performance [9–16].

A schematic of a typical cantilever and air-bridge type of MEMS switch is shown in Figure 11.1a and 1b, respectively. More specifically, Figure 11.1a depicts a cantilever ohmic contact switch, while Figure 11.1b depicts an air-bridge capacitive switch. The switch typically consists of a movable membrane suspended a few microns above the silicon substrate. With the application of an electrostatic voltage between the membrane and a pull-down electrode an electrostatic force is applied to the membrane, and as a result the membrane starts moving closer to the substrate. When the voltage and, thus, the electrostatic force go beyond a certain threshold value the membrane is pulled completely down. Typically the membrane is pulled down at height values around one third the original distance from the substrate.

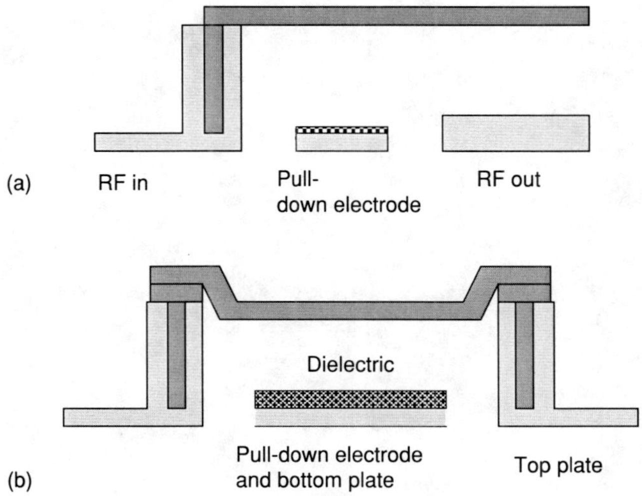

FIGURE 11.1 Schematic of (a) cantilever ohmic switch and (b) capacitive MEMS switch.

At Georgia Tech we have developed both ohmic and capacitive MEMS switches [17,18]. Photos of cantilever type switches are shown in Figure 11.2. The main difference between the two switches is that in the case of the ohmic switch the contact is metal-to-metal, whereas in the capacitive case there is a thin layer of dielectric between the two metals providing a capacitive short. Capacitive switches typically have lower loss at higher frequencies, whereas ohmic switches have smaller loss at lower frequencies. Both switches are activated using an electrostatic force that results from the application of a DC bias voltage. A bias electrode is typically used to apply the DC voltage separately from the RF signal. Typical fabrication steps for a MEMS switch include: (a) formation of the bottom electrode layer by depositing 2000 Å of gold with e-beam evaporation and patterning of the layer; (b) deposition of a thin layer of silicon nitride (typically 1000–2000 Å) by PECVD techniques and patterning by reactive ion etching (RIE) to cover the bottom pull-down electrode area and the RF contact area for the capacitive switch only; (c) patterning of the first metal layer and electroplating to a thickness of 1–3 μm to provide the contact point between the membrane and this layer; (d) patterning of 2-μm-thick photoresist layer ("sacrificial" layer) that will be used to form the movable membrane; (e) electroplating of the membrane areas to a total thickness of 1–3 μm depending on the switch design; (f) removal (release) of the sacrificial layer with solvents and drying of the switches with a critical point drier (step f can be replaced by RIE of the sacrificial layer).

The performance of the ohmic and capacitive switch can be seen in Figure 11.3 and Figure 11.4, respectively. From Figure 11.3 one can observe that when actuated the ohmic switch has a loss of 0.3 dB at 20 GHz, while when unactuated it yields an isolation of 17 dB at 20 GHz. The switch actuation voltage is varied between 20 and 40 V. The equivalent switch model consists of a resistor (R_S) in series with an inductor for the ON state, and a capacitor (C_{OFF}) for the OFF state. The modeled values are $R_S = 0.5\,\Omega$ and $C_{OFF} =$ 15 fF. The actuation time of the ohmic switch was measured to be around 5–10 μs. From Figure 11.4 it can be seen that the capacitive switch yields a loss of 0.15 dB at 18 GHz in the ON state and an isolation of 12 dB at 18 GHz in the OFF state. The equivalent switch model is a series combination of an inductor, capacitor, and a resistor with values of $L =$ 140 pH, $R = 0.5\,\Omega$, $C_{ON} = 900$ fF and $C_{OFF} = 20$ fF. The actuation voltage of the capacitive switch was 30–60 V. Capacitive switches typically have higher power handling capability than ohmic switches with values ranging between 2 and 5 W of a CW RF signal applied between switch actuation ("cold switching").

Reliability and low-cost packaging are probably the two most important remaining challenges for full commercialization of the RF MEMS switch technology. Several companies and universities have been

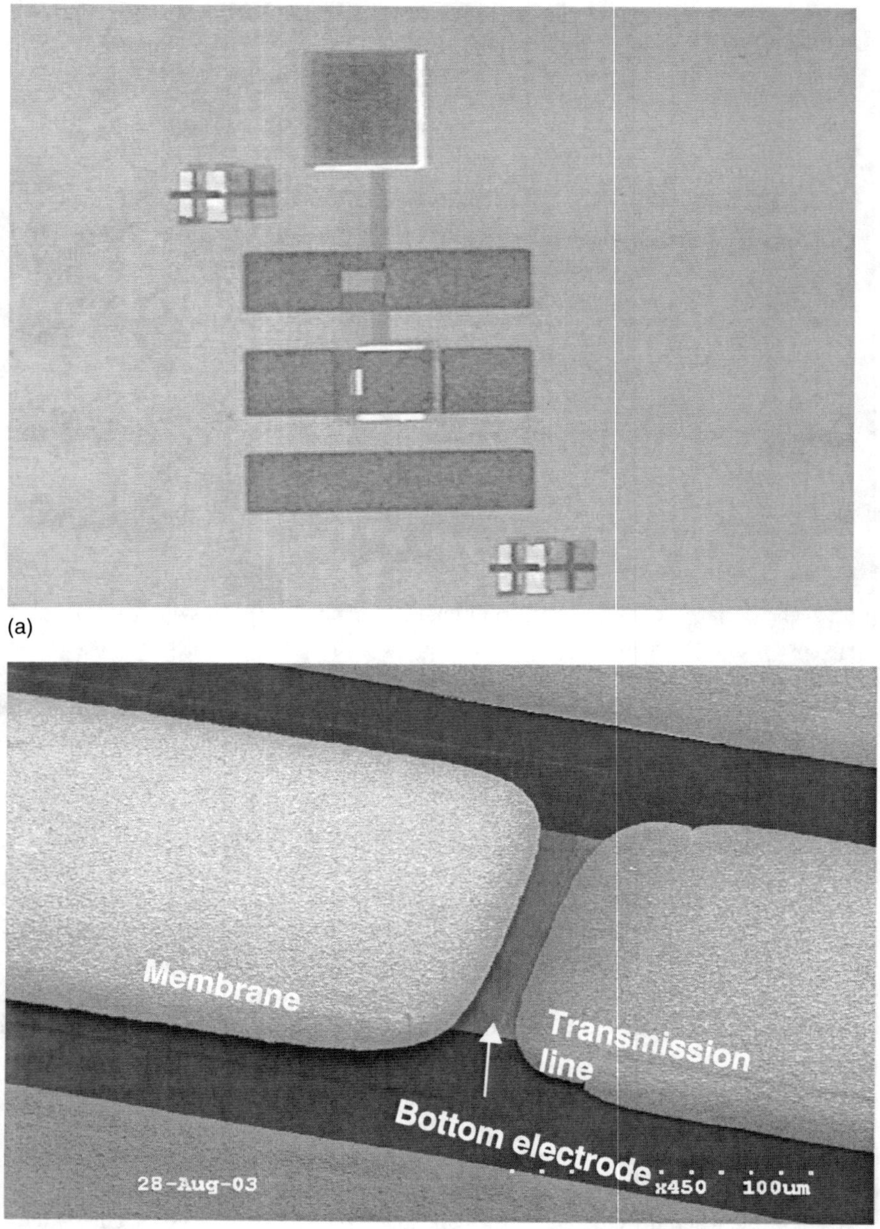

FIGURE 11.2 Photos of cantilever: (a) ohmic switch and (b) capacitive MEMS switch.

focusing during the last 4 to 5 years on addressing these two issues that are interdependent. New material technology, new switch designs, as well as new biasing algorithms have been implemented to improve switch lifetime and package performance. The mechanisms that lead to switch failure (e.g. stiction) are typically different in ohmic vs. capacitive switches. For example, microwelding can reduce the lifetime of ohmic switches, while dielectric charging can reduce it for capacitive switches. Debris in the contact area is a major consideration for both switching types, as well as environmental conditions such as increased levels of relative humidity. It is generally agreed in the research community that an appropriate package is absolutely necessary for increased switch reliability and, eventually, mass

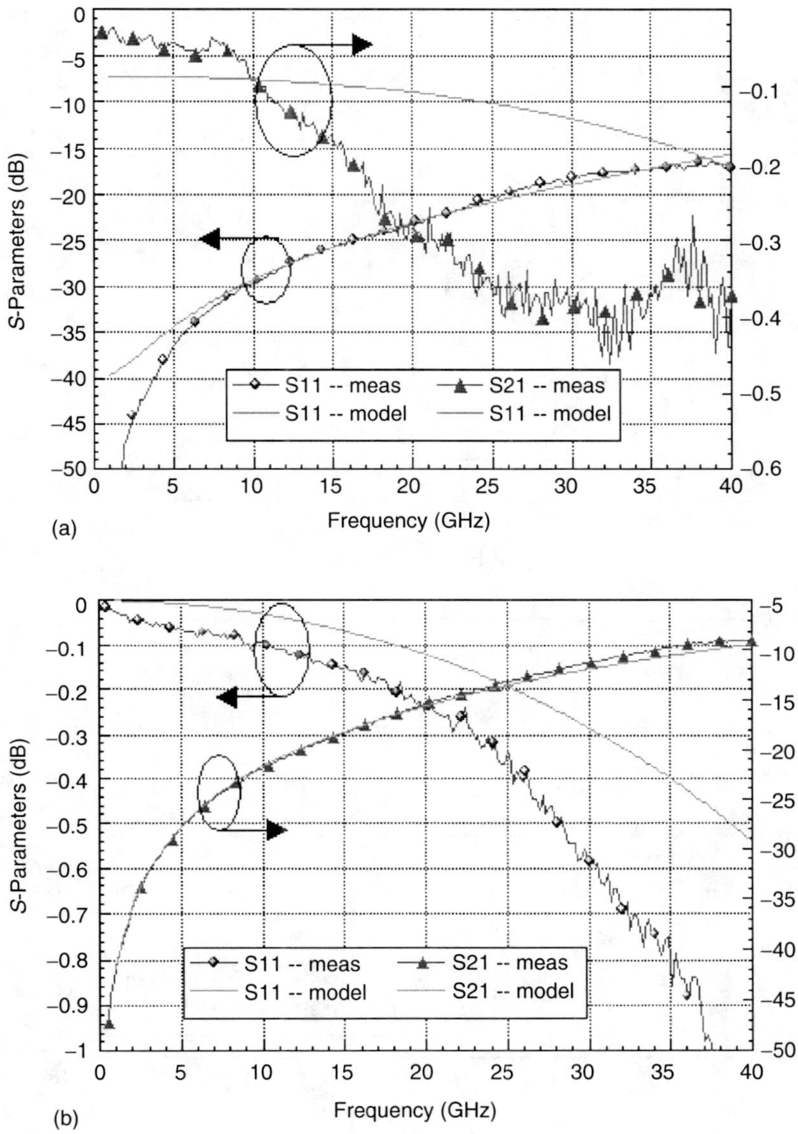

FIGURE 11.3 Measured and simulated *S*-parameters for an ohmic switch in the: (a) ON state and (b) OFF state. (From G. Zheng, P. Kirby, S. Pajic, A. Pothier, P. Blondy, J. Papapolymerou, and Z. Popovic, 2004 IEEE Topical Meeting on Si Monolithic Integrated Circuits in RF Systems, pp. 159–162, Atlanta, GA, September 2004. With permission.)

production and commercialization. Self-actuation under high RF power levels can also lead to reduced lifetimes for both types. At the time this chapter was written typical wafer capped lifetimes of ohmic switches for low RF power levels exceeded 20 billion cycles, while several devices with more than 100 billion cycles were also reported. Ohmic switches with 2 W of RF input power had achieved a lifetime of 2.5 billion cycles [19]. Capacitive switches had statistically demonstrated lifetimes higher than 35 billion cycles, while several devices had achieved more than 70 billion cycles [19]. Near hermetic or hermetic wafer-level packaging is also considered to be the most promising low-cost solution for packaging of RF MEMS switches with increased reliability and low loss. RF and DC interconnects that provide low-loss signal connections in and out of the capped switch are also under investigation. It is expected that within a short period of time RF MEMS switch reliability will statistically exceed values of 90 billion cycles for relatively low RF power levels (up to 2 W).

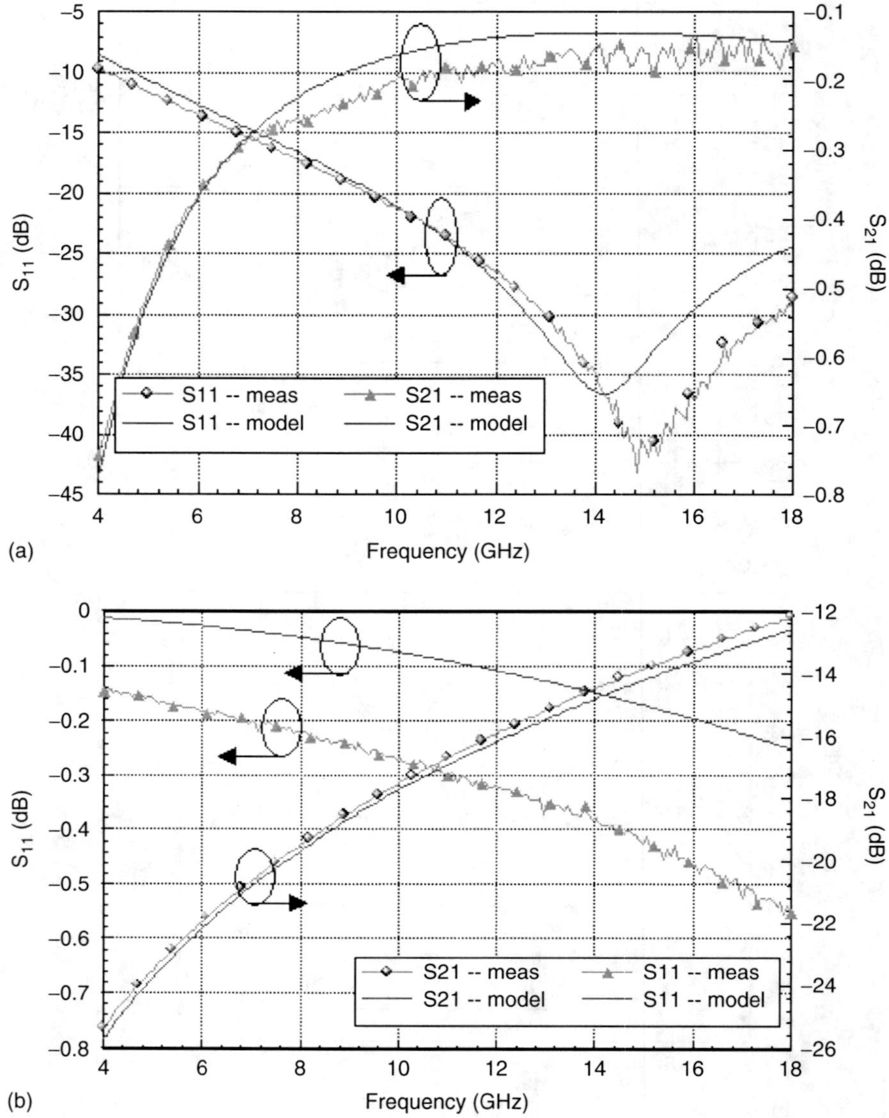

FIGURE 11.4 Measured and simulated *S*-parameters for a capacitive switch in the: (a) ON state and (b) OFF state. (From G. Zheng and J. Papapolymerou, *Int. J. RF Microwave Computer Aided Engineering (special issue on RF Applications of MEMS and Micromachining)*, 14(4), 373–382, 2004. With permission.)

11.3 Application Circuits

In this section we will present some examples of microwave circuits utilizing RF MEMS switches to achieve performance characteristics that are impossible with other microwave switching devices. It should be noted that there is a variety of RF MEMS based microwave circuits that cannot be covered in the extent of one chapter only.

Reconfigurable Low-Loss Resonators and Bandpass Filters

There is an increased interest in low loss, low power, small size tunable microwave filters for receiver applications to replace filter banks that traditionally take more space and use high-*Q* bulky structures.

Reconfigurable filters based on MEMS switches offer a viable alternative to filters using pin diodes or varactors that typically suffer from high losses. This section presents resonators and filters using ohmic contact switches combined with low loss digital capacitive loads to achieve reconfigurability. Their implementation in wide tuning range, low loss band pass filters is demonstrated, showing the interest of this technique [18].

A capacitance quality factor Q is defined as

$$Q = \frac{1}{2\pi f RC} \tag{11.1}$$

where f is the operation frequency, C the capacitance value, and R the series resistance of the capacitor. Figure 11.5a presents an electrical model of a digital capacitive load to ground, with a DC contact switch and its biasing network. Biasing is coupled to the capacitive load through the actuation pad capacitance $C_{act.}$ This capacitance is strongly increased when the switch is turned on. Consequently in the switch ON and OFF states, the biasing network has an important influence. Thus, for microstrip planar capacitors losses are not only due to the switch resistance R_{on} but also to the biasing network.

It can be proven that this effect can be highly reduced with the integration of high resistances R_{bias} in the biasing network. Indeed these resistances permit to isolate the RF device from its biasing network at microwave frequencies, while the electrostatic actuation of the switch is not affected by the presence of the resistance. In the proposed designs, a short open-ended microstrip transmission line is used as the high-Q capacitance to ground. In this configuration, the capacitance value depends on the line dimensions (length and width) and substrate properties (permittivity and thickness). Using this topology, one can obtain very high Q, reproducible capacitors. By careful selection of various capacitive values one can design a capacitor bank that can be switched between different states to achieve reconfiguration in terms of resonant frequency.

Using this topology, different digital capacitive loads have been developed, following the implementation presented in Figure 11.5a. These loads are fabricated on a 400-μm-thick high-resistivity silicon substrate and very high value resistors (0.8–1 MΩ) are integrated in the biasing network. Since a thick low loss metal (>3.5 μm thick gold) is being used for the switch and the load metallization, the switch is the main source of loss in this design. Each DC contact switch used here has two contacts resulting in 0.25 Ω series resistance per switch. Because of the parallel combination of several MEMS switches, these losses can be highly reduced. But, on the other hand, the total isolation is also degraded: a trade-off between ON state losses and OFF state isolation therefore exists. Hence, only two or three MEMS switches are associated to each digital load. Table 11.1 shows the computed quality factor for the digital loads implemented in this work. At 10 GHz, they exhibit a high Q value that relies on the capacitance value and the number of switches used. One can note that loads 3 and 4 use a reduced number of switches for a practical implementation in the circuits shown in this section.

The resonator developed has been designed by loading a planar resonator with two digital capacitive loads presented in Figure 11.5b and Table 11.1. The unloaded resonator is fabricated using a half wavelength microstrip line (3400 μm long and 400 μm wide). It presents a simulated quality factor of 105 at 15.6 GHz and uses an S-shaped line to facilitate the load integration. Each load decreases the resonator frequency, when it is turned on. With two capacitive loads, four combinations are possible and four different frequencies are obtained. A photo of the fabricated resonator along with measured results are shown in Figure 11.6. A summary of full-wave simulations and measured results can be found in Table 11.2. As seen, an unloaded quality factor of around 77 is achieved for all four different resonant frequencies.

Using the resonator presented previously, we have designed a bandpass filter based on a two-pole topology. Indeed, a two-pole filter is an easy way to demonstrate the digital capacitive load potential for filtering devices. Usually a two-pole filter frequency response depends on the coupling between each resonator and also on the coupling between the excitation lines and the resonators. These parameters determine the filter prototype. The main challenge with the tunable filter is to control the coupling

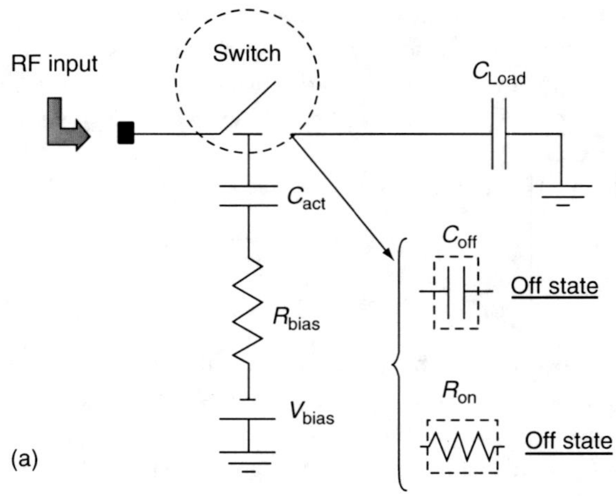

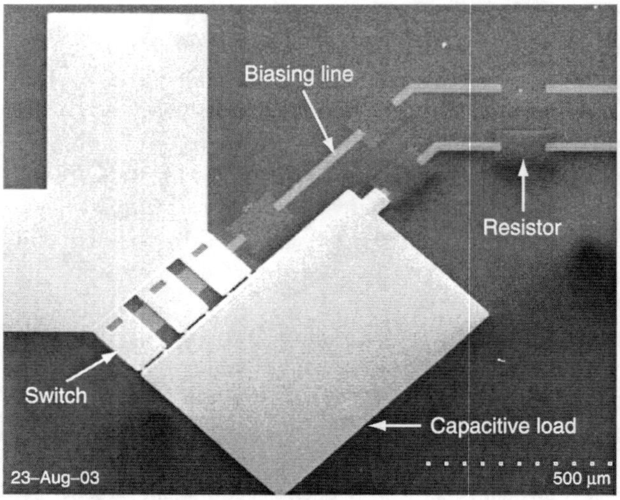

(b)

FIGURE 11.5 (a) Electrical equivalent circuit of a digital capacitive load to ground with its biasing network and (b) SEM picture of a MEMS digital capacitive load and its biasing network fabricated on a high-resistivity silicon substrate. (From A. Pothier, J.-C. Orlianges, E. Zheng, C. Champeaux, A. Catherinot, D. Cros, P. Blondy, and J. Papapolymerou, *IEEE Trans. Microwave Theory Techn.*, 53(1), 354–360, 2005. With permission.)

TABLE 11.1 Computed Performance of the Fabricated Capacitive Loads.

Capacitive Load	1	2	3	4
Length (μm)	320	500	500	900
Width (μM)	400	750	750	1300
Capacitance to ground (fF) @ 10GHz	135	260	260	570
Number of switches used	2	3	2	3
Quality factor (@ 10 GHz)	625	465	330	210

Source: A. Pothier, J.-C. Orlianges, E. Zheng, C. Champeaux, A. Catherinot, D. Cros, P. Blondy, and J. Papapolymerou, *IEEE Trans. Microwave Theory Techn.*, 53(1), 354–360, 2005. With permission.

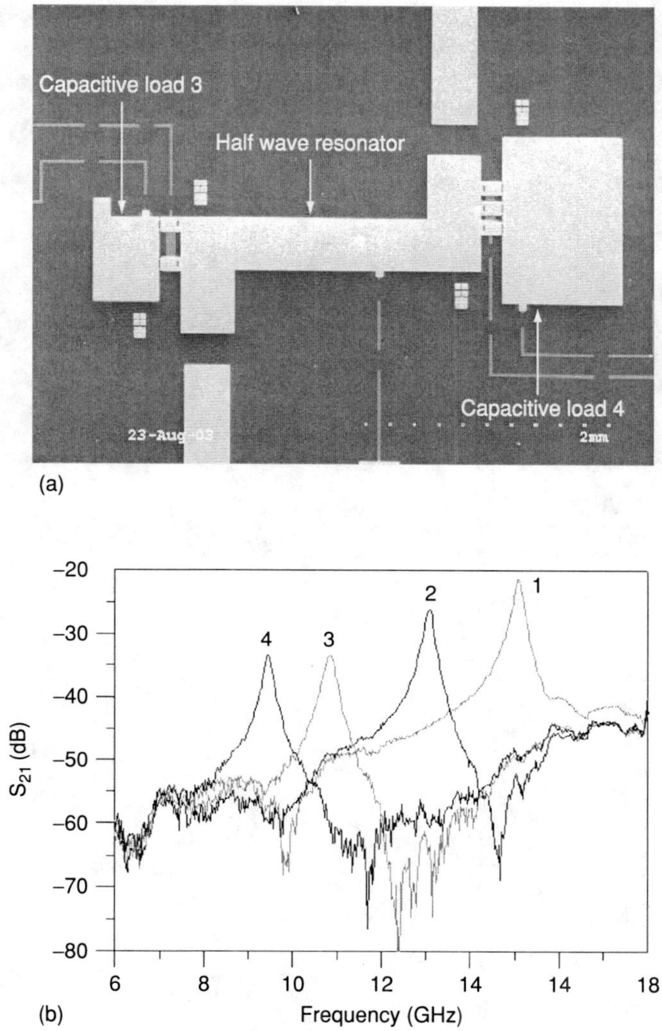

FIGURE 11.6 (a) SEM photo of reconfigurable resonator with MEMS switches on silicon substrate and (b) measured performance of the resonator. (From A. Pothier, J.-C. Orlianges, E. Zheng, C. Champeaux, A. Catherinot, D. Cros, P. Blondy, and J. Papapolymerou, *IEEE Trans. Microwave Theory Techn.*, 53(1), 354–360, 2005. With permission.)

TABLE 11.2 Simulated and Measured Results for Resonator with 43% Tuning Range.

Resonator state	1	2	3	4
Capacitance load turned on	None	#3	#4	#3 & #4
Simulated frequency (GHz)	15.18	13.13	11.19	9.69
Measured frequency (GHz)	15.03	13. 07	11.13	9.7
Simulated quality factor	102	101	90	89
Measured quality factor	79	77	80	75

Source: A. Pothier, J.-C. Orlianges, E. Zheng, C. Champeaux, A. Catherinot, D. Cros, P. Blondy, and J. Papapolymerou, *IEEE Trans. Microwave Theory Techn.*, 53(1), 354–360, 2005. With permission.

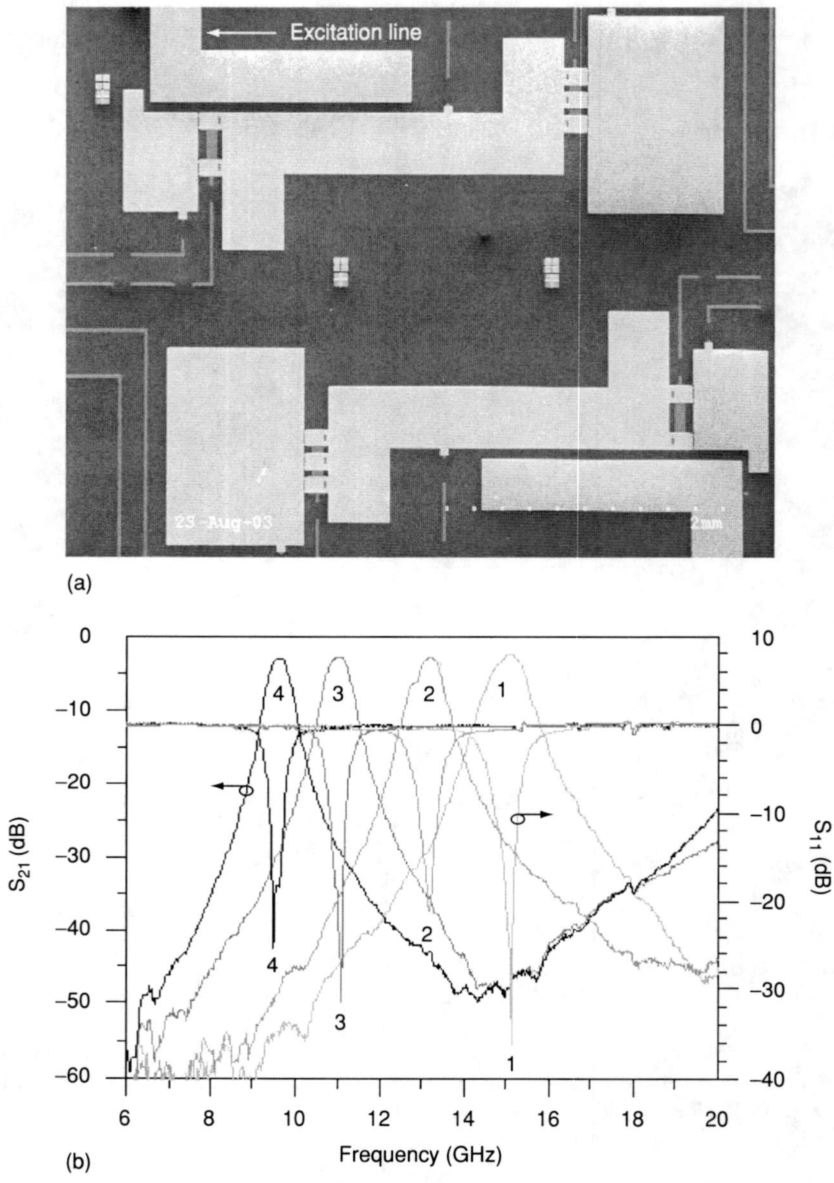

FIGURE 11.7 (a) SEM photo of two-pole MEMS filter with 44% tuning range and (b) measured performance of the filter. (From A. Pothier, J.-C. Orlianges, E. Zheng, C. Champeaux, A. Catherinot, D. Cros, P. Blondy, and J. Papapolymerou, *IEEE Trans. Microwave Theory Techn.*, 53(1), 354–360, 2005. With permission.)

variation when the frequency shifts in order to get the desired filter response. For large frequency shifts, the effect of not controlling the coupling is generally a degradation of the filter loss and the filter fractional bandwidth. In our case, the two microstrip resonators are coupled to each other using their magnetic fields while the capacitive loads maintain a correct input–output coupling. The two-pole filter topology chosen is tolerant to frequency shifting, and proves the concept of microstrip MEMS tunable filters using MEMS DC switches. However, a design methodology similar to the one described in Ref. [20] should be applied for multipole and more complex structures. A photo of the fabricated filter on silicon along with measured results is shown in Figure 11.7. The filter exhibits a low level of insertion loss (less

TABLE 11.3 Summary of Measured Results for Two-Pole Filter with 44% Tuning Range

Filter state	1	2	3	4
Capacitance load turned on	None	#1	#2	All
Measured frequency (GHz)	14.95	13.11	11.05	9.55
Measured bandwidth (MHz)	870	700	600	520
Fractional bandwidth %	5.8	5.3	5.4	5.4
Insertion losses (dB)	2.6	2.8	2.6	3.2

Source: A. Pothier, J.-C. Orlianges, E. Zheng, C. Champeaux, A. Catherinot, D. Cros, P. Blondy, and J. Papapolymerou, *IEEE Trans. Microwave Theory Techn.*, 53(1), 354–360, 2005. With permission.

than 3.2 dB) within the filter bandwidth. The return loss level is also very high (>15 dB), showing good filter matching for the four combinations. The full-wave simulations (method of moments), including metal losses and the biasing network influence, are in very good agreement with the measurements. For the measurements, a TRL calibration procedure has been used to de-embed the CPW to microstrip transition that excites the filters. The filter results are summarized in Table 11.3. The filter can be reconfigured from 9 to 15 GHz with a very small amount of loss. Of paramount importance is the DC biasing network as it can considerably increase losses due to RF leakage.

Reconfigurable Bandstop Filter

One type of tunable filter that is very important in RF receivers for communication and radar systems is the bandstop filter. This section presents a monolithic, reconfigurable, microstrip-based bandstop filter utilizing RF MEMS capacitive switches. The goal of this work is to develop a four-state or “2-bit” tunable bandstop filter from 8 to 15 GHz [17]. More specifically, the filter is designed to have a notch at 8, 10, 13 and 15 GHz, respectively, that can be selected electronically by activating the appropriate MEMS switches. One useful application of this filter is to provide a good image rejection if the microwave receiver experiences signal interferences. The cantilever beam capacitive MEMS switch with electrostatic actuation was chosen for the filter design based on its advantages as follows: first, comparing with ohmic contact switch, it can typically handle more RF power; second, it is easier to design and fabricate for microstrip circuits over the air-bridge capacitive switch.

The bandstop filter using quarter-wavelength open-circuited resonators is one type of configuration commonly used for this type of filtering [21,22]. The electrical length of the shunt stubs and the distance between the shunt stubs is $\lambda g/4$ at the center or design frequency. The four filters that resonate at 8, 10, 13, and 15 GHz were first designed as if a bandstop filter with a single resonant frequency were desired. The characteristic impedances of the filter stubs were calculated at each resonant frequency as given in the following eqution:

$$Z_{\text{on}} = \frac{4Z_0}{\pi g_n \Delta} \tag{11.2}$$

where $n = 1, 2, 3, \ldots$ Z_{on} is the characteristic impedance of each stub, Z_0 is the characteristic impedance of the feed line equal to 50 Ω, g_n is the low-pass prototype value, which can be found in Refs. [21,22] and Δ is the fractional bandwidth.

In order to develop one bandstop filter that can resonate at four different frequencies with some tunable devices, such as MEMS switches, these four individual filters needed to be combined together. To achieve this, the 50 Ω transmission line width at 10 GHz was chosen as the feedline for the tunable filter since it is around the center of the full frequency range. The difference between the stub lengths was first replaced with lumped components using the following equations:

$$\frac{1}{j\varpi C} = -Z_{\text{on}} \cot(\beta l) \tag{11.3}$$

$$j\varpi L = -Z_{\text{on}} \cot(\beta l) \tag{11.4}$$

where C is the capacitance, L is the inductance, Z_{on} is the characteristic impedance of the stub, β is the propagation constant, and l is the length of the microstrip stub.

Since the electrical lengths of the microstrip stubs are different at different resonant frequencies, the corresponding lumped element could be a capacitor (C) or an inductor (L). The equivalent values of the lumped elements at 8, 10, 13, and 15 GHz are: 1.12 nH, 45 fF, 264.5 fF, and 356 fF, respectively. These lumped components were then broken into two parallel subcircuits. In each subcircuit, there are two lumped components in series, which are a two-state capacitance with a fixed larger capacitance or larger inductance as shown in Figure 11.8. The two-state variable capacitance is used as the circuit model value of the MEMS switch, which is very small (~ 20–30 fF) at the off-state and very large (~0.9–1 pF) at the on-state; the other lumped component will be an appropriate fixed capacitance or inductance. By choosing the MEMS switch to be on- or off-state, the total reactance can be easily determined as it should be equal to the reactance before the lumped components were broken into two subcircuits. When the MEMS switch is at the off-state with a very small capacitance, its value will be the dominant one to determine the resonant frequency; but when the switch is at the on-state, its capacitance is relatively large, thus the lumped component which connects to the MEMS switch will be the dominant component to determine the resonant frequency. In other words, by selecting MEMS switches on or off, different resonant frequencies will be selected. To make the circuit monolithic, the fixed lumped components were replaced by microstrip lines with radial stubs. Based on the equivalent capacitances/inductances calculated above and the schematic diagram of Figure 11.8, the required capacitances and inductances were re-calculated and the resulting values are: $C_1 = C_2 = 20/900$ fF, $C_3 = 356$ fF, and $L_4 = 1.12$ nH. The tunable filter with lumped elements has now been designed, and the resonant frequency is determined by selecting the different combinations of capacitors C_1 and C_2.

The S-parameters of the fabricated filters were measured with an Agilent 8510 network analyzer and a Cascade probe station. A multiline Thru-Reflect-line (TRL) calibration technique was used to de-embed the probe pad-to-circuit transitions and move the reference planes to the locations shown in Figure 11.9. A summary of the measured results and a comparison with the same filter design utilizing perfect open and short circuits instead of MEMS switches are shown in Table 11.4. For the bandstop filter with the

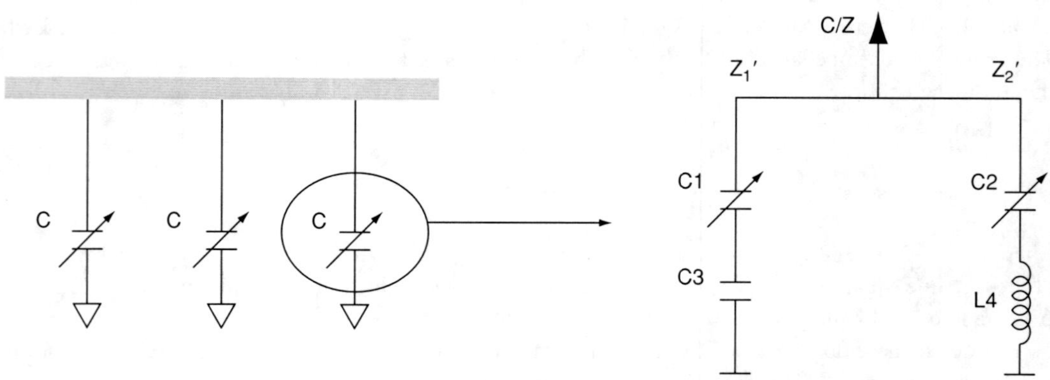

FIGURE 11.8 Schematic of the capacitor of each path breaking into two subcircuits. (From G. Zheng and J. Papapolymerou, *Int. J. RF Microwave Computer Aided Engineering (special issue on RF Applications of MEMS and Micromachining)*, 14(4), 373–382, 2004. With permission.)

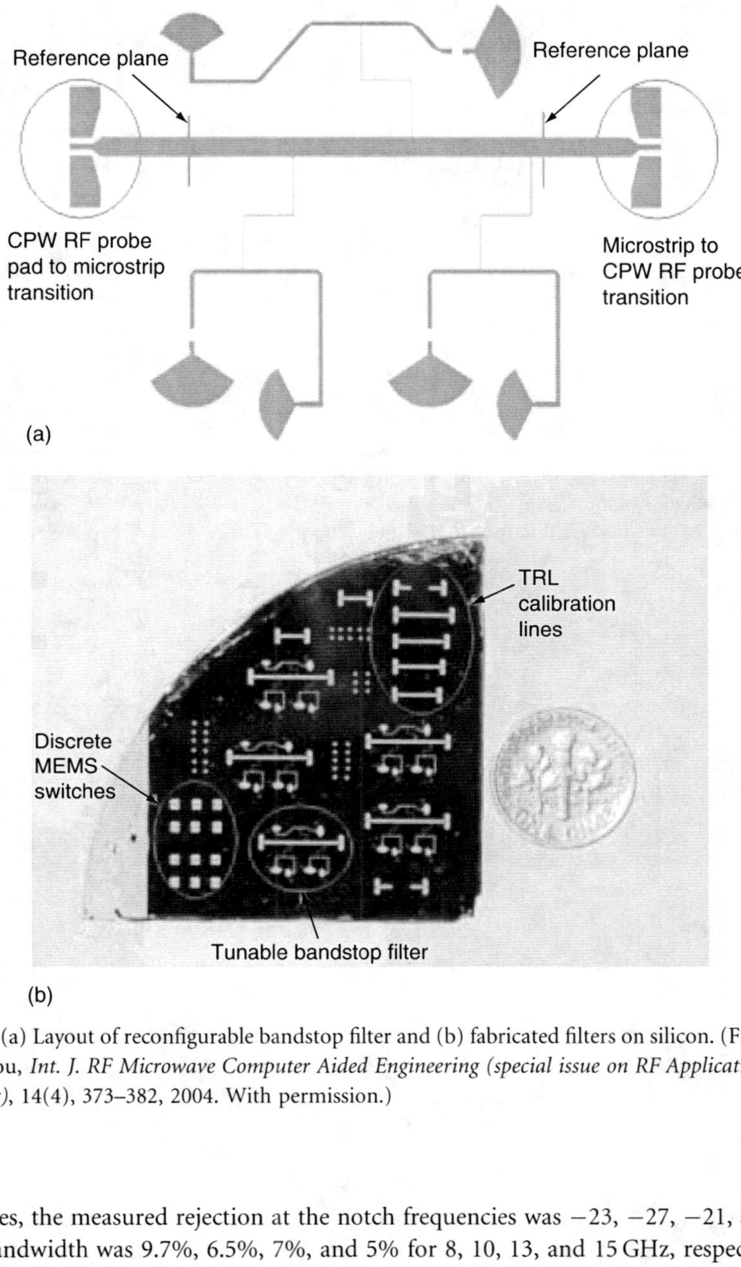

FIGURE 11.9 (a) Layout of reconfigurable bandstop filter and (b) fabricated filters on silicon. (From G. Zheng and J. Papapolymerou, *Int. J. RF Microwave Computer Aided Engineering (special issue on RF Applications of MEMS and Micromachining)*, 14(4), 373–382, 2004. With permission.)

MEMS switches, the measured rejection at the notch frequencies was −23, −27, −21, and −20 dB and the −10 dB bandwidth was 9.7%, 6.5%, 7%, and 5% for 8, 10, 13, and 15 GHz, respectively, while the insertion loss at the passband was measured to be around 0.5 dB.

The measured results were also compared with simulated results (Figure 11.10 to Figure 11.13) where the MEMS switches were modeled as 28 fF at off-state, and 1000 fF at the on-state. The simulation was done in two main steps: (1) the layout simulation was done in Agilent Momentum, but in the simulation, the MEMS switches were removed and six pairs of internal ports were inserted into the gaps; (2) the simulated results from Momentum were imported into the Agilent ADS schematic simulation and the six capacitors with variable values were then connected between these six different pairs of internal ports. As shown in Figure 11.10, Figure 11.11, and Figure 11.13 the measurements and the simulations agree very well at 8, 10, and 15 GHz. Figure 11.12 shows that the bandwidths of the measured and the simulated results at 13 GHz are different. This is caused by the fact that when all the MEMS switches are at the on state, there is RF leakage through the DC bias lines. It should be noted

TABLE 11.4 Center Frequency and Bandwidth Comparison between the Measurement with MEMS Switches and Perfect Open-/Short Circuits.

		MEMS switches	Perfect open or short
Off–off/open–open	Center Freq/Rejection (GHz/dB)	10.16/-27	11.5/-23
	Bandwidth (GHz)[a]	0.42	0.34
	Percentage BW (%)[a]	5	3
	Center Freq/Rejection (GHz/dB)	8.62/−23	9/−40
Off–on/open–short	Bandwidth (GHz)[a]	0.84	0.7
	Percentage BW (%)[a]	9.7	7.8
	Center Freq/Rejection (GHz/dB)	15.13/−21	16.015/−22
On–off/short–open	Bandwidth (GHz)[a]	0.97	0.89
	Percentage BW (%)[a]	6.5	5.6
	Center Freq/Rejection (GHz/dB)	13.275/−20	13.42/−20
On–on/short–short	Bandwidth (GHz)[a]	1.05	0.72
	Percentage BW (%)[a]	7	5.4

Source: G. Zheng and J. Papapolymerou, *Int. J. RF Microwave Computer Aided Engineering (special issue on RF Applications of MEMS and Micromachining)*, 14(4), 373–382, 2004. With permission.
[a]The bandwidth was calculated as the −10 dB bandwidth.

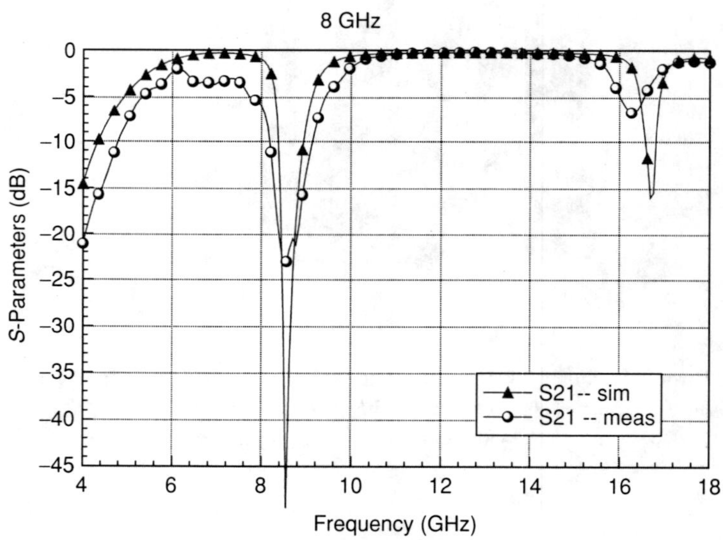

FIGURE 11.10 Measurement vs simulation results for the off–on state of the filter with MEMS switches (8 GHz response). (From G. Zheng and J. Papapolymerou, *Int. J. RF Microwave Computer Aided Engineering (special issue on RF Applications of MEMS and Micromachining)*, 14(4), 373–382, 2004. With permission.)

here that integrated resistors that provide a perfect open at the switch location were not used in this implementation. As a result of the RF leakage or loss, the larger bandwidth is more pronounced for the measurements of Figure 11.12 when compared to the other measurements. For Figure 11.11, where all the switches are at the off-state, an ideal agreement can be observed between simulated and measured data as the DC bias lines are not loading the circuit.

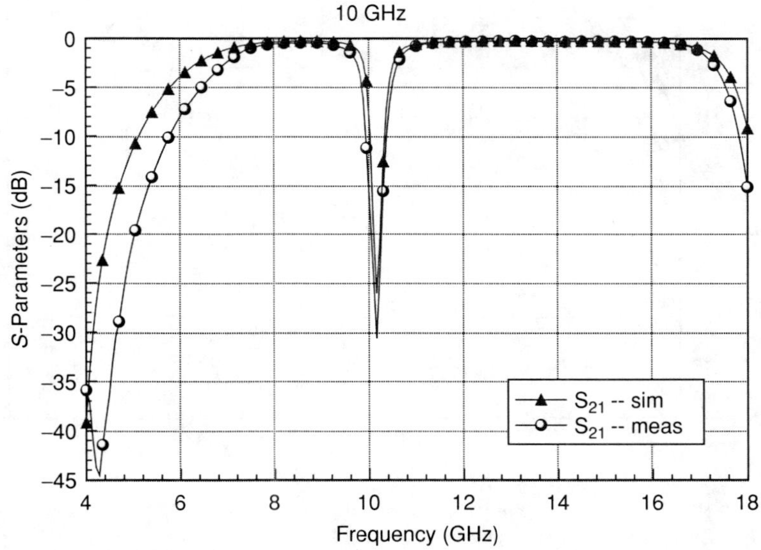

FIGURE 11.11 Measurement vs simulation results for the off–off state of the filter with MEMS switches (10 GHz response). (From G. Zheng and J. Papapolymerou, *Int. J. RF Microwave Computer Aided Engineering (special issue on RF Applications of MEMS and Micromachining)*, 14(4), 373–382, 2004. With permission.)

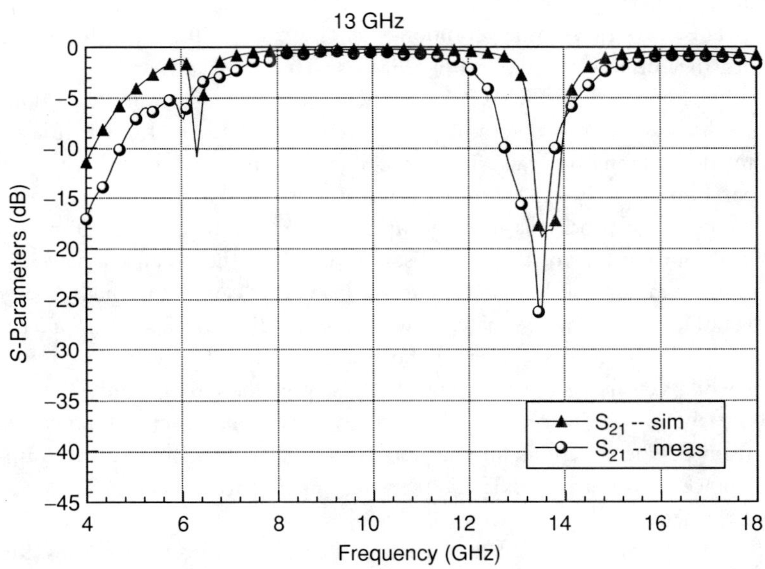

FIGURE 11.12 Measurement vs simulation results for the on–on state of the filter with MEMS switches (13 GHz response). (From G. Zheng and J. Papapolymerou, *Int. J. RF Microwave Computer Aided Engineering (special issue on RF Applications of MEMS and Micromachining)*, 14(4), 373–382, 2004. With permission.)

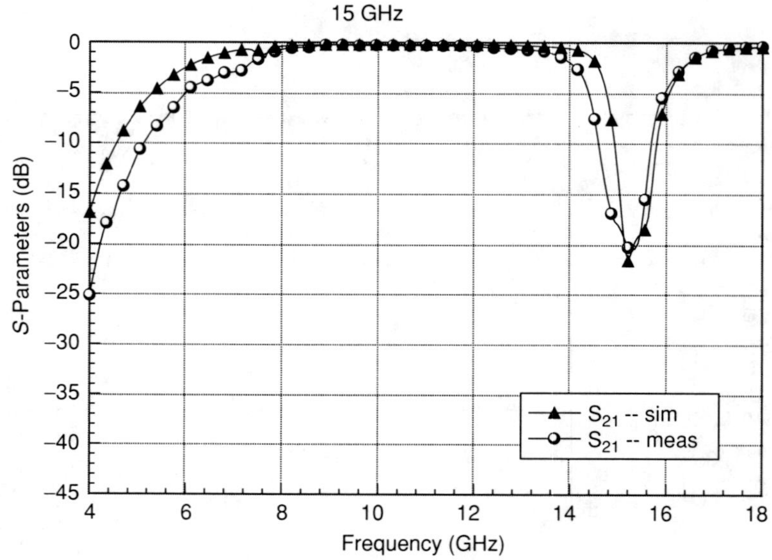

FIGURE 11.13 Measurement vs simulation results for the on–off state of the filter with MEMS switches (15 GHz response). (From G. Zheng and J. Papapolymerou, *Int. J. RF Microwave Computer Aided Engineering (special issue on RF Applications of MEMS and Micromachining)*, 14(4), 373–382, 2004. With permission.)

Reconfigurable Matching Networks (Tuners)

Matching networks are fundamental circuit components in RF front ends since they minimize reflection losses and maximize power transfer from one stage to the other. They are widely used in antenna and amplifier designs, as well as other components. This section presents a reconfigurable, planar double-stub tuner that utilizes capacitive RF MEMS switches and achieves the widest tuning range from 10 to 20 GHz with the highest VSWR [23]. The basic design procedure for a double-stub tuner, outlined in Ref. [23], was used to determine the susceptances of each stub and the lengths between them to match the desired range of loads. The range of loads to be matched was arbitrarily chosen to be 20 to 80 Ω real part and −150 to +150 Ω imaginary. It was assumed that the desired match was to 50 Ω. The first step in the design process is to determine the distance of the first stub to the load. It was assumed that this distance was zero. The second step is to choose the distance *d* between the two stubs. This distance is important since it limits the range of loads that can be matched. A large distance will decrease the range of loads that can be matched while a small distance may be impractical to fabricate. For this tuner, the distance *d* was chosen to be relatively small ($0.1\,\lambda$) to maximize the range of matchable loads while sacrificing the frequency sensitivity of the circuit. Once the distance between the stubs has been chosen, the susceptances of each stub that will match the range of loads to 50 Ω need to be determined. The equations for the susceptance of the two stubs can be found in Ref. [21]. Once the susceptance values are calculated, the stub capacitances can be evaluated at the design frequency ($C = B/2\pi f_d$).

The next step in the design is the implementation of the variable susceptance stubs. Knowing that RF MEMS switches would be used to make the circuit reconfigurable a "digital" approach was pursued, where different capacitances were realized by a combination of a switch and a fixed capacitor. For each stub, four such combinations were used (Figure 11.14), resulting in a "4bit × 4bit" double-stub tuner. The more the bits used, the wider the range of load impedances that can be matched. If fewer bits are used, the more simplistic the circuit is to design and use and the less area it occupies. For our

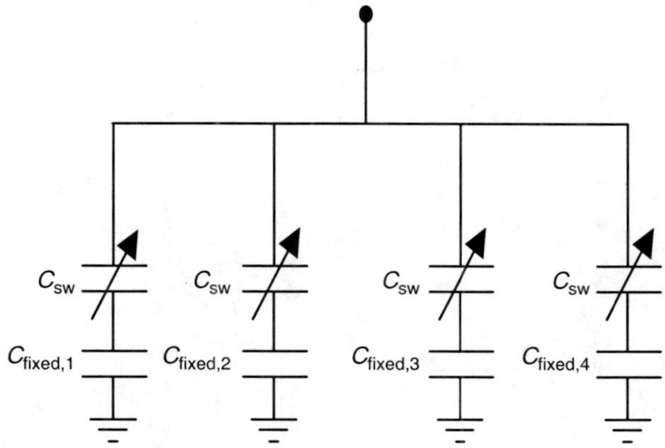

FIGURE 11.14 Schematic of "four-bit" reconfigurable tuner that consists of fixed capacitors and MEMS switches. (From J. Papapolymerou, K. Lange, C. Goldsmith, A. Malczewski, and J. Kleber, *IEEE Trans. Microwave Theory Techn.*, 51(1), 271–278, 2003. With permission.)

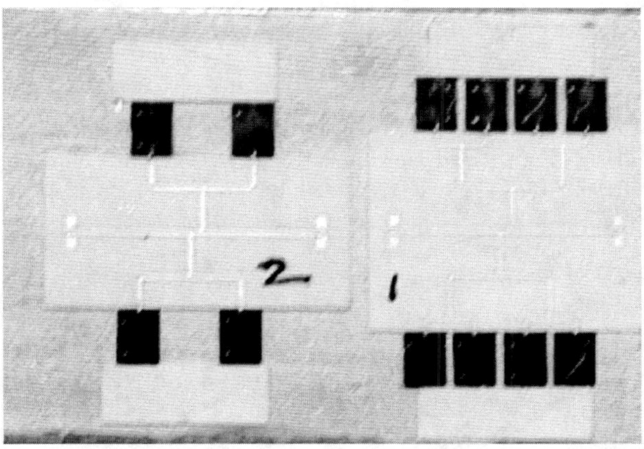

FIGURE 11.15 Photo of fabricated RF MEMS tuners. (From J. Papapolymerou, K. Lange, C. Goldsmith, A. Malczewski, and J. Kleber, *IEEE Trans. Microwave Theory Techn.*, 51(1), 271–278, 2003. With permission.)

implementation the design frequency was chosen at 20 GHz. The RF MEMS switches were capacitive switches, with an "off" capacitance of 35 fF and an "on" capacitance of 3 pF [9]. The fixed capacitors were realized as open stubs and were fabricated on high-resistivity silicon along with the switches. The remainder of the transmission lines was fabricated on low-loss alumina and ribbon bonds were used to connect the two substrates. A photo of the fabricated tuner is shown in Figure 11.15. The tuner size was approximately 18 mm × 11 mm. The tuner response was simulated using HFSS and Agilent ADS. The former tool was used to extract the response of the switch–stub combination that was incorporated in ADS for the calculation of the entire circuit response.

Measured and simulated results for the "4bit × 4bit" tuner at 20 and 15 GHz are shown in Figure 11.16 and Figure 11.17, respectively. As can be seen in Figure 11.16, the tuner can match at 20 GHz load impedances with $1.5\,\Omega < \mathrm{Re}\{Z_L\} < 109\,\Omega$ and $-107\,\Omega < \mathrm{Im}\{Z_L\} < 48\,\Omega$. In addition, the measured and simulated results agree very well. This circuit by far provides the largest range of matchable load impedances equivalent to three quadrants of the Smith chart. Furthermore, measurements show that a

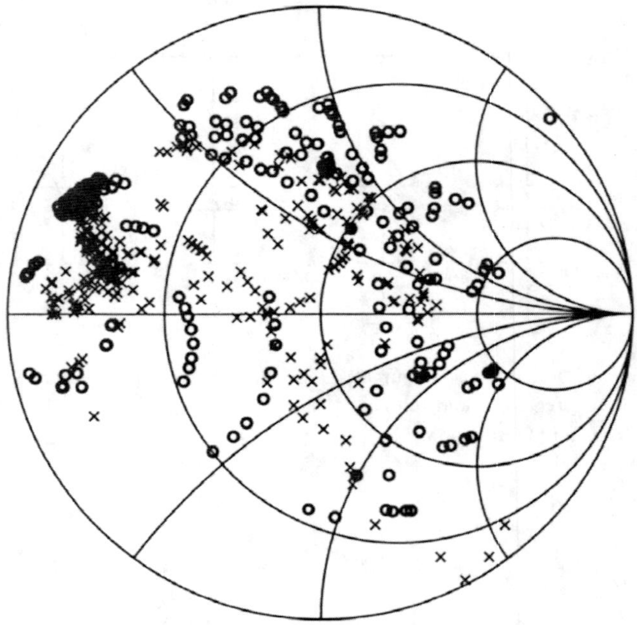

FIGURE 11.16 Simulated (O) and measured (X) range of matched load impedances for the 20 GHz-optimized, 4-bit by 4-bit tuner at 20 GHz. (From J. Papapolymerou, K. Lange, C. Goldsmith, A. Malczewski, and J. Kleber, *IEEE Trans. Microwave Theory Techn.*, 51(1), 271–278, 2003. With permission.)

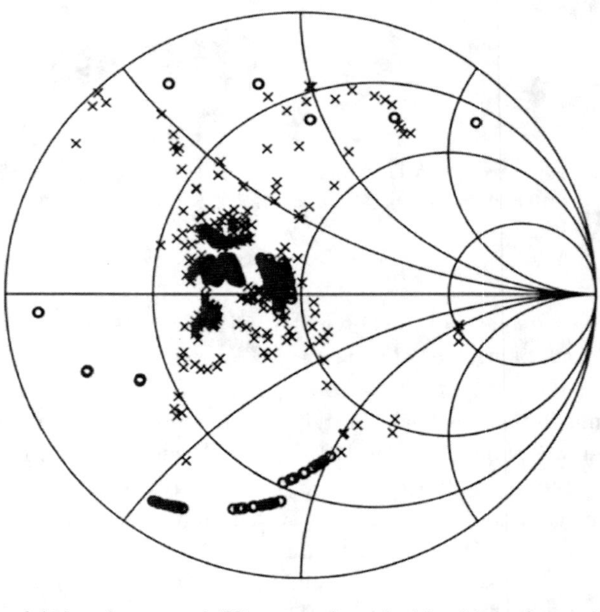

FIGURE 11.17 Simulated (O) and measured (X) range of matched load impedances for the 20 GHz-optimized, 4-bit by 4-bit tuner at 15 GHz. (From J. Papapolymerou, K. Lange, C. Goldsmith, A. Malczewski, and J. Kleber, *IEEE Trans. Microwave Theory Techn.*, 51(1), 271–278, 2003. With permission.)

maximum VSWR of 99 was achieved at the fourth quadrant and 15.7 at the second and third quadrants of the Smith chart. This is attributed to the low loss nature of the MEMS components. Thus, the MEMS tuner performance is superior compared to conventional tuners using FET devices that typically achieve VSWRs less than 20. At 15 GHz, this circuit can match load impedances with $1\,\Omega < \text{Re}\{Z_L\} < 155\,\Omega$ and $-68\,\Omega < \text{Im}\{Z_L\} < 79\,\Omega$, equivalent to three quadrants of the Smith chart. There are some minor discrepancies between the measured and simulated data that arise mostly from the variation of ribbon bond lengths and placements. In addition, the spacing between the alumina and the silicon substrates that were bonded together was not modeled in the HP-ADS simulations. Also, it should be noted that while the simulated and measured data agree fairly well at the design frequency, the two data sets agreed less at other frequencies.

To overcome some of the limitations and challenges of the hybrid MEMS tuner, such as shifting of the load range due to wire bonds and the relatively large size, a fully monolithic microstrip tuner has also been developed [24]. The latter is the first monolithic, microstrip-type reconfigurable X-band tuner on a silicon substrate using ohmic contact MEMS switches for maximizing the efficiency of X-band class-E power amplifiers [25]. Unlike the hybrid tuner, the monolithic one is designed to achieve a specific tuning matching range for the amplifier and not to maximize the overall load tuning range. In this mode of operation, the active device is driven hard into saturation, and small device-to-device and parasitic variations have a large effect on efficiency. A load-pull technique combined with measurements on around 30 hybrid class-E 10 GHz power amplifiers determines the optimal output impedances range required of the output tuner. The resulting output impedances for class-E operation are obtained to be $Z_{EMAX} = (34.1 + j39.3)\,\Omega$ and $Z_{EMIN} = (22.7 + j26.2)\,\Omega$, while the nominal impedance is $Z_{ENOM} = (27.3 + j31.5)\,\Omega$.

The microstrip-based double-stub tuner with the MEMS switches was designed to provide the impedance tuning range required above. The design procedure is as follows: first of all, a double-stub tuner was designed with the fixed stub length to provide the nominal impedance required by the power amplifier, which is $34+j37\,\Omega$ (this value takes into account the interconnects between the Si tuner and the amplifier circuit), and then each stub was split into three paths and the microstrip transmission lines were replaced with radial stubs as shown in Figure 11.18. Each radial stub is connected with the main RF line through an ohmic contact MEMS switch; this connection will enable the selection of different radial stubs depending on the different combinations (on/off states selection) of the MEMS switches. The dimensions of the radial stubs were chosen so that by selecting the different on and off states of the MEMS switches, they will provide the sufficient tuning range. In order to properly actuate the MEMS switches, the DC bias lines were designed to provide the DC path while blocking the RF path to keep the RF signal from leaking; the high resistive material indium tin oxide (ITO) was used to form the DC bias lines. The average sheet resistance measured was $10\,\text{K}\Omega$/square and the DC bias lines were designed as 100 squares. A photo of the fabricated monolithic tuners is shown in Figure 11.19.

Tuner measurements were performed with an Agilent 8510C network analyzer using GGB Picoprobes. The switches were activated using Cascade DC probes and one probe card. The actuation voltage of the MEMS switches was found to be around 60 V. The simulated and the measured results for the total of 64 states are shown in Figure 11.20 and Figure 11.21, respectively.

From Figure 11.20 and Figure 11.21, it can be seen that the simulated and the measured results have a good agreement. In Figure 11.20, marker 1 $(33.4+j33.8\,\Omega)$ is the required nominal impedance for the output of the power amplifier. Correspondingly, in Figure 11.21, marker 1 shows the measured result with exactly the same combinations of the MEMS switches. The marker reading in this case is $27.7 + 20.6\,\Omega$. The real and the imaginary parts of the impedance from the simulation and measurement in this case are slightly different. The reason for this might be from the simulation and measurement error, and slight shift of the reference plane during the measurement. Overall, the

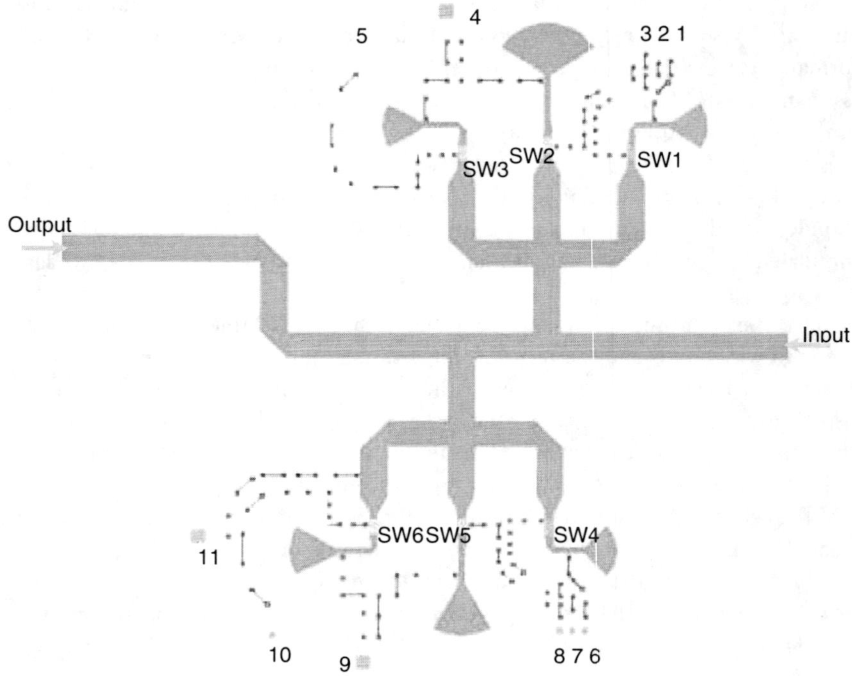

FIGURE 11.18 Layout of monolithic tuner with DC bias lines shown. (From G. Zheng, P. Kirby, S. Pajic, A. Pothier, P. Blondy, J. Papapolymerou, and Z. Popovic, 2004 IEEE Topical Meeting on Si Monolithic Integrated Circuits in RF Systems, pp. 159–162, Atlanta, GA, September 2004. With permission.)

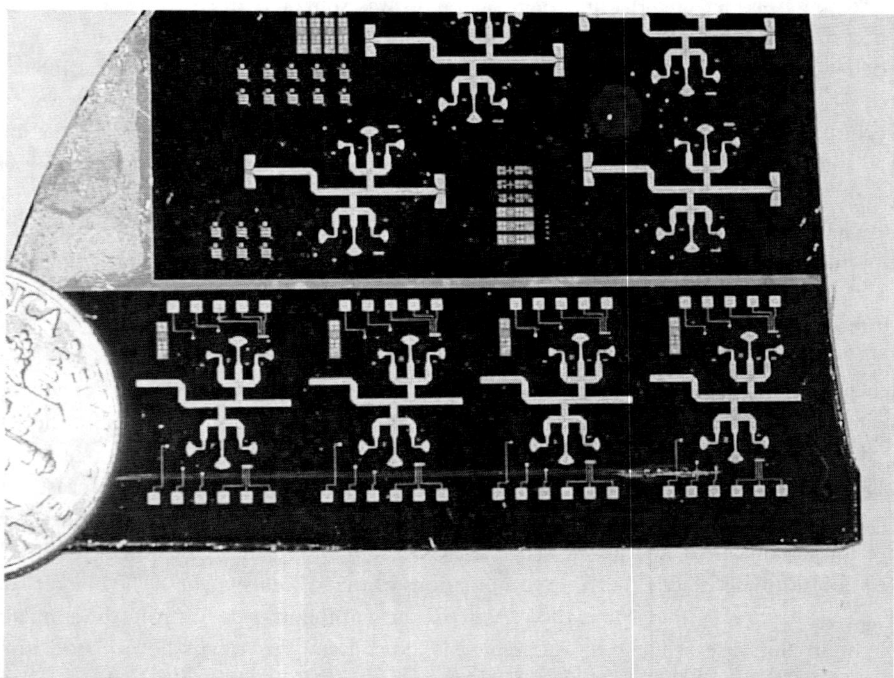

FIGURE 11.19 Photo of fabricated monolithic tuners. (From G. Zheng, P. Kirby, S. Pajic, A. Pothier, P. Blondy, J. Papapolymerou, and Z. Popovic, 2004 IEEE Topical Meeting on Si Monolithic Integrated Circuits in RF Systems, pp. 159–162, Atlanta, GA, September 2004. With permission.)

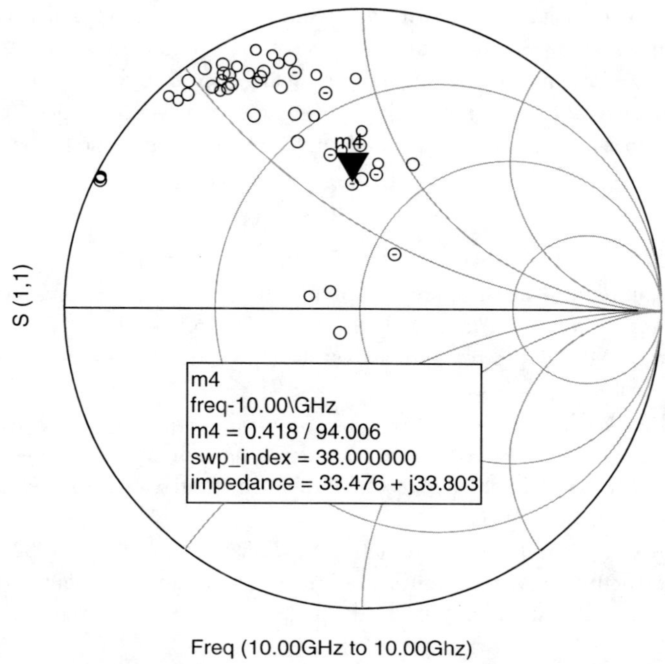

FIGURE 11.20 Simulated impedance tuning range with the same measured 64 different MEMS switch on/off states. (From G. Zheng, P. Kirby, S. Pajic, A. Pothier, P. Blondy, J. Papapolymerou, and Z. Popovic, 2004 IEEE Topical Meeting on Si Monolithic Integrated Circuits in RF Systems, pp. 159–162, Atlanta, GA, September 2004. With permission.)

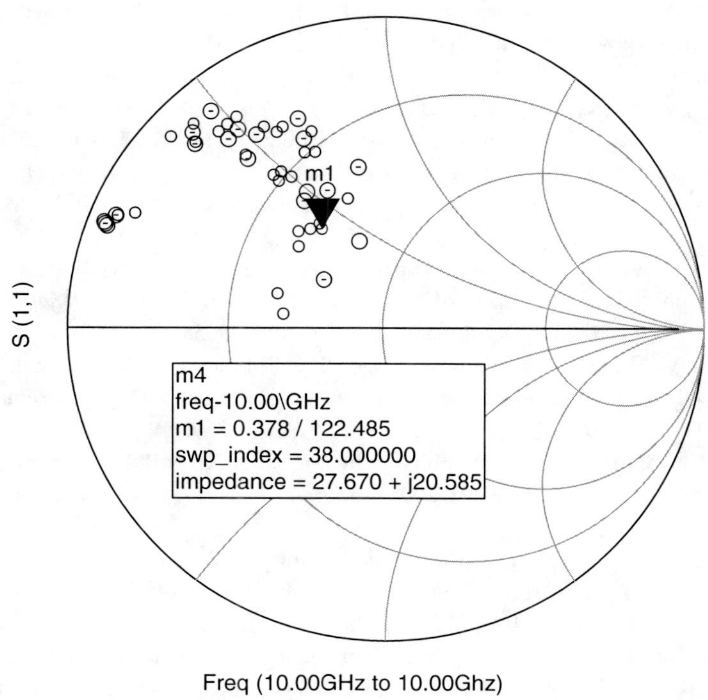

FIGURE 11.21 Measured impedance tuning range for the 64 different MEMS switch on/off states. (From G. Zheng, P. Kirby, S. Pajic, A. Pothier, P. Blondy, J. Papapolymerou, and Z. Popovic, 2004 IEEE Topical Meeting on Si Monolithic Integrated Circuits in RF Systems, pp. 159–162, Atlanta, GA, September 2004. With permission.)

monolithic tuner can match loads that are within the desired impedance range of a class-E amplifier. The measured maximum VSWR is 38 and the tuner size is 6 mm×7 mm. A max VSWR of 83 has also been achieved with the same design that had higher per square resistance for the DC bias lines. The size reduction is at least threefold compared to the hybrid tuner that was designed around 20 GHz. The insertion loss of the tuner was also calculated based on the measurements, and the average loss was found to be around 1 dB.

References

1. H.H. Meinel, Commercial applications of millimeter waves: history, present status and future trends, *IEEE Trans. Microwave Theory Techn.*, 43(7), 1639–1653, 1995.
2. I. Papapolymerou, MMIC passive and active structures, PhD dissertation, The University of Michigan, Ann Arbor, 1999.
3. K.J. Herrick, J.G. Yook, and L.P.B. Katehi, Microtechnology in the development of three-dimensional circuits, *IEEE Trans. Microwave Theory Techn.*, 46(11), 1832–1844, 1998.
4. E.R. Brown, RF MEMS switches for reconfigurable integrated circuits, *IEEE Trans. Microwave Theory Techn.*, 46(11), 1868–1880, 1998.
5. K. Petersen, Silicon as a mechanical material, *Proc. IEEE*, 70(5), 420–457, 1982.
6. J.H. Comtois, Structures and techniques for implementing and packaging complex, large scale microelectromechanical systems using foundry fabrication, PhD dissertation, Air Force Institute of Technology, 1996.
7. J.J. Sniegowski and E.J. Garcia, Microfabricated actuators and their application to optics, *Proc. SPIE—Int. Soc. Opt. Eng. (USA)*, Vol. 2383, San Jose, CA, February 7–9, 1995, pp. 46–64.
8. G.M. Rebeiz, *RF MEMS: Theory, Design and Technology*, Wiley Interscience, 2003.
9. C. Goldsmith, Z. Yao, S. Eshelman, and D. Denniston, Performance of low-loss RF MEMS capacitive switches, *IEEE Microwave Guided Wave Lett.*, 8(8), 269–271, 1998.
10. S. Pacheco, C. Nguyen, and L. Katehi, Micromechanical electrostatic K-band switches, *IEEE MTT-S International Microwave Symposium Digest*, pp. 1569–1572, Baltimore, MD, June 1998.
11. J. Muldavin and G. Rebeiz, Inline capacitive and DC-contact MEMS shunt switches, *IEEE Microwave Wireless Components Lett.*, 11(8), 334–336, 2001.
12. R. Mihailovich, M. Kim, J. Hacker, E. Sovero, J. Studer, J. Higgins, and J. DeNatale, MEM relay for reconfigurable RF circuits, *IEEE Microwave Wireless Components Lett.*, 11(2), 53–55, 2001.
13. D. Hyman and M. Mehregany, Contact physics of gold microcontacts for MEMS switches, *IEEE Transactions Comp. Packaging Technol.*, 22(3), 357–364, 1999.
14. G. Wang, S. Barstow, A. Jeyakumar, J. Papapolymerou, and C. Henderson, Low cost RF MEMS switches using photodefinable mixed oxide dielectrics, *2003 IEEE International Microwave Symposium Digest*, Vol. 3, pp. 1633–1636, Philadelphia, PA, June 2003.
15. S.C. Chen and M. Feng, Low actuation voltage RF MEMS switches with signal frequencies from 0.25 GHz to 40 GHz, *Proceedings IEEE International Electronics Device Meeting*, pp. 689–692, December 1999.
16. G. Ponchak, N. Varaljay, and J. Papapolymerou, Ka-band, RF MEMS switches on CMOS grade silicon with a polyimide interface layer, *2003 European Microwave Conference*, Vol. III, pp. 971–974, October 2003.
17. G. Zheng and J. Papapolymerou, Monolithic reconfigurable bandstop filter using RF MEMS switches, *Int. J. RF Microwave Computer Aided Engineering (special issue on RF Applications of MEMS and Micromachining)*, 14(4), 373–382, 2004.
18. A. Pothier, J.-C. Orlianges, E. Zheng, C. Champeaux, A. Catherinot, D. Cros, P. Blondy, and J. Papapolymerou, High tuning range bandpass filters using MEMS DC contact switches, *IEEE Trans. Microwave Theory Techn.*, 53(1), 354–360, 2005.
19. Reliability testing and reliability enhancement of RF MEMS switches, 2004 IEEE International Microwave Symposium Workshop Notes, Fort Worth, TX, June 2004.

20. G. Torregrosa-Penalva, G. Lopez-Risueno, J.I. Alonso, A simple method to design wide band electronically tunable comb-line filters, *IEEE Trans. Microwave Theory Techn.*, 50(1), 2002.
21. David M. Pozar, *Microwave Engineering*, 2nd edition, John Wiley & Sons, Inc., New York.
22. G. Matthaei, L. Young, and E. M. T Jones, *Microwave Filters, Impedance Matching Networks, and Coupling Structures*, Artech House, New Jersey, 1980.
23. J. Papapolymerou, K. Lange, C. Goldsmith, A. Malczewski, and J. Kleber, Reconfigurable double stub tuners using MEMS switches for intelligent RF front ends, *IEEE Trans. Microwave Theory Techn.*, 51(1), 271–278, 2003.
24. G. Zheng, P. Kirby, S. Pajic, A. Pothier, P. Blondy, J. Papapolymerou, and Z. Popovic, A monolithic reconfigurable tuner with ohmic contact MEMS switches for efficiency optimization of X-band power amplifiers, 2004 IEEE Topical Meeting on Si Monolithic Integrated Circuits in RF Systems, pp. 159–162, Atlanta, GA, September 2004.
25. S. Pajic and Z. Popovic, An efficient 16-element X-band spatial combiner of switched-mode power amplifiers, *IEEE Trans. Microwave Theory Techn.*, 51(73), 2003.

12

Wideband Antennas on Silicon

Manos Tentzeris and
Rong Lin Li
Georgia Institute of Technology

12.1 Microstrip Antennas on Cavity-Etched Silicon Substrate 12-2
SiO_2 • Polyimide • SiO_2 versus Polyimide
12.2 Microstrip Antennas on Hole-Perforated Silicon Substrate 12-10
12.3 Tapered Slot Antennas on Micromachined Silicon Substrates 12-11
12.4 Surface Micromachined Monopoles for W-Band Applications 12-13

In modern sensor and communication systems, there is an increasing demand for the monolithic integration of the antennas with the RF front-ends to avoid external transmission line connections or sophisticated packaging, thus reducing manufacturing cost and improving system performances [1,2]. Especially, microstrip antennas are used in a broad range of applications from communication systems (WLAN, radars, telemetry, and navigation) to biomedical systems, primarily due to their simplicity, conformability, and low manufacturing cost. They are commonly utilized in wireless applications due to the fact that can be planar or conformal, can be fed in numerous configurations and also are compact and suitable for antenna array designs. In general, they can be used in applications requiring high-performance compact low-cost planar antennas. With the recent development of microwave- and millimeter-wave integrated circuits and the trend to incorporate all microwave devices on a single chip for low-cost and high density, there is a need to fabricate microstrip antennas in a monolithic fashion with the rest of the circuitry on semiconducting materials such as silicon, GaAs, or InP. One of the main limitations of these microstrip antennas is the excitation of surface waves in the silicon substrate, especially in cases where the bandwidth and the radiation efficiency requirements demand large values of the substrate thickness, resulting in compromised efficiency, reduced bandwidth, and degradation of the radiation pattern [4,5]. Furthermore, in monolithic designs the feedlines share in most cases the same interface with the antennas and lead into parasitic radiation, which deforms the antennas' pattern and increases crosstalk.

The suitability of high-resistivity silicon as a substrate material for microwave- and millimeter-wave circuits has been demonstrated in numerous examples [6,7]. However, direct use of silicon substrates with patch antennas results in strong surface-wave modes and reduced bandwidth. The ideal solution requires the capability to integrate the antenna on electrically thick low-dielectric constant regions while the circuitry remains on the high-dielectric constant regions in the same substrate. This requirement can be satisfied by selecting the substrate that offers optimum component performance; unfortunately this leads to hybrid integration schemes and high development cost. As the frequency increases, moreover, this approach becomes increasingly difficult and costs are prohibitively high [8]. In addition, the low-resistivity Si wafer used for traditional microwave circuit is troublesome due to the high loss. Several methods have been recently used to overcome this problem, including drilling a well-designed cavity underneath the antenna [8–14], and etching few holes underneath the antenna to disturb the formation of substrate modes [15–17]. In this review, we first focus on two categories of micromachined antennas (i.e., micromachined cavity and micromachined holes' perforation) and their design methodologies. Micromachining of silicon is a process where the semiconductor substrate is mechanically altered, either by removing parts of the substrate (bulk micromachining) or by adding layers and structures to the top of the wafer (surface micromachining). Finally, we will introduce the development of surface micromachined monopoles for W-Band applications.

12.1 Microstrip Antennas on Cavity-Etched Silicon Substrate

When a microstrip patch antenna is fabricated on a cavity-etched silicon substrate (as shown in Figure 12.1), the resonant frequency of the antenna with length and width L and W is determined by [8,18]

$$f_r = \frac{c}{2(L + 2\Delta L)\sqrt{\varepsilon_{\text{reff}}}} \tag{12.1}$$

where

$$\varepsilon_{\text{reff}} = \varepsilon_{\text{cavity}} \left(\frac{L + 2\Delta L \frac{\varepsilon_{\text{fringe}}}{\varepsilon_{\text{cavity}}}}{L + 2\Delta L} \right) \tag{12.2}$$

$$\Delta L = 0.412t \frac{(\varepsilon_{\text{reff}} + 0.3)(W/t + 0.264)}{(\varepsilon_{\text{reff}} - 0.258)(W/t + 0.8)} \tag{12.3}$$

with

$$\frac{\varepsilon_{\text{fringe}}}{\varepsilon_{\text{cavity}}} = \frac{\varepsilon_{\text{air}} + (\varepsilon_{\text{sub}} - \varepsilon_{\text{air}})x_{\text{air}}}{\varepsilon_{\text{air}} + (\varepsilon_{\text{sub}} - \varepsilon_{\text{air}})x_{\text{fringe}}} \tag{12.4}$$

$$\varepsilon_{\text{cavity}} = \frac{\varepsilon_{\text{air}}\varepsilon_{\text{sub}}}{\varepsilon_{air} + (\varepsilon_{\text{sub}} - \varepsilon_{\text{air}})x_{\text{fringe}}} \tag{12.5}$$

The thickness parameters x_{air} and x_{fringe} are ratio of the air to full substrate thickness in the mixed and fringing field regions, respectively [8].

A plot of the theoretical and measured effective dielectric constant versus the air gap thickness for silicon substrate is shown in Figure 12.2, where an effective dielectric constant of approximately 2.2 is achieved for a mixed air–silicon ratio of 1:1 using the capacitor model (Equation 12.2 with $\Delta L = 0$) and 3:1 ratio for the capacitor model with ΔL extension length (Equation 12.2 with ΔL calculated from Equation 12.3). The dimensions for the micromachined patch that was used to produce the results of Figure 12.2 are listed in Table 12.1. The finite difference time-domain (FDTD) calculation was based on a three-dimensional (3D) full-wave scheme that yields the return loss of the micromachined antenna and the effective permittivity of a 1:1 air–silicon substrate for a range of frequencies. Once the resonant frequency is determined from the return loss, the effective permittivity of interest is found. Regarding the measured (micromachined) data point, the resonant frequency of the fabricated antenna was measured and was used to extract the dielectric constant of the substrate of a patch having the same dimensions as the measured one.

Two antennas were fabricated on silicon, with resonant frequencies in the K-band and air–substrate thickness ratios of 1:1. In the silicon micromachined patch antenna, the conductor has been electroplated to a metal thickness of approximately 3.2 μm and the substrate is chemically etched (EDP

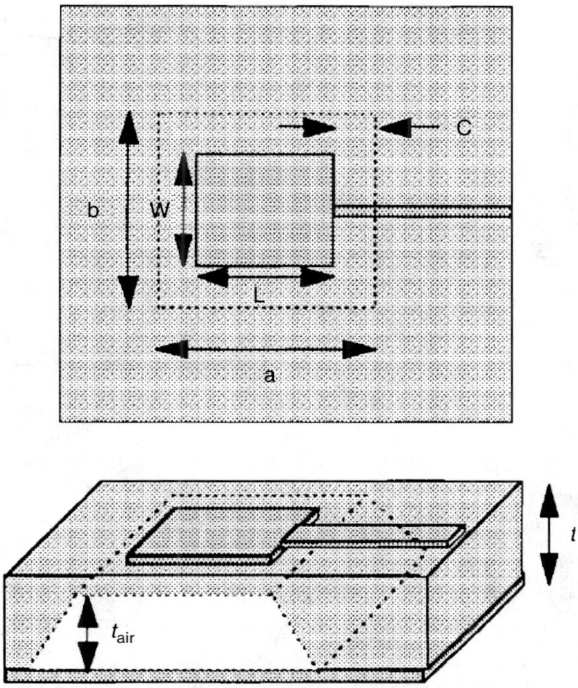

FIGURE 12.1 Microstrip patch antenna on a cavity-etched silicon substrate. (From J Papapolymerou, RF Drayton, and LPB Katehi. *IEEE Trans. Antennas Propagat.* 46:275–283, 1998. With permission. Copyright IEEE.)

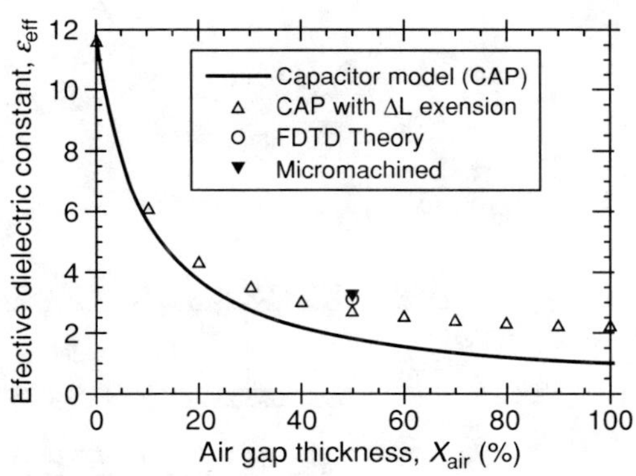

FIGURE 12.2 Effective dielectric constant versus air gap thickness (%) for the silicon-based micromachined patch using the cavity and FDTD model compared to the measured response. (From J Papapolymerou, RF Drayton, and LPB Katehi. *IEEE Trans. Antennas Propagat.* 46:275–283, 1998. With permission. Copyright IEEE.)

TABLE 12.1 Design Parameters for the Antennas on Silicon Substrates

Patch	t (mm)	t_{air} (mm)	L (mm)	w (mm)	a (mm)	b (mm)	c (mm)
Regular	0.355	0	2.019	4.08	0	0	0
Micromachined	0.355	0.165	3.616	3.445	3.616	8.108	0

Source: J Papapolymerou, RF Drayton, and LPB Katehi. *IEEE Trans. Antennas Propagat.* 46:275–283, 1998. With permission. Copyright IEEE.

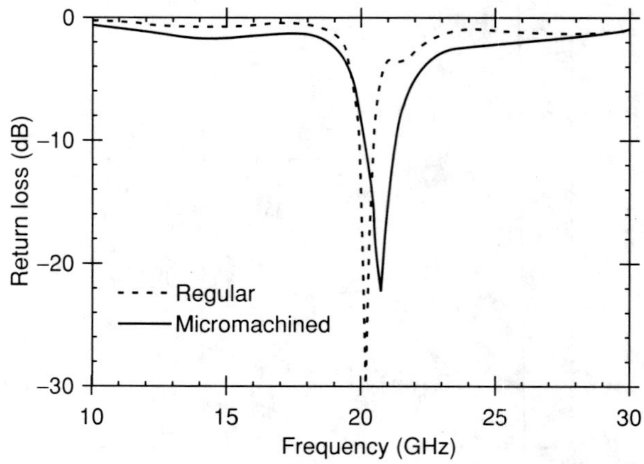

FIGURE 12.3 Measured return loss for the regular and micromachined patch antenna printed on a full thicknesss substrate and substrate with mixed air–silicon thickness ratio (1:1), respectively. (From J Papapolymerou, RF Drayton, and LPB Katehi. *IEEE Trans. Antennas Propagat.* 46:275–283, 1998. With permission. Copyright IEEE.)

process) in a single-etch step underneath the antenna. Return loss measurements are shown in Figure 12.3 where the −10 dB bandwidth increases from 2.9% for the regular antenna to 5% for the micromachined patch.

The fabrication process of a micromachined patch antenna is shown in Figure 12.4 [11] for an antenna used for satellite communications (around 16 GHz). The first 254 μm-thick high-resistivity silicon wafer is used to support the microstrip patch. The lift-off process shown in Figure 12.4a is employed to produce a required patch and feed line patterns on top of the silicon wafer. A second silicon

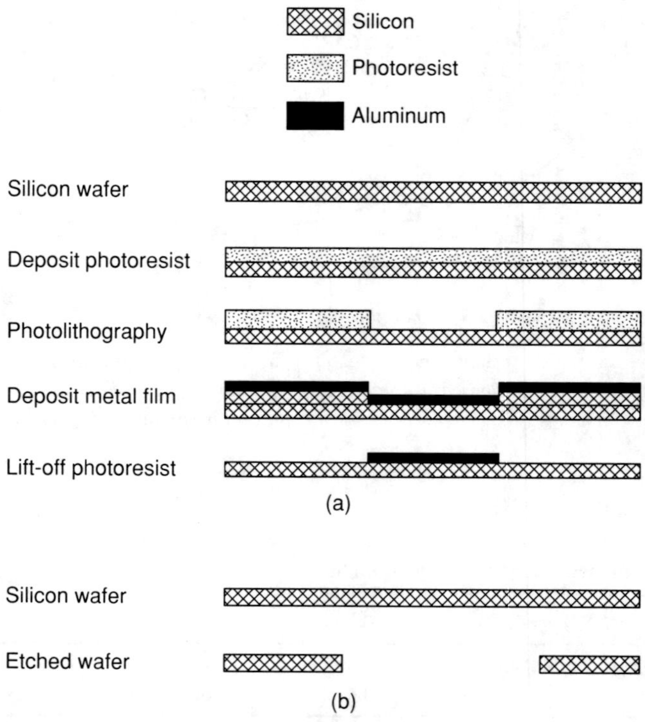

FIGURE 12.4 Fabrication of cavity-etched patch antenna: (a) lift off process and (b) etching process. (From M Zheng, Q Chen, PS Hall, and VF Fusco. *Electron. Lett.* 34:3–4, 1998. With permission. Copyright IEEE.)

wafer of the same thickness is used to produce a 0.254 mm × 12 mm × 12 mm cavity as shown in Figure 12.4(b) by using an abrasive etching method. Finally, two wafers and a polished brass carrier are assembled together using silver epoxy to form a complete air-filled-cavity-backed microstrip patch antenna, which is depicted in Figure 12.5 with $t_{air}/t = 50\%$ ($H_1 = H_2$).

The measured and calculated results for input return loss of the antenna are presented in Figure 12.6 and demonstrate good agreement. The calculation was carried out using a 3D FEM model. It is found that the effective dielectric constant of the patch antenna is only 2.8, compared to 9.7 of a noncavity-backed antenna of the same thickness. The drastic reduction in the permittivity of the patch antenna by partially etching out the silicon material underneath the patch, obviously results in the increase of the antenna's bandwidth and efficiency. The 10 dB impedance bandwidth of the antenna is 12.5%, which is much larger than the typical value of 2% and 4.4% for conventional microstrip patch antennas of dielectric constants of 10 and 2.2 with the same substrate thickness-to-wavelength ratio [18]. The measured radiation patterns are plotted in Figure 12.7, where the patterns in the E-plane are not symmetrical because of the disturbance from the feed line. The gain of the antenna is measured to be 6.6 dBi.

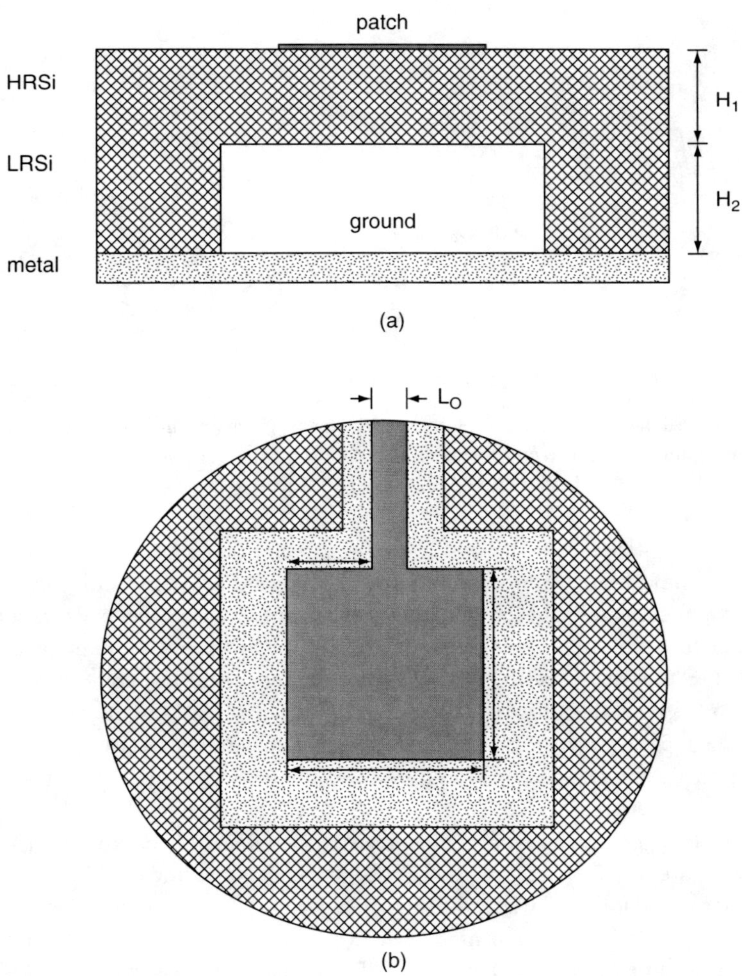

FIGURE 12.5 Cavity-backed microstrip patch antenna ($H_1 = H_2 = 254\,\mu$m, $L = 5$ mm, $W = 7$ mm, $L_1 = 2.95$ mm. $L_0 = 1.3$ mm; (a) side view, (b) top view). (From M Zheng, Q Chen, PS Hall, and VF Fusco. *Electron. Lett.* 34:3–4, 1998. With permission. Copyright IEEE.)

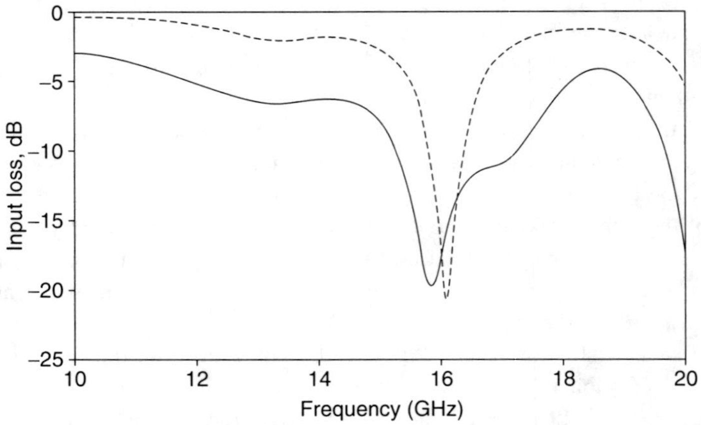

FIGURE 12.6 Input return of the cavity-etched patch antenna (—) measured, (- -) calculated. (From M Zheng, Q Chen, PS Hall, and VF Fusco. *Electron. Lett.* 34:3–4, 1998. With permission. Copyright IEEE.)

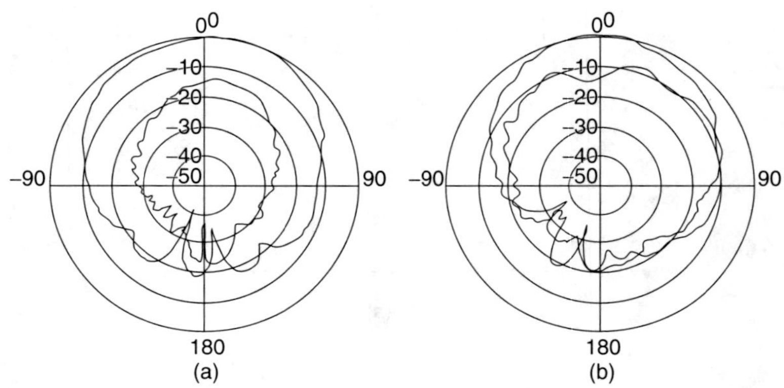

FIGURE 12.7 Measured radiation patterns of the cavity-etched patch antenna: (a) E-plane and (b) H-plane. E-plane, H-plane (—), and cross-polarization patterns (- -) are shown. (From M Zheng, Q Chen, PS Hall, and VF Fusco. *Electron. Lett.* 34:3–4, 1998. With permission. Copyright IEEE.)

In Ref. [12], a micromachining approach that allows etching a shallow cavity directly under the area of microstrip antenna, and then fills the cavity with lower permittivity materials, such as silicon dioxide or polyimide, is presented. A metal layer is deposited between the low permittivity material and the silicon underneath. The thin layer of low permittivity substrate can reduce the loss and the excitation of surface waves. The tradeoff is that the very small thickness of this layer makes the realization of practical bandwidths for this type of microstrip antennas very difficult. Finding the most efficient thickness, in terms of fabrication feasibility and bandwidth of the antenna, is the most critical problem of these designs.

One important factor of the antenna performance is the choice of the optimum feeding options. Microstrip antennas are usually fed with either probe, coplanar waveguide (CPW), or microstrip feed. The radiating structures of Ref. [12] use a coplanar CPW feed to the microstrip antenna [13], in order to minimize the interaction of the coplanar mode fields with the lossy Si substrate. The latter will prevent loss of power in the feeding network. In addition, CPW interfaces very well with the pad-probes of most measurement equipment. To integrate the microstrip antenna in the CPW feeding network, a simple, low-loss, compact and via-less CPW to microstrip transition was designed and optimized for operation around 30 GHz [14]. The material used was a silicon substrate with relative permittivity (ε_r) of 11.8, conductivity (σ) of 100 S/m, and thickness of 500 μm at 30 GHz. With these high values of thickness and conductivity, it was assumed that the substrate provided a virtual ground, below which the E-field would be less than 5% of the E-field on the antenna. The attenuation constant (α) for such a material can be calculated with

$$\alpha = \omega\sqrt{\frac{\mu\varepsilon'}{2}\left(\sqrt{1+\left(\frac{\varepsilon''}{\varepsilon'}\right)^2}-1\right)} \tag{12.6}$$

where $\varepsilon' = \varepsilon_r\varepsilon_0$ and $\varepsilon'' = \sigma/\omega$, giving a value of $\alpha = 3120.41$ Np/m for the above silicon substrate. Using this attenuation constant in the formula [19]

$$|E(z)| = |E_0|e^{-\alpha z} \tag{12.7}$$

the electric field left at the bottom of the silicon is approximately 21.01% of the electric field on top. This means that there is still energy left at the bottom of the substrate, which excites significant substrate modes and could potentially distract the radiation of the antenna, unless a metal plane is added to the bottom. To alleviate this problem, a shallow cavity was etched directly under the area of microstrip antenna, and then filled with lower permittivity materials, such as silicon dioxide or polyimide. A metal layer was added at the horizontal planar interface between the low permittivity material and the silicon underneath. Figure 12.8 shows the top view of this antenna. Figure 12.9 and Figure 12.10 are the side views of the antenna with SiO_2 and polyimide underneath, respectively. The entire structure, including CPW, microstrip line, and the patch antenna are on top of the cavity. The CPW and microstrip line are designed to match a 50 Ω characteristic impedence.

The thickness of the cavity depends on the material used to fill the cavity. The dimension of the antenna is 3300 μm × 2600 μm. To match the 50 Ω input impedence, the antenna is fed at 600 μm from the edge. The width of the microstrip transmission line is 452 μm and the distance from the edge of the antenna to the end of the cavity is 350 μm on the side and 511.3 μm in the front. The SiO_2 has $\varepsilon_r = 3.5$. The cavity depth with SiO_2 filling has been simulated at 50, 100, and 200 μm. The bottom of the cavity is metalized with gold.

The polyimide has $\varepsilon_r = 3.2$. During the fabrication, the polyimide is applied using a spin coating method that results in an extra layer of polyimide over the cavity and the silicon. The depths of the cavity that are simulated are equal to 50, 100, and 200 μm.

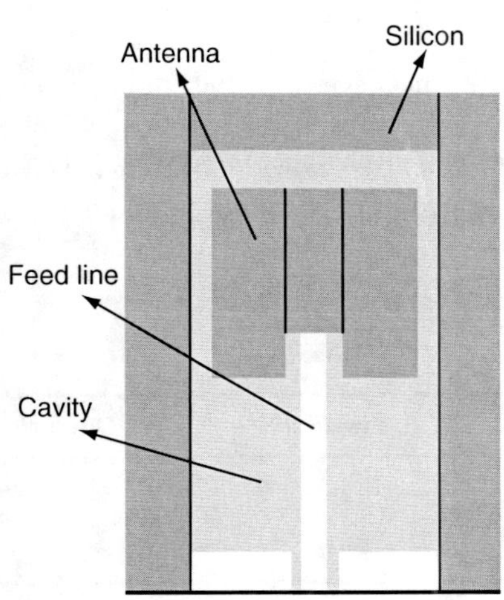

FIGURE 12.8 Top view of the antenna.

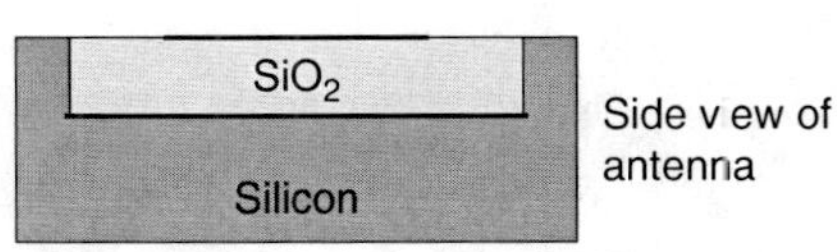

FIGURE 12.9 Side view of antenna with SiO_2.

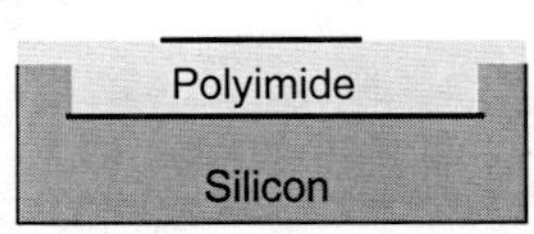

FIGURE 12.10 Side view of antenna with polyimide.

TABLE 12.2 SiO_2 Filled Cavity for Different SiO_2 Thickness Values

SiO_2 Thickness (μm)	Resonant Frequency (GHz)	Bandwidth (%)	Directivity (dBi)
50	28.946	1.4	6.736
100	29.165	1.7	6.766
200	29.430	2.4	6.809

Source: EY Tsai, AM Bacon, M Tentzeris, and J Papapolymerou. Proceedings of the 2002 Asian-Pacific Microwave Symposium, Kyoto, 2002, pp. 821–824. With permission.

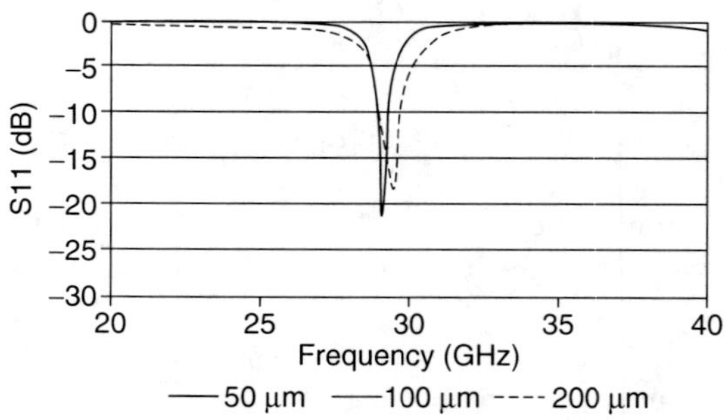

FIGURE 12.11 Simulated S_{11} for SiO_2 cavity. (From EY Tsai, AM Bacon, M Tentzeris, and J Papapolymerou. Proceedings of the 2002 Asian-Pacific Microwave Symposium, Kyoto, 2002, pp. 821–824. With permission.)

SiO_2

Table 12.2 summarizes the results and Figure 12.11 is a graphical presentation of the S_{11} values versus frequency for the three different simulated thicknesses of SiO_2 layers, 50, 100, and 200 μm.

All three designs show a resonant frequency around 30 GHz. The curves for 50 and 100 μm are almost identical with only 0.3% difference of the bandwidth. The 200 μm design shows more bandwidth, but the maximum return loss is slightly higher than for the thinner layers. The 200 μm design is the only one having a bandwidth close to the commonly required value of 3%. However, fabrication of the 200 μm design is difficult, since growing such large oxidation thicknesses on silicon is very time-consuming and technologically challenging.

Polyimide

The polyimide structure is also simulated with three different thicknesses: 50, 100, and 200 μm. Table 12.3 and Figure 12.12 summarize the results.

At 30.975 GHz, the 200 μm design has a return loss of 16.5 dB, and a bandwidth of 2.7% (83.63 MHz). As a conclusion, the simulation results show that with 200 μm of polyimide, both the resonant frequency and the bandwidth are in a desirable range.

TABLE 12.3 Polyimide Filled Cavity for Different Polythickness Values

Polymide Thickness (μm)	Resonant Frequency (GHz)	Bandwidth (%)	Directivity (dBi)
50	30.318	1.3	6.616
100	29.474	1.3	7.223
200	30.975	2.7	6.809

Source: EY Tsai, AM Bacon, M Tentzeris, and J Papapolymerou. Proceedings of the 2002 Asian-Pacific Microwave Symposium, Kyoto, 2002, pp. 821–824. With permission.

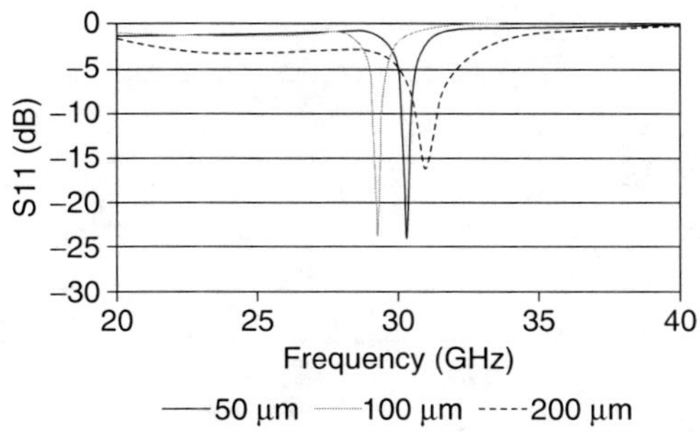

FIGURE 12.12 Simulated S_{11} for polyimide cavity. (From EY Tsai, AM Bacon, M Tentzeris, and J Papapolymerou. Proceedings of the 2002 Asian-Pacific Microwave Symposium, Kyoto, 2002, pp. 821–824. With permission.)

SiO_2 versus Polyimide

Figure 12.13 is a plot of S_{11} for the structures with 200 μm of material, both SiO_2 and polyimide. It can3 be easily observed that there is a shift of the resonant frequency for the two substrates. This is due to the fact that the polyimide has an extra layer on top of silicon that creates a cavity of a noncanonical shape. In addition, this introduces a dielectric interface of Z-shape in comparison to the dielectric interface of L-shape for SiO_2 (Figure 12.9 and Figure 12.10), leading to a slightly lower value of effective dielectric constant and thus to a higher resonant frequency for the same physical dimensions.

Both simulations demonstrate a bandwidth of approximately around 2.5%. SiO_2 shows better return loss at the resonant frequency, though it is more difficult to fabricate. On the contrary, polyimide-filled geometry is more feasible in terms of fabrication. In order to decrease the difference between the resonant frequencies of the SiO_2 and polyimide antennas, the size of the latter antenna was changed from 3300 μm × 2600 μm to 3300 μm × 2650 μm. The result shows a maximum return loss of 29.09 dB at 30.022 GHz, and the bandwidth is 2.7%. Figure 12.14 shows the comparison from this optimized dimension to the SiO_2 result. The central bandwidth is offset by 600 MHz, which is 2%. The difference can be attributed to different dielectric constants and numerical errors due to the simulation tools.

As a conclusion, preliminary designs of planar antenna structures on top of low-resistivity silicon with a layer of low-permittivity material in between were presented and evaluated in terms of return loss and bandwidth. The materials used are SiO_2 and polyimide and simulation results show that to achieve bandwidths above 2.5%, it is necessary to have at least 200 μm of either material. Polyimide is easier to fabricate, but SiO_2 could provide a better performance in terms of the return loss.

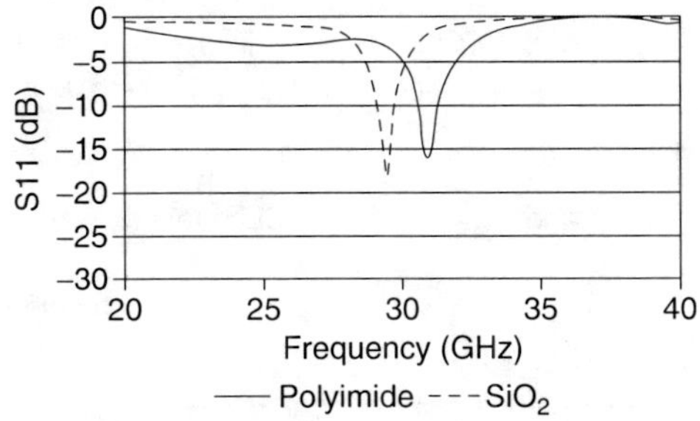

FIGURE 12.13 Simulated S_{11} for the same thickness (200 μm) of SiO_2 and polyimide. (From EY Tsai, AM Bacon, M Tentzeris, and J Papapolymerou. Proceedings of the 2002 Asian-Pacific Microwave Symposium, Kyoto, 2002, pp. 821–824. With permission.)

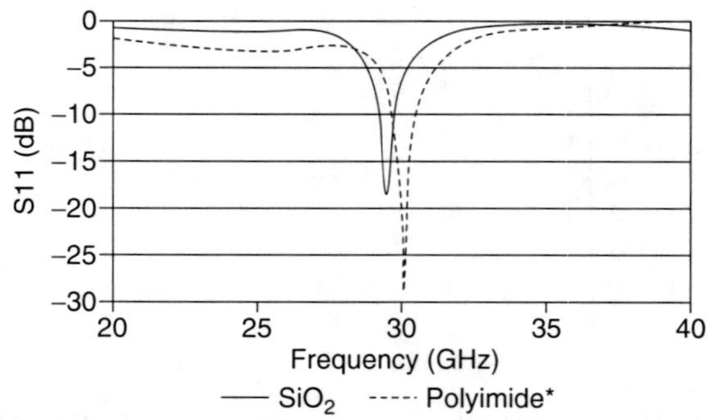

FIGURE 12.14 Simulated S11 for same thickness (200 μm) of SiO_2 revised polyimide. (From EY Tsai, AM Bacon, M Tentzeris, and J Papapolymerou. Proceedings of the 2002 Asian-Pacific Microwave Symposium, Kyoto, 2002, pp. 821–824. With permission.)

12.2 Microstrip Antennas on Hole-Perforated Silicon Substrate

Another commonly used technique to alleviate the problem of surface waves in Si-based antenna structures is the perforation of the silicon substrate with holes (Figure 12.15) [15,16], leading to an effective dielectric constant given by

$$\varepsilon_{\text{reff}} = \frac{\varepsilon_{\text{rhole}} + 1}{2} + \frac{\varepsilon_{\text{rhole}} - 1}{2}\left(1 + \frac{12t}{W}\right)^{-1/2} \tag{12.8}$$

The hole-perforated dielectric constant $\varepsilon_{\text{rhole}}$ was derived using

$$\varepsilon_{\text{rhole}} = \varepsilon_{\text{sub}}(1 - x) + x \tag{12.9}$$

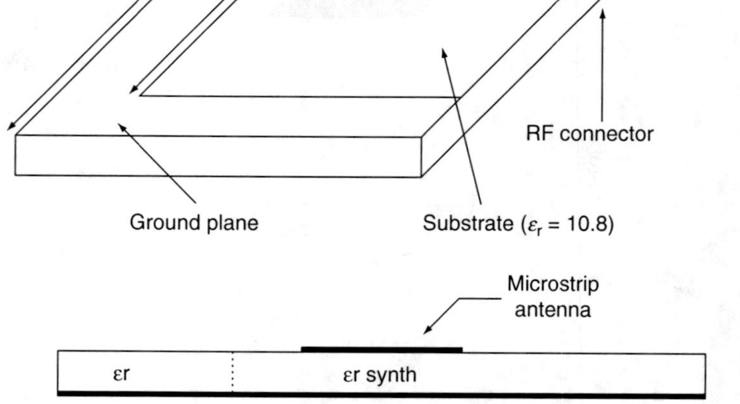

FIGURE 12.15 Microstrip patch antenna on a hole-perforated substrate. (From GP Gauthier, A Courtay, and GM Rebeiz. *IEEE Trans. Antennas Propagat.* 45:1310–1314, 1997. With permission. Copyright IEEE.)

$$x = \frac{\pi d_h^2}{2d_c^2} \tag{12.10}$$

where x is the proportion of volume removed, d_h the hole diameter, and d_c the center-to-center spacing in a triangular lattice grid.

12.3 Tapered Slot Antennas on Micromachined Silicon Substrates [17]

The tapered slot antenna (TSA) is a wideband antenna desirable for millimeter-wave applications, such as phased arrays and focal-plane imaging systems, due to its compact design and endfire radiation pattern. The main limitation of the TSA comes from its sensitivity to the thickness and dielectric constant of the supporting substrate. An *effective thickness*, which represents the electrical thickness of the substrate, has been defined as $t_{eff} = t(\sqrt{\varepsilon_r} - 1)$. An accepted range for good operation of a TSA has been experimentally determined by Yngvesson et al. [20] to be $0.005 < t_{eff}/\lambda_0 < 0.03$. For substrate thickness above the upper bound, unwanted substrate modes degrade the performance of the TSA, while antennas on thinner substrates suffer from decreased directivity.

The upper bound on the effective thickness $t_{eff}/\lambda_0 < 0.03$ requires mechanically thin substrates for millimeter-wave applications, especially for high dielectric constant substrate such as silicon, GaAs, or InP. In fact, to operate at 90 GHz, the maximum allowable thickness on silicon is 42 μm. This results in a mechanically fragile antenna and therefore is not practical in large arrays. Another approach is to micromachine holes in the substrate, resulting in a lower effective dielectric constant. In this section, the hole-micromachining technique at 90 to 100 GHz (W-band) on silicon substrate for the TSA applications is introduced.

The W-band TSA is designed and fabricated on 100-μm-thick silicon substrate ($\varepsilon_r = 11.7$, $\rho >$ 10,000 Ω cm) with (a) 300-μm and (b) 600-μm wide holes etched from the backside (as shown in Figure 12.16). A 100-μm thick silicon wafer with 4000 Å of oxide on both sides is first mounted on a 500-μm thick mechanical wafer, and the holes are patterned using a 500-μm thick photoresist film. The oxide is then etched in the holes using buffered HF and the silicon is exposed to the deep silicon RIE etcher, The etch rate is 4.5 to 5 μm/min, resulting in a total etch time of 20 min for 90- to 100-μm deep holes. The sidewall etch profile is 84° steep and the width of the etched hole increases by 10 to 15 μm. The wafer is then flipped and mounted again on a 500-μm thick mechanical wafer. The antenna metal layer is patterned using a lift-off process has a thickness of 5000 Å. The 100-μm wafer is released from the mechanical support wafer and attached at its edge to a glass microslide.

The holes' pattern shown in Figure 12.16 results in $\varepsilon_{eff} = 5.1$ and $t_{eff} / \lambda_0 = 0.037$ at 90 GHz as per Equation 12.11 and Equation 12.12 by choosing the spacing between the holes W is such that $D/W = 0.625$, where D is the diameter of the holes, which is chosen to be $D = 300$ and 600 μm, respectively. The effective dielectric constant does not depend on the hole diameter but only on the ratio D/W

$$\varepsilon_{reff} = \varepsilon_r \left(1 - \frac{\pi}{2}\left(\frac{D}{W}\right)^2\right) + \frac{\pi}{2}\left(\frac{D}{W}\right)^2 = 5.1 \tag{12.11}$$

for micromachined silicon wafers, and

$$t_{eff}/\lambda_0 = t(\sqrt{\varepsilon_{eff}} - 1)/\lambda_0 = 0.037 \tag{12.12}$$

for micromachined silicon at 90 GHz.

For reference, the TSA on silicon without any etched holes, $\varepsilon_{eff} = \varepsilon_r = 11.7$ gives $t_{eff} / \lambda_0 = 0.075$ at 90 GHz. The effective thickness is well above the upper limit of the Yngvesson condition and should

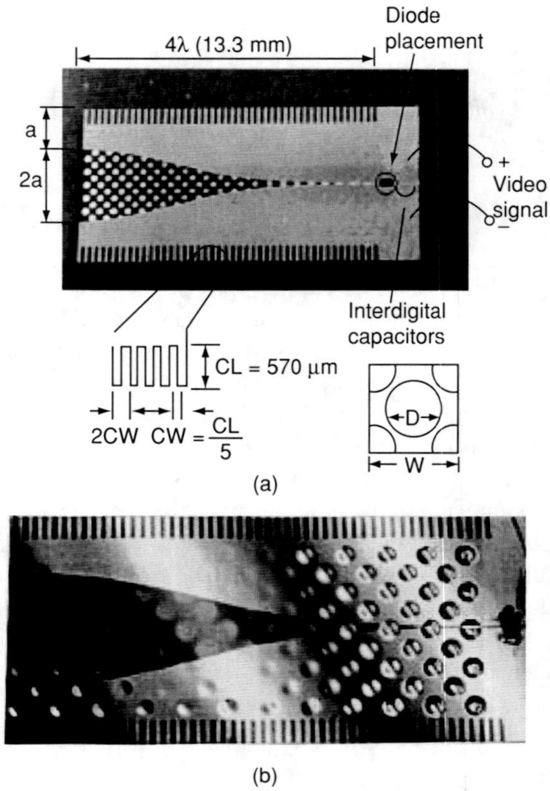

FIGURE 12.16 Picture of the W-band TSA fabricated on 100-μm thick silicon substrate with (a) 300-μm and (b) 600-μm wide holes etched from the backside. (From JB Rizk and GM Bebeiz. *IEEE Trans. Antennas Propagat.* 50:379–383, 2002. With permission. Copyright IEEE.)

result in bad radiation pattern. For comparison purpose, the same antenna is also fabricated on a quartz substrate ($\varepsilon_{eff} = \varepsilon_r = 3.78$) whose thickness (150 μm) was chosen to give an effective thickness (0.042) close to the one of the micromachined 100-μm thick silicon TSA.

Figure 12.17 to Figure 12.20 show the radiation pattern for the antennas at 90 GHz. It is clear that the TSA on 100-μm thick silicon does not perform well, and micromachining is essential to reduce the effective dielectric constant. It is important to notice the similar radiation patterns of the TSA on 150-μm quartz and the micromachined TSAs on silicon with 300- and 600-μm wide holes. These antennas have almost the same effective thickness ($t_{eff} / \lambda_0 = 0.042$), and therefore, the improvement in the micromachined silicon TSA is directly related to the effective dielectric constant of the substrate, and not to a substrate mode suppression phenomena due the presence of the periodic holes. The 300-μm wide hole antenna and the 150-μm quartz antenna resulted in very similar cross-polarization components. The improvement in the micromachined TSA starts to deteriorate for a hole diameter of 750 μm. In fact, in these cases, the hole diameter is so large that only two hole period are defined in the TSA aperture at the edge of the substrate. Therefore, the volumetric average of the dielectric constant cannot be used accurately. Still the E-plane pattern is quite acceptable, but the H-plane pattern starts to break down. The measured sidelobe at -45° is due to the large diameter hole and move to $+45^\circ$ if the antenna is flipped. The quart TSA and the micromachined silicon TSA with 300 μm holes were also measured at 105 GHz and shown in Figure 12.21. The measured patterns of the quartz and the micromachined TSA are very similar and are much better that the nonmicromachined case. This shows that the effective dielectric constant model is quite wideband and the design shown in Figure 12.16 can easily cover the 80 to 110 GHz range.

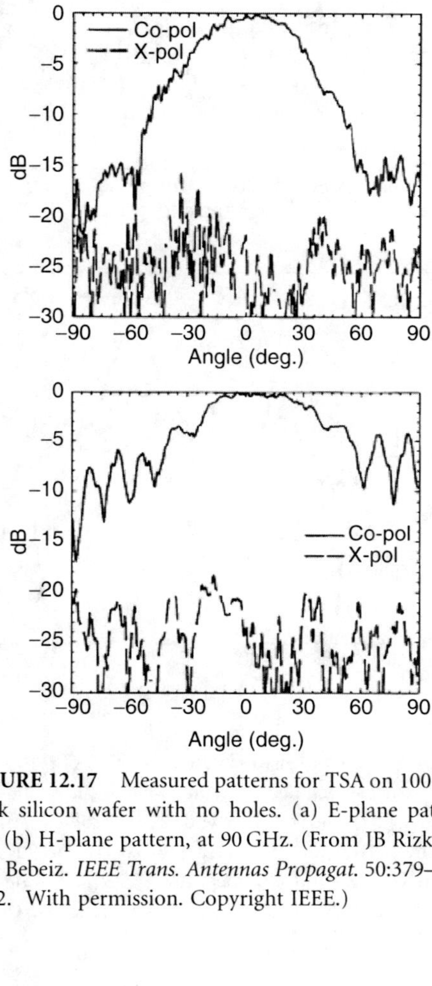

FIGURE 12.17 Measured patterns for TSA on 100-μm thick silicon wafer with no holes. (a) E-plane pattern and (b) H-plane pattern, at 90 GHz. (From JB Rizk and GM Bebeiz. *IEEE Trans. Antennas Propagat.* 50:379–383, 2002. With permission. Copyright IEEE.)

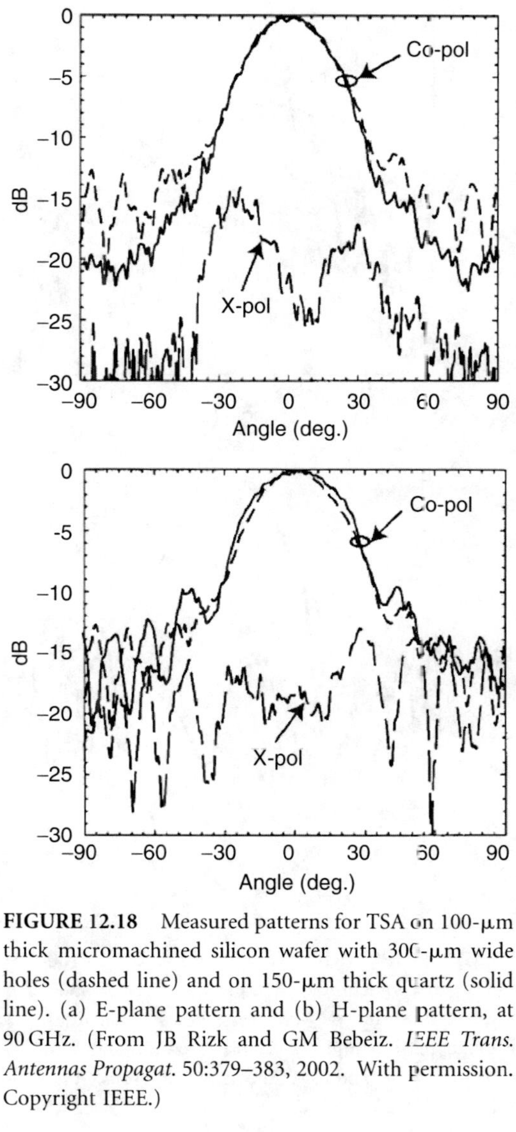

FIGURE 12.18 Measured patterns for TSA on 100-μm thick micromachined silicon wafer with 300-μm wide holes (dashed line) and on 150-μm thick quartz (solid line). (a) E-plane pattern and (b) H-plane pattern, at 90 GHz. (From JB Rizk and GM Bebeiz. *IEEE Trans. Antennas Propagat.* 50:379–383, 2002. With permission. Copyright IEEE.)

12.4 Surface Micromachined Monopoles for W-Band Applications [21]

Printed circuit antennas are often used in compact millimeter-wave systems as a low-cost solution [22]. Nevertheless, microstrip patch antennas often suffer from low bandwidth, high loss, and perturbation of radiation pattern caused by surface wave, which limit their application in broadband modules and have led to various bandwidth improvement techniques [23]. On the contrary, thin wire or cylindrical monopoles have broad impedance bandwidths and might be better candidates for broadband radiation. The major roadblock for their wide application is that the 3D transition from the preceding planar transmission lines to the monopole radiation structures is more complicated than those of printed circuit antennas. It is well known that in lower frequency systems, the cylindrical monopole is usually fed from the backside by a coaxial line, requiring the use of an etching process. In compact millimeter-wave

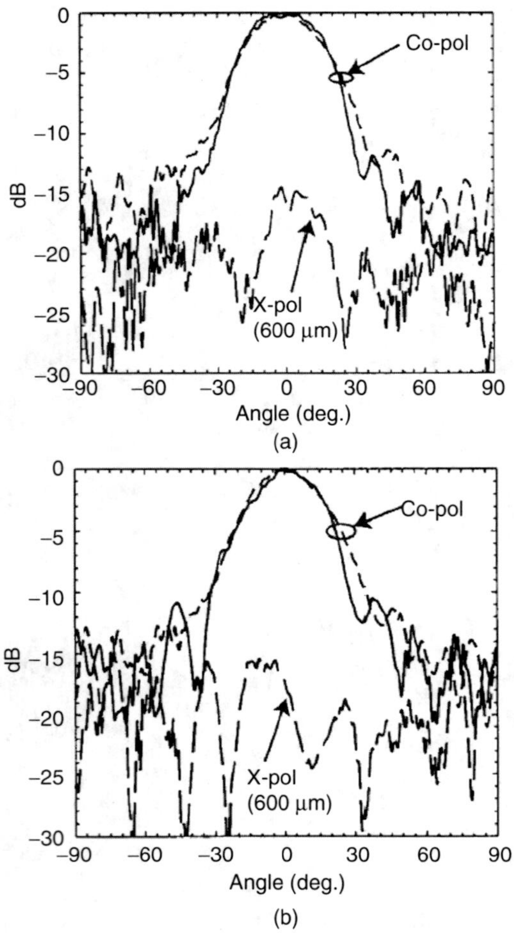

FIGURE 12.19 Measured patterns for TSA on 100-μm thick micromachined silicon wafer with 300-μm (dashed line) and 600-μm wide holes (solid line). (a) E-plane pattern and (b) H-plane pattern, at 90 GHz. (From JB Rizk and GM Bebeiz. *IEEE Trans. Antennas Propagat.* 50:379–383, 2002. With permission. Copyright IEEE.)

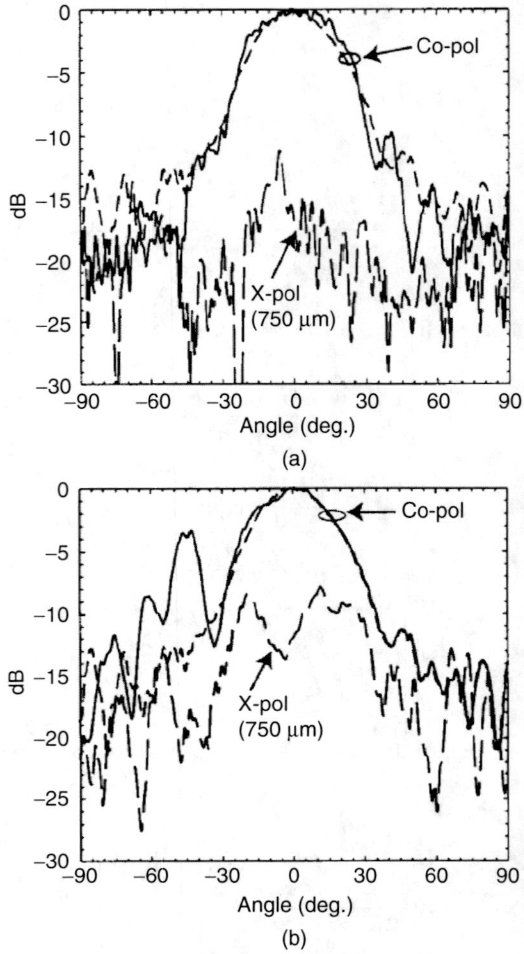

FIGURE 12.20 Measured patterns for TSA on 100-μm thick micromachined silicon wafer with 300-μm (dashed line) and 750-μm wide holes (solid line). (a) E-plane pattern and (b) H-plane pattern, at 90 GHz. (From JB Rizk and GM Bebeiz. *IEEE Trans. Antennas Propagat.* 50:379–383, 2002. With permission. Copyright IEEE.)

systems, this process should be avoided for the reasons of cost reduction and fabrication compatibility with coplanar transmission lines. Therefore, a transition from a two-dimensional CPW structure to a 3D monopole is essential for the effective realization of this concept [24].

A novel cylindrical monopole that is vertically mounted on silicon substrate operating in W-band is presented for the first time in Ref. [21]. The work has been enabled by surface micromachining technology and can be realized on other substrates as well. This scheme is proposed as an alternative for traditional leaky wave mode and microstrip radiators. It is via-free, and thus a low-cost solution. In the future, this structure is expected to be easily integrated with other key RF front-end components in millimeter-wave systems.

The general configuration and the illustration of key geometrical parameters are shown in Figure 12.22. The quarter-wavelength monopole is vertically mounted at the end of the center conductor of the CPW line. The two sides of the ground plane are connected radially to enclose the monopole

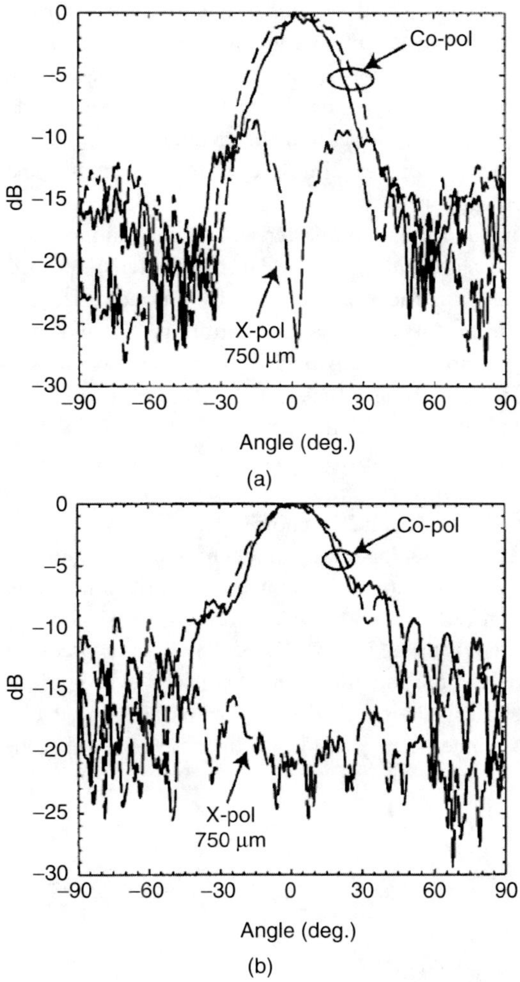

FIGURE 12.21 Measured patterns for TSA on 100-μm thick micromachined silicon wafer with 300-μm (dashed line) and 150-μm thick quartz (solid line). (a) E-plane pattern and (b) H-plane pattern, at 105 GHz. (From JB Rizk and GM Bebeiz. *IEEE Trans. Antennas Propagat.* 50:379–383, 2002. With permission. Copyright IEEE.)

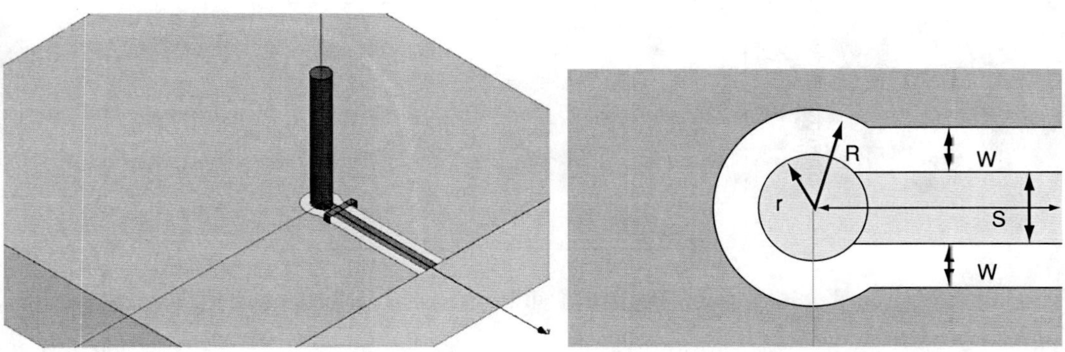

FIGURE 12.22 (a) Configuration of the structure and (b) illustration of the key geometrical parameters. (From B Pan, Y Yoon, J Papapolymerou, MM Tentzeris, and M Allen. Proceedings of the IEEE IMS symposium, 2004, Fort Worth, TX, June 2004, pp. 1935–1938. With permission.)

bottom [24]. The radius of monopole *r*, the radius of the ground aperture *R*, the center conductor line width *s*, and the gap width *w* between center conductor and ground have been optimized with Ansoft HFSS9.0. The characteristic impedance of CPW lines on silicon has been calculated as a function of normalized center conductor width, using LineCalc of Agilent ADS, and is depicted in Figure 12.23. The gap width is fixed to 50 μm for these curves and the ground is assumed to be infinite. In order to get compatibility with both fabrication process and measurement facilities, the center conductor width and gap width have been chosen appropriately.

The theoretical input impedance of the quarter-wave length monopole is 36.5 Ω at resonance. On some substrates, the width of center conductor for 50 Ω characteristic impedance will be comparable to and even significantly larger than the diameter of the monopole. On other side, in order to get a symmetrical radiation pattern, the transition discontinuity should be only a small portion of the circumference of the ground aperture. Considering this, we have two alternative choices for the feeding: (a) use a thinner line width to feed the monopole and transform it to wider lines with the aid of different types of matching networks; (b) from Figure 12.23, we can observe that the characteristic impedance is still around 50 Ω when using a narrow line width of 80 μm, if the diameter of the monopole is 100 μm. Therefore, simply connecting the CPW line without other impedance transforming techniques would also be a reasonable choice. These two schemes have been simulated and compared. In choice (b) the central operating frequency for the monopole was chosen as 85 GHz and the theoretical height of the monopole was 800 μm. Simulation results with Ansoft HFSS 9.0 are shown in Figure 12.3 and Figure 12.4. Figure 12.24 shows S_{11} for silicon and sapphire substrates, while radiation pattern for silicon given in Figure 12.25. The results predict greater than 10 dB return loss from 82 to 92 GHz. As for the radiation performance, a symmetrical far field pattern has been achieved for all three substrates. The far field radiation parameter calculation tool in HFSS 9.0 was utilized to estimate the radiation efficiency and led to the observation that most of the energy is radiated from the monopole.

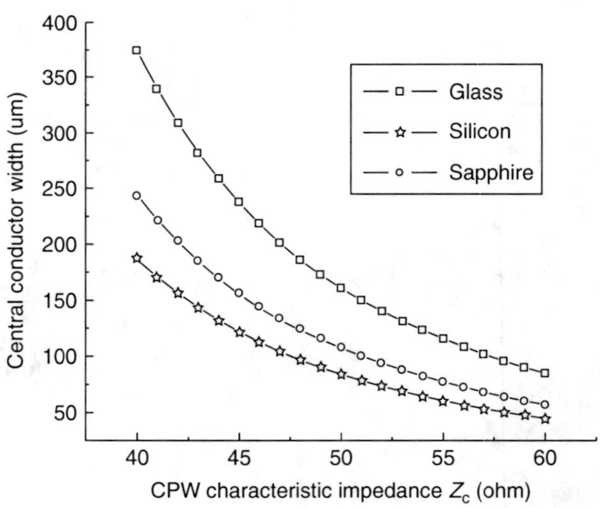

FIGURE 12.23 CPW dimensions versus characteristic impedances on different substrates (gap width fixed to 50 μm). (From B Pan, Y Yoon, J Papapolymerou, MM Tentzeris, and M Allen. Proceedings of the IEEE IMS symposium, 2004, Fort Worth, TX, June 2004, pp. 1935–1938. With permission.)

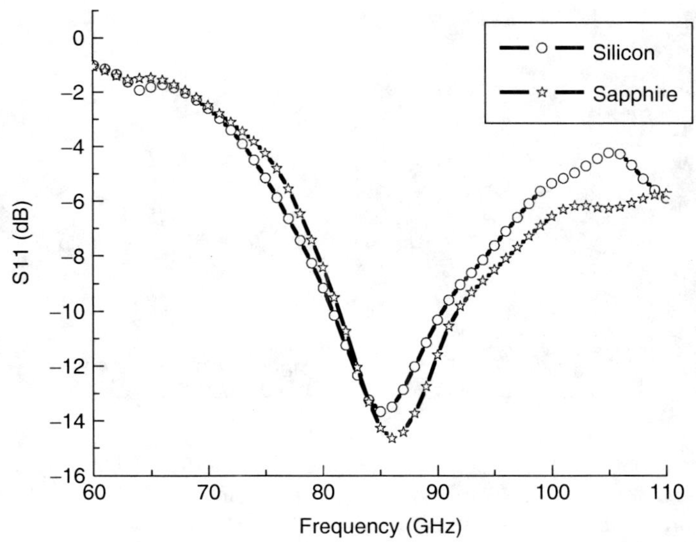

FIGURE 12.24 Simulated input return loss for the monopole on silicon substrate and sapphire substrate. (From B Pan, Y Yoon, J Papapolymerou, MM Tentzeris, and M Allen. Proceedings of the IEEE IMS symposium, 2004, Fort Worth, TX, June 2004, pp. 1935–1938. With permission.)

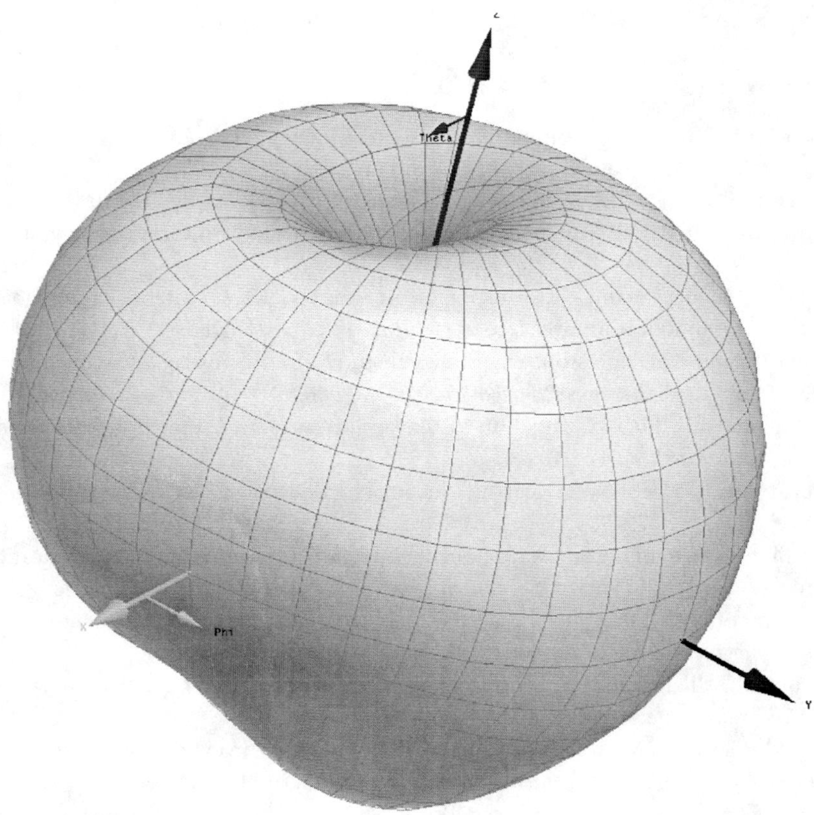

FIGURE 12.25 Simulated radiation pattern for the monopole on silicon substrate. (From B Pan, Y Yoon, J Papapolymerou, MM Tentzeris, and M Allen. Proceedings of the IEEE IMS symposium, 2004, Fort Worth, TX, June 2004, pp. 1935–1938. With permission.)

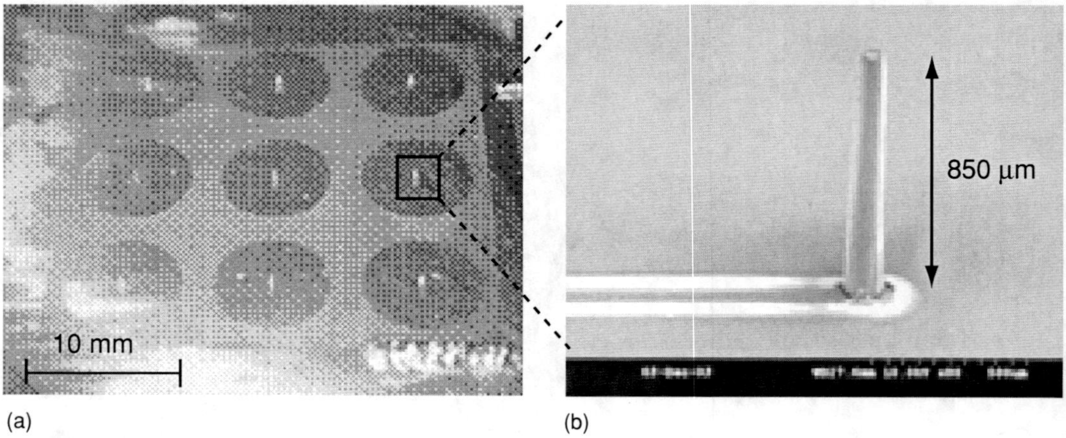

FIGURE 12.26 (a) A photomicrograph of 3 × 3 monopole array and (b) an SEM image of a single monopole antenna. (From B Pan, Y Yoon, J Papapolymerou, MM Tentzeris, and M Allen. Proceedings of the IEEE IMS symposium, 2004, Fort Worth, TX, June 2004, pp. 1935–1938. With permission.)

Figure 12.26a and b shows a photomicrograph of a fabricated monopole antenna array (3 × 3) and an SEM image of a single monopole antenna, respectively. The fabricated structures were measured to be approximately 850 μm high, which usually vary between 800 and 880 μm.

References

1. KK O, K Kim, et al. Wireless communications using integrated antenna. Proceedings of the IEEE Interconnect Technology Conference, 2003. pp. 111–113.
2. B Strassner and K Chang. Integrated antenna system for wireless RFID tag in monitoring oil drill pipe. Proceedings of IEEE Antenna and Propagation Society International Symposium, 2003, pp. 208–211.
3. JJ Lin, L Gao, A Sugavanam, X Guo, R Li, JE Brewer, and KK O. Integrated antennas on silicon substrates for communication over free space. *IEEE Electron Device Lett.* 25:1–3, 2004.
4. DH Schaubert and KS Yngvesson. Experimental study of a microstrip array on high permittivity substrate. *IEEE Trans. Antennas Propagat.* 34:72–97, 1986.
5. MO Thieme and EM Biebl. Calculation of the surface-wave excitation in multilayered structures. *IEEE Trans. Antennas Propagat.* 46:686–693, 1998.
6. EM Biebl. Integrated active antennas on silicon. Proceedings of Microwave and Optoelectronics Conference, 1997, pp. 279–284.
7. P Russer. Si and SiGe millimeter-wave integrated circuits for sensor and communications applications. Proceedings of 12th International Conference on Microwaves and Radar, 1998, pp. 330–344.
8. J Papapolymerou, RF Drayton, and LPB Katehi. Micromachined patch antennas. *IEEE Trans. Antennas Propagat.* 46:275–283, 1998.
9. DR Jackson, JT Williams, and AK Bhattacharyya. Microstrip patch designs that do not excite surface waves. *IEEE Trans. Antennas Propagat.* 41:1026–1037, 1993.
10. JG Yook and PB Katehi. Micromachined microstrip patch antenna with controlled mutual coupling and surface waves. *IEEE Trans. Antennas Propagat.* 49:1282–1289, 2001.
11. M Zheng, Q Chen, PS Hall, and VF Fusco. Broadband microstrip patch antenna on micromachined silicon substrates. *Electron. Lett.* 34:3–4, 1998.
12. EY Tsai, AM Bacon, M Tentzeris, and J Papapolymerou. Design and development of novel micromachined patch antennas for wireless applications. Proceedings of the 2002 Asian-Pacific Microwave Symposium, Kyoto, 2002, pp. 821–824.

13. G Gauthier, LP Katehi, and GM Rebeiz. W-band finite ground coplanar waveguide (FGCPW) to microstrip line transition. Proceedings of the 1998 IEEE IMS Symposium, Baltimore, MD, June 1998, pp. 107–109.
14. G Zheng, J Papapolymerou, and MM Tentzeris. Wideband coplanar waveguide RF probe pad to microstrip transitions without via holes. *IEEE Microwave Wireless Components Lett.* 13:544–546, 2003.
15. GP Gauthier, A Courtay, and GM Rebeiz. Microstrip antennas on synthesized low dielectric-constant substrates. *IEEE Trans. Antennas Propagat.* 45:1310–1314, 1997.
16. MJ Vaughan, KY Hur, and RC Compton. Improvement of microstrip patch antenna radiation patterns. *IEEE Trans. Antennas Propagat.* 42:882–885, 1994.
17. JB Rizk and GM Bebeiz. Millimeter-wave Fermi tapered slot antennas on micromachined silicon substrates. *IEEE Trans. Antennas Propagat.* 50:379–383, 2002.
18. R Garg, P Bhartia, I Bahl, and A Ittipiboon. *Microstrip Antenna Design Handbook.* Norwood, MA, Artech House, 2001.
19. F Ulaby. *Fundamentals of Applied Electromagnetics.* Prentice Hall, Upper Saddle River, NJ, 1999.
20. KS Yngvesson, DH Schaubert, TL Korzeniowski, EL Kollberg, T Thungren, and JF Joansson. Endfire tapered slot antennas on dielectric substrate. *IEEE Trans. Antennas Propagat.* 33:1392–1400, 1985.
21. B Pan, Y Yoon, J Papapolymerou, MM Tentzeris, and M Allen. A W-band surface micromachined monopole for low-cost wireless communication systems. Proceedings of the IEEE IMS symposium, 2004, Fort Worth, TX, June 2004, pp. 1935–1938.
22. FK Schwering. Millimeter wave antennas. *Proc. IEEE* 80:92–102, 1992.
23. DM Pozar. Considerations for millimeter wave printed antennas. *IEEE Trans. Antenna Propagation* 31:740–747, 1983.
24. D Staiculescu, J Laskar, and M Tentzeris. Flip chip design rule development for multiple signal and ground bump configurations. Proceedings of the 2000 Asian-Pacific Microwave Conference, 2000, pp. 136–139.

13

Packaging Issues for SiGe Circuits

Kyutae Lim, Stephane Pinel, and Joy Laskar
Georgia Institute of Technology

13.1 Background 13-1
13.2 IC Package Evolution 13-2
13.3 Recent Package Technology Trend 13-2
Ball Grid Arrays (BGA) • Flip Chip Technology • Multichip Module
13.4 Future Direction 13-5
13.5 Package Design Consideration 13-6
13.6 Summary 13-7

13.1 Background

The role of the SiGe package is more important than other IC packages, since the SiGe devices make it possible to integrate digital, analog, and radio frequency (RF) functions on the same die and process. More care should be taken to minimize the noise and parasitic in the package of the SiGe die than the usual integrated circuit (IC) package. Packaging is a method for allowing electrical connection to an integrated circuit while maintaining and regulating its operating environment, and achieving performance, reliability, and cost requirements. The primary function of a package is to provide a means for electrical connectivity from the semiconductor device to a printed wiring board (PWB), also known as a printed circuit board (PCB). It provides a path for power to be applied to the chip as well as a way for the data signals to be transmitted into and out of the chip. Its secondary function is to house and protect the fragile chip from harsh environmental conditions, like moisture, light, and dust, that might hinder its performance. Finally, the package provides a pathway for dissipating the heat generated by the semiconductor device [1].

In its early evolution, the influence of the package on performance was limited; however, as the systems evolve to provide increasing performance and operation frequencies, the package must evolve to keep up, and packaging design must ensure that it optimally enables the systems. From the traditional role of a protective mechanical enclosure, the modern package has been transformed into a sophisticated thermal and electrical management platform. Furthermore, system architecture and design techniques can have significant impact on the complexity and cost of packaging. The need to optimize the total solution (chip, package, board, and assembly) has never been more important to maximize performance and minimize cost.

Recent advances in packaging indicate a migration from wirebond (where the chip or die is interconnected to the package only on the periphery of the die) to flip chip (where the die is interconnected to the package using the entire die area) [2]; and from ceramic to organic packages, with cartridge and multichip technologies emerging as key form-factors. With the emergence of the "segmented" market (mobile, desktop, server, and associated subsegments), we see a significant proliferation of packaging types tailoring functionality and costs to the different application specifications.

13.2 IC Package Evolution

The original packaging devices were typically large. They connected to the outside world by means of long leads or pins that needed to be inserted through holes in the PCBs. These devices came to be known as "through-hole" devices for that reason and were made of metal, ceramic, and plastic. The most predominant of these packages in the 1970s and 1980s was the dual-in-line package (DIP) [3], which is still in use today. The DIP's main limitation is the lead count with an upper limit of 64 pins. After that point, the package simply becomes too large for any practical application. A solution to this is the pin grid array (PGA), which uses a two-dimensional array of pins protruding from the bottom of the package. The PGA package can have around 200 pins on average [4].

The main problem with through-hole devices was their size. Furthermore, their performance in high frequencies was quite limited. The trend in PCB manufacturing was to increase density while decreasing board area and increasing signal frequencies. Through-hole devices did not easily allow this to happen. In the 1980s, through-hole devices began to give way to a new packaging technology called surface mount technology (SMT) [5]. The leads from these packages mount directly to rectangular pads on the surface of the PCBs. They did not require holes to be drilled into the PWBs. As a result, the width of the leads could be smaller and the spacing between the leads (lead pitch) could be decreased. This allowed for a package with the same number of pins as a traditional through-hole device to be significantly smaller, despite the fact that these devices were made with the same materials as the through-hole devices. Common surface mount devices included the plastic leaded chip carrier (PLCC) and the small outline integrated circuit (SOIC). In recent years, the quad flat pack (QFP) has become a predominant fine pitch, high lead count package solution. Figure 13.1 illustrates most of the previous topologies and Figure 13.2 shows some popular packaging devices used through the 1980s and 1990s [1].

Today, as semiconductor technology continues the minimization trend, the level of complexity on a single silicon chip is increasing. This leads to more functionality in a smaller area, higher I/O counts, higher frequencies, and higher heat dissipation requirements. All of the technologies listed up to this point are inadequate and impractical to satisfy all of these requirements. This is not to say that they do not have a place anymore in modern integrated circuit design, but they will not meet the needs of leading-edge technologies and, in particular, in the case of very high-performance systems, such as RF and millimeter waves integrated systems.

New packaging technologies have emerged in the last 10 years that are aimed at solving the I/O and heat challenges. The first of these solutions continue to use the same wire bonding technology used in traditional packages. However, these packages can also accommodate advances in interconnect technology. Flip chip technology (also called direct chip attach [DCA]) has emerged as a possible alternative to wire bonding [6]. Regardless of the interconnect technology, the most promising packaging technology being pursued at the present is the ball grid array (BGA) package [7], which is a descendant of the PGA package discussed earlier. Instead of through-hole pins, the BGAs have small conductive balls that are soldered directly to the surface of PCBs. As the balls are located at the bottom of the package, an obvious disadvantage is the inability to visually inspect the connections between the package and the PWB. As a result, the assembly process for PWBs using BGA technology must be very precise with very low tolerance for error. Its advantages in I/O count, I/O density, and heat dissipation easily outweigh this somewhat minor disadvantage.

13.3 Recent Package Technology Trend

Ball Grid Arrays (BGA)

BGA packages allow for PWB space savings since an array of solder bumps (or balls) are used in place of traditional package pins. An example of a BGA applied to a 3D integrated module concept is presented in Figure 13.3 [8].

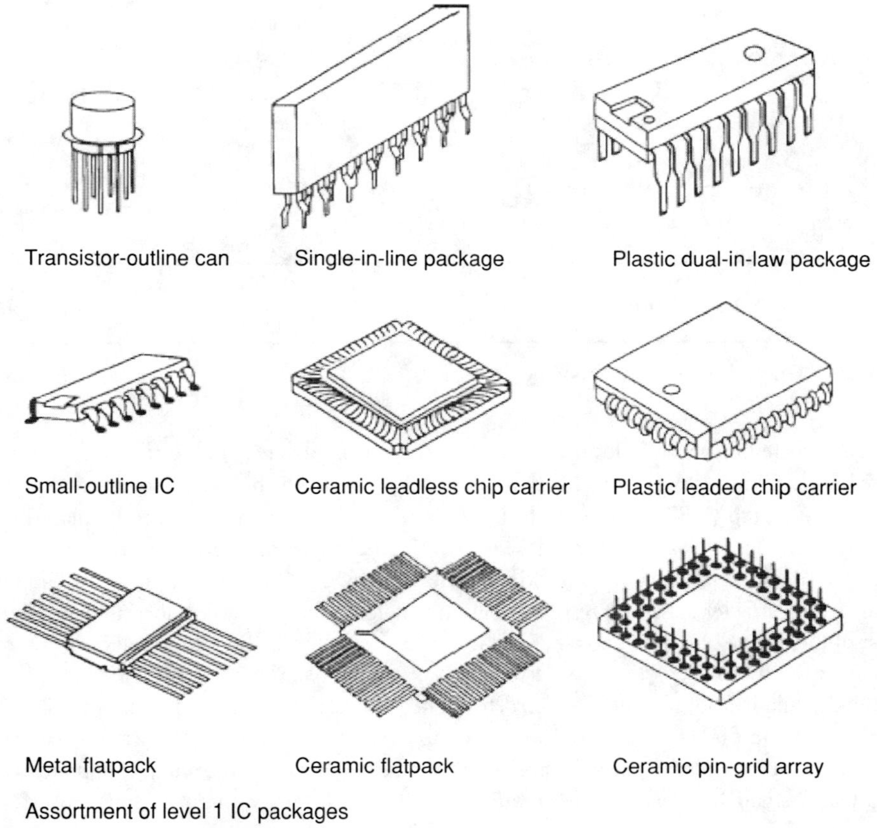

FIGURE 13.1 Various types of IC packaging through 1970 to 1980.

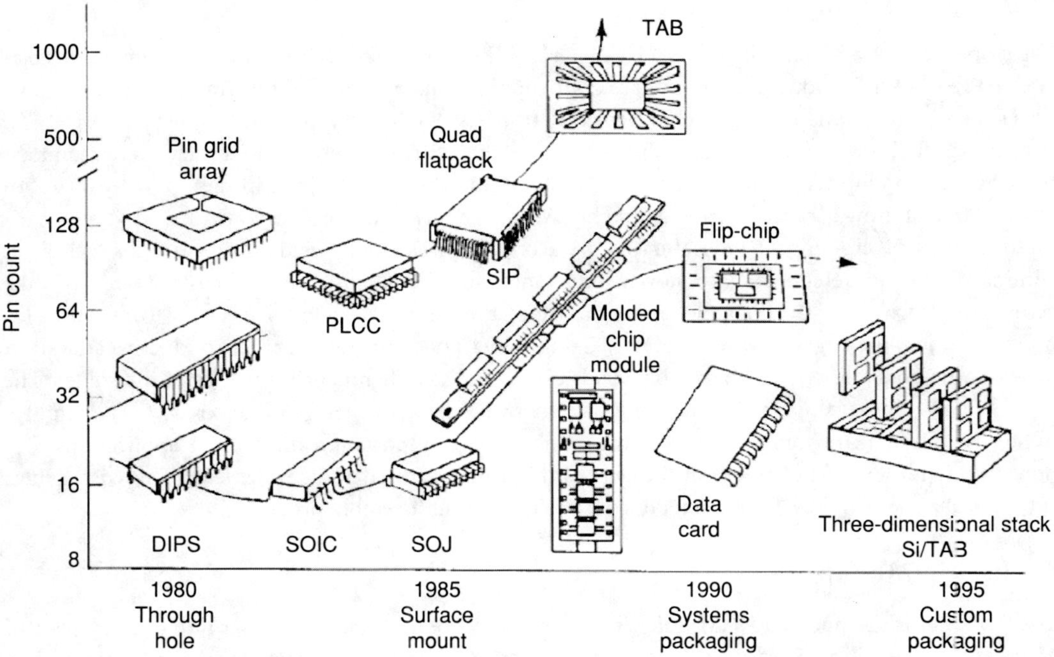

FIGURE 13.2 Packaging evolution through 1980 to 1990s.

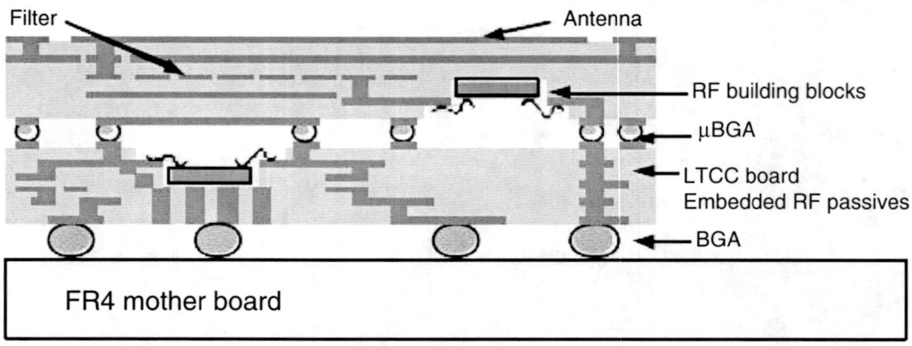

FIGURE 13.3 BGA package-based 3D module.

BGAs are expected to be the solutions for packages requiring over 200 pins and will be implemented using wire bonding in the lower I/O density parts and flip chip in the higher I/O count and higher power devices. Ultrafine pitch (UFP) wire bond technology will allow for a larger die to be placed on smaller substrates and an effective pad pitches below 60 μm. Increasing circuit density can accommodate more than 1000 I/O devices and enables shrinking of die size by 20–50%. Placing a larger die with more functionality on a smaller substrate results in significant cost savings, as the substrate is a driving factor in the cost of the device. Recent trends indicate that I/O counts for leading-edge devices grow at a rate of approximately ten times every 14 years.

Liquid-encapsulant underfills are expected to be needed to relieve the stress due to CTE differences between the chip and the package substrate in the flip chip devices. Chip scale packages using fine pitch BGAs with a size in the order of the chip size will be the next level of advancement for applications where low weight and small package size are required. As the technology matures and processes become cheaper, BGAs have the potential to become the workhorse of packaging technologies, like the DIP was for the 1970s and 1980s.

Flip Chip Technology

Flip chip is actually a 30-year-old technology that has only recently matured to be widely accepted and cost-effective for the semiconductor industry. It refers to flipping a silicon die or chip and mounting it face down on a substrate. A schematic of a flip chip transition from CPW to CPW is presented in Figure 13.4 [9].

Of the 60 billion integrated circuits produced in 1998, approximately 1.5% of them were manufactured using flip chip technology. From 1997 to 1998, the number of flip chip dies grew by 40%. An average annual growth rate of 48% for flip chip is expected over the next several years.

Advantages of flip chip include efficient die access, high assembly yields using solder attachment, elimination of an interconnect layer, lower inductance, and the potential for low cost. Efficiency of die access comes from the fact that the entire surface of the die is available for electrical connection (as opposed to just the edges in wire bonding.) Using solder attachment, instead of conductive adhesives, provides the possibility of high assembly yields due to the self-alignment properties of the solder. The wire bond is eliminated, providing higher reliability by reducing one level of interconnect. The removal of the “long-lead” wire bond also reduces the inductance, an important feature for RF applications. The opportunity for low cost comes from the fact that flip chip is created at the wafer level. This cost savings is only realizable with a well-understood, controlled, efficient assembly process.

Multichip Module

The term multichip module (MCM) refers to the packaging of multiple silicon dies into one device [10]. MCMs offer the ability to reduce package pin count by combining two or more high pin count devices that would normally connect to each other at the board level into one package where the interconnect is

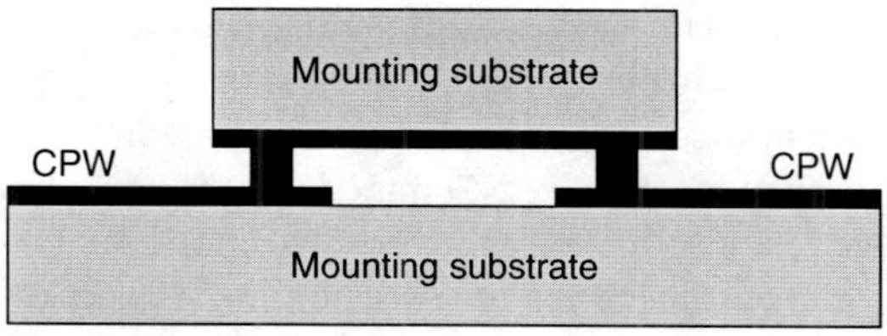

FIGURE 13.4 Flip chip configuration.

performed at the chip level. The resulting packaged system only needs a reduced set of pin to connect power and signals from the outside world. It can also serve to reduce PWB real estate (an expensive commodity) and presents the following advantages:

- Improved performance, such as shorter interconnect lengths between die (resulting in reduced delay time, lower RF parasitic and losses), lower power supply inductance, lower capacitance loading, less cross talk, and lower off-chip driver power.
- Miniaturization, since MCMs result in a smaller overall package when compared to packaged components performing the same function, hence resulting I/O to the system board is significantly reduced.
- Shorter time-to-market, making them attractive alternatives to ASICs, especially for products with short life cycles.
- Low-cost silicon sweep, allowing integration of mixed semiconductor technology, such as SiGe or GaAs.
- Hybrid configurations, including surface mount devices in the form of chip scale or micro-BGA packages and discrete chip capacitors and resistors.
- Simplification of board complexity by integrating several devices onto one package, thereby reducing total opportunities for error at the board assembly level.
- Capability of accommodating a variety of second-level interconnects. While BGA are the most popular ones, lead-frame solutions can be employed for plugability, enabling modularity for upgrades.

The widespread use of MCMs has been hurt by a few factors. It is a relatively high-cost process. Since not all of the signals are connected to the outside world, package level testing is difficult, so effort must be made to insure that the die being put into the chip is good. Obtaining such "known good die" (KGD) is still a challenge. Also, the proliferation of flip chip, BGAs, and chip scale packages for single-chip package solutions has reduced the need for chip-level system integration.

13.4 Future Direction

The needs of assembly and packaging are driven equally by silicon technology and marketing requirements. As always, cost will be the driving factor. It is expected that packaging costs per pin will decrease in the coming years, but the overall packaging pin count is expected to increase at a faster rate than the cost decrease. The increase in pin count is also expected to affect the substrate and the system-level costs. The roadmap identifies thermal management as a significant challenge. Handheld devices that do not use forced air and rely on the operator's hand to dissipate the heat from the unit will need new heat sink technologies and materials with better thermal conductivity.

The cost-performance market (desktop processors) requires forced air-cooling. Flip chip could be a possible enhancement to the forced air-cooling as the front-side of the silicon chip provides a "direct, efficient heat path from the chip to the heat sink." Existing heat sink solutions are predicted to be ineffective above 50 W in applications where forced air is not a viable solution due to market requirements. A reduction of internal thermal resistance and better air-cooling techniques will be critical for future thermal solutions. The high-end market, with predicted power consumption between 110 and 120 W per chip will pose an even greater challenge. A closed-loop cooling system that meets market and customer requirements will be necessary. In general, flip chip is predicted to become the predominant technology for chip-to-next levels interconnect. Commodity products will continue to use advanced forms of wire bonding until the cost of flip chip becomes affordable for that market. Flip chip is especially desirable because it is possible to reduce a level of interconnect when using this technology.

Another important RF packaging challenge is matching the coefficient of thermal expansion (CTE) between the silicon chip and the substrate. When using organic substrates, underfills will be required with high reliability, ease of manufacturability, stronger attachment at the interface, and higher resistance to moisture. Liquid-crystal polymers (LCP) with engineered CTE could be a substrate solution [11].

High-frequency packaging geometries are often sorted in two independent areas of focus, single-chip and multichip packages. The single-chip packages include the technologies discussed above in the "Background" section. Devices like QFPs will reach a maximum lead count of 304 and lead pitch will reach a minimum of 0.5 mm. After this point, the package body size and the surface mount assembly complexity become cost ineffective and multichip solutions have to be investigated.

The RF front-end module is the core of these systems and its integration poses a great challenge. Microelectronics technology, since the invention of the transistor, has revolutionized many aspect of electronics product. This integration and cost path has led the microelectronics industry to believe that this kind of progress can go on forever, leading to the so-called "system-on-chip" (SOC) for all applications [12]. But it is becoming clear that it is still a dream to produce a complete on-chip solution for the novel wireless communication front-ends.

Considering the characteristics of the RF front-end modules, such as high performance up to 100 GHz operating frequency, large number of high-performance discrete passive components, design flexibility, reconfigurable architecture, low power consumption, compactness, customized product, short time to market and low cost, the "system-on-package" (SOP) approach, has emerged as the most effective way to provide a realistic integration solution [13].

13.5 Package Design Consideration

The major issues are manufacturing cost, size and weight, signal integrity, low high-frequency loss, heat dissipation, mechanical, stability, testability, reliability:

- *Manufacturability and cost*: Materials, fabrication steps, and IC Costs incurred from testing, rework, yield loss. Manufacturability depends on process control, cycle time, repirability, equipment downtime, design tolerances
- *Electrical design*: Interconnect speed now plays a dominant role in determining performance limits. Each connection has parasitic capacitance, resistance, and inductance that limit speed, potentially distort signals, and add noise

  Leads for connections are also a source of reliability problems

  Several factors need to be considered, including:

  —Lead length

  —Matched impedances

  —Low ground resistance

  —Simultaneous switching and power supply spiking
- *Thermal design*:

  Objective: remove heat from the junctions of the ICs (to keep dopants from moving and avoid self heating effects)

  Techniques: forced air, liquid cooling, monophase cooling, dual phase cooling

  Considerations:

  1. How to remove heat (from the front or backside of the IC)?
  2. Air or liquid? mono or dual phase?

3. Thermal conductivity of the substrate
4. Stresses induced due to CTE mismatches

- *Mechanical design*:
  Susceptibility to thermal stresses must be considered in design
  Tensile modulus ("stiffness") also a consideration

13.6 Summary

In this section, IC package technology was reviewed and the future direction has been discussed. Package requirement for the SiGe IC package is more stringent than that for the conventional package, especially when the IC deals with mixed signal applications. In designing of those packages, the designers should take more care to minimize the high-frequency noise as well as $1/f$ noise. Also Ohmic loss and parasitic effect should be considered together since the I/Os deal with the low frequency to RF frequency at the same time. Recent progress in the system shows SOC and SOP approaches are the direction for the future system. In other words, a full system function can be realized in a single die thanks to SiGe process. Or for the more demanding system, a system can be realized in a single package. So it becomes very important to decide how the system would be realized. By properly allocating the functions to the IC or package, we can optimize the system in performance and cost.

In the package technology itself, significant challenges lie ahead to achieve the package I/O densities, lead pitch, power dissipation, and circuit speeds required of today's and tomorrow's high-performance integrated circuits. Packaging solutions are being pursued to meet these challenges. Each option has its own advantages and associated challenges to make it a feasible, cost-effective solution. BGAs utilizing wire bonding is a short-term solution. The proliferation of flip chip and its application into BGAs provides the next level solution. MCMs offer even further solutions. Of course, these are not all the answers. There are other packaging technologies that have not been discussed here. These are, however, the solutions for the foreseeable future. Advances beyond today's technology will produce new packaging options and with these options, more challenges as well.

References

1. R. Tummala, E. Rymaszewski, and A. Klopfenstein, *Microelectronics Packaging Handbook – Part II*, Second Edition, Kluwer Academic Publishers, Dordrecht, 1997.
2. D. Staiculescu, K. Lim, A. Sutono, H. Liang, M. Tentzeris, and J. Laskar, Flip chip vs. wirebond, Printed Circuit Design Magazine, June 2002, pp. 12–16.
3. M. DiOrio and S. Pinamaneni, Material effects on the performance and reliability of high-power molded dual-in-line packages, Proceedings of the 38th Electronics Components Conference, May 1988, pp. 406–410.
4. G.C. Phillips, Jr. Planar pin grid array (PGA) ceramic packaging, Proceedings of the 38th Electronics Components Conference, May 1988, pp. 350–354.
5. R. Chroneos, D. Mallik, and S. Prough, Packaging alternatives for high lead count, fine pitch, surface mount technology, Proceedings of Electronics Manufacturing Technology Symposium, September 1991, pp. 181–186.
6. M. Christensen, *Flip Chip: A Technology Reborn.* Solid State Technology; Pennwell Publishing Co., 1999.
7. D. Dunn, BGAs given a couple of boosts. Electronic Buyer's News. CMP Media, Inc. November 29, 1999.
8. M.J. Kuzawinski, IBM's Ultra Fine Pitch Wire Bond PBGA Modules, IBM Microelectronics Web Site, http://www.chips.ibm.com/micronews/vol4_no4/pbga.html, 1998.
9. D. Staiculescu, J. Laskar, and M.M. Tentzeris, Design rule development for microwave flip chip applications, *IEEE MTT Journal*, 48(9), 2000, 1476–1481.

10. C. Mitchell, Assembly and reliability study for the micro-ball grid array, Proceedings of the 16th IEEE/CPMT IEMT Symposium, September 1994, pp. 344–346.
11. S. Pinel, M.F. Davis, V. Sundaram, K. Lim, J. Laskar, G. White, and R. Tummala, High *Q* passives on Liquid Crystal Polymer substrates and μBGA technology for 3D integrated RF front-end module, *IEICE Transactions on Electronics.* E86–C(8), 2003, 158–1592.
12. A. Matsuzawa, RF-SoC—expectations and required conditions, Invited Paper, *IEEE Transactions on MTT,* 50(1), 2002, 245–253.
13. K. Lim, S. Pinel, M.F. Davis, A. Sutono, C-H. Lee, D. Heo, A. Obatoynbo, J. Laskar, M. Tentzeris, and R. Tummala, RF-SOP for wireless communications, *IEEE-Microwave Magazine,* 3(1), 2002, 88–99.

14

Industry Examples of the State-of-the-Art: IBM—High-Speed Circuits for Data Communications Applications

Daniel A. Friedman and
Mounir Meghelli
IBM Thomas J. Watson Research Center

14.1 Introduction **14**-1
14.2 Transmitter and Receiver Chip Set at 43 Gb/sec **14**-2
Optical Link Overview • Technology Overview • Receiver Circuit • 40 Gb/sec Transmitter Implementations Using Half-Rate and Full-Rate Architectures
14.3 Summary **14**-14

14.1 Introduction

The advent of advanced SiGe technologies in the 0.18 [1] and 0.13 μm [2] nodes has enabled the implementation of silicon-based wired data communications circuits operating at data rates of 40 Gb/sec and beyond [3–8] targeting standards such as SONET OC-768. Two key classes of circuits for high data rate communications are serializers, built from a clock multiplying PLL and a multiplexer, and deserializers, built from a clock and data recovery PLL and a demultiplexer. The raw multiplexing and demultiplexing functions are critical not only for communications systems, but also for high-speed test equipment such as pattern generators, which typically generate output patterns in parallel form at low data rates and multiplex these to generate the high-speed output, and bit error rate testers, which typically check input patterns in parallel form after a demultiplexing stage.

There are many challenges that must be addressed in the execution of very high data rate circuit designs. Specifications for random and deterministic jitter tend to be very stringent and must be satisfied over fairly broad process, temperature, and supply regimes. Signal integrity must be addressed not only at the package and board level, but must be considered on-chip as well. High-speed clock distribution is difficult and consumes large amounts of power, making architectures that enable minimal, efficient clock distribution extremely desirable. Finally, testing of such circuits is very demanding, in part because small problems like loose cable connections can significantly degrade results, but also because when data rates become sufficiently high, there is no standard test equipment available.

The design of a 40 Gb/sec serializer and deserializer chip set implemented in a 0.18 μm SiGe BiCMOS process will be the focus of this chapter, with emphasis placed on recently reported technical results [7–10]. In the course of describing the chip set, key building blocks will be presented, as well as an architecture and performance comparison of two approaches to the implementation of the serializer.

14.2 Transmitter and Receiver Chip Set at 43 Gb/sec

Optical Link Overview

A key application for advanced SiGe technologies is optical data communication. In a typical optical link and associated electronics (Figure 14.1) the transmit subsystem takes relatively low-speed parallel data, multiplexes it onto a high-speed serial line, and converts these data from an electrical to an optical format for transmission over an optical fiber. The receive subsystem converts the serial optical signal at its input to an electrical format and demultiplexes the resulting serial data stream to create multiple lower speed parallel outputs. The transmit electrical serialization function is implemented by a multiplexer, which converts the multiple parallel data inputs to a single serial output, and an associated clock multiplier unit, which generates the clock frequencies and phases necessary for multiplexer operation. The multiplexer output is typically translated to the optical domain via a laser diode driver driving a laser diode for many 10 Gb/sec and slower systems, and via a modulator driver driving an electro-absorption modulator used to modulate a continuous wave laser output for 40 Gb/sec systems. On the receive side, a photodetector is used to convert the optical signal to an electrical current and a combination of a transimpedance amplifier and a limiting amplifier is typically used to generate an input signal with sufficient amplitude to be used by the deserializer. Because no explicit clock is sent with the data over the optical link, the deserializer must execute a clock and data recovery function in addition to the demultiplexing and thus consists of a clock and data recovery (CDR) phase locked loop (PLL) and a demultiplexer.

In the case of 40 Gb/sec optical data transmission, the SONET OC-768 standard effectively provides a specification for the performance requirements of the elements of a link. The data rate range of 39.8 to 43 Gb/sec is set by the base OC-768 transmission rate at the low end and by the line rate when forward error correction (ITU-T G.709) is used at the high end. For the serializer and deserializer circuits that are the main topics of this chapter, transmit jitter generation specifications and receive jitter tolerance specifications can also be derived from this standard.

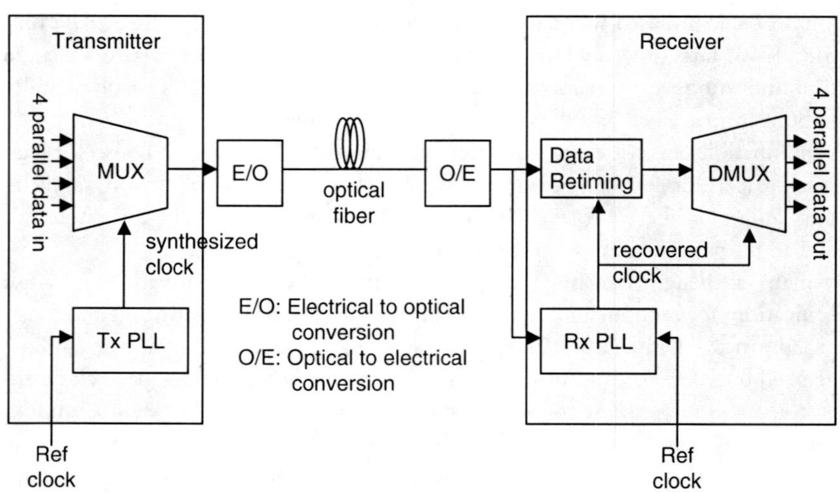

FIGURE 14.1 Typical high data rate optical link block diagram.

Technology Overview

The circuits described in this section were implemented in a production-level 0.18 μm SiGe BiCMOS process. This technology [1] features 120 GHz f_T, 100 GHz f_{max} HBTs, thick metal enabling inductor implementations, MIM capacitors, and a full suite of 0.18 μm CMOS devices. The bipolar logic family chosen for this work was emitter-coupled logic (ECL) because this choice enabled the use of a lower supply voltage, hence lower power consumption logic blocks, without sacrificing significant performance as compared to alternate, more headroom-hungry approaches such as emitter–emitter coupled logic (E^2CL). Differential internal swing levels were chosen to be between 500 and 600 mV in these designs, here reflecting a compromise between signal-to-noise ratio and limiting headroom (and thus supply voltage) requirements.

Receiver Circuit

The task of the receiver circuit is to take an input 40 Gb/sec serial data stream and demultiplex it. The data is sent with an implicit clock that must be recovered from the incoming data to enable proper retiming of the input. The serial data will be corrupted by deterministic and random jitter accumulated in its journey from the transmitter, through electrical to optical conversion, through a fiber, and through optical to electrical conversion. A key measure of the quality of the receiver is thus its jitter tolerance, or the degree of data impairment for which the receiver is still able to correctly recover the original data. Furthermore, the amplitude of the input signal may be small, particularly if the signal is taken directly from a transimpedance amplifier without passing through a limiting amplifier. Because of the extremely high data rate of the input, details of physical design and clock distribution have a significant impact on architecture, along with traditional factors like device performance. In the receiver design, a half-rate architecture was chosen over a full-rate architecture, thus enabling a lower power implementation that did not unduly stress the limits of the high-speed HBTs in the technology. This choice did incur costs, however, with main drawbacks of using a half-rate architecture in the receiver being the degradation of jitter tolerance due to the effects of duty cycle distortion and this architecture's requirement for quadrature clock generation. In the section covering the transmitter, the half-rate versus full-rate architecture question will be revisited for that circuit block.

Design Details of Key Receiver Circuit Blocks

The receiver itself (Figure 14.2) includes an integrated limiting amplifier, a half-rate clock and data recovery unit, a 1:4 demultiplexer, a frequency acquisition aid, and a frequency lock detector [7]. The intent of executing this design was to demonstrate the high-speed core of the deserializer function while keeping the input and output counts at a manageable level for a test site implementation. Note that a common implementation choice for a SONET product-level receiver would integrate the clock and data recovery function with a 1:16 demultiplexer [4].

As a precursor to implementing the receiver, a 1:4 demultiplexer was implemented as a stand-alone design, with the architecture of the stand-alone block used in large measure in the full receiver. High-speed building blocks of this type, as well as multiplexers, dividers, and latches, have been reported at high data rates and in multiple technologies [9,12–18]. Such blocks serve as technology demonstration vehicles, as parts of home-grown high data rate test environments for more complex circuits, and as elements in testers themselves. The stand-alone demultiplexer used in the receiver described in this chapter (Figure 14.3) uses a tree architecture with a recursive series of 1:2 demultiplexer stages, with a half-rate clock input used for the first 1:2 demultiplexing stages. A quarter-rate clock is generated from the half-rate input clock using a static divider to perform the last 1:2 demultiplexing stages. In the stand-alone implementation, input clock and data are received with double stage, wide bandwidth Cherry–Hooper [19] amplifiers, improving input sensitivity. The output buffers for the parallel data are implemented as differential pair stages with on-chip resistor terminations. In the physical design of the demultiplexer, transmission lines are used on long on-chip runs, a design approach followed in all high data rate designs described in this chapter. The stand-alone demultiplexer was demonstrated

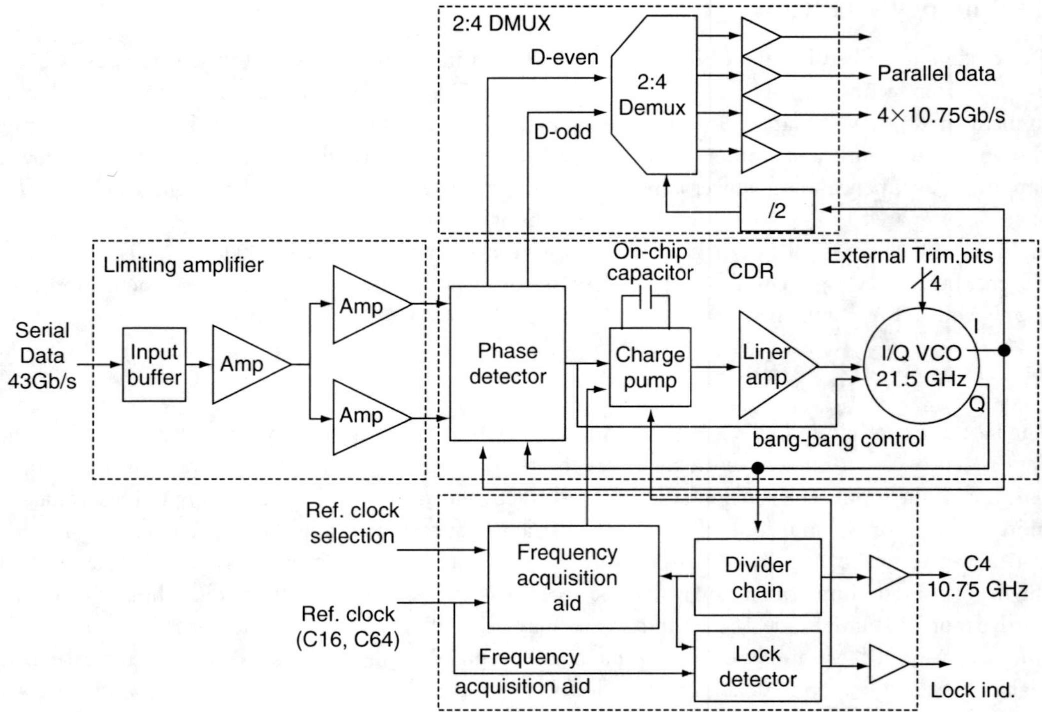

FIGURE 14.2 Block diagram of 40 Gb/sec receiver, including CDR and 1:4 demultiplexer. (From M Meghelli, A Rylyakov, S Zier, M Sorna, and D Friedman. *IEEE Journal of Solid-State Circuits* 38:2147–2154, 2003. With permission.)

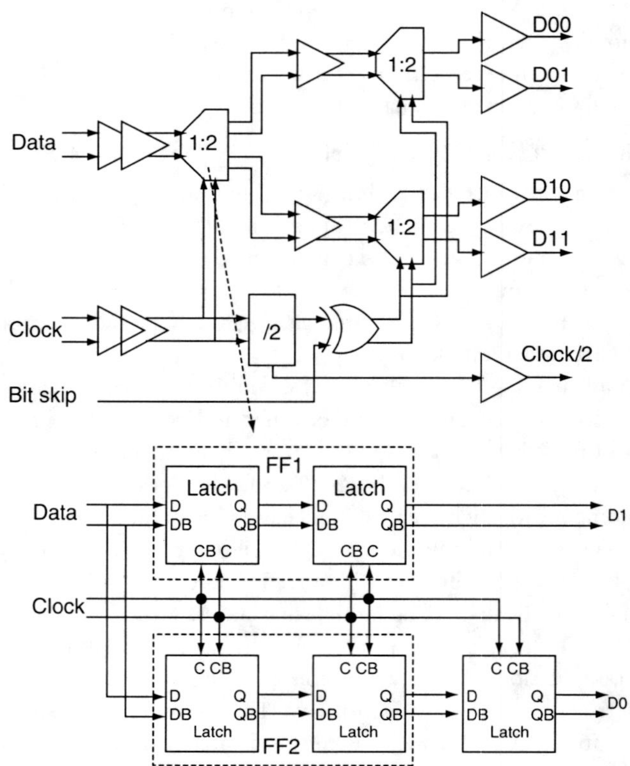

FIGURE 14.3 Simplified block diagram of the 1:4 demultiplexer circuit. (From M Meghelli, A Rylyakov, and L Shan. *IEEE Journal of Solid-State Circuits* 37:1790–1794, 2002. With permission.)

operational to input serial data rates exceeding 60 Gb/sec, more than ample performance for use as a building block in the 40 Gb/sec receiver implementation.

In the full receiver (CDR plus 1:4 demultiplexer) implementation, the Cherry–Hooper stage was included at the receiver input, effectively serving as a limiting amplifier. In this application, the Cherry–Hooper architecture was realized using a pair of emitter followers with 50 Ω on-chip termination resistors for impedance matching as an input buffer, followed by a single amplifier stage driving two parallel amplifier stages. The tree drive architecture reduces loading on the high-speed nets, enabling improved input sensitivity while maintaining bandwidth. Common approaches for band-limited signals, inductive peaking and shunt peaking, were not followed in this design due to the risk of incurring a deterministic jitter penalty in the signal provided to the CDR for retiming.

The CDR recovers half-rate quadrature clocks from the input random data stream, using these to recover the incoming 40 Gb/sec data. The capture range of the CDR is narrow, so a dual-loop architecture is used to extend the PLL capture range to an acceptable level. The second loop, used for frequency acquisition, locks the VCO to an input reference clock [20,21]. An on-chip lock detector counts reference clock pulses against divided VCO output pulses and indicates frequency lock when these count values agree to within $\pm 0.1\%$. Once the loop has achieved frequency lock, the data recovery loop is activated. Because the reference clock is set such that the VCO frequency will be close to half the data rate, at switchover the data rate falls within the capture range of the CDR and the loop locks very quickly. The lock detector continues to monitor the recovered clock against the reference clock and will switch the PLL back to frequency acquisition mode if the detector flags an unlocked condition during operation. The lock detector itself is built using on-chip standard-cell CMOS. Because a supply voltage of −3.6 V was used for the analog circuits in this design and the standard-cell CMOS library was designed for a nominal 1.8 V ground-referenced supply, it was decided to use a −1.8 V supply voltage (1.8 V above the −3.6 V analog supply) for the CMOS. This approach not only enabled the use of standard CMOS library elements, but also as the lock detector consumes less than 3 mA, power for the detector could be supplied by an on-chip voltage regulator with virtually no power impact on the scale of the complete design, although it was provided via a separate pin in this implementation. In the full-rate transmitter described later in this chapter, the lock detector was supplied from an on-chip regulator.

The half-rate architecture of the receiver demanded the use of a quadrature VCO (Figure 14.4). Several implementation approaches for this circuit are possible, including ring oscillators, coupled bipolar LC oscillators, and coupled CMOS LC oscillators. In this design, coupled CMOS LC oscillators [22] were chosen, enabling the combination of the phase noise performance of an LC implementation with an easy coupling scheme using FET switches. Each of the coupled CMOS LC-VCOs is implemented with cross-coupled inverters creating the negative resistance required for oscillation. In order to further improve the phase noise performance of the VCO, the gain of the VCO in the control loop was reduced, with overall tuning range requirements addressed by the use of digitally controlled band-switching varactors. The total tuning range for this VCO is 2.3 GHz, taking into account band-switching and loop control, with overlap between adjacent bands sufficient to ensure that once the proper initial band of the 16 available is chosen, the PLL stays locked under temperature and supply drift conditions without requiring a change to a new band. The VCO frequency control is provided in two paths. A proportional path provides low-latency update pulses directly to the VCO, bypassing the charge pump and linear amplifier and enabling quick effective phase adjustments by the VCO [23]. The integral path passes through the charge pump and linear amplifier and acts to set the center frequency of the VCO. In the design, an external control to set the amplitude of the proportional bang–bang pulses can be used to adjust PLL loop bandwidth.

The phase detector (Figure 14.5) chosen for this implementation has advantages in its low gate count and simple clock distribution [20]. When adapted for half-rate operation, the complexity of this block increases as compared to a full-rate implementation, but the benefits accrued in the logic gate performance by going to a slower clock rate more than compensate for this effect. The phase detector receives the quadrature clock inputs from the VCO in the form of an in-phase (I) clock and a quadrature phase (Q) clock, as well as the buffered data. The phase detector is built from three double-edge triggered

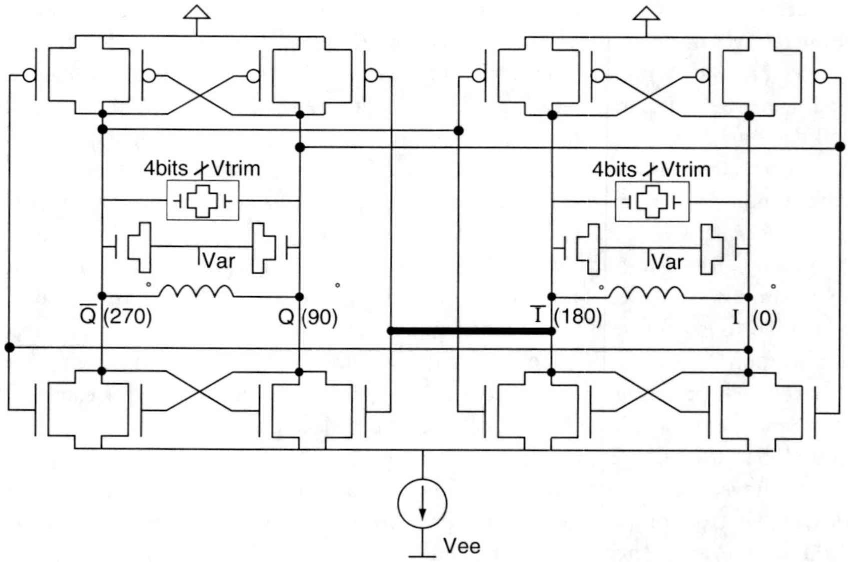

FIGURE 14.4 Quadrature CMOS LC-VCO. (From M Meghelli, A Rylyakov, S Zier, M Sorna, and D Friedman. *IEEE Journal of Solid-State Circuits* 38:2147–2154, 2003. With permission.)

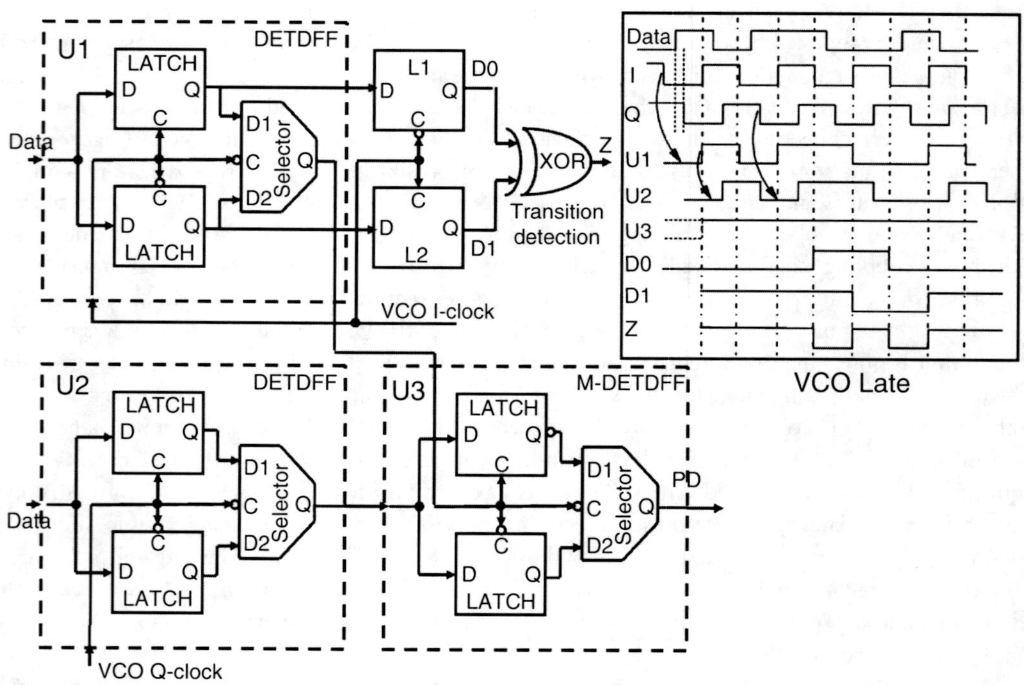

FIGURE 14.5 Phase detector block diagram and timing example for VCO lagging data. (From M Meghelli, A Rylyakov, S Zier, M Sorna, and D Friedman. *IEEE Journal of Solid-State Circuits* 38:2147–2154, 2003. With permission.)

master-slave flip-flops U1, U2, and U3, latches L1 and L2, and a XOR gate, all implemented with ECL logic. U1 samples the incoming data stream on the rising and falling edges of the I clock; similarly, U2 samples the incoming data stream on the rising and falling edges of the Q clock. Depending on the relative phase of the clock with respect to the data, the U1 output will lead or lag the U2 output by 90°. U3 samples the output of U2 on every transition of U1. If the U1 output leads the U2 output, on every rising edge of the U1 output, the U2 output will still be low, while on every falling edge of the U1 output, the U2 output will still be high. By inverting either the rising edge or falling edge sampled data (in U3, the rising edge data), therefore, a constant output from U3 will be generated as long as the U1 output leads the U2 output. Similarly, if the U1 output lags the U2 output, U3 will generate a constant output with the opposite polarity of that generated in the lead case. The U3 output thus indicates whether the VCO leads or lags the data (the latter is illustrated in Figure 14.5) and can be used to drive the loop into phase lock. Under locked conditions, the Q clock edges will be aligned with the data transitions and the I clock transitions will be centered within the data eye, independent of process, temperature, and supply voltage variation, provided the quadrature relationship between I and Q is robust and the duty cycle distortion in these half-rate clocks is small.

Another desirable feature in a phase detector–charge pump combination is the ability to provide correction signals only when transitions in the data have occurred and to produce a tri-state or neutral output otherwise. In this way, loop drift during long runs of consecutive ones or zeros in the input data stream is minimized. In the phase detector described here, the outputs of latches L1 and L2, which are generated 90° out of phase and correspond to the even and odd 1:2 demultiplexed data, are compared in the XOR gate. The XOR output is used in a tri-level logic gate (Figure 14.6) to enable the creation of positive, negative, and neutral control signals for the charge pump. When the XOR gate output is 1, a transition has occurred and the normal differential path in the tri-level gate is selected, resulting in a positive or negative differential output signal. When the XOR gate output is 0, however, no transition has occurred and the balanced output path is chosen, resulting in a neutral differential output signal.

Receiver Measurement Results

The receiver was measured on-wafer and in packaged form using a 43 Gb/sec $2^{31} - 1$ pseudo-random bit sequence (PRBS) input data pattern [7]. Error-free operation (bit error rate [BER] $< 10^{-15}$) was

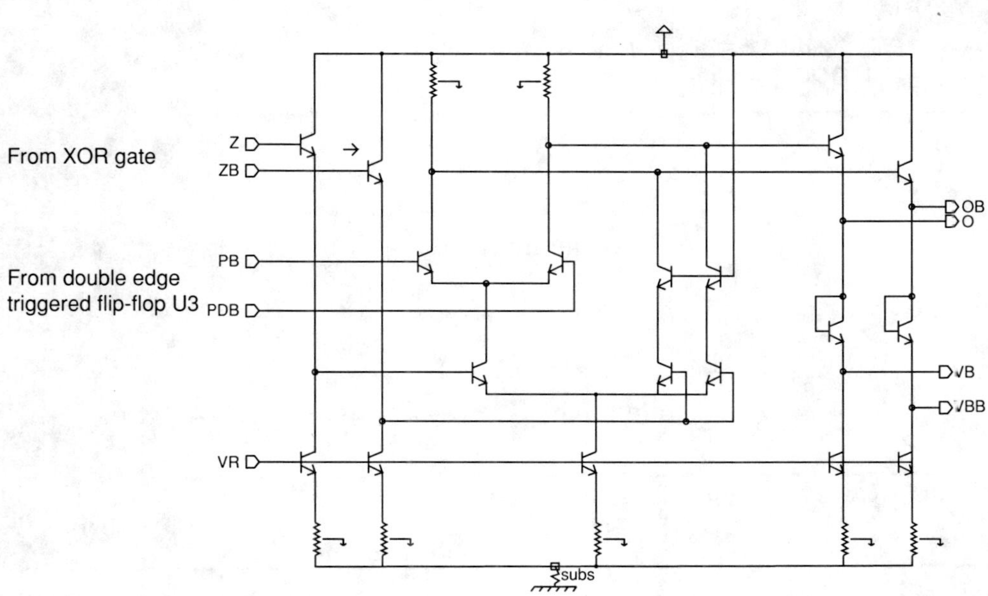

FIGURE 14.6 Gate generating three-level phase detector final output. (From M Meghelli, A Rylyakov, S Zier, M Sorna, and D Friedman. *IEEE Journal of Solid-State Circuits* 38:2147–2154, 2003. With permission.)

measured for temperatures from 25 to 100°C and supply voltages from −3.3 to −3.9 V. To evaluate the success of the tri-state feature of the phase detector in enabling the loop to handle long strings of consecutive bits, data patterns including up to 192 consecutive ones or zeros followed by 64 bits of alternating "10" were presented to the receiver at 43 Gb/sec; the receiver was able to recover such patterns without errors. The performance of the Cherry–Hooper input stage was tested by providing low-amplitude single-ended input to the receiver. For a BER $< 10^{-9}$, less than 40 mV single-ended input was successfully recovered. The VCO lock range within a given band was typically 700 MHz, which, since the VCO is running at half-rate, is equivalent to a data rate range of 1.4 Gb/sec. The overall frequency band over which the CDR was able to operate successfully was 39.5 to 44 Gb/sec. The free-running phase noise of the VCO itself was measured to be −103 dBc/Hz at a 1 MHz offset from a 21.5 GHz operating frequency. VCO temperature sensitivity was less than 1.9 MHz/°C, while VCO supply voltage sensitivity was less than 133 MHz/V.

The jitter generation of the recovered clock of interest in SONET repeater applications can be assessed from the phase noise plot (Figure 14.7) of a clock-divided-by-2 output of the chip. In the figure, the divided CDR clock output is shown along with the corresponding free-running divided-by-2 output of the VCO and the clock source for the input data, also divided by 2 to match the VCO frequency. The jitter generation was measured by integration over a 10 kHz to 1 GHz bandwidth, resulting in a recovered clock jitter generation as low as 188 fsec rms under typical conditions and less than 230 fsec rms under worst case temperature and supply voltage conditions. This low level of generated jitter validates the design strategy of using coupled LC VCOs combined with band-switching within the VCO.

The jitter tolerance of the receiver for SONET compliance would typically be measured by evaluating BER performance of the CDR against input phase modulated with sine jitter of various amplitudes corresponding to a defined jitter mask. When this part was evaluated, integrated test equipment to measure jitter tolerance in this manner did not exist, although some researchers developing alternate receiver chips built a custom test environment to make such measurements [5]. In our testing, tolerance to both sinusoidal and deterministic jitter was measured. In the case of the sinusoidal jitter, however, the

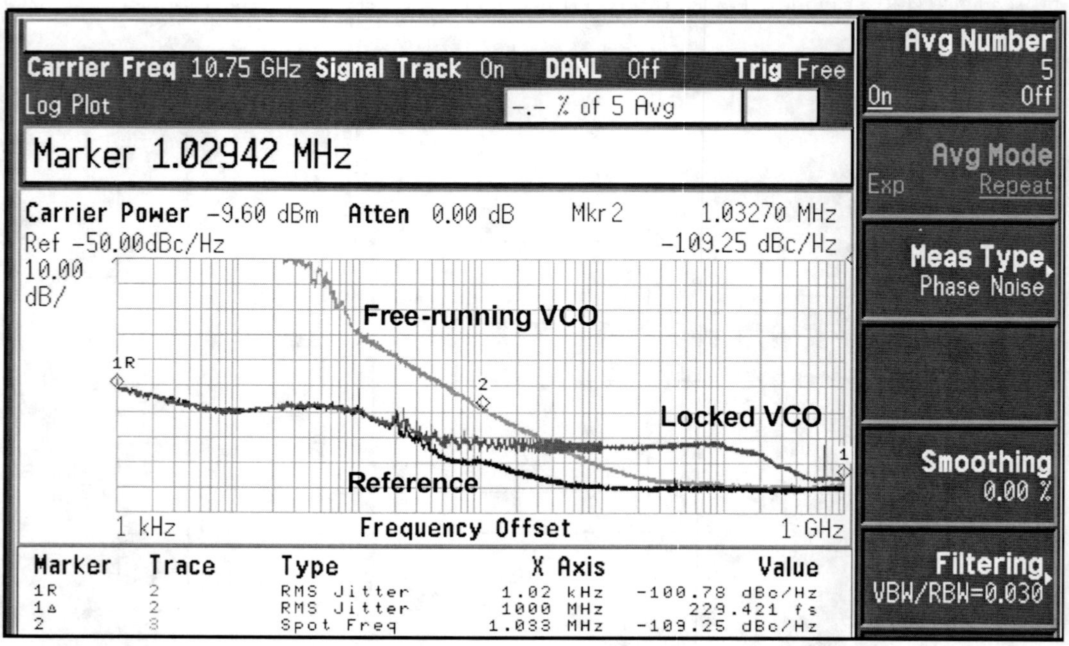

FIGURE 14.7 Phase noise performance of reference, free-running VCO, and locked VCO in CDR. (From M Meghelli, A Rylyakov, S Zier, M Sorna, and D Friedman. *IEEE Journal of Solid-State Circuits* 38:2147–2154, 2003. With permission.)

modulation bandwidth of the clock source was unfortunately much less than the bandwidth of the PLL, so while the CDR passed the test, this result merely served to verify the tracking capabilities of the PLL. The deterministic jitter testing was done by providing an input data stream that had been passed through a low bandwidth, high loss cable, thus introducing significant input signal degradation. The testing was performed on a CDR–demultiplexer chip packaged on a 1 in. × 1.2 in. Arlon substrate with 50 Ω microstrip transmission lines connected to GPPO connectors within an overall brass housing [24]; an eye diagram of the degraded input and one of the four recovered demultiplexed outputs shown in Figure 14.8. The power consumption of the receiver chip was quite low at 2.4 W under nominal operating conditions; the power consumption of similar designs reported in the same time frame as this result in similar or identical technologies was far higher, in part because of more conservative supply voltage and logic family choices made in those implementations [4–6].

40 Gb/sec Transmitter Implementations Using Half-Rate and Full-Rate Architectures

The function of each of the two transmitter circuits described in this section is to serialize four input parallel data streams, creating a 40 Gb/sec output data stream minimally corrupted by jitter, whether

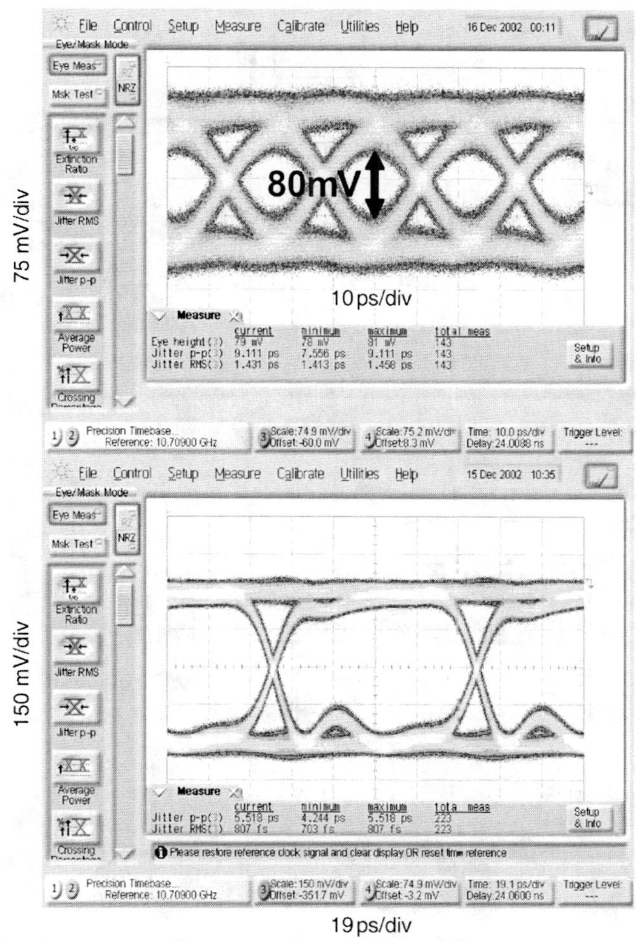

FIGURE 14.8 Receiver 43 Gb/sec input eye diagram (top) and one demultiplexed eye diagram (bottom), triggered with BERT clock. (From M Meghelli, A Rylyakov, S Zier, M Sorna, and D Friedman. *IEEE Journal of Solid-State Circuits* 38:2147–2154, 2003. With permission.)

random or deterministic. A key question in the implementation of this block is whether or not a full-rate architecture should be used. A full-rate approach enables the use of a baud-rate retiming latch just prior to the output driver, creating a very favorable design point for minimizing deterministic jitter at the cost of a very high-performance latch, the need to generate and distribute a full-rate clock to the full-rate latch, and the requirement that a very precise timing relationship between data and clock for that latch be satisfied. The half-rate approach reduces the stress on technology and timing introduced by the final latching stage while easing the VCO design challenge, at the cost of potential additional deterministic jitter introduced by duty cycle distortion. The technology used for these designs is the same production-level 0.18 μm SiGe BiCMOS process as that used for the receiver design. In both the full-rate and half-rate designs, the design target was a combined clock multiplier unit and a 4:1 multiplexer (Figure 14.9).

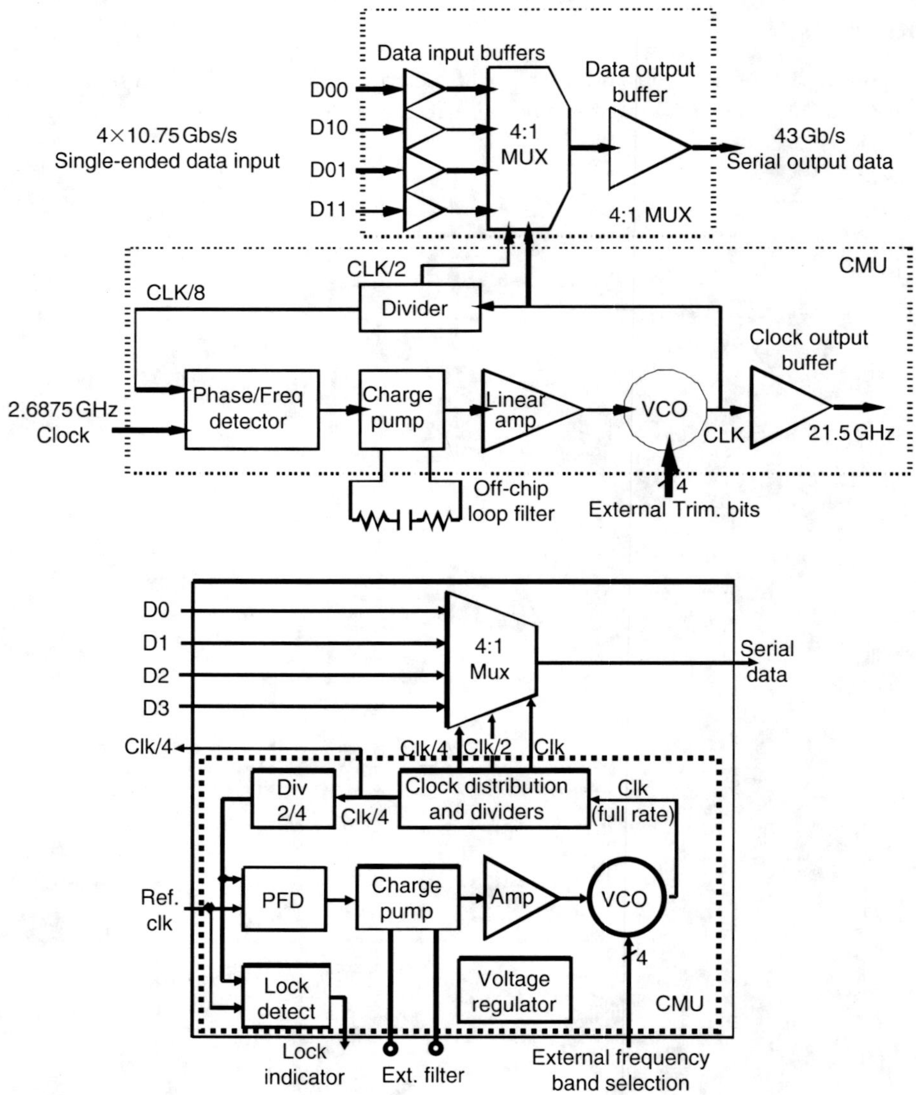

FIGURE 14.9 Block diagrams of half-rate (top) and full-rate (bottom) transmitters, each including a clock multiplier unit and a 4:1 multiplexer. (From M Meghelli, A Rylyakov, S Zier, M Sorna, and D Friedman. *IEEE Journal of Solid-State Circuits* 38:2147–2154, 2003; M Meghelli. A IEEE Proceedings of the Bipolar Circuits and Technology Meeting, 2004, pp. 289–292. With permission.)

Half-Rate and Full-Rate Transmitter Designs

The half-rate transmitter design features a clock multiplier unit (CMU) generating an output frequency in the 20 to 22 GHz range, supporting 40 to 44 Gb/sec data rates, as well as a multiplexer designed to work with the half-rate clock. A key design decision that must be made is the loop bandwidth of the CMU PLL, if not set by standard requirements. Inside the loop bandwidth, random jitter will be dominated by noise from the reference and subcircuits contributing to in-band noise such as the charge pump, phase-frequency detector, and divider; outside the loop bandwidth, random jitter is dominated by the performance of the VCO. In the half-rate CMU design, the loop bandwidth was set to 3 MHz. The reference multiplication factor chosen for the CMU PLL was 8.

As in the case of the 1:4 demultiplexer used in the receiver, the 4:1 multiplexer design was implemented as a stand-alone circuit prior to its inclusion in the transmitter design. A tree architecture with a recursive series of 2:1 multiplexer stages was used for this design (Figure 14.10). The stand-alone

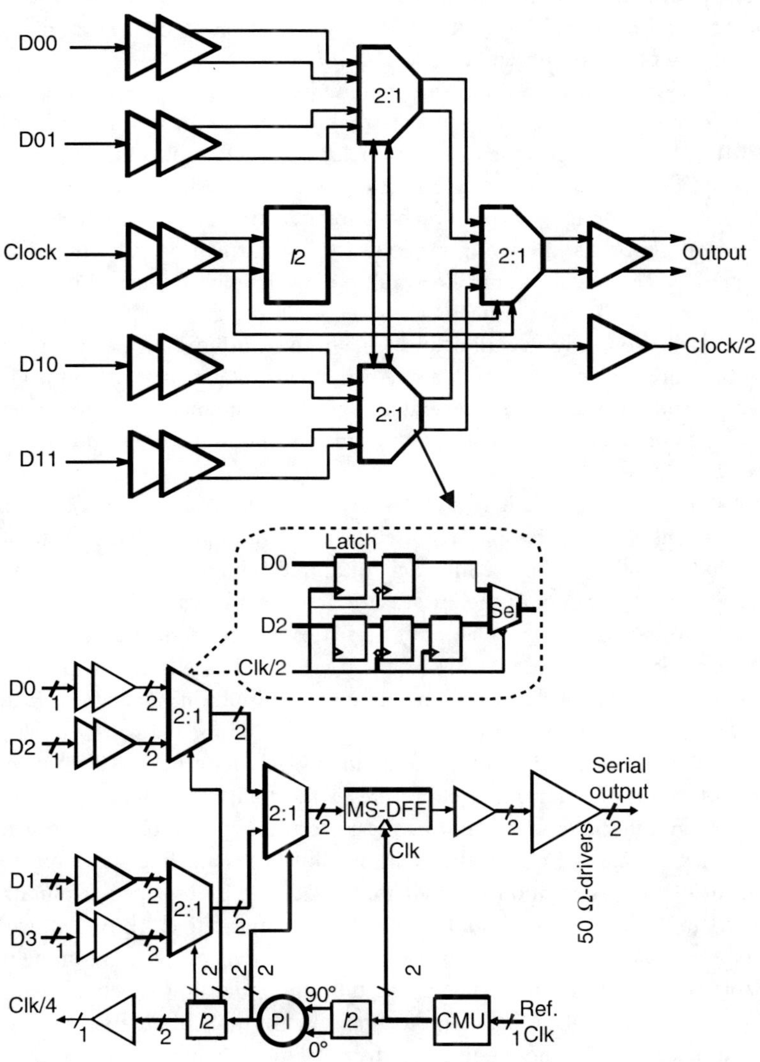

FIGURE 14.10 Simplified block diagram of the 4:1 multiplexer as used in the half-rate (top) and full-rate (bottom) transmitter implementations. (From M Meghelli, A Rylyakov, and L Shan. *IEEE Journal of Solid-State Circuits* 37:1790–1794, 2002; M Meghelli. A IEEE Proceedings of the Bipolar Circuits and Technology Meeting, 2004, pp. 289–292. With permission.)

multiplexer takes four parallel single-ended data as input and uses a half-rate clock with internal clock divider to perform input data latching, 4:2 multiplexing, and 2:1 multiplexing. The clock receiver in this multiplexer was designed with a Cherry–Hooper amplifier. The output driver was a critical design as it must support broadband operation up to a bandwidth sufficient for target data rates. The broadband requirement limits the amount of peaking that can be applied to the design problem, while maximizing the output data rate stresses the capabilities of the technology; the driver must operate at the same rate regardless of half-rate or full-rate architecture choice. The output driver was built using a simple differential pair with 50 Ω termination resistors for impedance matching. The first implementation of this block included series peaking in the termination that created undesirable, jitter-inducing excessive peaking in the driver's transfer function; the second implementation of this circuit did not include series peaking and worked with very low jitter to far higher data rates than did the first. Feedback emitter resistors were used to reduce output ringing, shunted with a capacitor to enhance bandwidth. The packaged version of the second implementation of this block operated to 70 Gb/sec output data rates, far in excess of the performance required in the 40 Gb/sec transmitter implementation. A subsequent implementation of a half-rate architecture multiplexer implemented in a 0.13 μm SiGe bipolar process was demonstrated with open output data eyes to 132 Gb/sec [8].

In the complete half-rate transmitter implementation, the multiplexer architecture and output driver design were taken from the stand-alone design, with the clock source for the multiplexer block being the output of the CMU. The oscillator in the PLL is a bipolar LC-VCO (Figure 14.11), employing a positive feedback cross-coupled differential pair to generate the negative resistance necessary to sustain oscillation. As in the receiver VCO design, a band-switching architecture was used to enable lower VCO gain in the fine tune control loop, thus improving random jitter performance while still supporting a broad tuning range capability via digitally controlled multiple overlapping frequency bands. The phase and frequency detector used in the transmitter is a conventional design built from a NOR gate and two synchronously set, asynchronously reset flip-flops. One of the flip-flops is set by the reference clock, the other by the feedback divider clock, and both are reset by the NOR gate output. In the locked condition, narrow, matched up and down pulses are generated. The NOR gate includes dummy devices that allow the flip-flops to drive nominally identical loads, thus minimizing sources of static phase error from this circuit block.

While the full-rate transmitter design retains much of the design used for the half-rate version, significant modifications to key blocks are required. The 4:1 multiplexer block in the full-rate design (Figure 14.10) includes a full-rate clock retiming circuit implemented with a master–slave data flip-flop. In order to ensure proper operation of the 2:1 selector gate, data retiming latches are used by each 2:1 multiplexing stage to offset the two parallel input data streams with respect to each other by half a bit time. The full-rate clock is generated on-chip by the CMU and is used at that frequency to clock the retiming flip-flop, and at divided frequencies to time the multiplexing operation. In this design, the reference clock was chosen to be the same frequency as that used for the half-rate design, demanding a reference multiplication factor of 16 and thus a divide-by-16 circuit in the CMU loop.

A primary benefit in full-rate transmit architectures is the ability to include full-rate retiming of the data just prior to the output driver, but achieving the correct timing relationship between full-rate clock arrival and multiplexed data arrival at the retiming flip-flop data and clock ports is nontrivial. Effectively, the sum of the propagation delay of the first divide by two circuit plus the propagation delay of the last 2:1 multiplexer stage should be less than one full-rate clock period, which is less than 24 psec at a data rate of 43 Gb/sec. The clock phase margin requirements of the retiming flip-flop further erode the available time. To ease this problem, the full-rate design implemented here (Figure 14.10) introduces a block which interpolates between the 0° and 90° outputs of the first divider stage, with the interpolator output used to drive the multiplexing tree. In this way, the data phase can be moved over 90° of half-rate clock phase, corresponding to 180° of full-rate clock phase. Control of the interpolator setting will thus enable the establishment of an optimal timing relationship between the multiplexed data and the full-rate clock at the retiming flop-flop inputs. Because this approach allows phase adjustment in the less power-hungry and less technology-stressful half-rate clock domain, it was chosen

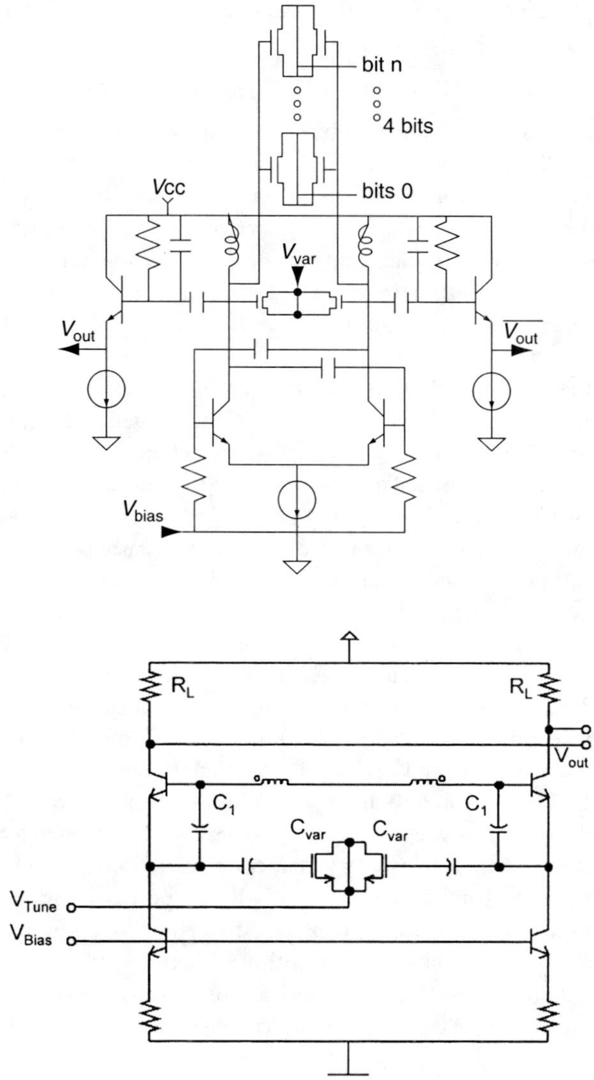

FIGURE 14.11 VCOs used in half-rate (top) and full-rate (bottom) transmitter implementations. (From M Meghelli, A Rylyakov, S Zier, M Sorna, and D Friedman. *IEEE Journal of Solid-State Circuits* 38:2147–2154, 2003; M Meghelli. A IEEE Proceedings of the Bipolar Circuits and Technology Meeting, 2004, pp. 289–292. With permission.)

over alternatives [25] that use extra delay stages operating at the full clock and data rates to establish appropriate timing.

The second circuit for which significant changes were required by the full-rate architecture as compared to the half-rate one is, of course, the VCO. The cross-coupled HBT-based VCO used in the half-rate design was replaced by a differential bipolar LC-VCO based on the Colpitts architecture in the full-rate VCO design (Figure 14.11). Varactors for band switching (not shown in the figure) and fine-tuning in this design are composed of nFETs operated in inversion mode. A key consideration in the half-rate versus full-rate decision is that if the phase noise performance of the VCO degrades by more than 6 dB as compared to the half-rate VCO, the RJ performance of the full-rate design will not be able to match that of the half-rate design. This is a difficult challenge to meet as the Q factor of the varactors available in a given technology tend to degrade significantly as the frequency increases from that required for half-rate to that required for full-rate. Because the 4 MHz PLL bandwidth of the full-rate design is similar to the 3 MHz PLL bandwidth of the half-rate design (while the feedback divider value is doubled), the in-band phase noise performance of the full-rate PLL must match or exceed that of the half-rate PLL to maintain in-band RJ performance.

Measurement Results from Half-Rate and Full-Rate Transmitters

The half-rate design was measured on-wafer and in packaged form, using parallel 10.75 Gb/sec $2^{31}-1$ PRBS inputs and a 2.6875 GHz reference clock, yielding error-free operation (BER $< 10^{-15}$) measured at the multiplexer serial output port at temperatures to 100°C over supply voltages ranging from −3.3 to −3.9 V. The full-rate design was measured only on-wafer under similar conditions, but above 50°C the on-chip divider began to fail at the high end of the VCO tuning range. Half-rate and full-rate implementations of the stand-alone multiplexer block show the same qualitative behavior; at 100°C, the half-rate multiplexer ran to 60 Gb/sec output data rates as compared to 44 Gb/sec output data rates for the full-rate design.

The tuning range of the PLL in the half-rate design was 20.2 to 22 GHz, whereas that of the full-rate design was 41.8 to 45.8 GHz. While neither implementation succeeded in covering the target frequency range exactly in hardware, no problem is anticipated in achieving this range from a theoretical perspective for either architecture choice. The phase noise of the full-rate VCO is significantly degraded as compared to that of the half-rate design (Figure 14.12), a performance decline of 14 dB, 8 dB in excess of the 6 dB budget that would enable similar phase noise performance between the half- and full-rate VCOs. As a result, the 260 fsec rms measured jitter integrated over a 10 kHz to 1 GHz bandwidth for the full-rate CMU is far greater than the 140 fsec rms value achieved by the half-rate design.

Recall that a key reason for pursuing a full-rate design was the opportunity to reduce duty cycle distortion. In the output eye diagrams (Figure 14.13), this benefit, as well as a reduction in the clock feed through, is clearly visible, although the half-rate design's performance still yields wide open data eyes with a signal-to-noise ratio greater than 12. The price for this benefit is high, however; in addition to the reduced robustness over temperature and generated jitter performance, the power consumption of the full-rate design was 2.31 W as compared to an expected 1.61 W for a half-rate design using logic gates powered commensurately with those used in the full-rate design (the actual half-rate design, done first, conservatively used higher power per logic gate targeting operation at a given data rate than the subsequent full-rate design), with much of the extra full-rate power absorbed in the added current needed for clock distribution and first two divider stages. In our estimation, the half-rate architecture represents the best compromise for 40 Gb/sec transmitter serializers implemented in the 0.18 μm technology used for this design. A performance summary of the two transmitters is provided in Table 14.1. Die photos of the three designs discussed in this chapter can be found in Figure 14.14.

14.3 Summary

In this chapter, we presented a brief overview of an optical data link at 40 Gb/sec and then focused on the high-speed serializer and deserializer circuitry components of such a link implemented in a 0.13 μm

TABLE 14.1 Performance Summary for Half-Rate and Full-Rate Transmitter Implementations

Measured Key Parameter	Full-Rate Tx	Half-Rate Tx [8]
Nominal data rate	41.8 to 45.8 Gb/sec	40.5 to 44 Gb/sec
Clock rms jitter generation	260 fsec	140 fsec
Data jitter	600 fsec rms, 4.6 psec pp	540 fsec rms, 3.4 psec pp
Data SNR	12.3	12.7
Data duty cycle distortion	1.4%	2.3%
Free-running VCO phase noise	−98 dBc/Hz at 1 MHz, 10.75 GHz	−100 dBc/Hz at 1 MHz, 21.5 GHz
Maximum chip temperature	50°C	100°C
Supply voltage	−3.3 to −3.9 V	−3.3 to −3.9 V

Source: M Meghelli. IEEE Proceedings of the Bipolar Circuits and Technology Meeting, 2004, pp. 289–292. With permission.

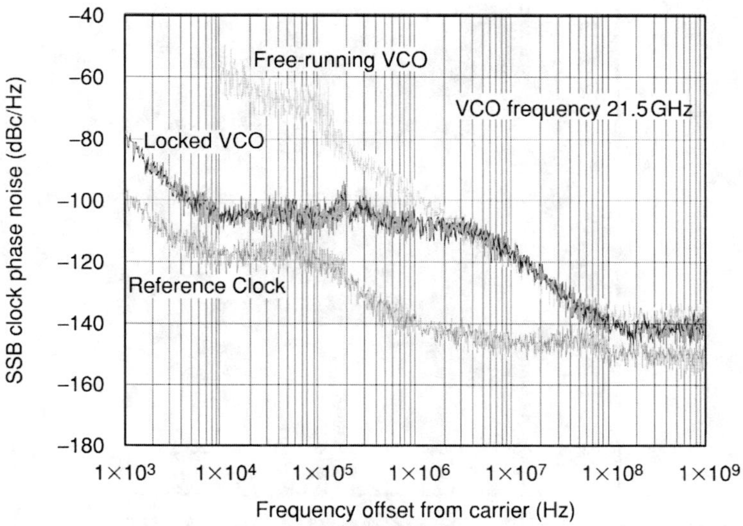

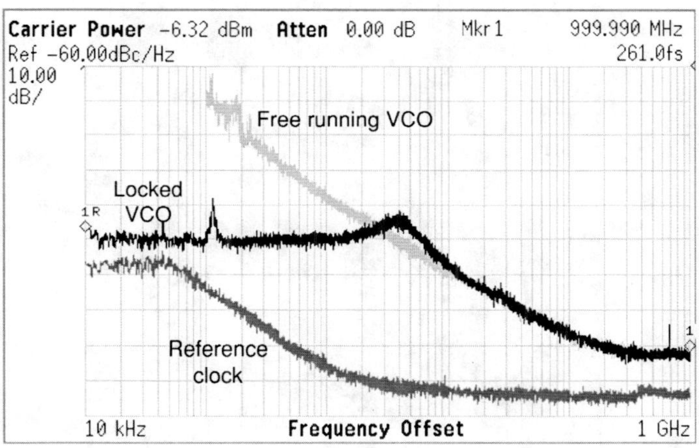

FIGURE 14.12 Phase noise plots for half-rate (top) and full-rate (bottom) transmitter implementations. (From M Meghelli, A Rylyakov, S Zier, M Sorna, and D Friedman. *IEEE Journal of Solid-State Circuits* 38:2147–2154, 2003; M Meghelli. A IEEE Proceedings of the Bipolar Circuits and Technology Meeting, 2004, pp. 289–292. With permission.)

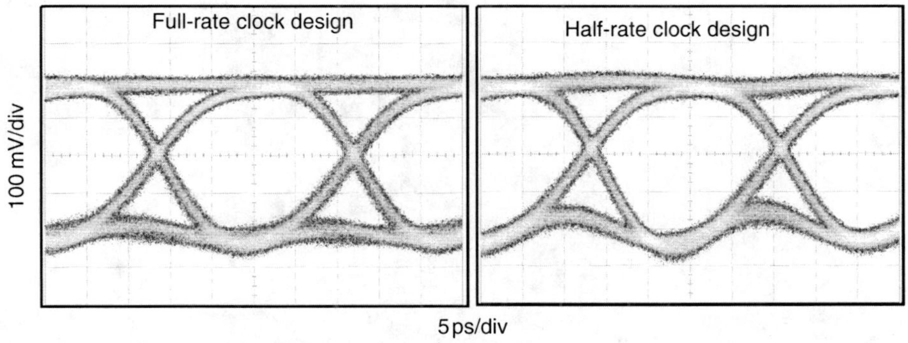

FIGURE 14.13 Output eye diagrams from full-rate (left) and half-rate (right) transmitter implementations. (From M Meghelli. A IEEE Proceedings of the Bipolar Circuits and Technology Meeting, 2004, pp. 289–292. With permission.)

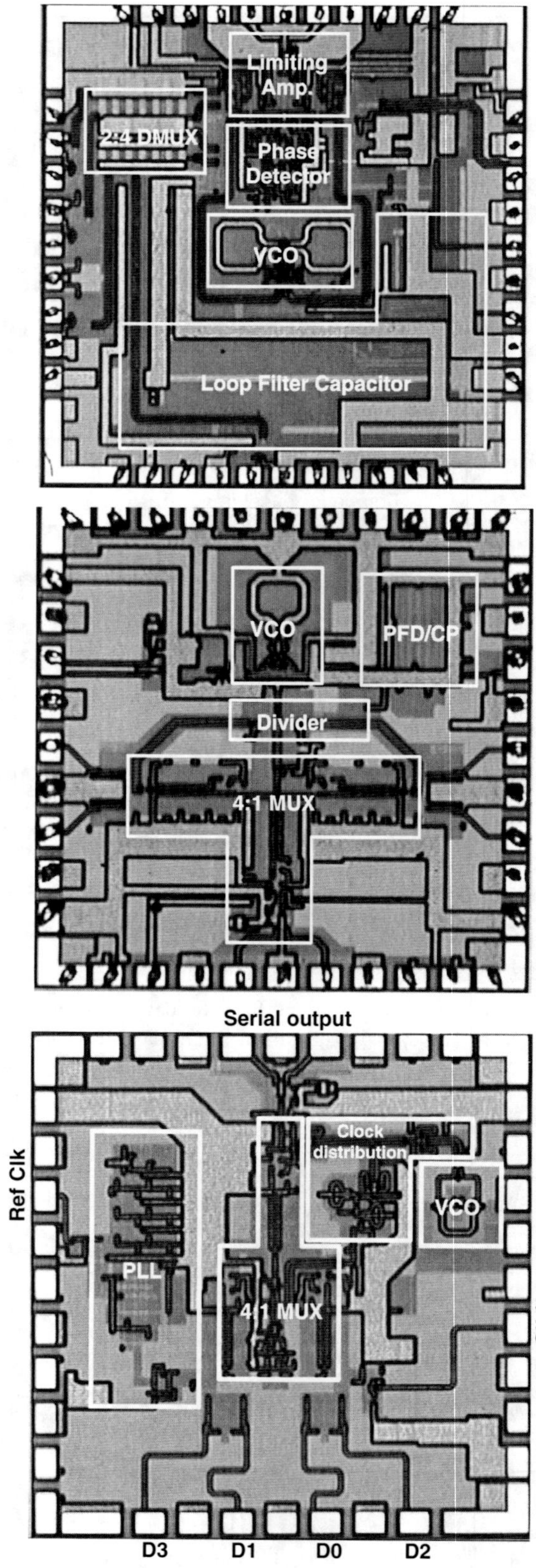

FIGURE 14.14 Die photos of the receiver (top), the half-rate transmitter implementation (middle), and the full-rate transmitter implementation (bottom). (From M Meghelli, A Rylyakov, S Zier, M Sorna, and D Friedman. *IEEE Journal of Solid-State Circuits* 38:2147–2154, 2003; M Meghelli. A IEEE Proceedings of the Bipolar Circuits and Technology Meeting, 2004, pp. 289–292. With permission.)

SiGe BiCMOS process. Design details for a half-rate receiver–demultiplexer as well as half-rate and full-rate transmitter–multiplexer were presented, including key performance metrics, circuit approaches, and hardware results. In order to develop the high speed, relatively complex designs described in this chapter, we first implemented and validated key smaller high-speed pieces of the design, specifically, the 1:4 demultiplexer and the 4:1 multiplexer. This approach enabled us to correlate hardware performance against models, identify and correct unexpected problems, and was critical in the success achieved in the first-pass designs of the full receiver and transmitters presented here. In addition, because they could be clocked from an external source, these blocks allowed us to explore technology performance limits more effectively than was possible in the complete transmitter and receiver designs. The half-rate receive architecture achieved excellent jitter generation and tolerance performance at very low power, validating supply voltage, architecture, and building block choices for such designs. The full-rate and half-rate architecture transmitter designs also achieved excellent jitter generation at low power, with the realized expected duty cycle distortion improvement from the full-rate design outweighed by its degraded phase noise performance and narrower range of operating temperature as compared to the half-rate design. These design examples explore circuit and architectural tradeoffs in high-speed serializer and deserializer design. Ultimately, the results achieved in this work confirm the viability of half-rate architectural approaches for 40 Gb/sec serializer and deserializer design in technologies with device performance like that of the chosen technology.

Acknowledgments

The authors would like to acknowledge the SiGe technology group of IBM Microelectronics for the fabrication of all chips described in this chapter. The authors would also like to thank Alexander Rylyakov, Michael Sorna, Steven Zier, Lei Shan, Mehmet Soyuer, and Modest Oprysko for their contributions and support.

References

1. A Joseph, D Coolbaugh, M Zierak, R Wuthrich, P Geiss, Z He, X Liu, B Orner, J Johnson, G Freeman, D Alhgren, B Jagannathan, L Lanzerotti, V Ramachandran, J Malinowski, H Chen, J Chu, P Gray, R Johnson, J Dunn, S Subbanna, K Schonenberg, D Harame, R Groves, K Watson, D Jadus, M Meghelli, and A Rylyakov. A 0.18 μm BiCMOS technology featuring 120/100 GHz (f_T/f_{max}) HBT and ASIC-compatible CMOS using copper interconnect. IEEE Proceedings of the Bipolar Circuits and Technology Meeting, 2001, pp. 143–146.
2. A Joseph, D Coolbaugh, D Harame, G Freeman, S Subbanna, M Doherty, J Dunn, C Dickey, D Greenberg, R Groves, M Meghelli, A Rylyakov, M Sorna, O Schreiber, D Herman, and T Tanji. 0.13 μm 210 GHz f_T SiGe HBTs—Expanding the horizons of SiGe BiCMOS. International Solid-State Circuits Conference Technical Digest of Papers, 2002, pp. 181–182.
3. M Reinhold, C Dorschky, E Rose, R Pullela, P Mayer, F Kunz, Y Baeyens, T Link, and J-P Mattia. A fully integrated 40-Gb/s clock and data recovery IC with 1:4 DEMUX in SiGe technology. *IEEE Journal of Solid-State Circuits* 36:1937–1944, 2001.
4. A Ong, S Benyamin, V Condito, Q Lee, J-P Mattia, D Shaeffer, A Shahani, X Si, H Tao, M Tarsia, W Wong, and M Xu. A 40–43 Gb/s clock and data recovery IC with integrated SFI-5 1:16 demultiplexer in SiGe technology. *IEEE Journal of Solid-State Circuits* 38:2155–2168, 2003.
5. D Shaeffer, H Tao, Q Lee, A Ong, V Condito, S Benyamin, W Wong, X Si, S Kudszus, and M Tarsia. 40/43 Gb/sec SONET OC-768 SiGe 4:1 MUX/CMU chipset with SFI-5 compliance. *IEEE Journal of Solid-State Circuits* 38:2169–2180, 2003.
6. A Koyama, T Harada, H Yamashita, R Takeyari, N Shiramizu, K Ishakawa, M Ito, S Suzuki, T Yamashita, S Yabuki, H Ando, T Aida, K Watanabe, K Ohhata, S Takeuchi, H Chiba, A Ito, H Yoshioka, A Kubota, T Takahashi, and H Nii. 43 Gb/s full-rate-clock 16:1 multiplexer and 1:16 demultiplexer with SFI-5 interface in SiGe BiCMOS technology. International Solid-State Circuits Conference Technical Digest of Papers, 2003, pp. 232–233.

7. M Meghelli, A Rylyakov, S Zier, M Sorna, and D Friedman. A 0.18-μm SiGe BiCMOS receiver and transmitter chipset for SONET OC-768 transmission systems. *IEEE Journal of Solid-State Circuits* 38:2147–2154, 2003.
8. M Meghelli. A 108 Gb/s 4:1 multiplexer in 0.13 μm SiGe-bipolar technology. International Solid-State Circuits Conference Technical Digest of Papers, 2004, pp. 236–237.
9. M Meghelli, A Rylyakov, and L Shan. A 50-Gb/s SiGe BiCMOS 4:1 multiplexer and 1:4 demultiplexer for serial communication systems. *IEEE Journal of Solid-State Circuits* 37:1790–1794, 2002.
10. M Meghelli. A 43 Gb/s full-rate clock transmitter in 0.18 μm SiGe BiCMOS technology. IEEE Proceedings of the Bipolar Circuits and Technology Meeting, 2004, pp. 289–292.
11. H Rein and M Moller. Design considerations for very-high-speed Si-Bipolar IC's operating up to 50 Gb/s. *IEEE Journal of Solid-State Circuits* 31:1076–1090, 1996.
12. A Rylyakov. A 51 GHz master-slave latch and static frequency divider in 0.18 μm SiGe BiCMOS. IEEE Proceedings of the Bipolar Circuits and Technology Meeting, 2003, pp. 75–77.
13. A Rylyakov, L Klapproth, B Jagannathan, and G Freeman. 100 GHz dynamic frequency divider in SiGe bipolar technology. *Electronics Letters* 39:217–218, 2003.
14. Y Suzuki, Y Amamiya, Z Yamazaki, S Wada, H Uchida, C Kurioka, S Tanaka, and H Hida. 110 Gb/s multiplexing and demultiplexing ICs. International Solid-State Circuits Conference Technical Digest of Papers, 2004, pp. 232–233.
15. T Suzuki, T Takahashi, K Makiyama, K Sawada, Y Nakasha, T Hirose, and M Takikawa. Under 0.5 W 50 Gb/s full-rate 4:1 MUX and 1:4 DEMUX in 0.13 μm InP HEMT technology. International Solid-State Circuits Conference Technical Digest of Papers, 2004, pp. 234–235.
16. T Yamamoto, KM Horinaka, D Yamazaki, H Nomura, K Hashimoto, and H Onodera. A 43 Gb/s 2:1 selector IC in 90nm CMOS technology. International Solid-State Circuits Conference Technical Digest of Papers, 2004, pp. 236–237.
17. A Felder, M Moller, M Wurzer, M Rest, T Meister, and H Rein. 60 Gbit/s regenerating demultiplexer in SiGe bipolar technology. *Electronics Letters* 33:1984–1985, 1997.
18. M Muller, H Rein, A Felder, and T Meister. 60 Gb/s time-division multiplexer in SiGe bipolar technology with special regard to mounting and measuring technique. *Electronics Letters* 33:679–680, 1997.
19. E Cherry and D Hooper. The design of wide-band transistor feedback amplifiers. *Proceedings of the IEEE* 110:375–389, 1963.
20. M Meghelli, B Parker, H Ainspan, and M Soyuer. SiGe BiCMOS 3.3 V clock and data recovery circuits for 10 Gb/s serial transmission systems. International Solid-State Circuits Conference Digest of Technical Papers, 2000, pp. 56–57.
21. D Friedman, M Meghelli, B Parker, J Yang, H Ainspan, and M Soyuer. A single-chip 12.5 Gbaud transceiver for serial data communication. IEEE Symposium on VLSI Circuits Digest of Technical Papers, 2001, pp. 145–148.
22. A Rofougaran, J Rael, M Rofougaran, and A Abidi. A 900 MHz CMOS LC oscillator with quadrature outputs. International Solid-State Circuits Conference Digest of Technical Papers, 1996, pp. 392–393.
23. R Walker, C Stout, and C-S Yen. A 2.488 Gb/s Si bipolar clock and data recovery IC with robust loss of signal detection. International Solid-State Circuits Conference Digest of Technical Papers, 1997, pp. 246–247.
24. L Shan, M Meghelli, J Kim, and J Trewhella. Millimeter wave package design: a comparison of simulation and measurement results. IEEE 10th Topical Meeting on Electrical Performance of Electrical Packaging, 2001, pp. 29–34.
25. Y Nakasha, T Suzuki, H Kano, A Ohya, K Sawada, K Makiyama, T Takahashi, M Nishi, T Hirose, M Takikawa, and Y Watanabe. A 43 Gb/s full-rate-clock 4:1 multiplexer in InP-based HEMT technology. International Solid-State Circuits Conference Digest of Technical Papers, 2002, pp. 190–191.

15

Industry Examples at the State-of-the-Art: Hitachi

15.1 Introduction 15-1
15.2 IC and LSI for Optical Transmission Systems 15-1
IC Chipset and LSIs for 40-Gb/s Optical-Fiber-Links • Single-Chip 10 Gb/s Transceiver LSI
15.3 IC for Wireless Communication Systems 15-4
Frequency-Divider ICs • 5.8-GHz Electronic-Toll-Collection Transceiver IC • Other ICs
15.4 Summary 15-8

Katsuyoshi Washio
Hitachi, Ltd.

15.1 Introduction

High-speed monolithic integrated circuits (ICs) and large-scale ICs (LSIs) are the key components for multi-gigabit data communication systems and wide-bandwidth radio communication systems. These systems include backbone networks, intercity communication networks, local area networks, and Ethernet for data communications, and microwave and millimeter-wave mobile networks, fixed wireless access (FWA), and intelligent transport systems (ITS) for radio communications. As applications of the SiGe HBT and BiCMOS technologies described in Chapter 9 (see *Fabrication of SiGe HBT BiCMOS Technology*), an IC chipset and LSIs for 40-Gb/s optical-fiber links, a single-chip 10-Gb/s transceiver LSI, frequency-divider ICs, a 5.8-GHz electronic-toll-collection (ETC) transceiver IC, and other ICs that are applicable to optical transmission and microwave/millimeter-wave wireless communication systems have been implemented.

15.2 IC and LSI for Optical Transmission Systems

IC Chipset and LSIs for 40-Gb/s Optical-Fiber-Links

To meet the demand for an expansion of the transmission capacity that accompanies the rapid growth of multimedia communications, the development of a 40 Gb/s optical transmission system for backbone networks is an effective solution. The IC chipset should be capable of operation at up to about 50 Gb/s or 50 GHz for practical use; it must also be sufficiently inexpensive to receive widespread commercial approval. The SiGe HBT can therefore be considered a promising candidate. Several ICs for 40 Gb/s optical-fiber-link communication systems have been developed by using self-aligned SEG SiGe HBTs [1–4]. A block diagram of a transmitter and a receiver built with the fabricated IC chipset for 40 Gb/s optical-fiber-link communication systems is shown in Figure 15.1. The IC chipset includes a 4:1 multiplexer in the transmitter, and a transimpedance preamplifier, an automatic gain-control (AGC) amplifier, a full-wave rectifier, a limiting amplifier, and a 1:4 demultiplexer with a decision circuit in the

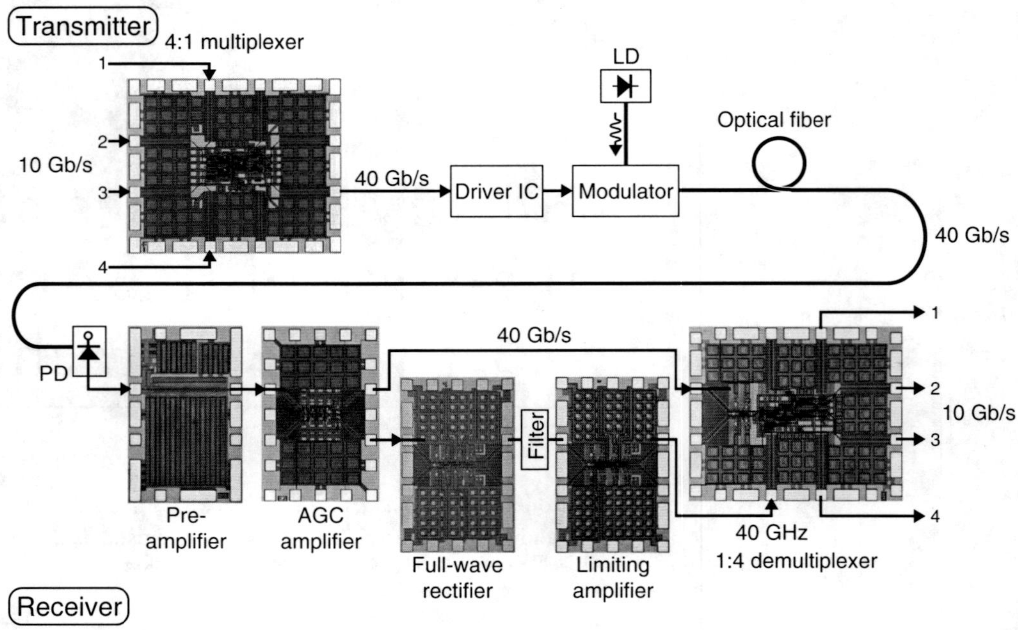

FIGURE 15.1 Block diagram of a transmitter and a receiver built with the IC chipset fabricated by using self-aligned SEG SiGe HBTs. The system is for 40-Gb/s optical-fiber-link communication systems. The IC chipset includes a 4:1 multiplexer in the transmitter, and a transimpedance preamplifier, an automatic gain-control (AGC) amplifier, a full-wave rectifier, a limiting amplifier, and a 1:4 demultiplexer with a decision circuit in the receiver. (From K. Washio. *IEEE Trans. Electron Devices* 50:656–668, 2003. With permission.)

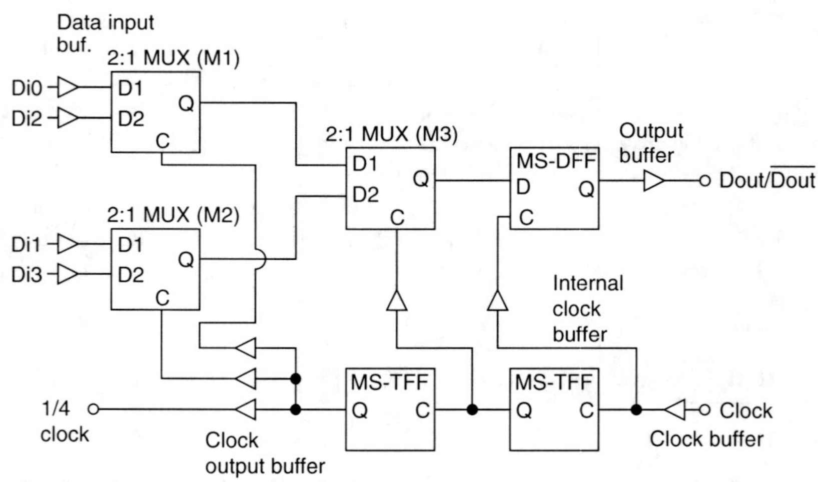

FIGURE 15.2 Block diagram of a 4:1 multiplexer. (From T. Masuda, K. Ohhata, N. Shiramizu, E. Ohue, K. Oda, R. Hayami, H. Shimamoto, M. Kondo, T. Harada, and K. Washio. Digest of the IEEE International Microwave Symposium, Phoenix, 2001, pp. 1697–1700. With permission.)

receiver. The block diagram of the 4:1 mutiplexer (4:1 MUX) is shown in Figure 15.2. The 2:1 multiplexer (2:1 MUX), the core circuit of the 4:1 MUX, is designed to maximize the phase margin between the input data transitions and the select clock. The 4:1 MUX consists of three 2:1 MUXs connected in a tree structure, an output master–slave delayed flip-flop (MS-DFF), an output

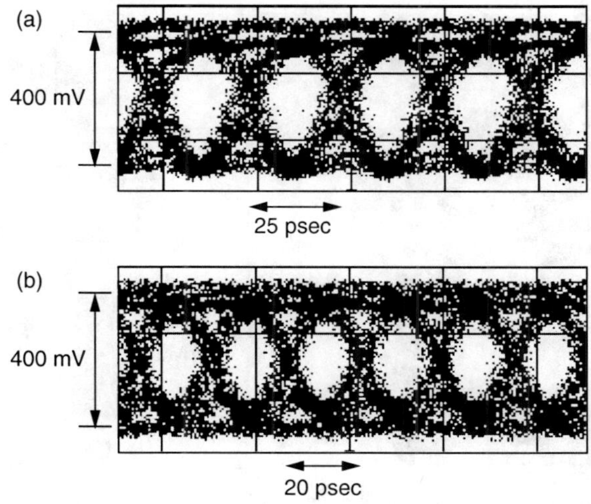

FIGURE 15.3 Output waveform of a 4:1 multiplexer at (a) 40 Gb/s and (b) 50 Gb/s. (From T. Masuda, K. Ohhata, N. Shiramizu, E. Ohue, K. Oda, R. Hayami, H. Shimamoto, M. Kondo, T. Harada, and K. Washio. Digest of the IEEE International Microwave Symposium, Phoenix, 2001, pp. 1697–1700. With permission.)

TABLE 15.1 Performance of the IC Chipset for 40 Gb/s Optical-Fiber-Link Communication Systems Developed by Using Self-Aligned SEG SiGe HBTs.

IC	Performance	Remarks
Multiplexer	>50 Gb/s (probe), 48 Gb/s (module)	4:1, full-rate
Preamplifier	45 GHz, 50 dBΩ	$C_{in} = 0.1$ pF
AGC amplifier	48 GHz, 0.5–21.5 dB	40 Gb/s (module)
Limiting amplifier	32 dB, 500 mVpp @ 40 GHz	BW = 49 GHz
Full-wave rectifier	90 mVpp (filter output) @ 40 GHz	
Demultiplexer	48 Gb/s (probe), 45 Gb/s (module)	1:4, full-rate

Source: K. Washio. *IEEE Trans. Electron Devices* 50:656–668, 2003. With permission.

buffer, and two master–slave toggle flip-flops (MS-TFFs) in series. The output eye diagrams of the 4:1 MUX IC for a 40-Gb/s and a 50-Gb/s pseudorandom bit sequence (PRBS) measured on-wafer probes are shown in Figure 15.3. Well-opened eye diagrams with output voltage swing of 400 mVpp were obtained. The performance of the IC chipset is summarized in Table 15.1. These excellent results indicate that the self-aligned SEG SiGe HBT technology, which offers high reliability and cost-effectiveness, will play an important role in optical-fiber-link systems, operating at a data rate of 40 Gb/s, for global communications.

To meet recent demand for high functionality, 43-Gb/s full-rate-clock 16:1 MUX and 1:16 DEMUX LSIs with a Serdes Framer Interface Level 5 (SFI-5) interface [6], and a fully integrated 39.8 to 43-Gb/s 16:1 MUX and 1:16 DEMUX chipset [7], have been developed.

Single-Chip 10 Gb/s Transceiver LSI

Optical transmissions at a data rate of 10 Gb/s are the highest serial data links commercially available for use in backbone networks, and are expected as candidates for use in the next generation of Ethernet, local area networks, and intercity communication networks for data communication. The demand for low-cost, high-productivity, small physical structure, and low power dissipation is therefore very strong.

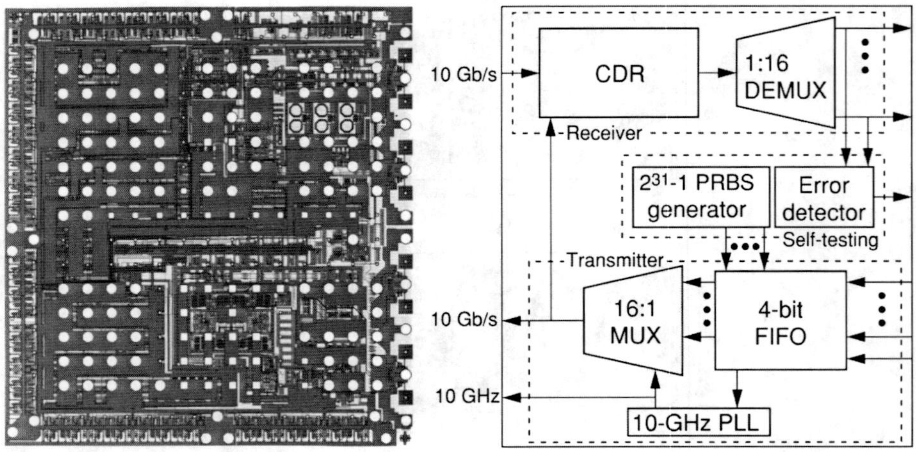

FIGURE 15.4 Photomicrograph and rough block diagram of a single-chip 10-Gb/s transceiver LSI developed by using the SiGe BiCMOS technology. A 1:16 DEMUX (demultiplexer) and a CDR (clock and data recovery) circuit in the receiver, a 4-bit 16-channel data input FIFO (fast-in fast-out) memory, a 16:1 MUX (multiplexer), and a 10-GHz PLL circuit with a 155-MHz external voltage control oscillator (VCO) control circuit in the transmitter, and a 2^{31}-1 pseudorandom bit sequence (PRBS) generator and an error detector for self-testing are all integrated in a single chip. (From S. Ueno, K. Watanabe, T. Kato, T. Shinohara, K. Mikami, T. Hashimoto, A. Takai, K. Washio, R. Takeyari, and T. Harada. Digest of Technical Papers of the IEEE International Solid-State Circuits Conference, San Francisco, 2001, pp. 82–83. With permission.)

The SiGe BiCMOS technology can be used to provide high-speed and sophisticated-function system LSIs at low power because of its combination of high-f_T and f_{max} SiGe HBTs and highly integrated CMOS. A photomicrograph and rough block diagram of a single-chip 10-Gb/s transceiver LSI developed by using SiGe BiCMOS technology are shown in Figure 15.4 [5]. This all-in-one transceiver LSI integrates: a 1:16 DEMUX (demultiplexer) and a clock and data recovery (CDR) circuit in the receiver; a 4-bit 16-channel data input fast-in fast-out (FIFO) memory, a 16:1 MUX (multiplexer), and a 10-GHz phase-locked loop (PLL) circuit with a 155-MHz external voltage control oscillator (VCO) control circuit in the transmitter; and a 2^{31}-1 PRBS generator and an error detector for self-testing. It occupies an area of 5.6 × 5.3 mm and contains 26,000 active elements, i.e., 16,000 SiGe HBTs and 10,000 CMOS transistors. CMOS transistors are used for the interfaces and test circuit to get the best characteristics in terms of operation at low supply voltage for low power consumption. This chip demonstrates that SiGe BiCMOS is a suitable technology for producing high-level integrated system LSIs to be used in optical transmission networks.

15.3 IC for Wireless Communication Systems

Frequency-Divider ICs

To satisfy the explosively growing demand for wide-bandwidth radio communication systems that accompany the availability of large-bandwidth spectra and the allocation of such frequency bands for commercial use, the development of monolithic millimeter-wave ICs has been advancing rapidly. Such systems include wireless local area networks (WLAN), local multiple-distribution services (LMDS), and consumer radar systems (automotive radars). Here, the availability and ease of use of low-cost monolithic ICs are essential for the penetration of millimeter-wave systems into the fields of consumer and commercial electronics. The high-speed frequency divider (FD) is a key circuit for many applications. Two types of FD ICs have been developed by using self-aligned SEG SiGe HBTs [8–11]. The circuits are a 1/4 pre-tracking static FD and a 1/4 dynamic FD (DFD), and have respective maximum operating frequencies of 81 and 92.4 GHz. Chip micrographs of the DFD, zooming in to the main circuit region in two steps, are shown in Figure 15.5. It consists of a 50-Ω-terminated three-emitter–follower input buffer, a dynamic toggle flip-flop (D-TFF) based on the regenerative frequency division principle as the first divide-by-two stage, a static MS-TFF as the second divide-by-two stage, internal buffers to reform the output signal of each TFF, and an output buffer driving 50-Ω lines. The internal single-ended voltage swing for the D-TFF was designed to be 500 mV, which was optimized by taking into account the gain cutoff frequency of a Gilbert multiplier. The single-ended voltage swing of the MS-TFF is 250 mV. To increase the operating speed, two emitter followers were used and the emitter size of each emitter follower transistor was optimized to reduce the loading of the flip-flops. An input sensitivity of the DFD for $f_{\text{toggle}}^{\text{max}}$ up to 92.4 GHz is shown in Figure 15.6a. The operational bandwidth of the dynamic FD was a broad 57 GHz (35–92 GHz) without tuning. Figure 15.6b shows the 23.1 GHz divided-by-four output

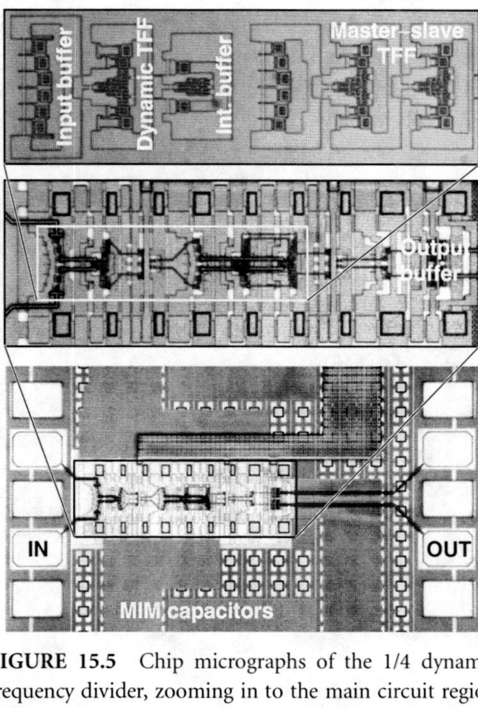

FIGURE 15.5 Chip micrographs of the 1/4 dynamic frequency divider, zooming in to the main circuit region in two steps. The DC bias terminals are connected via MIM capacitors to obtain a stable voltage supply. The area of the D-TFF is 40 × 50 μm, and that of the main circuit region is 420 × 150 μm. (From K. Washio, E. Ohue, K. Oda, R. Hayami, M. Tanabe, H. Shimamoto, T. Harada, and M. Kondo, Digest of Technical Papers of the IEEE International Solid-State Circuits Conference, San Francisco, 2000, pp. 210–211. With permission.)

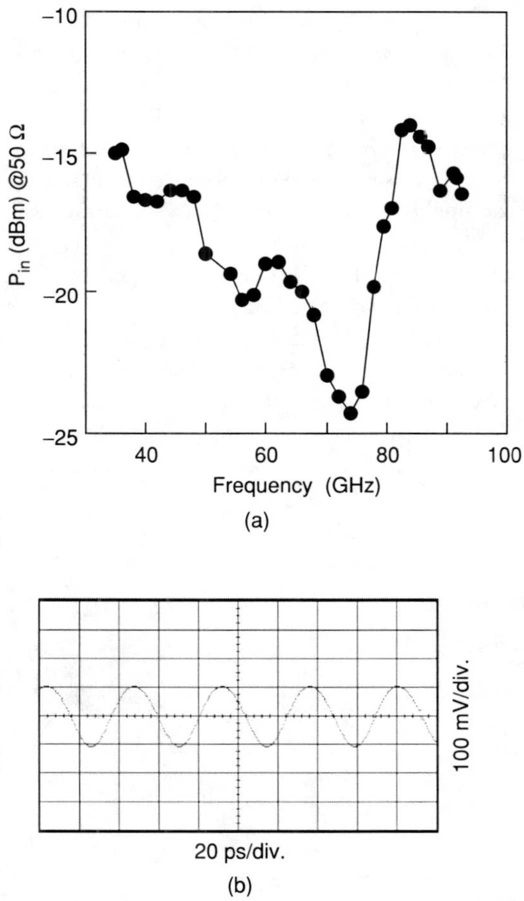

FIGURE 15.6 (a) Input sensitivity of the 1/4 dynamic frequency divider up to its maximum operating frequency of 92.4 GHz. (b) The 23.1-GHz divided-by-four output waveform for a 92.4-GHz input. (From K. Washio, E. Ohue, K. Oda, R. Hayami, M. Tanabe, and H. Shimamoto. *IEEE Trans. Electron Devices* 49:1755–1760, 2002. With permission.)

waveform for a 92.4 GHz input to the dynamic FD. These results indicate that this SiGe HBT technology will play an important role in the monolithic millimeter-wave communication systems of the near future.

5.8-GHz Electronic-Toll-Collection Transceiver IC

High-performance wireless communication systems in the 5-GHz band, such as FWA, microwave mobile networks, and ITS, have become increasingly popular. For these systems to enter commercial service, the necessary items of equipment must be small and cheap. When several separate chips are used to implement a necessary operation, extra space and cost for assembly is incurred, so a fully integrated single-chip solution is most desirable. A single-chip 5.8-GHz ETC transceiver IC has been developed by using the SiGe BiCMOS technology as a demonstrator for 5-GHz wireless communication systems [12]. A chip photomicrograph is shown in Figure 15.7; the chip occupies an area of 2.65×2.5 mm. This fully integrated ETC chip consists of four blocks, i.e., a receiver (RX), a demodulator (DEMO), a PLL, and a transmitter (TX). The chip includes a matching network circuit (MC), a low-noise amplifier (LNA), a down-conversion mixer, and RF detector circuits for the RX, a received signal strength indicator (RSSI) circuit and an amplitude-shift-keying (ASK) demodulator for the DEMO, a varactor-tuned LC-VCO and a synthesizer for the PLL, and a modulator, a variable-gain amplifier (VGA), and power amplifier (PA) for the TX (Figure 15.8). The receiver transfer characteristic is shown in Figure 15.9. The receiver is composed of an LC-matching circuit, two-stage LNAs, and a Gilbert-type double-balanced mixer. The load resistance is 1 kΩ, equal to input impedance of an external band-pass filter. The single-ended gain, the input 1-dB compression point, and the double-side-band noise figure were 31 dB, -40 dBm, and 8 dB, respectively. The -54 dBm local oscillation signal leakage at RF input terminal is attributed to low parasitic capacitance of SiGe HBT. The input VSWR of 1.2 at 5.835 GHz results from well-matched MC considering the parasitic effect of bond-wire and the package. This transceiver chip is intended for use in ETC systems, but all of the circuits and the implementation technique have many other potential applications. The good measurement results for the IC in a plastic package indicate the suitability of SiGe BiCMOS as the base technology of single-chip transceiver ICs for 5-GHz-band wireless communication systems.

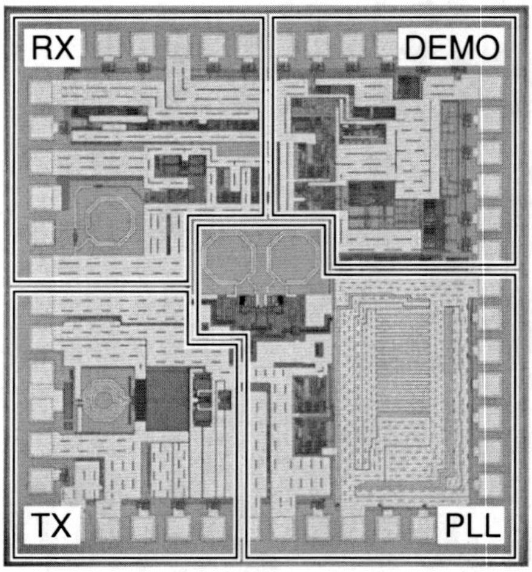

FIGURE 15.7 Chip photomicrograph of a single-chip 5.8-GHz electronic-toll-collection (ETC) transceiver IC using SiGe BiCMOS technology. This fully integrated ETC chip consists of four blocks: a receiver (RX), a demodulator (DEMO), a phase-locked loop (PLL), and a transmitter (TX). (From T. Masuda, K. Ohhata, N. Shiramizu, S. Hanazawa, M. Kudoh, Y. Tanba, Y. Takeuchi, H. Shimamoto, T. Nagashima, and K. Washio. Digest of Technical Papers of the IEEE International Solid-State Circuits Conference, San Francisco, 2002, pp. 96–97. With permission.)

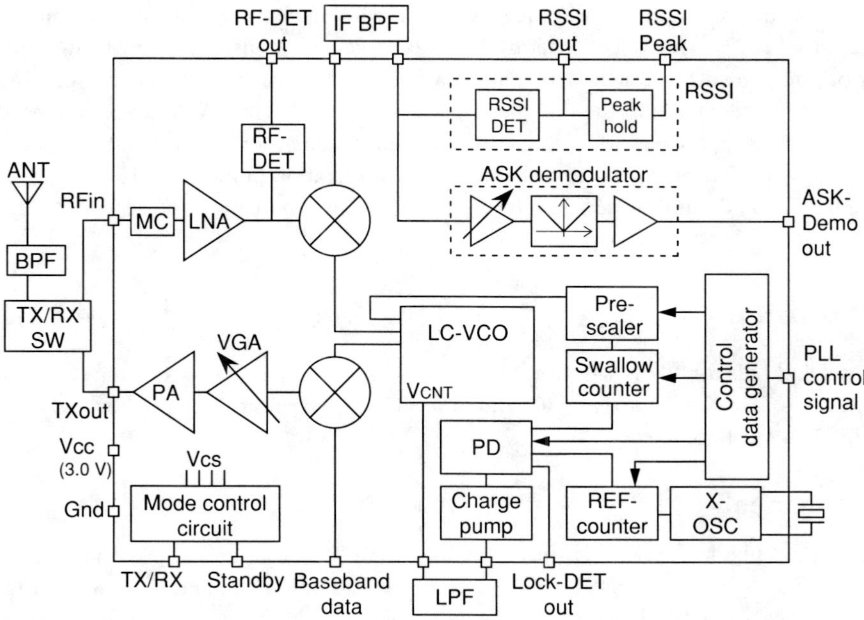

FIGURE 15.8 Block diagram of a single-chip 5.8-GHz ETC transceiver IC. (From T. Masuda, K. Ohhata, N. Shiramizu, S. Hanazawa, M. Kudoh, Y. Tanba, Y. Takeuchi, H. Shimamoto, T. Nagashima, and K. Washio. Digest of Technical Papers of the IEEE International Solid-State Circuits Conference, San Francisco, 2002, pp. 96–97. With permission.)

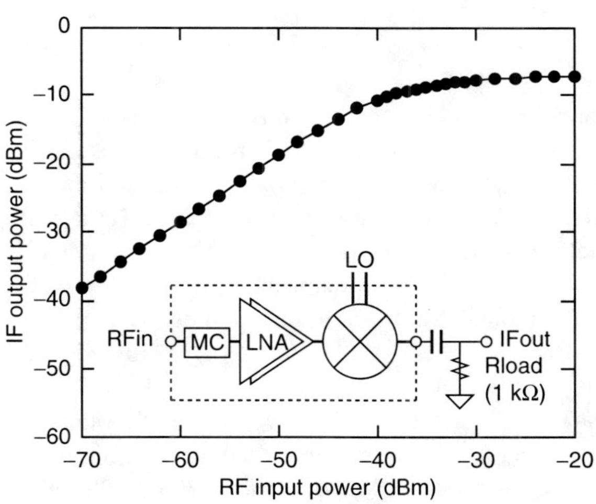

FIGURE 15.9 Receiver transfer characteristics of a single-chip 5.8-GHz ETC transceiver IC. (From T. Masuda, K. Ohhata, N. Shiramizu, S. Hanazawa, M. Kudoh, Y. Tanba, Y. Takeuchi, H. Shimamoto, T. Nagashima, and K. Washio. Digest of Technical Papers of the IEEE International Solid-State Circuits Conference, San Francisco, 2002, pp. 96–97. With permission.)

Other ICs

A PA IC for wideband code-division multiaccess (WCDMA) modulation cellular phone systems has been developed [13]. By optimizing total emitter area, this PA exhibited 44% power-added efficiency and 27.3-dBm output power with an adjacent-channel power ratio of less than −40 dBc at 1.95 GHz and 3.4-V bias voltage.

Two other microwave/millimeter-wave ICs have been developed for wireless applications, specifically microwave link, WLAN, and radar systems. These ICs again demonstrated the strong potential and suitability of self-aligned SEG SiGe HBT. One is a one-dimensional travelling-wave amplifier (TWA) with a bandwidth extending into the millimeter-wave region [14]. This TWA exhibited a bandwidth of about 67 GHz with a gain of 6 dB. The chip occupies approximately an area of 1×1 mm and consumes 23 mA at a supply voltage of 3.3 V. The other is a LNA for operation in the 23-GHz band [15]. The LNA exhibited a gain of 21 dB and noise figure of 4.1 dB at 23 GHz, and consumed 20 mA at a 2.5-V single supply voltage.

15.4 Summary

An IC chipset and LSIs for 40-Gb/s optical-fiber-links, a single-chip 10-Gb/s transceiver LSI, frequency-divider ICs, a 5.8-GHz ETC transceiver IC, and other ICs have been implemented by applying the SiGe HBT and BiCMOS technologies.

Acknowledgments

The author would like to express his sincere thanks to Dr. A. Anzai and Dr. Y. Hatta at the Hitachi Device Development Center (HDDC), and Dr. O. Kanehisa, Dr. K. Seki, and K. Kimura at the Hitachi Central Research Laboratory (HCRL) for their encouragement. The author would also like to express his sincere thanks to T. Masuda, Dr. K. Ohhata, N. Shiramizu, F. Arakawa, T. Nakamura, Dr. M. Kondo, R. Hayami, R. Takeyari at HCRL, and to T. Harada, K. Mikami, S. Ueno, A. Koyama, K. Watanabe at HDDC, to T. Nagashima at Hitachi Digital Media Systems R&D Division, and to Dr. M. Mokhtari, Dr. Y. Li, Dr. B. Hansson, and Dr. T. Lewin at Ericsson Microwave Systems for their extensive contributions throughout this work.

References

1. T. Masuda, K. Ohhata, F. Arakawa, N. Shiramizu, E. Ohue, K. Oda, R. Hayami, M. Tanabe, H. Shimamoto, M. Kondo, T. Harada, and K. Washio: 45 GHz transimpedance 32 dB limiting amplifier and 40 Gb/s 1:4 high-sensitivity demultiplexer with decision circuit using SiGe HBTs for 40 Gb/s optical receiver. Digest of Technical Papers of the IEEE International Solid-State Circuits Conference, San Francisco, 2000, pp. 60–61.
2. K. Ohhata, F. Arakawa, T. Masuda, N. Shiramizu, and K. Washio: 40-Gb/s analog IC chipset for optical receivers—AGC amplifier, full-wave rectifier and decision circuit—implemented using self-aligned SiGe HBTs. Digest of the IEEE International Microwave Symposium, Phoenix, 2001, pp. 1701–1704.
3. T. Masuda, K. Ohhata, N. Shiramizu, E. Ohue, K. Oda, R. Hayami, H. Shimamoto, M. Kondo, T. Harada, and K. Washio. 40 Gb/s 4:1 multiplexer and 1:4 demultiplexer IC module using SiGe HBTs. Digest of the IEEE International Microwave Symposium, Phoenix, 2001, pp. 1697–1700.
4. K. Washio. SiGe HBT and BiCMOS technologies for optical transmission and wireless communication systems. *IEEE Trans. Electron Devices* 50:656–668, 2003.
5. S. Ueno, K. Watanabe, T. Kato, T. Shinohara, K. Mikami, T. Hashimoto, A. Takai, K. Washio, R. Takeyari, and T. Harada. Single-chip 10 Gb/s transceiver LSI using SiGe SOI/BiCMOS. Digest of Technical Papers of the IEEE International Solid-State Circuits Conference, San Francisco, 2001, pp. 82–83.
6. A. Koyama, T. Harada, H. Yamashita, R. Taketari, N. Shiramizu, K. Ishikawa, M. Ito, S. Suzuki, T. Yamashita, S. Yabuki, H. Ando, T. Aida, K. Watanabe, K. Ohhata, S. Takeuchi, H. Chiba, A. Ito, H. Yoshioka, A. Kubota, T. Takahashi, and H. Nii. 43 Gb/s full-rate-clock 16:1 multiplexer and 1:16 demultiplexer with SFI-5 interface in SiGe BiCMOS technology. Digest of Technical Papers of the IEEE International Solid-State Circuits Conference, San Francisco, 2003, pp. 232–233.

7. K. Watanabe, A. Koyama, T. Harada, R. Sotomura, T. Aida, A. Ito, T. Murata, H. Yoshioka, M. Sonehara, H. Yamashita, K. Ishikawa, M. Ito, N. Shiramizu, T. Nakamura, K. Ohhata, F. Arakawa, T. Kusunoki, H. Chiba, T. Kurihara, and M. Kuraishi. A low-jitter 16:1 MUX and a high-sensitivity 1:16 DEMUX with integrated 39.8 to 43GHz VCO for OC-768 communication systems. Digest of Technical Papers of the IEEE International Solid-State Circuits Conference, San Francisco, 2004, pp. 166–167.
8. K. Washio, E. Ohue, K. Oda, R. Hayami, M. Tanabe, H. Shimamoto, T. Harada, and M. Kondo, 82 GHz dynamic frequency divider in 5.5 ps ECL SiGe HBTs. Digest of Technical Papers of the IEEE International Solid-State Circuits Conference, San Francisco, 2000, pp. 210–211.
9. K. Washio, R. Hayami, E. Ohue, K. Oda, M. Tanabe, H. Shimamoto, and M. Kondo. 67-GHz static frequency divider using 0.2-μm self-aligned SiGe HBTs. *IEEE Trans. Microwave Theory Tech.* 49: 31–34, 2000.
10. K. Washio, E. Ohue, K. Oda, R. Hayami, M. Tanabe, and H. Shimamoto. Optimization of characteristics related to the emitter-base junction in self-aligned SEG SiGe HBTs and their application in 72-GHz-static/92-GHz-dynamic Frequency Dividers. *IEEE Trans. Electron Devices* 49:1755–1760, 2002.
11. K. Washio, E. Ohue, R. Hayami, A. Kodama, H. Shimamoto, M. Miura, K. Oda, I. Suzumura, T. Tominari, and T. Hashimoto. High-speed scaled-down self-aligned SEG SiGe HBTs. *IEEE Trans. Electron Devices* 50:2417–2424, 2003.
12. T. Masuda, K. Ohhata, N. Shiramizu, S. Hanazawa, M. Kudoh, Y. Tanba, Y. Takeuchi, H. Shimamoto, T. Nagashima, and K. Washio. Single-chip 5.8 GHz ETC transceiver IC with PLL and demodulation circuits using SiGe HBT/CMOS. Digest of Technical Papers of the IEEE International Solid-State Circuits Conference, San Francisco, 2002, pp. 96–97.
13. M. Kondo, I. Miyashita, M. Koshimizu, Y. Kagotoshi, H. Nagai, and K. Washio. High-efficiency power characteristics for WCDMA applications of SiGe HBT devices using a novel form of base-bias resistance. Digest of the IEEE International Microwave Symposium, Philadelphia, 2003, pp. 2205–2208.
14. B. Kerzar, M. Mokhtari, Y. Li, B. Hansson, K. Washio, T. Harada, and T. Lewin. Millimeter-wave bandwidth, SiGe-HBT travelling wave amplifier. Technical Digest of the IEEE Gallium Arsenide Integrated Circuit Symposium, Seattle, 2000, pp. 57–59.
15. G. Schuppener, T. Harada, and Y. Li. A 23-GHz low-noise amplifier in SiGe heterojunction bipolar technology. Digest of the IEEE Radio Frequency Integrated Circuits Symposium, Phoenix, 2001, pp. 177–180.

16

Industry Examples at the State-of-the-Art: ST

16.1 Introduction 16-1
16.2 Process Definition 16-3
16.3 Switched Gain LNA Implementation 16-6
16.4 Down Mixers and PMA Implementation 16-8
16.5 Variable Gain Amplifier (VGA) and Low-Pass Filter (LPF) 16-10
16.6 LNA-Mixer Isolation Strategy 16-12
16.7 Receiver Implementation 16-13
16.8 Validation Strategy 16-13
16.9 RF and Analog Blocks Validation 16-14
16.10 RF and Analog Receiver Validation 16-17
16.11 WCDMA Receiver Global Test 16-18
16.12 Summary 16-19

Didier Belot
ST Microelectronics

16.1 Introduction

Whereas the trends for wireless transceivers design are in the direction of integration of most of the blocks to reduce the cost, two main terminal families are identified:

- The low-end terminals should serve voice centric standards like GSM, and should be time duplex mode operation (TDD). These standards permit global integration, RF–analog–digital, approach. This means that the best process candidate for this family is the VLSI CMOS if the cointegration between RF, analog, and digital is demonstrated.
- The high-end terminals will serve multimode, multimedia applications; they will integrate frequency domain mode operation (FDD). Due to the complexity of the RF–analog and the complexity of the DSP, it seems that, in a first time at least, two main circuits must be developed, a pure state-of-the-art VLS CMOS for the DSP, and an RF–analog one. The BiCMOS SiGe-carbon process is the best candidate to help the integration of such an analog–RF parts of this kind of wireless transceivers on silicon.

To be ready for the next 3GPP generation of the handset, a focus on the WCDMA standard was made, taking into account the specificity of the GSM, in order to have the possibility to develop a reconfigurable multimode receiver in a second time. This approach imposes several challenges for the integration instead of developing multimode reconfigurable receiver [1]. The main issue is the full duplex mode operation (FDD) of the receiver and the transmit paths. This leads to aggressive requirements of the receiver due to the strong modulated TX leakage of −23 dBm at the low-noise amplifier (LNA) input while the receiver sensitivity must be met with a wanted modulated channel of −110 dBm. Noise factor, conversion gain, and out of band second- and third-order linearity are the major parameters to be guaranteed to fulfil the WCDMA requirements. For this receiver, the direct conversion architecture has been chosen for its capacities to integrate both analog and RF blocks, and the flexibility offered by this kind of architecture in order to implement more than one standard in the receiver. However, an external interstage SAW filter centered at 1.95 GHz must be added off-chip to reduce the TX power at the down mixer input.

This choice introduces the discussion about the process selection and the partitioning. A zero IF architecture is very sensitive to the DC offset, the matching especially on IIP2 linearity parameter, and the $1/f$ noise, especially for narrow band applications. A silicon germanium bipolar is the best answer to solve these issues. Another issue to take into account is the substrate noise immunity, we will see in the process paragraph how we can easily modulate an MOS transistor by the bulk, and this is not the case with the bipolar collector.

The receive path is shown in Figure 16.1. An external duplexer filter is connected to the antenna to separate the RX and the TX bands. Then, the signal feeds into the single-ended switched gain LNA. After the LNA, the TX leakage is attenuated by −50 dB to relax the IIP2 linearity requirements of the Quadrature downconverter mixers (DCM) [1]. The Quadrature signals are generated by a divide-by-two based on a cascaded D-latch flip flop. As the frequency synthesizer was not integrated, the 4-GHz local oscillator (LO) differential signals are coming from external inputs. The DCM output current is then converted into voltage by the postmixer amplifier (PMA). This PMA performs a common mode voltage shifting and 4 MHz pre-filtering of the adjacent channels to relax the linearity requirements of the analog base-band (ABB) filter. The analog base-band filter includes a first-stage VGA followed by a fifth-order Butterworth low-pass filter and exhibits −22 dB/+38 dB variable gain by 2 dB step. The overall low-pass −3 dB cutoff frequency 2.4 MHz is fixed by the unitary resistor and capacitor (RC) module, which is tuned by the calibrator to keep the RC product constant in case of process and temperature spreading.

To compensate the static DC offset on each I and Q paths coming from the PMA and LPF mismatches, 8 bits current DAC are connected at the PMA inputs. The dynamic range of the DAC allows the cancellation of the ±5 V DC-offset at the receiver outputs in the maximum gain.

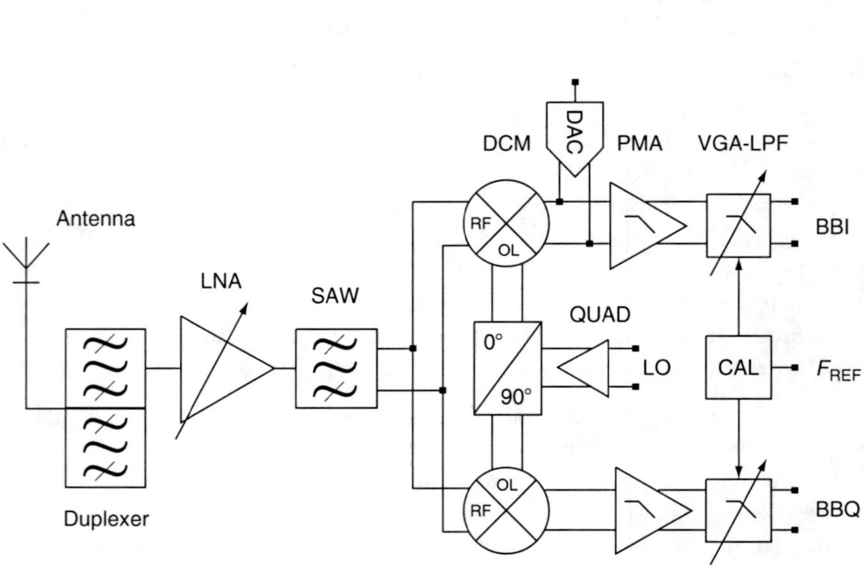

FIGURE 16.1 Zero-IF WCDMA architecture.

A read/write three wires serial bus allows the set-up of all the functionalities such as power-down, LNA and VGA gain modes, DC-offset calibration, and PLF cutoff frequency calibration.

This chip also includes separate currents biasing (with external reference resistor) and voltage band-gap, respectively, used for RF parts (LNA, mixers, LO Quadrature and DAC) and for the ABB filters.

16.2 Process Definition

The zero IF architecture permits a reconfigurable multimode approach, and is cost effective in terms of external components such as SAW filters but it is very sensitive to the Mixer IIP2, the DC offset, and the $1/f$ noise in base band.

A way to increase the IIP2 is to design a pure differential mixer, in that case the second-order linearity depends on the mismatching between the LO switching transistors and the LO phase mismatching at the input of these transistors, in a first order. The use of bipolar transistors, in the switching pair, due to their better matching than MOS transistors, allows to increase this second-order linearity parameter.

The DC offset is shared by the down-converter mixer offset, the postmixer amplifier offset, the voltage gain amplifier and the analog base-band filter offset. It is well known that bipolar offset is, at least, ten times lower than CMOS one at same operating point.

The $1/f$ noise is another issue of CMOS transistors; pink noise frequency corner is in the 10–100 kHz range when the bipolar pink range is in the 10–100 Hz range, which gives a significant advantage to the bipolar.

If we want to go deeper in the integration toward a transceiver, we have to take into account the substrate coupling effects, and to analyze the substrate noise immunity of the devices [2,3]. The substrate modulation mechanism in an MOS transistor is described in the Figure 16.2, the recombination of positive loads with the substrate increases the negative loads collected by the drain, which modifies the threshold voltage of the transistor, as given by the equation included in Figure 16.2

$$V_T = V_{T0} + \gamma\lfloor\sqrt{(2\varphi_F - V_{BS})} - \sqrt{2\varphi_F}\rfloor$$

This V_T variation creates new intermodulation products inside the RF amplifiers and the RF mixers, which can affect dramatically the receiver performances. A solution to reduce this effect is the possibility of using an extra-cost N buried layer, in order to create an isolated bulk; this is efficient low frequency, but it behaves as a transmit capacitor at high frequency, thus LO harmonics, RF harmonics can create intermodulation products. It seems that this limitation is one of the main issues of using CMOS processes in an FDD standard.

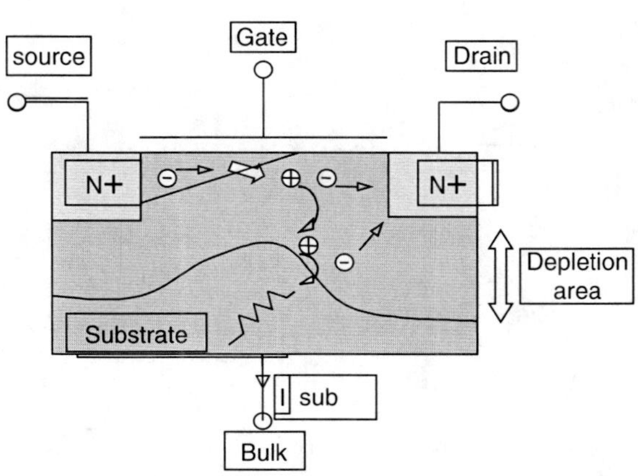

FIGURE 16.2 MOS substrate modulation.

We have identified some issues, which define why a bipolar transistor is well suited for FDD standards, but we have to continue this process in order to define why we need SiGeC transistors instead of classic Si bipolar ones.

The global receiver noise figure (NF), in order to satisfy the IEEE-3GPP requirement, has been estimated to be 7.5 dB, which is very challenging and needs to design an LNA with an NF less than 2 dB at 2.2 GHz in the worst case. The silicon–germanium–carbon transistor allows to maintain the transition frequency, particularly reducing the base collector capacitor and reducing the base resistance, which reduces the intrinsic noise figure of the transistor, and increases its intrinsic linearity as it is described in the following study.

The objective of this study is to understand the nonlinear behavior of the heterojunction bipolar transistor (HBT) at 2 GHz. A quantitative method validating an equivalent nonlinear HBT [4] allowed to identify the nonlinearity sources (mainly C_{bc}) and to observe the second and third order distortion compensations. However, it presents limitation factors, which do not permit to fulfill, understand the HBT implications.

In order to accurately observe the compensation an analytic method based on Volterra series has been developed. In this transfer functions theory [5], the second and third order intermodulation currents (IMD2 and IMD3, at respective frequencies (f_1-f_2, f_1+f_2) and ($2f_1-f_2$, $2f_2-f_1$)) and first- and second-order harmonics (H1 and HD2, at respective frequencies f_1 and f_2) are defined through HBT parameters and the source and load impedances. (Figure 16.3).

In order to validate this approach, a common emitter LNA was studied. The main results are given in Figure 16.4 where the entire contributor to the IMD3 is presented.

The main contributors are the C_{bc} capacitor and the transconductance G_m. Taking into account the high compensation between second and third order nonlinearity of C_π and G_m, the third order of Cbc is mainly implicated.

This nonlinearity is given by the following equation:

$$I_{\text{Cbc\_IMD3}} = C_2 V_{\text{bc},\text{w2}} V_{\text{bc},\text{w2}-\text{w1}} + C_2 V_{\text{bc},-\text{w1}} V_{\text{bc},2\text{w2}} + (3/4) C_3 V_{\text{bc},\text{w2}}^2 V_{\text{bc},-\text{w1}}$$

The amplitude of this one is due for 83% to the third order term $C_3 V^2_{\text{bc, w2Vbc},-\text{w1}}$. A 20% reduction of C_3 reduces directly the IMD3 by 13%. The SiGeC HBT, reducing this capacitor, increases its third order linearity.

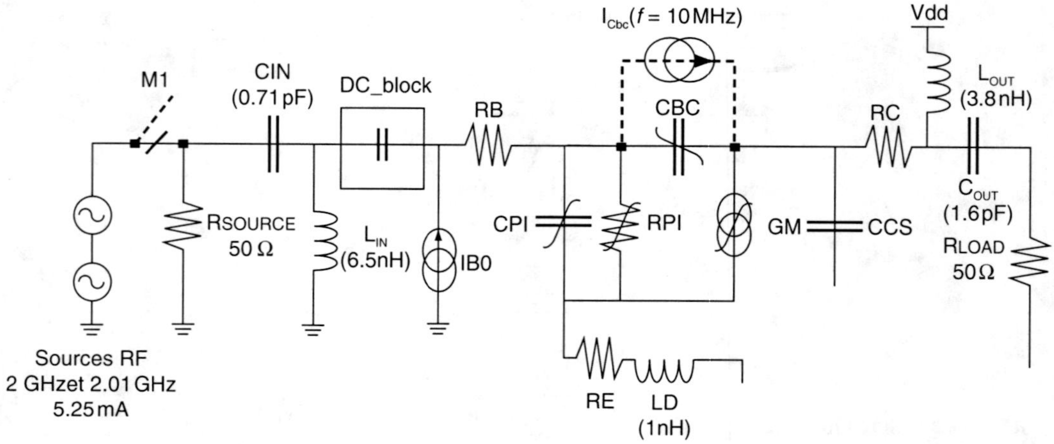

FIGURE 16.3 Transfer function HBT.

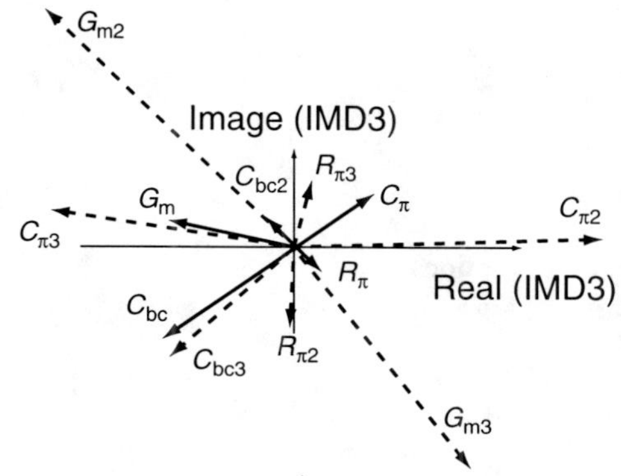

FIGURE 16.4 Main contributors to IMD3 in a HBT.

TABLE 16.1 Noise Figure Improvements

Criteria	NR671A10 SiGeC HBT 0.25 μm	NR671A20 SiGe HBT 0.35 μm	NR671A10 SiGe HBT 0.25 μm	LNA kit part with NR671A20 SiGe HBT
Vcc (V)	2.5	2.7	2.5	2.7
NFmin	0.405	0.95	0.9	NF = 1.1 dB
Icoll (mA)	4.8	5.5	3.5	5.5
Rn (Ω)	7.5	9	15.53	
Mod (Γ opt)	0.18	0.47	0.267	
Ph(Γ opt) (°)	3	70	−3	
S21 (dB)	21.65	15.7	19.5	14.7
S12 (dB)	−26	−22.15	−22.72	−19.1
S11 / 50 Ω	0.21–j1.63	0.1–j0.48	0.3–j1.54	−14.5 dB
S22 / 50 Ω	1.25–j3.5	0.93–j1.06	1.12–j2.71	−18.5 dB
IIP1 (dBm)				−0.2
IIP3 (dBm)				10.5

Finally, the noise figure, mainly due to the base resistor, is reduced in a SiGeC HBT, due to the higher level of doping profile in the germanium base; Table 16.1 gives, examples of noise figure improvement between a SiGe HBT process and a SiGeC one. At a same level of generation, we can see an improvement of noise figure by 0.5 dB at the same collector current, mainly due by a dramatic reduction of Rn.

In conclusion, for these different topics, it has been decided to design the circuit with a SiGeC BiCMOS process based on a 0.25 μm CMOS. The main characteristics of this process are given below.

50 A as CMOS base process
—5 nm gate oxide thickness
—0.25 μm gate lengths
—2.5 V supply voltage
—Isolated NMOS (optional)

SiGeC heterojunction bipolar transistor (HBT)
—Deep trench isolation
—LV and HV with 3 and 6 V BVCE0, respectively
—Low noise characteristics (Nfmin <0.5 dB)

Other active devices
—NLDEMOS (2.5 V supply voltage, 15 V BVDS)
—Isolated Vertical PNP BJT (Optional)
—Lateral PNP

Resistors
—Silicided N + Poly
—Unsilicided N + Active
—Unsilicided N + Poly
—Unsilicided P + Poly
—Nwell under STI
—High-value Poly 1 kohms/sq (option)
—35 ohms/sq TaN resistor (option)

Capacitors
—5 fF/μm^2 MIM Capacitor
—High-value polyresistor
—N+Poly/NWell capacitor
—N+Poly/Sinker capacitor
—35 fF/μm^2 HIDT capacitor (option)

Varactors
—P+/Nwell diode
—P+/Nwell diode with differential structure
—MOS transistor

Diodes
—N+/Psubstrate
—P+/Nwell

Thick metal inductors
—Single-ended inductors
—Symmetrical inductors (high *Q*, low area, large *L*-range)

Metal
Five metal levels
—M1 in Tungsten
—M2–M5 in aluminum
—Thick M5 (2.5 μm)
—Copper M5 level (option)
—50 ohms-cm substrate

A cut of the HBT transistor is given in Figure 16.5.

We will go now more deeper in the design of each block beginning by the LNA, and then will have a discussion on the integration issues, we will also discuss about the opportunity of the use of specific technology for module approach such micromechanical devices (MEMs) and integrated passive devices on glass (IPAD).

16.3 Switched Gain LNA Implementation

A single-ended LNA was chosen to minimize the duplexer losses (compared with a single to differential one), which is inserted at the antenna output, the substrate noise sensitivity is managed using bipolar transistors, but it is obvious that this should be an issue in case of a transmit, receive application. This LNA, described in Figure 16.6, is based on a cascode topology with degenerated emitter inductor L1 to get the best trade-off between matching and noise factor.

As explained in the "Introduction," the circuit must be, in the final version, flip chipped on a glass module, which creates a constraint for L1 inductor: instead of using bonding inductor, thick metal inductor L1 was integrated.

The well-known cascode topology was selected for its abilities to reach higher gain. This helps designers to finally achieve the requested minimum power gain when taking into account all the

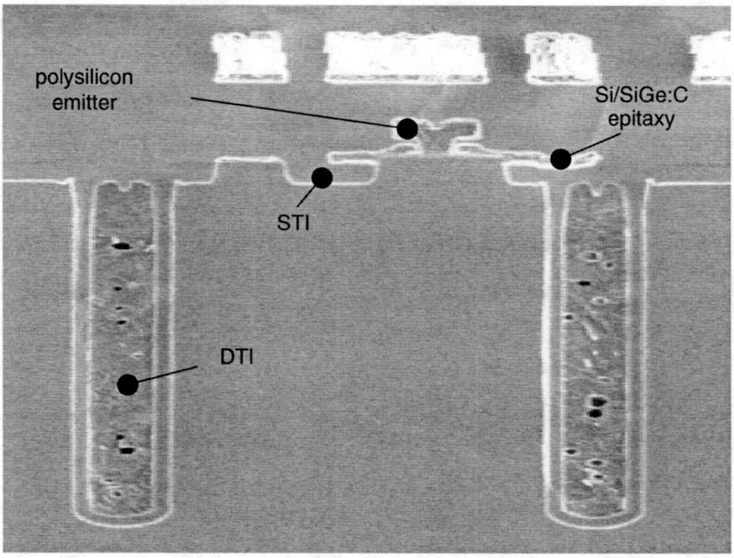

FIGURE 16.5 SiGeC cross section HBT.

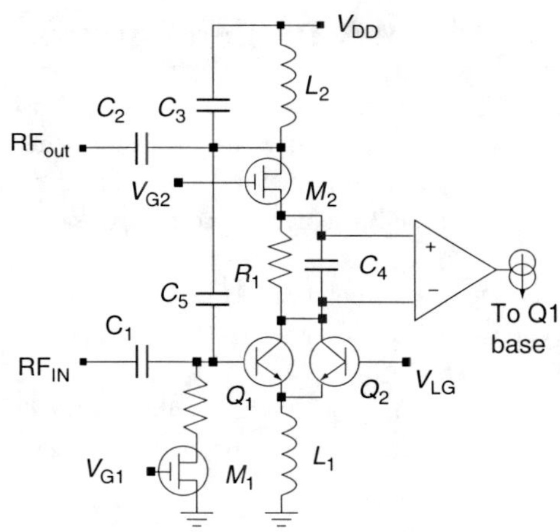

FIGURE 16.6 Switched gain LNA simplified schematic.

undesired elements involved in gain reduction, such as RF pads, ESD protection diodes, layout parasitic, and wire bonding.

The common gate M2 MOS transistor replaces advantageously the bipolar transistor to enhance the output compression point. As the output compression is limited by the voltage swing across the L2 load inductor, the MOS allows saving precious mV of swing by its lower V_{gs} than the V_{ce} bipolar.

A feedback capacitor array is inserted between the input and the output nodes to increase the overall third order linearity to the detriment of a power gain reduction.

In the low gain (LG) mode, the 4.5 mA collector current flowing through Q1 is switched off. Then the Q2 transistor is biased at 50 μA to reduce the power gain to −16 dB. In that configuration, the input matching in the LG mode is done by the input MOS M1 operating in the triode region.

The LNA is biased using a DC feedback loop A1, which compares the Q1 collector current with a current reference coming from the bandgap current source. Compared to the mirror current source, this biasing method gives more flexibility to set the Q1 transistor area.

Finally, the LNA is simulated with its RF pads including ESD protection diodes and the bonding models associated with the PCB lines extracted with Ansoft HFSS 3D field solver. Separately, the layout parasitic elements were extracted with Cadence ASSURA-RF extractor, which permits a full RLC+K extraction of the selected RF critical nets.

Thanks to the 0.25 μm SiGe-carbon process and the lower base–collector capacitance of the BJT, the LNA had some margin to fulfil the 3G specifications requirements at low current. The 5 fF/μm^2 metal–metal capacitor at the last metal level exhibits lower bottom to substrate parasitic capacitor and helps to save power gain. At 2.14 GHz, this LNA achieves +3 dBm IIP3 and 14.7 dB of power gain with 1.3 dB noise factor. The input compression point of −10 dBm is more critical within the 2.7 V supply voltage and the 4.5 mA collector current. In the LG mode, the power gain is reduced to −15 dB with 200 μA consumption.

Table 16.2 summarizes the main simulated results obtained on this block. The IPAD simulations are dedicated to the flip chipped on glass version, which is the final product, the COB simulations are dedicated to the chip on board, which is the tested one in this chapter, and AST spec columns give the specifications for this LNA.

Blocker = −20 dBm at 2 GHz
Simulations including: COB, ESD, Layout RC estimated parasitic

Figure 16.7 gives a layout view of the LNA, the adaptation self-inductance being the smaller, and the bigger one is the output load. The LNA core area is 500×510 μm.

16.4 Down Mixers and PMA Implementation

The direct conversion mixer (DCM), described in the Figure 16.8, is based on a differential input transconductor Q1 and Q2 with an emitter degenerating inductor. The 3.5 mA tail current source I0 is shared between the I and Q paths; this biasing method is helpful to compensate the symmetry error of the incoming signals (specially when the RF signals are coming from outside the chip) that contribute to increase IIP2 parameter.

A 200-Ohm input differential impedance matching requested by the interstage SAW filter is achieved at 2.14 GHz with 10 nH series external inductors. The Q1, Q4 . . . Q8 bases voltage comes from a bandgap reference and an operational amplifier.

The output base-band current at the Q3 · · · Q6 collector enters in the differential postmixer amplifier (PMA) based on the Rauch topology, which converts current to voltage. The transimpedance R1 value

TABLE 16.2 Simulate Results

	IPAD		COB		AST SPEC	
	HG	LG	HG	LG	HG	LG
ICC (mA)	4.84	0.65	T.B.D	T.B.D	5	0.5
S11 (dB)	−15	−12.8	−10	−11	−12	−9.5
S21 (dB)	15.40	−16.8	14.1	−16	N.A	N.A
S12 (dB)	−22.6	−22.8	−21	−20.6	−18	−18
S22 (dB)	−10	−4	−7.5	−4.6	−12	−9.5
NF (dB)	1.60	21	1.56	20	2	20
GP (dBm)	15.60	−16.8	14.1	−15.9	14	−17
IIP3 (dBm)	0.00	−13	+3	−11	−1	−12
IIP1	−12.1	−11	−10[a]	−11.4[a]	−8	−13
K	1.20	18.00	1.13	15	N.A	N.A
Settling time (μs)	4.00	4.00	N.A	N.A	10	10
ΔNF	0.10	−0.80	N.A	N.A	N.A	N.A
GP-1 dB	−12	T.B.D	N.A	N.A		

[a]Process = Typ.

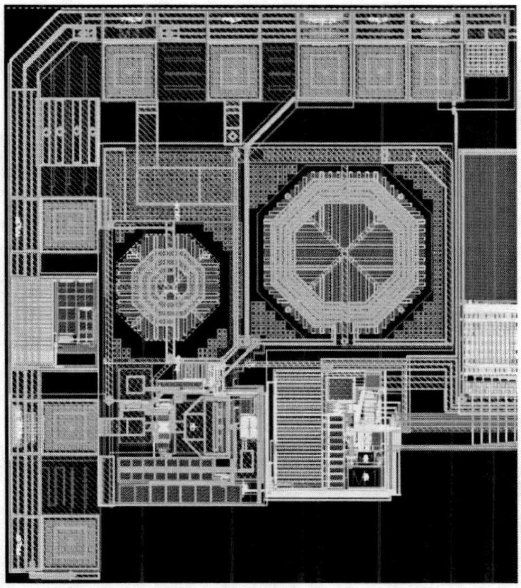

FIGURE 16.7 Switched gain LNA layout view.

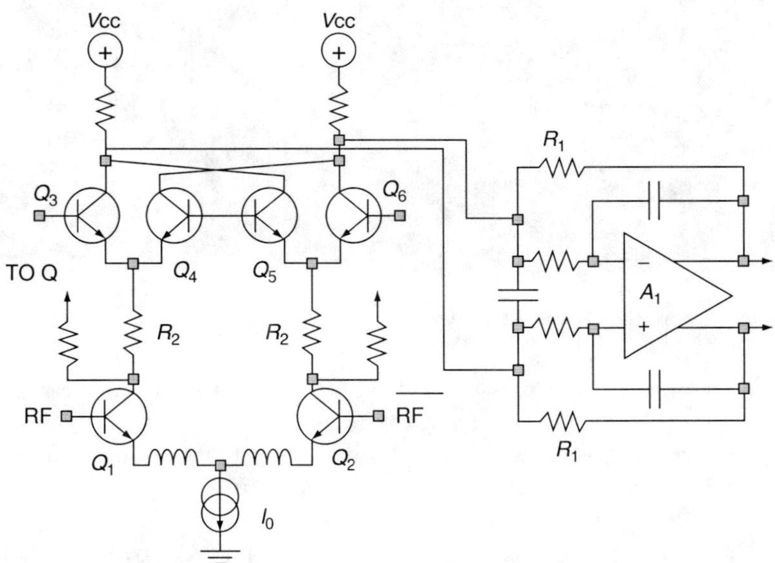

FIGURE 16.8 Down-converter mixer and PMA simplified schematic.

was chosen as 2400 ohm in order to reach 22 dB of typical gain when PMA and DCM are cascaded. The PMA (Figure 16.3) also exhibits a second-order low-pass Butterworth frequency response tc filter out the 10 and 20 MHz adjacent blocking channels by, respectively, 10 and 20 dB. The 5.8 MHz cutoff frequency is high enough to avoid phase distortion of the receive channel.

The PMA fully differential operational amplifier A1 uses $f_T = 6$ GHz vertical isolated PNP transistors into the input pairs. These transistors bring the lowest $1/f$ noise available in the process offer. The PMA achieves 380 MHz of gain–bandwidth product in order to present good linearity behavior when the 10 and 20 MHz adjacent spurious signals are received.

The I/Q down-converters and PMA consumption are, respectively, 3.5 and 3 mA for a voltage gain of 22 dB with an IIP3 equal to 0 dBm and 4.3 nV/sq.Hz of equivalent input noise.

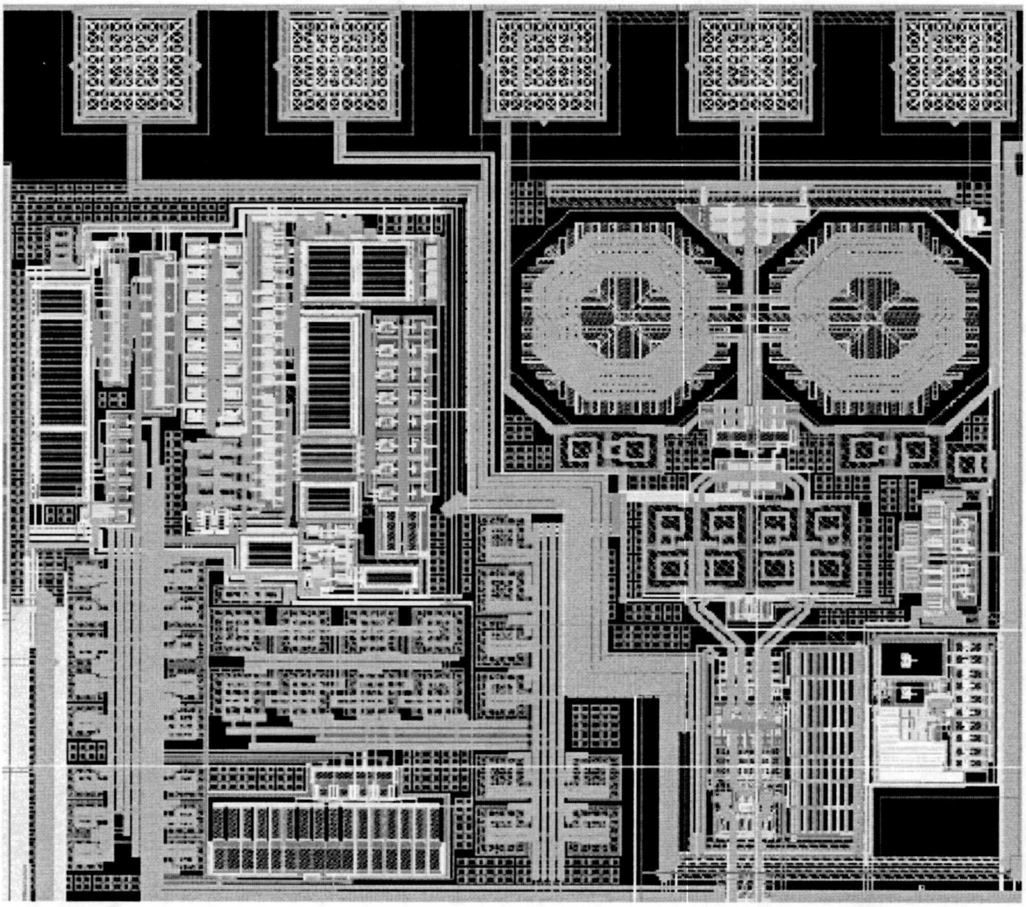

FIGURE 16.9 Down-converter mixer and PMA layout.

The next gives the simulated results of the DCM+PMA obtained in the worst case in temperature/process/power supply.

Figure 16.9 gives a layout view of the down-converter mixer with the postmixer amplifier, the mixer being on the right. The total area of this block is 500 μm by 500 μm.

16.5 Variable Gain Amplifier (VGA) and Low-Pass Filter (LPF)

Following the postmixer amplifier, the VGA ensures that the signal remains in an acceptable dynamic. A classical inverting amplifier structure has been selected; a description is given in Figure 16.10. The VGA is located before the low-pass filter to satisfy the noise requirement; on the other hand, it must be more linear to support adjacent blockers.

The low-pass filter is an active RC fifth-order Butterworth filter. The main parameter of the filter is the accuracy of the 3 dB cutoff frequency, maximum specification being 5% deviation. In order to achieve this performance an internal RC calibrator has been designed. This calibrator compares the time constant of an RC structure to a reference period and provides a 6-bit digital control word, which adjusts the capacitor to correct process and temperature drift.

The RC structure of the calibrator cell is then adjusted with 3.2% resolution. A specific care is given to the layout in order to guarantee this accuracy; the capacitor layout in the filter should be exactly the same as the one of the calibrator.

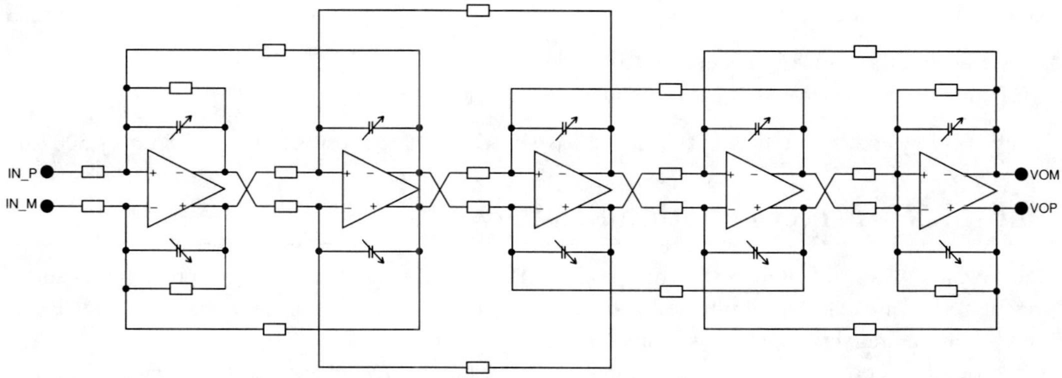

FIGURE 16.10 Fifth-order low-pass filter schematic.

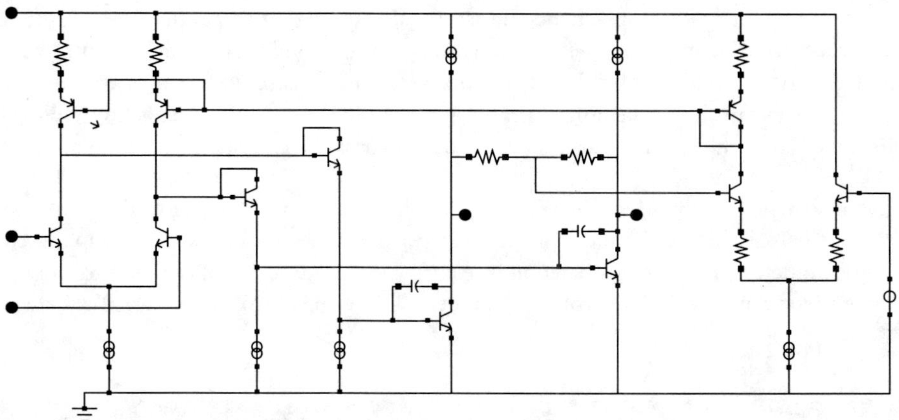

FIGURE 16.11 Low-pass filter operational amplifier schematic.

By taking resistor and capacitor matching between calibrator and filter into account (1% and 0.5% respectively) it leads to 3.5% RC product accuracy in the filter. If we want to guarantee the 5% accuracy of the band pass filter, we should avoid any parasitic effect that may degrade the cutoff frequency.

As a consequence, the gain bandwidth product of the filter's operational amplifier should be high enough, in order to maintain the 3 dB band pass. The gain–bandwidth (GBW) product of the operational amplifier is 300 MHz for a 2.4 MHz filter bandwidth. The linearity (IIP3) and the group delay specifications also require a high GBW in order to get a high gain in the feedback loop. The use of npn bipolar input stage and pnp current mirrors is very helpful to achieve the wanted trade-off between GBW, DC offset, and power consumption. Figure 16.11 describes the operational amplifier schematic. Moreover, bipolar transistor allows a huge reduction of the flicker noise contribution of analog blocks regarding to a pure CMOS design [6]. It is not so important for WCDMA application, but it becomes a tremendous advantage when making the analog blocks design compatible with both WCDMA and GSM standards at low power. The GSM bandwidth in zero-IF architecture being around 100 KHz, it is not possible to integrate a kind of DC blocker, like a capacitor path.

The specifications of this Analog base-band are given below:

First (VGA + FILTER) stage:

Gain: −15 dB to +15 dB in steps of 2 dB
Filter: Third order—with a cutoff frequency of 2.68 MHz ± 5%
Input noise < 28 nV/SQR(Hz)
IIP3 > 3 dBVp out of band

Second (VGA + FILTER) stage

Gain: −7 dB to +23 dB in steps of 2 dB
Filter: Second-order with a cutoff frequency of 2.94 MHz ± 5%

Figure 16.12 gives a layout view of the analog base-band, the block area is around 700 μm by 500 μm.

16.6 LNA-Mixer Isolation Strategy

Since WCDMA is a full duplex system, Receive (RX) and Transmit (TX) signals are simultaneously present at the antenna. Even if filtered by the front-end duplexer, the TX signal remains at −23 dBm at the LNA input and very much constraints its IIP3 requirement.

In order to filter out this TX leakage before mixer, an external SAW filter is used between the LNA and the mixer used is as shown in Figure 16.1. This SAW filter rejects out TX signal by 40 dB.

An LNA and Mixer floor planning has been studied in order not to bypass these 40 dB attenuation by internal coupling. There are both magnetic and electrical coupling on chip, after Cadence ASSURA-RF and Cadence SubstrateStorm postextracted simulations; it was found that the main contribution to the coupling effect between these two blocks was the magnetic one. Indeed, inductors are used in the mixer input stage in order to optimize the noise-linearity trade-off in the cells, and magnetic coupling between LNA output inductor and Mixer input inductors could bypass the external filter attenuation of the TX signal.

Differential structure of the Mixer increases its coupling immunity but due to the layout constraint the position of LNA inductors is not symmetrical with the mixer positions. The coupling coefficient between these inductors has been estimated to be 6×10^{-4} that leads to 90 dB isolation from LNA output to Mixer input (simulated with ESD protection diodes). This value means that internally there is not any

FIGURE 16.12 Analog base-band layout.

risk to bypass the SAW filter, but the circuit, in the validation module is bonded to board module, which means that the LNA output should be coupled to the Mixer inputs through the wire bonding. A specific model of the wire bonding ring was made using Ansoft HFSS, in order to define the coupling factor between the different inputs and outputs. After extraction of these coupling factors, a solution was found with −50 dB bonding coupling between the LNA outputs and the Mixer inputs, to guarantee the 40 dB rejection of TX signal by the external filter.

The drawback of this isolation constraint is that we had to increase the distance between LNA and Mixer. This free space was used to place the offset compensation DACs and the postmixer amplifier that leads to a nonoptimal base-band signal path.

16.7 Receiver Implementation

Figure 16.13 gives a layout view of the WCDMA receiver, the core area of this function is 1.8 mm^2, and the total power consumption is 25 mA in high gain and 20 mA in low gain. The LNA is on the right side, its input is on the right, and its output is on the top, in order to respect a 90° angle between the two-wire bonding, which allows to reduce the coupling effect drastically. Between the LNA and the Mixer we can see the PMA at the right of the Mixer and the DACs in order to increase the distance between the two RF blocks. This floor plan constraint creates a distance to manage between the PMA and VGA-Filter. On the bottom of this layout, we can discover the serial bus, which drives this function and in the final product; the global transceiver, this die being the receiver prototype of the global transceiver program.

16.8 Validation Strategy

The validation strategy was thought out early in the project, and was at the same level as the process selection, with the architecture and the designs as the main concerns for success of the layout. First of all, this circuit, being a step to a transceiver, has to be validated in chip on board (COB) configuration, and will be used in an intermediate version in flip chipped (FC) on glass. That means that the validation strategy must cover these two cases. In a second time, this circuit must be validated at an RF–analog level, before going to a chip-set application; these considerations have guided our approach.

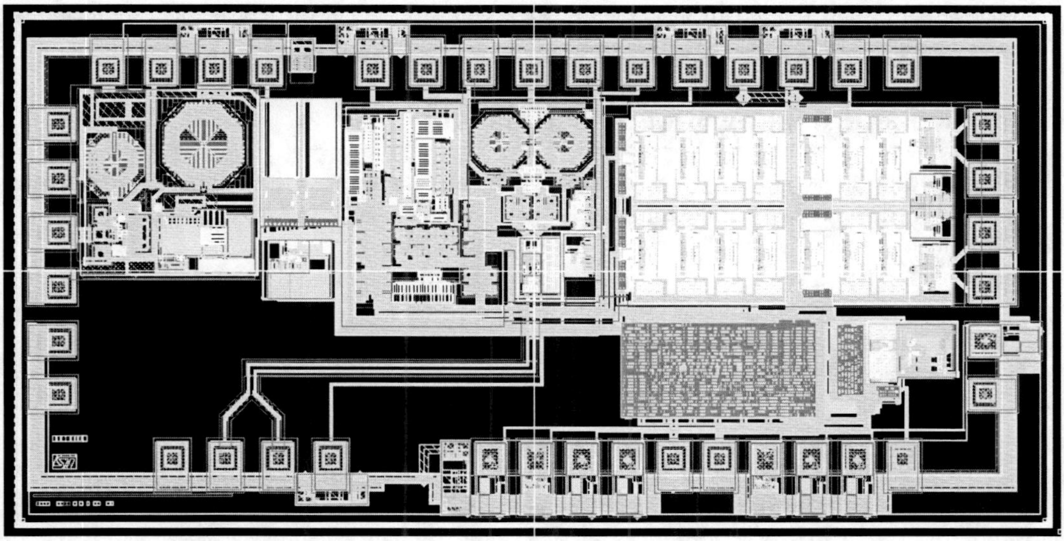

FIGURE 16.13 Receiver layout: 2.9 mm × 1.4 mm.

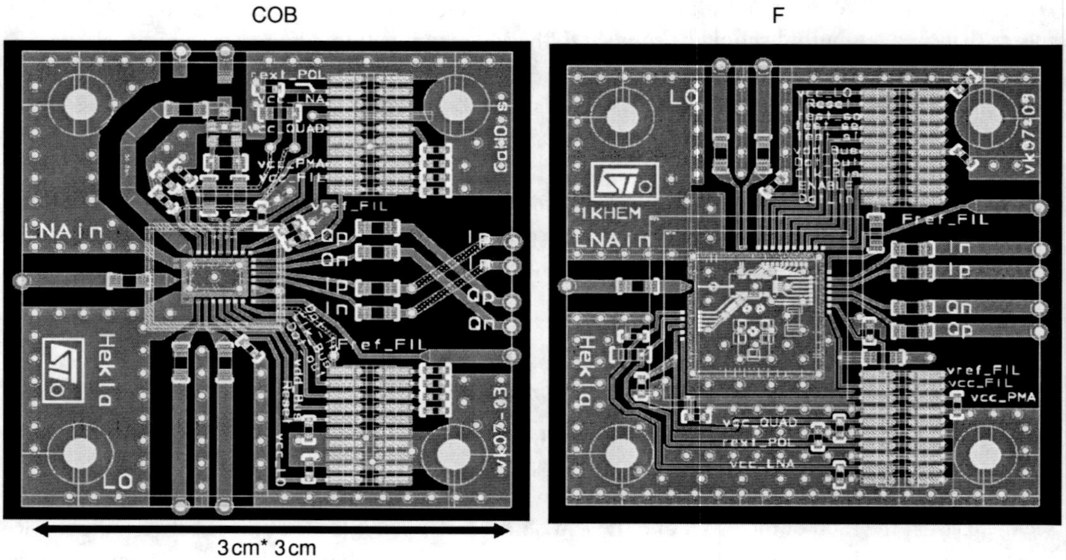

FIGURE 16.14 COB module and FC module.

In order to satisfy the first constraint, a module approach was defined, which allows to develop two compatible modules, one for the COB version and one for the FC configuration. In order to satisfy to the second constraint, the two modules interface with the same RF board, which is a part of the global chip-set board. Figure 16.14 shows the two modules.

On the left side, the COB module exhibits the external SAW filter between the LNA and the Mixer at the top with its adaptation network, while it is integrated in the integrated passive devices on glass in the FC version. The two modules are symmetric, due the fact that the circuit is flip chipped in the FC version.

The targeted objective of this approach is to guarantee the global RF analog behavior of the module to the customer, which needs to validate the path from the module inputs/outputs to the internal circuit. In order to aim this target, a modelization of board lines was made, and a global simulation was performed as shown in Figure 16.15.

Finally, the RF validation board was designed to be compatible with the global application chip-set board; a description of this board is given in Figure 16.16.

16.9 RF and Analog Blocks Validation

To validate the functions separately, three different versions were processed: the stand-alone analog baseband, the PMA-down-mixer, and the complete RX path shown in Figure 16.1. The integrated LNA could also be validated separately as the RF output enters the SAW filter.

The validation board was split into two parts: the IC is wire bounded on an RF module, which is reported on a motherboard; this allows more flexibility to test several dies. The RF module contains the RF external elements such as impedance matching, supply capacitors, and SAW filter. This module also includes the de-embedding footprints to validate separately the scattering parameters of the LNA, SAW filter, and input mixer in order to help impedance matching between these RF blocks.

The LNA S11 and S22 parameters show good agreements with the simulations: 2.2 nH series inductor is added in series at the LNA input while the output is matched to 50 ohms without external elements. At 2.14 GHz, the measured power gain is 13.4 dB and the noise factor is 1.85 dB. In band IIP3 was measured at 0 dBm with the 10 MHz offset two tones interferers for 5 mA current supply at 2.7 V. A summary of the measured results is given in Table 16.3.

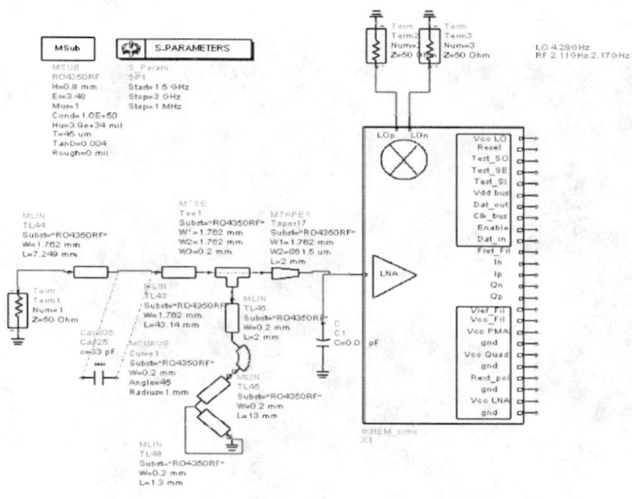

Mother board

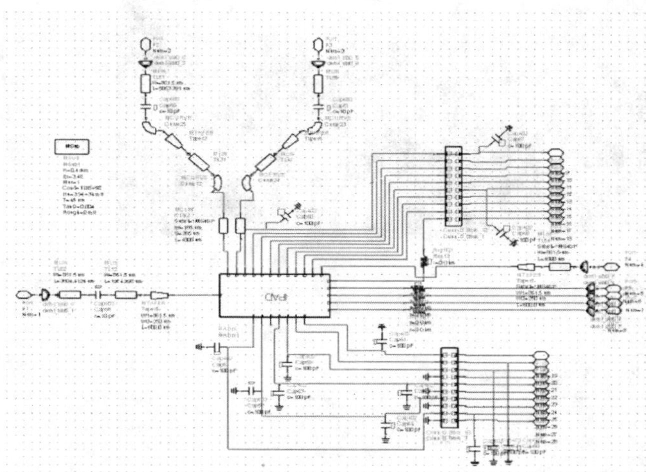

Module

FIGURE 16.15 Board to chip simulation flow.

The I/Q PMA down-converter and LO Quadrature generator were validated together with the SAW filter (measurements have been done at the analog–base-band outputs with 0 dB of VGA gain). The input return loss seen at the SAW filter input is around −6 dB in the middle of the RX band. With 4 nH series inductors between the SAW filter and the mixer, the power gain reaches 16.7 dB, with an in band IIP3 of 4 dBm. The noise factor is around 14.5 dB when the VGA gain is set to 16 dB. On this sample, we report +47 dBm IIP2 for the cascaded mixer-PMA-VGA with 3° and 0.06 dB of Quadrature phase error and gain error, respectively, at the base-band outputs. These measured results are in line with the simulations except for the in band IIP3 that is 3 dB lower than predicted. A summary of measurement results is given in Table 16.4.

The analog VGA low-pass filter exhibits +38 dB maximum gain with 22 nV/Hz input referred noise. The frequency response shows 26 and 84 dB attenuation at 5 and 20 MHz, respectively, with 62 nsec group delay variation (from 1 kHz to 1.92 MHz). At the maximum VGA gain, the linearity reached 7 dBVp IIP3 for 6 mA of current consumption (I and Q).

A summary of the measurement results is given in Table 16.5.

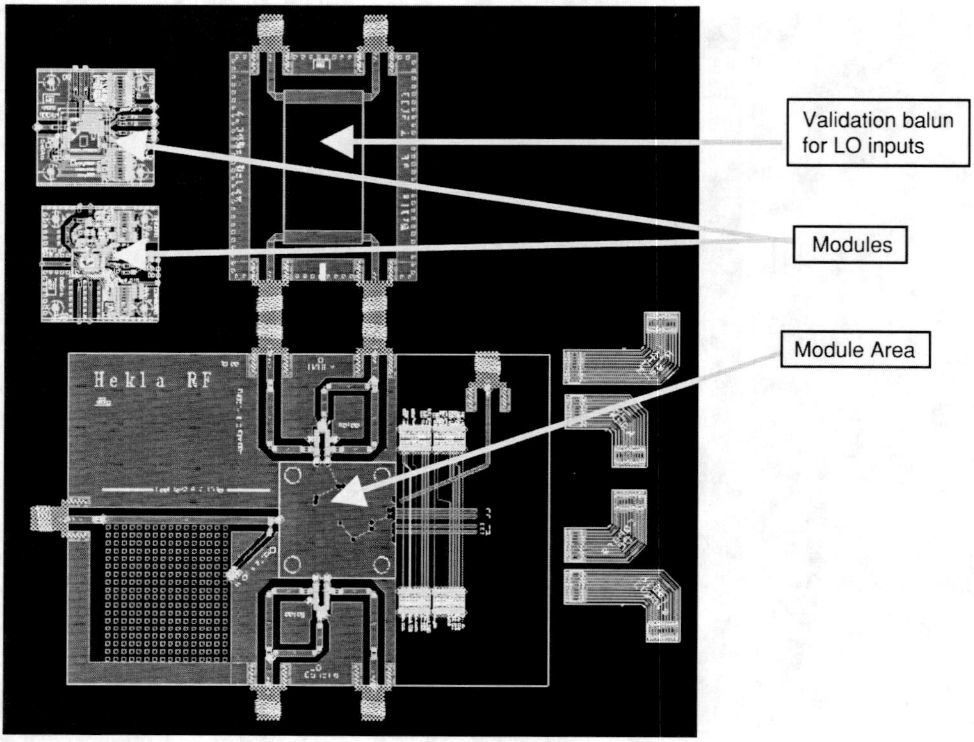

FIGURE 16.16 RF analog board picture.

TABLE 16.3 Measurement Results

	Simulated Performances
Gain (dBm to dBVp)	54.5
Input linear operation (dBm)	−13
IIP3 (dBm)	7
SSB noise figure (dB)	14.8

TABLE 16.4 Measurement Results

	HG	LG
ICC	5.00	0.20
S11	−15	−23
S21	13.2	−18.5
S12	−22	−22
S22	−17	−12
NF	1.85	N.A
IIP3	0.00	N.A
IIP1	−12.5	> −10

HG means high-gain configuration, while LG means low gain, ICC is given in mA, *S* parameters and NF in dB, while IIP3 and IIP2 are given in dBm.

TABLE 16.5 Specifications vs Performance

Worst Case Electrical Performances	AST Specifications	Performances
Current consumption (mA)	6	6.3
Gv (dB)	21	20
PMA cutoff frequency (MHz)	3.8	3.6
Input linear operation (dBVp)	−28	−23
IIP3 (dBVp)	−2	−1.5
SSB noise figure (dB)	13	13.3

16.10 RF and Analog Receiver Validation

The I/Q signals at the complete received outputs were sampled at $F_s = 38.4$ MHz and filtered with a 0.13 μm HCMOS ΣΔ ADC-FIR companion chip [7], offering 55 dB equivalent input dynamic range.

A first level of test was made in single tone configuration in order to define the sensibility and the selectability of the global receiver integrating the two circuits, the RF–Analog circuit developed in this chapter and its companion chip. The main results are given below.

Reference Sensitivity Test

RF In −117 dBm	Minimum Requested	Measured
SNR [dB]	10.5	11.5

Note: TX Power at antenna Connector is 22 dBm instead of 24 dBm.

Maximum Input Level

RF In −25 dBm	Minimum Requested	Measured
SNR [dB]	10.5	>30 dB

Adjacent Channel Selectivity

RF In −103 dBm Blk @ 5 MHz −52 dBm	Minimum Requested	Measured
SNR [dB]	10.5	20

Intermodulation Characteristics

RF In −114 dBm Blk @ 10 MHz −46 dBm Blk @ 20.5 MHz −46 dBm	Minimum Requested	Measured
SNR [dB]	12	13.2
SIR	–	7.8 dB
Input IP3 [dBm]	−17	−9
	PA off (not enough RF generators)	

Blocking Characteristics RFin: −114 dBm—TX: 22 dBm

SNR [dB]	Spec.	Measured	Comment
RFin ± 10 MHz	11	13	
RFin ± 15 MHz	11	13	
F = 2045 MHz −30 dBm	11	12	2xBlk-Tx (in spec despite off out spec LNA IIP3 due to not worst case Duplexer Isolation)
F = 4090 MHz −15 dBm	11	12	Blk-Tx
F = 1760 MHz −15 dBm	11	11	2xTx-Blk
F = 4280 MHz −15 dBm	11	12	2xRx
F = 1070 MHz −15 dBm	11	–	Rx/2 Not done due to generator harmonics
F = 713 MHz −15 dBm	11	–	Rx/3 Not done due to generator harmonics

Spurious Emission

Up to 7GHz ? 12 GHz	Spec.	Measured
268.8 MHz (H7 Fref.)	−57 dBm	−80 dBm
4.284 MHz (H2 LO)	−47 dBm	−64 dBm

These results show that the complete receiver is in the specifications and allow receiving modulated signals, this last step is detailed in the next chapter.

16.11 WCDMA Receiver Global Test

This section presents the performance of the receiver platform including the SiGeC RF–Analog receiver, the VLSI-CMOS ADC-FIR, a software modem running in a PC on a Linux real-time platform. The full performance against specification 25.101 by 3GPP was evidenced.

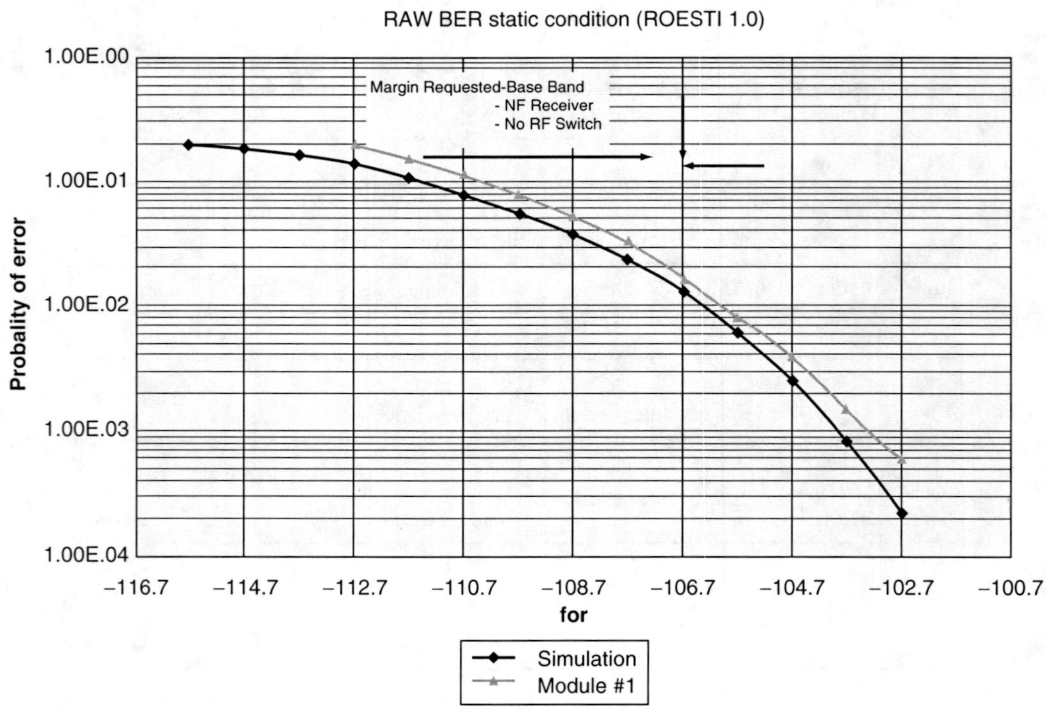

FIGURE 16.17 Simulated bit error rate performance.

In order to illustrate this operation, a RAW BER is given in Figure 16.17. The circuit exhibits a 5 dB margin with the diplexer, without RF switches, which means that the complete receiver should have a 3 dB margin at least.

16.12 Summary

The first BiCMOS SiGe-carbon 0.25 μm/60 GHz f_T process has been validated with a complete WCDMA receive chain including switched gain LNA, I/Q down mixers, LO Quadrature generator and VGA fifth-order Butterworth low-pass filter in 4 mm^2 silicon area including pads. Thanks to the duplexer TX isolation [8] and to the careful estimation of the coupling (between cells inside the IC and between RF lines on the validation board), the receiver meets all the SNR requirements with +22 dBm TX power measured at the antenna connector, even if the measured LNA IIP3 is slightly below than expected.

This receiver in a WCDMA chip-set including a digital–RF interface, ADC plus FIR, and a software modem, meets the full performance against specification 25.101 by 3GPP.

Acknowledgments

The author thanks especially the teams of ST CR&D, which have designed, implemented, founded the circuit and made the electrical validation. The team of ST AST, which was involved in the system definition, the software design, and the application evaluation. The IXL laboratory of the University of Bordeaux has made specific study on the MOS bulk modulation.

References

1. J. Ryynänen et al., A single-chip multimode receiver for GSM900, DCS1800, PCS1900, and WCDMA, *IEEE Journal of Solid-State Circuits*, 38(4), 2003, 594–602.
2. T. Taris et al., A 1 V 2 GHz VLSI CMOZS low noise amplifier, RFIC 2003 Conference Proceeding, p. 123.
3. T. Taris et al., A 0.9 V body effect feedback 2 GHz low noise amplifier, ESSCIRC 2003 Conference Proceeding, p. 659.
4. R. Paulin, Etude des causes de non-linéarité du transistor bipolaire SiGe 0.25 μm, JNRDM, mai 2003.
5. P. Wambacq and W. Sansen, *Distortion Analysis of Analog Integrated Circuits*, Kluwer Academic Publishers, Dordrecht, 1998.
6. K-Y. Lee et al., Full-CMOS 2-GHz WCDMA direct conversion transmitter and receiver, *IEEE Journal of Solid-State Circuits*, 38(1), 2003, 43–53.
7. A. Dezzani and E. Andre, A 1.2-V dual-mode WCDMA/GPRS ΣΔ modulator, IEEE International Solid-State Circuits Conference, Session 3, 2003.
8. S. Reynolds et al., A direct-conversion receiver IC for WCDMA mobile systems, IEEE Bipolar/BiCMOS Circuits and Technology Meeting, 2002, pp. 61–64.

A.1

Properties of Silicon and Germanium

John D. Cressler
Georgia Institute of Technology

The energy band structures of Si and Ge are depicted in Figure A.1.1, together with (1) their carrier effective mass parameters (Table A.1.1) and (2) their bulk structural, mechanical, optical, and electrical properties (Table A.1.2) [1–3].

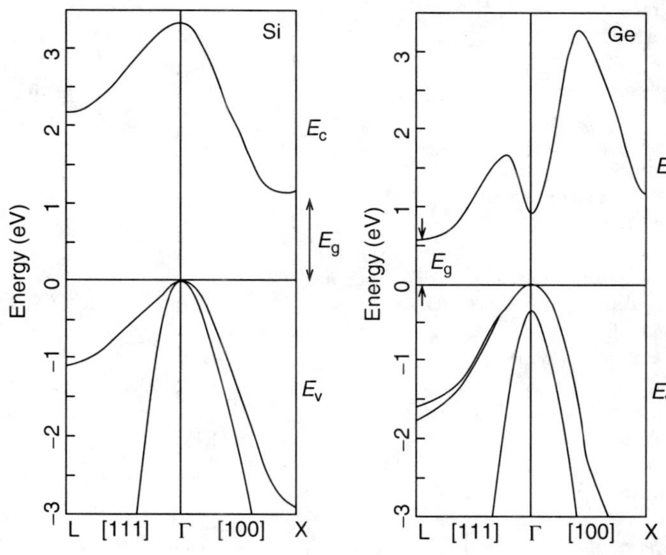

FIGURE A.1.1 Energy band structure, showing the principal conduction and valence bands of Si and Ge as a function of *k*-space direction. (From M Shur. *Physics of Semiconductor Devices.* Englewood Cliffs, NJ: Prentice-Hall, 1990. With permission.)

TABLE A.1.1 Carrier Effective Mass Parameters for Si and Ge

Parameter	Units	Silicon	Germanium
Effective electron mass (m_n^*)	($\times m_o$)		
Longitudinal (4.2 K)		0.9163	1.58
Transverse (4.2 K)		0.1905	0.082
Density-of-states (4.2 K)		1.062	—
Density-of-states (300 K)		1.090	—
Effective hole mass (m_p^*)	($\times m_o$)		
Heavy hole (4.2 K)		0.537	0.28
Light hole (4.2 K)		0.153	0.044
Density-of-states (4.2 K)		0.59	—
Density-of-states (300 K)		1.15	—

TABLE A.1.2 Properties of Bulk Si and Ge

Parameter	Units	Silicon	Germanium
Atomic number	—	14	32
Atomic density	(atoms/cm^3)	5.02×10^{22}	4.42×10^{22}
Atomic weight	(g/mole)	28.09	72.6
Density	(g/cm^3)	2.329	5.323
Electronic orbital configuration	—	(Ne) $3s^2 3p^2$	(Ar) $3d^{10} 4s^2 4p^2$
Crystal structure	—	Diamond	Diamond
Lattice constant (298 K)	(Å)	5.43107	5.65791
Dielectric constant	—	11.7	16.2
Breakdown strength	(V/cm)	3×10^5	1×10^5
Electron affinity	(V)	4.05	4.00
Specific heat	(J/g-°C)	0.7	0.31
Melting point	(°C)	1412	1240
Intrinsic Debye length (300 K)	(μm)	24	0.68
Index of refraction	—	3.42	3.98
Transparency region	(μm)	1.1–6.5	1.8–15
Thermal conductivity (300 K)	(W/cm-°C)	1.31	0.60
Thermal expansion coefficient (300 K)	(°C^{-1})	2.6×10^{-6}	5.9×10^{-6}
Young's modulus	(dyne/cm^2)	1.9×10^{12}	—
Energy bandgap (low doping)	(eV)	1.12 (300 K)	0.664 (291 K)
		1.17 (77 K)	0.741 (4.2 K)
Equivalent conduction band minima	—	6	8
Effective electron mass (300 K)	($\times m_o$)	1.18	—
Effective hole mass (300 K)	($\times m_o$)	0.81	—
Intrinsic carrier density (300 K)	(cm^{-3})	1.02×10^{10}	2.33×10^{13}
Effective conduction band DoS (300 K)	(cm^{-3})	2.8×10^{19}	1.04×10^{19}
Effective valence band DoS (300 K)	(cm^{-3})	1.04×10^{19}	6.00×10^{18}
Electron mobility (300 K)	(cm^2/V-sec)	1450	3900
Hole mobility (300 K)	(cm^2/V-sec)	500	1900
Electron diffusivity (300 K)	(cm^2/sec)	37.5	100
Hole diffusivity (300 K)	(cm^2/sec)	13	49
Optical phonon energy	(meV)	63	37
Phonon mean free path length	Å	76	105
Intrinsic resistivity (300 K)	(Ω cm)	3.16×10^5	47.62

References

1. JD Cressler and G Niu. *Silicon–Germanium Heterojunction Bipolar Transistors.* Boston, MA: Artech House, 2003.
2. M Shur. *Physics of Semiconductor Devices.* Englewood Cliffs, NJ: Prentice-Hall, 1990.
3. R Hull, editor. *Properties of Crystalline Silicon.* London: EMIS Datareviews Series, Number 20, INPSEC, 1999.

A.2
The Generalized Moll–Ross Relations

John D. Cressler
Georgia Institute of Technology

The classical solution for the collector current density in a Si BJT, derived by Shockley, necessarily assumes a constant base doping profile. In this case, for low-injection conditions, the drift component of the minority carrier transport equation can be neglected, and the minority carrier diffusion equation solved under the Shockley boundary conditions. The resultant equation, under the assumptions of negligible neutral base recombination and forward-active bias, is the well-known expression

$$J_C = \frac{qD_{nb}}{N_{ab}^{-} W_b} n_{io}^2 e^{\Delta E_{gb}^{app}/kT} \left(e^{qV_{BE}/kT} - 1 \right). \tag{A.2.1}$$

Here, D_{nb} is the minority electron diffusivity, N_{ab}^{-} is the ionized base doping level, n_{io}^2 is the low-doping intrinsic carrier density, given by,

$$n_{io}^2 = N_C N_V e^{-E_{go}/kT}, \tag{A.2.2}$$

and $\Delta E_{gb}{}^{app}$ is the heavy-doping induced bandgap narrowing.

The path to the generalization of this result to the "real-world" case of a nonconstant base doping profile (Figure A.2.1) is nonobvious, and even a cursory glance at the problem is enough to convince one that it cannot follow the original path in Shockley's approach. The complexity of this problem results from the addition of the field-driven transport, which is now no longer negligible due to the doping-gradient-induced field. The clever solution to this problem was first presented in the classic paper by Moll and Ross in 1956, the so-called "Moll–Ross relation" [1]. Unfortunately, that solution made two undesirable assumptions: (1) that the minority electron mobility (hence, diffusivity) is constant across the quasineutral base and (2) that the intrinsic carrier density is constant across the quasineutral base. The latter assumption, in particular, fails in the presence of a heavily doped base (i.e., real life), since the apparent bandgap narrowing is inherently position dependent across the base, and hence the effective bandgap in the base is also position dependent.* In essence, then, the problem becomes one of solving for the collector current density in the presence of *both* nonconstant base doping and nonconstant base bandgap, and is particularly relevant to the graded-base SiGe HBT. This problem remained unsolved for

*In fairness, Moll and Ross cannot be blamed for the second assumption since doping-induced bandgap narrowing had not yet been discovered.

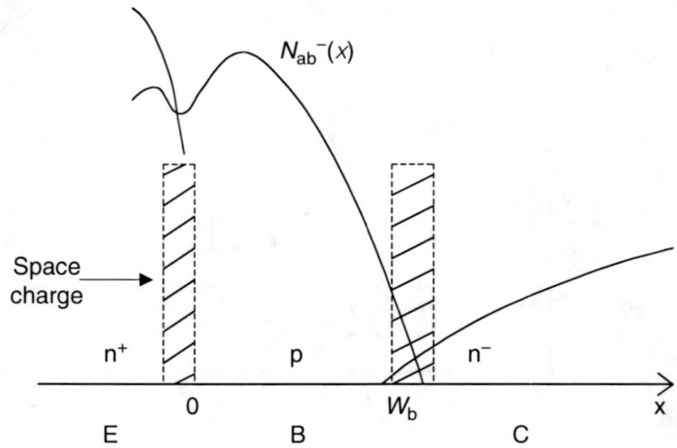

FIGURE A.2.1 Schematic nonconstant base doping profile used in the derivations.

almost 30 years until the seminal paper by Kroemer in 1985 [2]. Since Kroemer's "generalized Moll–Ross relations" are the starting point for both the dc and ac analysis of the graded-base SiGe HBT (Chapter 4, see *Silicon Heterostructure Devices*), we present that elegant derivation here (showing all of the mathematical steps that Kroemer neglected to include in his paper).

The assumptions in Kroemer's solution include: (1) 1-D transport, (2) transport by both drift and diffusion, (3) low-injection conditions (i.e., $n_b(x) \ll N_{ab}^-(x)$ for all x across the base), (4) negligible neutral base recombination, and (5) forward-active bias. Importantly, however, there are no assumptions on the position dependence of the base doping profile or the base bandgap.*

We begin from the generalized drift–diffusion minority electron transport equation, as expressed in terms of the minority electron quasi-Fermi potential

$$J_n = q\mu_n n \nabla \phi_n, \tag{A.2.3}$$

which for our 1-D Si BJT problem reduces to

$$J_C = q\mu_{nb}(x) n_b(x) \frac{d\phi_n(x)}{dx}. \tag{A.2.4}$$

In the quasineutral base, the majority carrier (hole) quasi-Fermi potential (ϕ_p) in low-injection is constant, such that

$$J_C = q\mu_{nb}(x) n_b(x) \frac{d}{dx}\left(\phi_n(x) - \phi_p\right), \tag{A.2.5}$$

and from the generalized Shockley boundary condition

$$n_b(x)p_b(x) = n_{ib}^2(x) e^{q(\phi_n(x)-\phi_p)/kT}, \tag{A.2.6}$$

which can be rewritten as

$$\frac{kT}{q} \ln\left\{\frac{n_b(x)p_b(x)}{n_{ib}^2(x)}\right\} = \phi_n(x) - \phi_p. \tag{A.2.7}$$

*Interestingly, additional generalizations to Kroemer's result have been recently offered [3]. Let it never be said that the final word in device physics is ever in.

Taking the derivative of both sides we have

$$\frac{kT}{q}\left\{\frac{n_{\mathrm{ib}}^2(x)}{n_{\mathrm{b}}(x)p_{\mathrm{b}}(x)}\right\}\frac{\mathrm{d}}{\mathrm{d}x}\left\{\frac{n_{\mathrm{b}}(x)p_{\mathrm{b}}(x)}{n_{ib}^2(x)}\right\}=\frac{\mathrm{d}}{\mathrm{d}x}\left(\phi_{\mathrm{n}}(x)-\phi_{\mathrm{p}}\right). \tag{A.2.8}$$

Substituting this result back into Equation (A.2.5), we obtain

$$J_{\mathrm{C}}=q\mu_{\mathrm{nb}}(x)n_{\mathrm{b}}(x)\frac{kT}{q}\frac{n_{\mathrm{ib}}^2(x)}{n_{\mathrm{b}}(x)p_{\mathrm{b}}(x)}\frac{\mathrm{d}}{\mathrm{d}x}\left\{\frac{n_{\mathrm{b}}(x)p_{\mathrm{b}}(x)}{n_{\mathrm{ib}}^2(x)}\right\} \tag{A.2.9}$$

We now integrate this expression from some arbitrary point in the base profile to the neutral base boundary (W_{b}) to obtain

$$\int_x^{W_{\mathrm{b}}}\frac{J_{\mathrm{C}}}{qD_{\mathrm{nb}}(x')}\frac{p_{\mathrm{b}}(x')}{n_{\mathrm{ib}}^2(x')}\mathrm{d}x'=\left.\frac{n_{\mathrm{b}}(x')p_{\mathrm{b}}(x')}{n_{\mathrm{ib}}^2(x')}\right|_x^{W_{\mathrm{b}}} \tag{A.2.10}$$

Under the assumptions of negligible neutral base recombination (i.e., J_{C} is a constant to the integration), and using the fact that in forward-active bias,

$$p_{\mathrm{b}}(W_{\mathrm{b}})n_{\mathrm{b}}(W_{\mathrm{b}})\simeq n_{\mathrm{ib}}^2(W_{\mathrm{b}}) \tag{A.2.11}$$

we find

$$\frac{J_{\mathrm{C}}}{q}\int_x^{W_{\mathrm{b}}}\frac{p_{\mathrm{b}}(x')\mathrm{d}x'}{D_{\mathrm{nb}}(x')n_{\mathrm{ib}}^2(x')}=1-\frac{n_{\mathrm{b}}(x)p_{\mathrm{b}}(x)}{n_{\mathrm{ib}}^2(x)} \tag{A.2.12}$$

At the emitter–base boundary ($x=0$), we know from the generalized Shockley boundary condition that

$$n_{\mathrm{b}}(0)p_{\mathrm{b}}(0)=n_{\mathrm{ib}}^2(0)\mathrm{e}^{q(\phi_{\mathrm{n}}(0)-\phi_{\mathrm{p}}(0))/kT}, \tag{A.2.13}$$

and

$$\phi_{\mathrm{n}}(0)-\phi_{\mathrm{p}}(0)=V_{\mathrm{BE}}, \tag{A.2.14}$$

so that we obtain

$$\frac{J_{\mathrm{C}}}{q}\int_0^{W_{\mathrm{b}}}\frac{p_{\mathrm{b}}(x)\mathrm{d}x}{D_{\mathrm{nb}}(x)n_{\mathrm{ib}}^2(x)}=1-e^{qV_{\mathrm{BE}}/kT}, \tag{A.2.15}$$

and thus finally,

$$J_{\mathrm{C}}=-\frac{q(e^{qV_{\mathrm{BE}}/kT}-1)}{\int_0^{W_{\mathrm{b}}}\frac{p_{\mathrm{b}}(x)\mathrm{d}x}{D_{\mathrm{nb}}(x)n_{\mathrm{ib}}^2(x)}}, \tag{A.2.16}$$

This is the "generalized Moll–Ross relation"* for the collector current density in a bipolar transistor with nonconstant base doping and arbitrary position-dependence of the base bandgap. Observe that if

*I personally would have no problem calling this elegant result the "Kroemer relation."

we allow $p_b(x) = N_{ab}^-(x) = N_{ab}^- = \text{constant}$, then we obtain Equation A.2.1, as expected (the extra negative sign simply accounts for the fact that the electron flow is in the opposite direction of the positive current flow).

As detailed in Chapter 4 (see *Silicon Heterostructure Devices*), this fundamental result is the starting point of the derivations for collector current density, the current gain, and the output conductance in a graded-base SiGe HBT. In this case, in addition to the bandgap-narrowing-induced position dependence in the base bandgap, we have an additional contribution from the Ge-strained layer (Figure A.2.2). This Ge contribution easily enters the generalized Moll–Ross relation via n_{ib}^2 in Equation A.2.16. For more detail on the resultant derivations and the assumptions and approximations involved, the reader is referred to Ref. [4].

An additional desirable feature of Kroemer's approach is that we can also easily obtain an analytical expression for the base transit time in a device with nonconstant base doping and bandgap. Under a quasistatic assumption we can generally define the base transit time as

$$\tau_b = \frac{-q}{J_C} \int_0^{W_b} n_b(x)\mathrm{d}x. \tag{A.2.17}$$

From Equation A.2.12 and neglecting the unity factor, we can solve for $n_b(x)$ as

$$n_b(x) = \frac{-J_C}{q} \frac{n_{ib}^2(x)}{p_b(x)} \int_x^{W_b} \frac{p_b(x')\mathrm{d}x'}{D_{nb}(x')n_{ib}^2(x')} \tag{A.2.18}$$

Substituting this result into Equation A.2.17, we finally obtain

$$\tau_b = \int_0^{W_b} \frac{n_{ib}^2(x)}{p_b(x)} \left\{ \int_x^{W_b} \frac{p_b(y)\mathrm{d}y}{D_{nb}(y)n_{ib}^2(y)} \right\} \mathrm{d}x \tag{A.2.19}$$

Again, observe that if we allow $p_b(x) = N_{ab}^-(x) = N_{ab}^- = \text{constant}$, we obtain the classical result for a BJT with constant base doping

$$\tau_b = \frac{W_b^2}{2D_{nb}}, \tag{A.2.20}$$

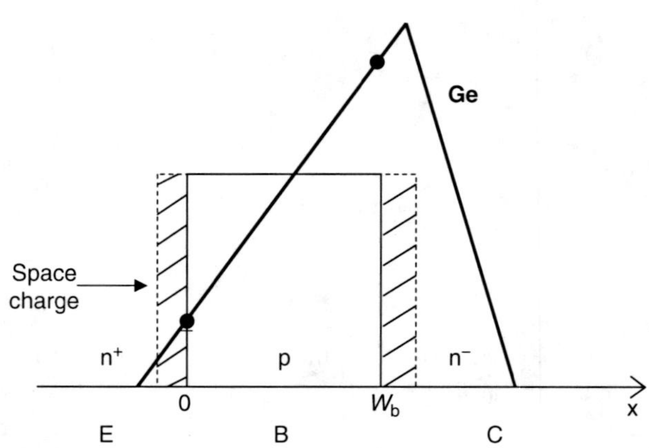

FIGURE A.2.2 Schematic drawing of a position-dependent, graded Ge base profile in a SiGe HBT.

as expected. This second generalized Moll–Ross relation is the starting point for the derivation of base transit in a graded-base SiGe HBT (Figure A.2.2), as detailed in Chapter 4 (see *Silicon Heterostructure Devices*) and Ref. [4].

References

1. J. Moll and I. Ross. The dependence of transistor parameters on the distribution of base layer resistivity. *Proceedings of the Institute of Radio Engineers* 44:72, 1956.
2. H Kroemer. Two integral relations pertaining to electron transport through a bipolar transistor with a nonuniform energy gap in the base region. *Solid-State Electronics* 28:1101–1103, 1985.
3. SN Mohammad. Generalization of the Moll–Ross relations for heterojunction bipolar transistors. *Solid-State Electronics* 46:589–591, 2002.
4. JD Cressler and G Niu. *Silicon–Germanium Heterojunction Bipolar Transistors.* Boston, MA: Artech House, 2003.

A.3

Integral Charge-Control Relations

Michael Schröter
University of California at San Diego

A.3.1 Introduction **A.3**-1
A.3.2 Derivation of a General Relationship **A.3**-2
A.3.3 Homojunction Transistors **A.3**-6
A.3.4 Heterojunction Transistors **A.3**-9
Box Ge Profile in the (Metallurgical) Base Region • Trapezoidal Profile in the (Metallurgical) Base Region
A.3.5 Further Extensions **A.3**-12
A.3.6 Summary **A.3**-13

A.3.1 Introduction

One of the most important requirements for compact models is an accurate description of the devices' main current. In an npn bipolar transistor, this is the (generally time dependent) collector current i_C which is given by the transport of electrons from emitter to collector. For time (and frequency) dependent *quasistatic* (q.s.) operation, which is the case in the vast majority of practical applications, the current flowing at the collector terminal can be partitioned into a q.s. transfer current i_T and a charging current supplying all charge storage elements connected to the collector terminal. Since the latter elements are described and represented separately in a model, a formulation of i_T, which equals I_C under d.c. conditions, is of major interest. Before a general relation for i_T is derived, a brief historical perspective of the respective theory is given below.

The obvious starting point for such an equation is the transport and continuity equation. Since the time derivative term in the latter is taken into account separately by the charge storage elements, and the impact of recombination on i_T in modern bipolar transistors is negligibly small, a spatially constant electron current density j_{nx} results (and is observed in device simulation) throughout the structure of a one-dimensional (1D) transistor. This fact can be used to solve the transport equation first for the carrier density and associated charge at a *given transfer current*, and then reformulate the result to obtain the transfer current at a *given charge*. The first solution of this kind was published by Moll and Ross in 1956 [1, Equation (13)],

$$i_T = q\mu_{nB} n_{iB}^2 V_T \frac{\exp(V_{B'E'}/V_T)}{\int_{x_l}^{x_u} h(x)p(x)\mathrm{d}x} \tag{A.3.1}$$

with $p(x) = N_B(x)$, $h(x) = 1$, $V_{B'E'}$ as internal base–emitter voltage, and $[x_l, x_u]$ as base region. The relation provided a great deal of insight into the dependence of i_T (and also the transit time) on doping profile and material parameters in the base region. An improved and more complete expression was later used for one of the first TCAD papers by Ghosh et al. in 1967 [2]. Here, the material dependent function $h(x) = (\mu_{nB}n_{iB}^2) / (\mu_n n_i^2)$ was included for the first time, but the integral was evaluated numerically, thus not providing a formulation suitable for compact modeling. Only shortly thereafter, in 1970 Gummel published his well-known paper about the integral charge-control relation (ICCR) [3]. For the first time, a compact formulation for i_T was derived, that contained in a consistent relation between i_T and stored charge and with all bias-dependent nonideal effects that were relevant for transistors at that time. In one way or the other, the ICCR has been used until today in any reasonably physics-based compact model. Note that the Scharfetter–Gummel discretization, which was first published and applied to a bipolar transistor, also assumes a constant current density between discrete points, which is the main reason for its numerical stability and wide use for device simulation.

An extension of the Moll–Ross relation with special emphasis on a *spatially variable bandgap* in the base layer was derived by Kroemer in 1985 [4]. Here also the doping density was replaced by the bias-dependent hole density as in Equation A.3.1, but no equation suitable for compact modeling was given. In 1993, for both the latter equation and the ICCR a generalization was presented in Ref. [5], with application to compact modeling. The respective generalized ICCR (GICCR) will be discussed in more detail in subsequent sections. During its derivation, the various results mentioned above will be obtained and be referred to again.

Since the theories above all apply only to a 1D transistor structure with reasonably smooth material and bandgap changes, and to quasistatic operation, extensions for practical cases are of interest, which will be briefly discussed in A.3.5. The considerations in this chapter apply to Si and SiGe technologies. In the latter, two fundamentally different types of doping profiles can be distinguished which are presently being manufactured and illustrated schematically in Figure A.3.1. The conventional emitter doping (CED) type has a similar profile shape as BJTs, but contains in addition a Ge distribution that increases from the BE junction to the BC junction. In contrast, the low-emitter concentration (LEC) type contains a box Ge profile (or at least a sufficiently high step at the BE junction) that allows to significantly increase the base doping and to lower the emitter doping.

A.3.2 Derivation of a General Relationship

Although the GICCR can be derived for the 3D case, the considerations here will be restricted to the 1D case in order to demonstrate the concept. The assumptions required for the derivation are summarized below:

(a) A one-dimensional transistor structure (cf. Figure A.3.1) is considered, with the emitter contact ($x = 0$) at the mono-silicon surface and the collector contact ($x = x_C$) at the transition from the lightly-doped collector region to the buried layer. This 1D structure does not include the emitter and *external* collector series resistance.

(b) Volume recombination within the above 1D transistor region is negligible.

(c) The time derivative is negligible, which corresponds to quasistatic operation.

(d) Effects such as thermionic emission and tunneling across the junctions are neglected. They can be accounted for though by separate terms and can be combined with the GICCR solution.

Assumptions (a), (b), and (c) together with the electron continuity equation lead to a spatially independent electron current density,

$$J_n = \text{const}(x) = -J_T = \frac{-I_T}{A_E}, \qquad \text{(A.3.2)}$$

which can be expressed by the transfer current and area (emitter) A_E of the 1D-transistor. The derivation starts with the electron transport equation,

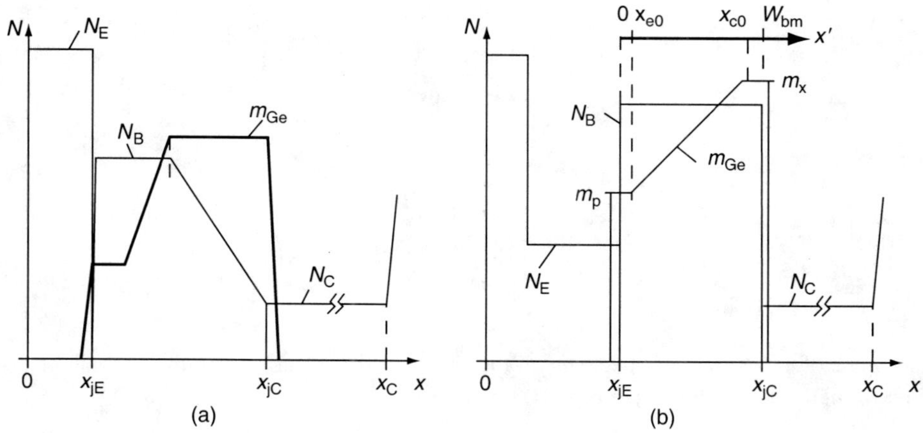

FIGURE A.3.1 Sketch of doping and Ge profiles for (a) CED transistors and (b) LEC transistors.

$$J_n = -q\mu_n n \frac{d\varphi_n}{dx} \tag{A.3.3}$$

where μ_n, n and φ_n are the electron mobility, density, and quasifermi potential. Inserting $n = n_i \exp[(\psi - \varphi_n)/V_T]$ into (A.3.3) gives

$$J_n = -q\mu_n n_i \exp\left(\frac{\psi}{V_T}\right) \exp\left(-\frac{\varphi_n}{V_T}\right) \frac{d\phi_n}{dx}. \tag{A.3.4}$$

Note that the effective intrinsic carrier density n_i accounts for bandgap differences caused by high-doping effects and bandgap-engineering. This topic will be discussed later. Using the transformation

$$\frac{d\exp(-\varphi_n/V_T)}{dx} = -\frac{1}{V_T} \exp\left(-\frac{\varphi_n}{V_T}\right) \frac{d\varphi_n}{dx} \tag{A.3.5}$$

leads to

$$J_n = qV_T\mu_n n_i \exp\left(\frac{\psi}{V_T}\right) \frac{d\exp(-\varphi_n/V_T)}{dx}. \tag{A.3.6}$$

Separation of the differential variables and rearranging terms gives

$$\frac{J_n}{qV_T\mu_n n_i} \exp\left(-\frac{\psi}{V_T}\right) dx = d\left(\exp\left(-\frac{\varphi_n}{V_T}\right)\right). \tag{A.3.7}$$

Extension of the l.h.s. by the product $n_i \exp(\varphi_p/V_T)$ allows to replace the inconvenient term $\exp(-\psi/V_T)$ by the hole density, yielding

$$\frac{J_n}{qV_T} \frac{p}{\mu_n n_i^2 \exp(\varphi_p/V_T)} dx = d\left(\exp\left(-\frac{\varphi_n}{V_T}\right)\right). \tag{A.3.8}$$

Integration of the above equation over the general spatial interval $[x_l, x_u]$ gives

$$\int_{x_l}^{x_u} \frac{J_n}{qV_T} \frac{\exp(-\varphi_p/V_T)}{\mu_n n_i^2} p \, dx = \int_{\exp(\varphi_n(x_l)/V_T)}^{\exp(\varphi_n(x_u)/V_T)} d\left(\exp\left(-\frac{\varphi_n}{V_T}\right)\right). \tag{A.3.9}$$

The result for the r.h.s. is

$$\int_{\exp(\varphi_n(x_l)/V_T)}^{\exp(\varphi_n(x_u)/V_T)} d\left(\exp\left(-\frac{\varphi_n}{V_T}\right)\right) = \left[\exp\left(-\frac{\varphi_n(x_u)}{V_T}\right) - \exp\left(-\frac{\varphi_n(x_l)}{V_T}\right)\right]. \tag{A.3.10}$$

The exact value of φ_n on the r.h.s. depends on the choice of the integration limits and will be discussed later. First though, the left-hand-side of (A.3.9) is integrated,

$$\int_{x_l}^{x_u} \frac{J_n}{qV_T} \frac{\exp(-\varphi_p/V_T)}{\mu_n n_i^2} p \, dx = \frac{-J_T \exp(-V_{B'E'}/V_T)}{qV_T \mu_{nr} n_{ir}^2} \int_{x_l}^{x_u} h(x) p(x) dx, \tag{A.3.11}$$

with $h = h_g h_i h_\varphi$ as the *normalized* weighting function of the hole density, and its components

$$h_g = \frac{\mu_{nr} n_{ir}^2}{\mu_n(x) n_i^2(x)}, \quad h_i = \frac{-J_n(x)}{J_T}, \quad h_\varphi = \exp\left(\frac{V_{B'E'} - \varphi_p(x)}{V_T}\right). \tag{A.3.12}$$

μ_{nr} and n_{ir} are the mobility and intrinsic carrier density, respectively, of a reference material.

Equating (A.3.10) with (A.3.11), and solving for the desired transfer current yields the "master" equation

$$J_T = qV_T^2 \mu_{nr} n_{ir}^2 \frac{\exp\left(\frac{V_{B'E'}}{V_T}\right)\left[\exp\left(-\frac{\varphi_n(x_l)}{V_T}\right) - \exp\left(-\frac{\varphi_n(x_u)}{V_T}\right)\right]}{\int_{x_l}^{x_u} h_g h_i h_\varphi \, p dx}, \tag{A.3.13}$$

from which different analytical formulations can be derived. The choice of the reference material (or transistor region) and its associated values for $\mu_{nr} n_{ir}^2$ is arbitrary and will be discussed later.

The integration interval is undefined yet, and so is the impact of the various weighting functions on the integral. An attractive choice for the integration limits is $[0, x_C]$, which corresponds to the entire 1D transistor region. As a result of this choice, the electron quasifermi potentials in the numerator assume their known internal terminal values (e.g., for common-emitter configuration),

$$\varphi_n(x_l) = \varphi_n(0) = 0 \quad \text{and} \quad \varphi_n(x_u) = \varphi_n(x_C) = V_{C'E'}, \tag{A.3.14}$$

with $V_{C'E'}$ as (internal) collector–emitter voltage.

The denominator integral is more difficult to oversee. Thus, Figure A.3.2 shows the spatial dependence of the weighting functions with the hole density superimposed. Since the hole quasifermi potential equals $V_{B'E'}$ over the entire base and adjacent space–charge regions, h_φ equals 1 where $p(x)$ is significant. Similarly, the electron current density is constant and equals J_T even in a larger interval, except in a small region close to the emitter contact, where it increases slightly due to the back injection of holes and the corresponding recombination; the maximum deviation can be $1/B_f$. Therefore, without introducing a significant error, it is possible for both BJT and HBT to assume an average value for the functions h_i and h_φ that is very close to 1. According to Figure A.3.2, the main spatial dependence of the weighting function h is caused by h_g via the bandgap variation of n_i. A smaller contribution to the spatial dependence comes from the doping and field dependence of the mobility.

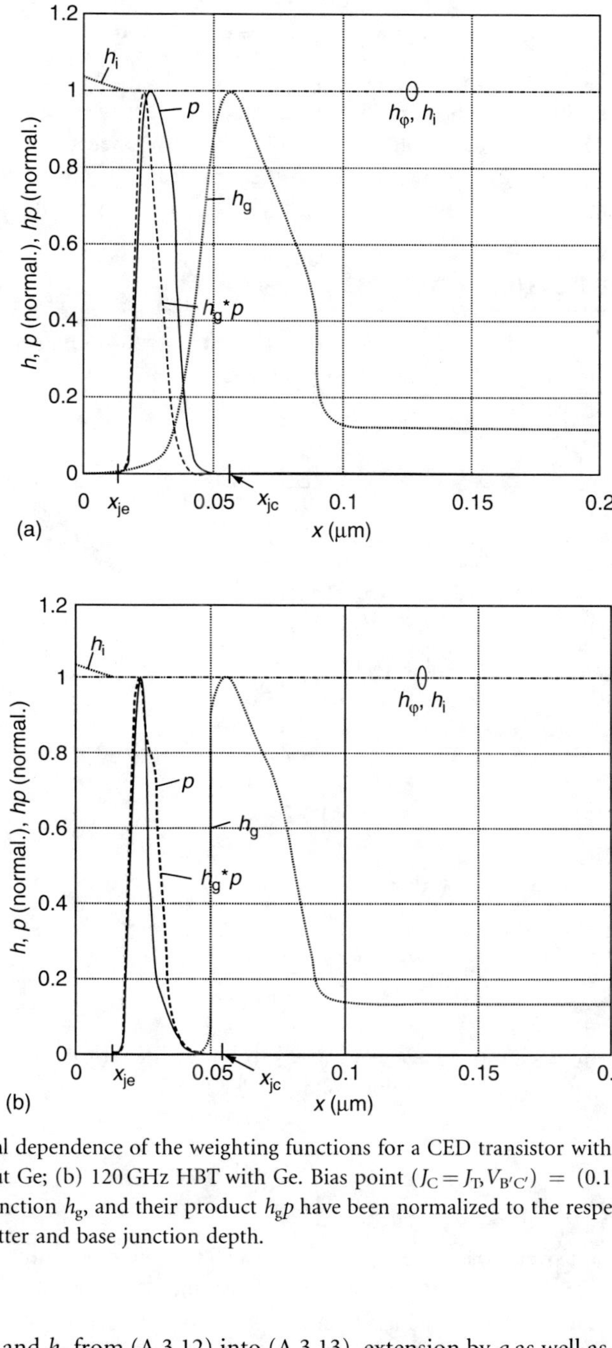

FIGURE A.3.2 Spatial dependence of the weighting functions for a CED transistor with the same doping profile: (a) 70 GHz BJT without Ge; (b) 120 GHz HBT with Ge. Bias point ($J_C = J_T, V_{B'C'}$) = (0.1 mA/μm^2, 0 V). The hole density p, weighting function h_g, and their product $h_g p$ have been normalized to the respective maximum value. x_{je} and x_{jc} denote the emitter and base junction depth.

Inserting (A.3.14) and h_g from (A.3.12) into (A.3.13), extension by q as well as using average values $\bar{h}_i$ and $\bar{h}_\varphi$ yields the basic formulation

$$J_T = q^2 V_T \frac{\mu_{nr} n_{ir}^2}{\bar{h}_i \bar{h}_\varphi} \frac{\exp(V_{B'E'}/V_T) - \exp(V_{B'C'}/V_T)}{q \int_0^{x_C} h_g p \mathrm{d}x}. \tag{A.3.15}$$

from which practically relevant equations can be derived as shown later. The denominator is not yet suitable for (compact) modeling, but at this point can be prepared for further evaluation by partitioning it into a bias-independent and a bias-dependent portion

$$q\int_0^{x_C} h_g p \mathrm{d}x = q\int_0^{x_C} h_g p_0 \mathrm{d}x + q\int_0^{x_C} h_g \Delta p \mathrm{d}x, \tag{A.3.16}$$

where p_0 is the hole density in equilibrium and Δp is the hole density *change* in the transistor with respect to equilibrium. Δp consists of depletion and minority components. While the latter density is distributed over the whole base region, the former densities are located around the junctions and related to the (ionized) base doping.

A.3.3 Homojunction Transistors

In homojunction transistors, the variation of h_g is only caused by high-doping effects. Thus, h_g is fairly constant in the region around the peak hole density (cf. Figure A.3.2a), which contributes most to the integral. Defining an average value

$$\bar{h}_g = \frac{\mu_{nr} n_{ir}^2}{\mu_n n_i^2}, \tag{A.3.17}$$

the denominator of (A.3.15) becomes

$$q\int_{x_E}^{x_C} \frac{\mu_{nr} n_{ir}^2}{\mu_n n_i^2} p(x)\mathrm{d}x = \bar{h}_g \bar{Q}_p \tag{A.3.18}$$

with the hole charge *per area* stored in the 1D transistor structure,

$$\bar{Q}_p = q\int_0^{x_C} p(x)\mathrm{d}x. \tag{A.3.19}$$

Inserting (A.3.18) and (A.3.19) into (A.3.15) yields

$$J_T = c\frac{\exp(V_{B'E'}/V_T) - \exp(V_{B'C'}/V_T)}{\bar{Q}_p} \tag{A.3.20}$$

which has the same form as Gummel's ICCR in Ref. [3], but a different definition of the integration region and, thus, of the charge and controlling voltages. The factor

$$c = q^2 V_T \frac{\mu_{nr} n_{ir}^2}{\bar{h}_i \bar{h}_\varphi \bar{h}_g} \cong q^2 V_T \overline{\mu_n n_i^2} \tag{A.3.21}$$

is assumed to be constant over bias. Usually, the most right term is used only, which is obtained by setting $\bar{h}_i = \bar{h}_\varphi = 1$ and is an excellent assumption for the 1D case. Since the hole density is concentrated mostly in the base, the value for $\bar{h}_g$ and $\overline{\mu_n n_i^2}$ is close to that for the base material.

The charge in (A.3.19) can be divided into a bias-independent and a bias-dependent component, $\overline{Q}_p = \overline{Q}_{p0} + \Delta\overline{Q}_p$. The bias-independent charge $\overline{Q}_{p0}$ is defined at $V_{B'E'} = V_{B'C'} = 0$ and consists of holes stored mostly in the neutral base region $0 \leq x' \leq w_{B0}$ at equilibrium:

$$\bar{Q}_{p0} = q\int_0^{x_C} p_0(x)\mathrm{d}x \cong q\int_0^{w_{B0}} p_0(x')\mathrm{d}x' \approx q\int_0^{w_{B0}} N_B(x')\mathrm{d}x'. \tag{A.3.22}$$

(Note the use of a different coordinate x' in the neutral base region, cf. Figure A.3.1b.) The bias-dependent charge component represents all holes, Δp, that for a nonequilibrium bias condition enter the base region (i) to charge the depletion regions, (ii) to compensate the electrons injected into the (neutral) base, and (iii) for the injection into emitter and collector,

$$\Delta\bar{Q}_p = q\int_0^{x_C} \Delta p(x)\mathrm{d}x = \bar{Q}_{pE} + \bar{Q}_{jE} + \bar{Q}_{nB} + \bar{Q}_{jC} + \bar{Q}_{pC} \qquad \text{(A.3.23)}$$

Here, $\overline{Q}_{jE}$ and $\overline{Q}_{jC}$ are the BE and BC depletion charge (per unit area), respectively, and the other components are the minority charges (per unit area) stored in E, B, and C. As a consequence of the choice of the chosen integration limits, $\Delta\overline{Q}_p$ can actually be measured via the terminals. From a physics-based point of view, the biggest advantage of the extended ICCR (A.3.20) is its firm link between the quasistatic transfer current and the *total* hole charge, which resembles the charge-controlled operation principle of a bipolar transistor in a very compact form.

Examples for the accuracy of various solutions are shown in Figure A.3.3 for a typical Gummel plot. In all cases, the denominator integral (e.g. $\overline{Q}_p$ in (A.3.20)) has been taken directly from device simulation. As expected, the basic Equation A.3.15 follows exactly the simulated behavior of J_T (crosses). For the older generation profile ("25 GHz transistor" in Figure A.3.3a), the practically interesting approximation (A.3.20) is very accurate at low and medium current densities, and shows only slight deviations at high current densities, which are tolerable though and typical for homojunction bipolar processes (e.g. [6,7]). The reason for these small deviations is caused by the assumption of a bias-independent value for $\overline{h}_g$. However, if the minority charge contributions of emitter and collector, $\overline{Q}_{pE}$ and $\overline{Q}_{pC}$, are neglected in (A.3.20), the model current becomes far too large. Thus, including only the base charge is not a feasible approach.

For the profile of a modern generation Si(Ge) transistor ("70 GHz BJT" in Figure A.3.3b), Equation A.3.20 starts to deviate already at medium current densities, affecting the accuracy of the transconductance modeling. This is caused by the increased difference in doping and associated variation of h_g across the base region, which requires a different weighting factor for Q_{p0} and Q_{jE} along the same lines as discussed in the next section on HBTs, applying the corresponding Equation A.3.28 gives again a much higher accuracy.

The assumption of a bias-independent value for $\overline{h}_g$ deserves some consideration. High-doping effects cause both n_i and μ_n (which is also field dependent) to depend on x within the base region, with the change of n_i partially being compensated by the change of μ_n. Obviously though, as long as the spatial distributions of p and $\mu_n n_i^2$ do not change with bias, $\overline{h}_g$ remains bias independent. This holds quite well for low and medium current densities where p is concentrated in the base region, and the mobility depends only very slightly on bias. Toward high current densities, however, p spreads out into the collector and partially also into the emitter region. In the collector, for instance, high-doping effects can be neglected but the field (and thus bias) dependence of μ_n is significant, leading to a different weighting of the hole density in this region. As a consequence, the average value $\overline{h}_g$ defined at low current densities is not quite correct anymore at high current densities and causes the observed small deviation from the exact current. Note though that the field dependence of the mobility around the BC junction decreases at high current densities.

At this point, the choice of the integration limits shall be briefly discussed. Fundamentally, there is an infinite number of possible integration intervals. Besides the one selected above (i.e., entire 1D transistor), the other most important choice in literature (e.g., Refs. [3,4]) is the base region, i.e., $x \in [x_{jE},x_{jC}]$. In this case, $\overline{h}_g$ is only a very weak function of bias, but there are serious disadvantages of this choice: (i) the unknown values for the electron quasi-fermi potentials at the base region boundaries, i.e., $\varphi_n(x_{jE})$ and $\varphi_n(x_{jC})$ in the numerator of (A.3.13), are not easily accessible by measurements; (ii) a separate determination (i.e., measurement) of the *base* hole charge, in particular the base minority charge, is difficult at medium to high current densities.

The original and most famous expression of the ICCR as given by Gummel in Ref. [3], which has the same form as (A.3.20), is in fact based on considering only the base region for the integration and assuming contributions from outside that region to be negligible. Although this was a reasonable assumption at that time for the charge, it was not for the voltage drop toward the contacts.

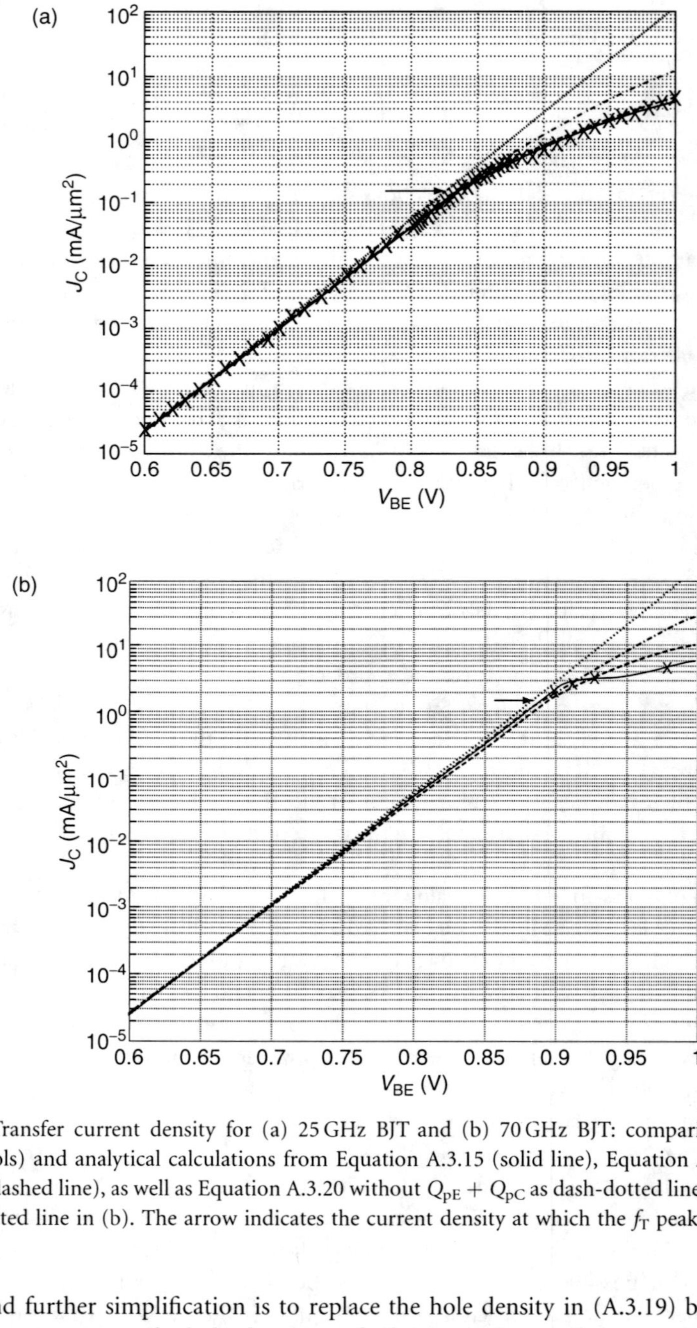

FIGURE A.3.3 Transfer current density for (a) 25 GHz BJT and (b) 70 GHz BJT: comparison between device simulation (symbols) and analytical calculations from Equation A.3.15 (solid line), Equation A.3.24 (dotted line), Equation A.3.20 (dashed line), as well as Equation A.3.20 without $Q_{pE} + Q_{pC}$ as dash-dotted line in (a) and Equation A.3.28 as dash-dotted line in (b). The arrow indicates the current density at which the f_T peak occurs.

An often found further simplification is to replace the hole density in (A.3.19) by the base doping density N_B. As a consequence, the hole charge equals the expression on the most r.h.s. of (A.3.22) and does not contain any bias-dependent "nonideal" effects anymore. Neglecting the voltage drops toward the contacts, Moll and Ross [1] first derived the corresponding expression for the transfer current

$$J_T = qV_T\,\overline{\mu_{nB}n_{iB}^2}\,\frac{\exp\left(V_{B'E}/V_T\right) - \exp\left(V_{B'C'}/V_T\right)}{\int_0^{W_B} N_B\,dx'} \qquad \text{(A.3.24)}$$

This is a more general relation compared to the classical transistor theory along the lines of Ebers and Moll [9], and aided the understanding that the transfer current is linked to the *integral* of the base doping and not to some *spot* value. As shown in Figure A.3.2, Equation (A.3.24) produces an almost ideal characteristic due to the missing bias-dependent charge components.

A.3.4 Heterojunction Transistors

A comparison for the weighting function h_g between a SiGe and a Si transistor with the same doping profile was already shown in Figure A.3.2. While the spatial dependence of the hole QFP and electron current density and, thus, of h_φ and h_i is still similar to BJTs, the intentional bandgap differences in HBTs lead to additional variations of n_i^2 within the integration interval $[0,x_C]$. As a consequence, it is impossible to define a single average value $\bar{h}_g$ over this whole interval. Thus, the denominator integral in (A.3.15) now has to be formulated in terms of charge components in *separate transistor regions that are multiplied with an average value of the weighting function* h_g taken over the respective region. In HBTs the transition from one bandgap value to another usually takes place close to the junctions. Therefore, partitioning of the respective integral $q\int_0^{x_c} h_g p dx$ into neutral and space–charge regions is *one* reasonable choice in order to obtain separate expressions, in which the weighting function is sufficiently independent on location and bias. This leads generally to

$$q\int_0^{x_C} h_g p dx = \bar{h}_{g,B0}\bar{Q}_{p0} + \bar{h}_{g,E}\bar{Q}_{pE} + \bar{h}_{g,jE}\bar{Q}_{jE} + \bar{h}_{g,B}\bar{Q}_{nB} + \bar{h}_{g,jC}\bar{Q}_{jC} + \bar{h}_{g,C}\bar{Q}_{pC} \tag{A.3.25}$$

$$\text{with} \quad \bar{h}_{g,v} = \frac{\overline{\mu_{n0r} n_{ir}^2}}{\overline{\mu_{nv} n_{iv}^2}} \quad \text{and} \quad v = \{\text{B0, E, jE, B, jC, C}\}. \tag{A.3.26}$$

Choosing the base region as reference and assuming $\bar{h}_{g,B} = \bar{h}_{g,B0}$, all weighting factors can be divided by $\bar{h}_{g,B}$ giving for the normalized remaining weighting factors

$$h_v = \frac{\bar{h}_{g,v}}{\bar{h}_{g,B}} = \frac{\overline{\mu_{nB} n_{iB}^2}}{\overline{\mu_{nv} n_{iv}^2}} \quad \text{with} \quad v = \{\text{E, jE, jC, C}\}, \tag{A.3.27}$$

which are considered to be model parameters. Inserting the above expression and charge components into (A.3.15), and making again the valid assumption $\bar{h}_i = \bar{h}_\varphi = 1$ yields the GICCR as the final expression for 1D-HBTs [5]

$$J_T = q^2 V_T \overline{\mu_{nB} n_{iB}^2} \frac{\exp(V_{B'E'}/V_T) - \exp(V_{B'C'}/V_T)}{\bar{Q}_{p,T}} \tag{A.3.28}$$

with the modified ("transfer current related") charge density

$$\bar{Q}_{p,T} = \bar{Q}_{p0} + \Delta\bar{Q}_{p,T} = \bar{Q}_{p0} + h_E\bar{Q}_{pE} + h_{jE}\bar{Q}_{jE} + \bar{Q}_{nB} + h_{jC}\bar{Q}_{jC} + h_C\bar{Q}_{pC}, \tag{A.3.29}$$

in which—as for BJTs—still a bias-independent hole charge, Q_{p0}, and a bias-dependent portion can be distinguished. According to Refs. [5,8], and also as Figure A.3.4 shows, (A.3.28) with (A.3.29) using bias-independent values for h_v leads to a significant improvement over the conventional ICCR (A.3.20). Thus, the GICCR results in an accurate description of the transfer current characteristics and the respective derivatives over the entire bias range of interest (up to high current densities). Again, as in Figure A.3.3, the charges and weighting factors have been calculated directly from device simulation results to avoid any errors introduced by analytical approximations.

For "ideal" doping profiles like in Figure A.3.1, the weighting factors h_v in (A.3.2) can be calculated analytically using (A.3.15). This aids the physical understanding of the impact of the Ge profile on device characteristics and shall be demonstrated below for two Ge profile examples. However, the calculation can be extended to other profiles but just becomes more elaborate.

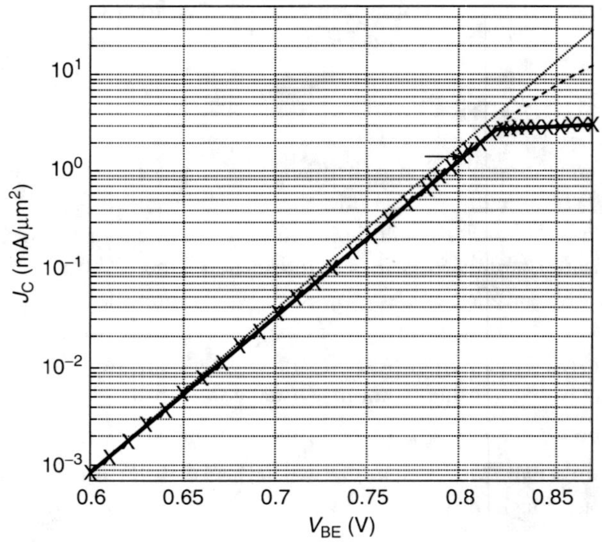

FIGURE A.3.4 Transfer current density for the 120 GHz HBT: comparison between device simulation (symbols) and analytical calculations from Equation A.3.15 (solid line), Equation A.3.28 (dashed line), Equation A.3.24 (dotted line), and Equation A.3.28 without $Q_{pE} + Q_{pC}$ (dash-dotted line).

Box Ge Profile in the (Metallurgical) Base Region

In this case, $\overline{\mu_n n_i^2}$ is constant over the base region and the portion of the space–charge regions that are associated with the base. For the sake of simplicity, low injection is assumed first. Setting $\overline{\mu_{nr} n_{ir}^2} = \mu_{nB} n_{iB}^2$ yields $h_g = 1$ already in the integral over the base region and, thus, $h_{jE} = h_{jC} = 1$. In the emitter and collector region, $h_E = \mu_{nB} n_{iB}^2 / \mu_{nE} n_{iE}^2 \gg 1$ and $h_C = \mu_{nB} n_{iB}^2 / \mu_{nC} n_{iC}^2 \gg 1$, respectively, due to the much larger bandgap in those regions. As a consequence of this bandgap, the hole charges $\overline{Q}_{pE}$ and $\overline{Q}_{pC}$ are very small and do not significantly impact the dynamic transistor behavior, regardless of the respective doping profile. However, according to (A.3.28) these charges can have a significant impact on the transfer current due to the large weighting factors h_E and h_C. The respective terms $h_E \overline{Q}_{pE}$ and $h_C \overline{Q}_{pC}$ in (A.3.28) actually cause the "saturation" of the $I_C(V_{BE})$ characteristics at high injection observed in Figure A.3.2b. In HBTs with a box Ge profile, Equation A.3.28 can be simplified to [5]

$$J_T = q^2 V_T \overline{\mu_{nB} n_{iB}^2} \frac{\exp(V_{B'E'}/V_T) - \exp(V_{B'C'}/V_T)}{\bar{Q}_p + (h_E - 1)\bar{Q}_{pE} + (h_C - 1)\bar{Q}_{pC}}, \qquad \text{(A.3.30)}$$

which contains the total hole charge as a lumped variable and correction factors in the denominator. In Ref. [8], the product $(h_C - 1)\bar{Q}_{pC}$ is described directly by a compact expression rather than separately by the weighting factor and charge component.

Trapezoidal Profile in the (Metallurgical) Base Region

Consider the Ge profile in Figure A.3.1b. For the sake of simplicity it is assumed that the Ge mole fraction m_{Ge} and the respective bandgap voltage increase over the width of the neutral base ($x' \in [x_{e0}, x_{c0}]$) only, but stay constant across the space–charge regions. Choosing the Si-base without Ge contents as reference material, the bandgap voltage differences $\Delta V_{Gp} = \Delta V_G(x' = x_{e0})$ and $\Delta V_{Gx} = \Delta V_G(x' = x_{c0})$, respectively, can be defined. Hence, the intrinsic carrier density within the neutral base with the width $w_{B0} = x_{c0} - x_{e0}$ can be written as

$$n_i^2 = n_{iB,Si}^2 \exp\left(\frac{\Delta V_{Gp} + a_G(x' - x_{e0})}{V_T}\right) \tag{A.3.31}$$

with the slope factor $a_G = (\Delta V_{Gx} - \Delta V_{Gp})/w_{B0}$. Neglecting the (much smaller) dependence of the mobility on field and Ge contents, and applying the above relation to (A.3.16) with $p_0 = N_B$ yields for the bias-independent term

$$q\int_0^{x_C} \frac{\mu_{nr} n_{ir}^2}{\mu_n n_i^2} p_0 \mathrm{d}x' \cong q\int_{x_{e0}}^{x_{c0}} \exp\left(-\frac{\Delta V_{Gp} + a_G(x' - x_{e0})}{V_T}\right) N_B \mathrm{d}x' \tag{A.3.32}$$

which gives after evaluating the r.h.s. integral

$$q\int_0^{x_C} \frac{\mu_{nr} n_{ir}^2}{\mu_n n_i^2} p_0 \mathrm{d}x' \cong \frac{\frac{V_T}{a_G}\left[\exp\left(-\frac{\Delta V_{Gp}}{V_T}\right) - \exp\left(-\frac{\Delta V_{Gx}}{V_T}\right)\right]}{w_{B0}} \bar{Q}_{p0} = \bar{h}_{g,B0}\bar{Q}_{p0} \tag{A.3.33}$$

with $\bar{Q}_{p0} = qN_B w_{B0}$. As can be seen, the average weighting factor depends exponentially on the bandgap voltages at the *beginning* and the *end* of the neutral base.

For the bias-dependent portion in (A.3.16), one can write at low current densities

$$q\int_0^{x_C} h_g \Delta p \mathrm{d}x \cong q\left[\int_{x_e}^{x_{e0}} \frac{N_B}{\exp\left(\frac{\Delta V_{Gp}}{V_T}\right)} \mathrm{d}x' + \int_{x_{c0}}^{x_c} \frac{N_B}{\exp\left(\frac{\Delta V_{Gx}}{V_T}\right)} \mathrm{d}x' + \int_{x_{jE}}^{x_{jC}} \frac{n}{\exp\left(\tilde{n}\frac{\Delta V_{Gp} + a_G(x'\tilde{n}x_{e0})}{V_T}\right)} \mathrm{d}x'\right] \tag{A.3.34}$$

The first two terms represent the depletion components that are only to be evaluated between the SCR boundary (i.e., x_e, x_c) at the given bias point and the respective equilibrium SCR boundaries (i.e., x_{e0}, x_{c0}); it also has been assumed that the bandgap (i.e., Ge mole) change within $(x_{e0}-x_e)$ and $(x_{c0}-x_c)$ is still negligible. The resulting depletion charges are $\bar{Q}_{jE} = qN_B(x_{e0} - x_e)$ and $\bar{Q}_{jC} = qN_B(x_c - x_{c0})$. For the case that the electric field in the base due to Ge grading causes the electrons to travel with saturation drift velocity v_s, i.e. $n = J_T/(qv_s)$ does not depend on x', the resulting base minority charge is then $\overline{Q}_{nB} = J_T w_B/v_s$. With these charge expressions, one obtains after evaluating all terms on the r.h.s. of (A.3.34)

$$q\int_0^{x_C} h_g \Delta p \mathrm{d}x \cong \exp\left(-\frac{\Delta V_{Gp}}{V_T}\right)\bar{Q}_{jE} + \exp\left(-\frac{\Delta V_{Gx}}{V_T}\right)\bar{Q}_{jC} + \bar{h}_{g,B0}\bar{Q}_{nB}, \tag{A.3.35}$$

where the last term follows the same evaluation as (A.3.32). The final step is to insert the components in (A.3.33) and (A.3.35) back into (A.3.15) and to normalize the denominator to the base weighting factor, $\overline{h}_{gB0}$ given by Equation A.3.33. The resulting expression then reads

$$J_T = \underbrace{q^2 V_T \frac{\mu_{nB,Si} n_{iB,Si}^2}{\bar{h}_i \bar{h}_\varphi \bar{h}_{g,B0}}}_{c_{10}} \frac{\exp(V_{B'E'}/V_T) - \exp(V_{B'C'}/V_T)}{\bar{Q}_{p0} + h_{jE}\bar{Q}_{jE} + h_{jC}\bar{Q}_{jC} + \bar{Q}_{nB}}. \tag{A.3.36}$$

which has the same form as (A.3.28), but for low current densities and with known analytical expressions of the weighting factors from the above analysis:

$$h_{jE} = \frac{\exp\left(-\frac{\Delta V_{Gp}}{V_T}\right)}{\bar{h}_{g,B0}} = \frac{v\exp(v)}{\exp(v) - 1}, \quad h_{jC} = \frac{\exp\left(-\frac{\Delta V_{Gx}}{V_T}\right)}{\bar{h}_{g,B0}} = \frac{v}{\exp(v) - 1}. \tag{A.3.37}$$

with $v = [\Delta V_{GX} - \Delta V_{GP}]/V_T$ as normalized bandgap difference between beginning and end of the base region (cf. Figure A.3.1b) . The dependence of the weighting factors as a function of v is shown in Figure A.3.5 for a practically relevant range.

From the above results, the forward Early voltage at low injection, can be calculated:

$$V_{Ef} = \frac{\bar{Q}_p}{h_{jC}\bar{C}_{jCi}} \approx \frac{\bar{Q}_{p0}}{\bar{C}_{jCi}} \frac{\exp(v) - 1}{v}. \tag{A.3.38}$$

According to (A.3.37), a 20% difference in Ge across the base region corresponds to an about 40 times increase in Early voltage, which is a significant enhancement factor over a Si-BJT or a SiGe HBT with a Ge box profile ($v = 0$).

In addition to the strong variation in n_i, the mobility varies within the transistor as a function of both doping and bias (via the electric field). The variation caused by the latter is most pronounced in the BC junction and collector region. In general, μ_n and n_i possess an opposite dependence on doping, leading to a partial compensation within h_g. However, the influence of n_i still remains much stronger than that of μ_n. As a consequence, the weighting function h_g always deviates strongly from 1 and has to be considered for all processes.

A.3.5 Further Extensions

All of the considerations so far apply to a 1D transistor structure and quasistatic operation. Extensions in both directions have been investigated and proposed. A solution of the time-dependent continuity equation led to the transient ICCR (TICC) [10], in which the "in-phase" component gives the same expression as the ICCR for the for q.s. transfer current, while the "out-of-phase" solution yields a physical definition of the charging currents flowing through the E and C contact. Hence, the out-of-phase solution defines a physics-based capacitance matrix associated with the E and C terminals, that includes the case of non-quasistatic operation. Extensions of the TICC towards including recombination and non-1D effects were presented in, for example, Ref. [11]. The application of the TICC results in a compact model, however, is quite challenging due to the bias-dependent weighting functions in the integrals defining the charging components.

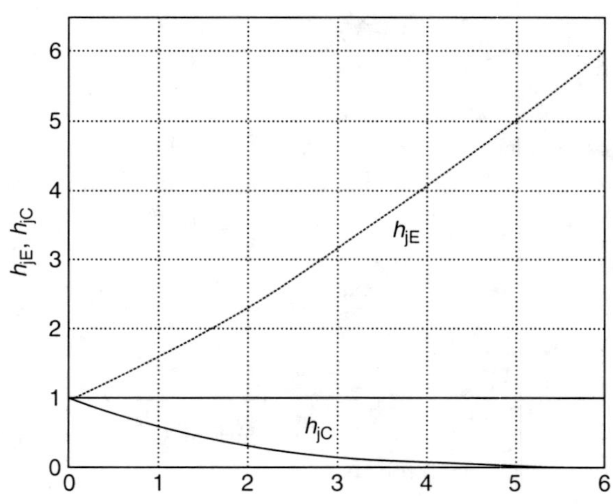

FIGURE A.3.5 Weighting factors of the depletion charges according to Equation A.3.37 as a function of the normalized bandgap difference $v = (\Delta V_{Gx} - \Delta V_{Gp})/V_T$.

As shown in Ref. [7], it is possible to extend the GICCR to two- and three-dimensional transistor structures. The respective derivation is beyond the scope of this chapter, but the result shall be briefly discussed. For instance, the resulting 2D-GICCR reads

$$I_T = c_{10} \frac{\exp\left(V_{\mathrm{B'E'}}/V_{\mathrm{T}}\right) - \left[\exp\left(V_{\mathrm{B'C'}}/V_{\mathrm{T}} \frac{2(y_{\mathrm{Bcon}}+b_{\mathrm{Bcon}})}{b_{\mathrm{E}}} + \frac{2b_{\mathrm{Bcon}}}{b_{\mathrm{E}}}\right]}{Q_{\mathrm{p0,T}} + \Delta Q_{\mathrm{p,T}}}, \tag{A.3.39}$$

where the constant c_{10} depends on an enlarged (effective) emitter width

$$b_{\mathrm{E}} = b_{\mathrm{E0}} + 2\int_{b_{\mathrm{E0}}/2}^{y_{\mathrm{Bcon}}} \exp\left(-\frac{\varphi_n(x=0,y)}{V_{\mathrm{T}}}\right) \mathrm{d}y. \tag{A.3.40}$$

b_{E0} is the emitter window width, y_{Bcon} is the edge of the base contact or polysilicon next to the emitter, and b_{Bcon} is the base contact or polysilicon width on mon-silicon. $\Delta Q_{\mathrm{p,T}}$ is defined as in (A.3.29), but now includes, among others, the impact of electron current crowding. This also applies to $Q_{\mathrm{pC,T}}$, which introduces a bias-dependent geometry dependence at higher current densities. In practice, $Q_{\mathrm{pC,T}}$ can be approximated by a constant value to first order.

A.3.6 Summary

A set of integral charge-control relations has been derived and put in perspective to the (classical) literature. It was shown that a "master" equation exists, from which integral charge-control relations of different complexity and accuracy can be derived. The most general form, that is suitable for accurately describing the transfer current in a compact model for HBTs and BJTs, is the GICCR, which includes bandgap differences in the various device regions and also contains the weakest assumptions among the known theories for the transfer current.

The GICCR is a powerful tool to analytically derive the relationship between transfer current, stored charges, and physical as well as structural parameters of a transistor. The GICCR can be very accurate, provided that the respective weigthing factors as a function of device structure and the hole charge as a function of bias are accurately modeled. Notice that the latter is a prerequisite for the description of high-speed applications in any way. Also, since the hole charge has to be continuously differentiable with respect to bias, the transfer current is also automatically continuously differentiable over all bias regions and, hence, is modeled via a single-piece formulation. This is a very desirable feature of the (G)ICCR for compact models.

Applying the "master" equation to compact modeling requires partitioning of the hole charge and analytical approximations for its various components. These measures as well as the determination of charge model parameters and appropriate weighting factor values introduce additional inaccuracies with respect to the results shown here, which are unavoidable though for any compact model equation.

Acknowledgment

The author would like to thank H. Tran for performing simulations and model calculations.

References

1. J.L. Moll and J.M. Ross, The dependence of transistor parameters on the distribution of base layer resistivity, *Proc. IRE*, 44, 1956, S72–S78.
2. H.N. Ghosh, F.H. De La Moneda, and N.R. Dono, Computer-aided transistor design, characterization, and optimization, *Proc. IEEE*, 55, 1967, 1897–1912.
3. H.K. Gummel, A charge-control relation for bipolar transistors", *Bell System Technical J.*, 49, 1970, S115–S120.

4. H. Kroemer, Two-Integral equations pertaining to the electron transport through a bipolar transistor with nonuniform energy gap in the base region, *Solid-State Electron.*, 28, 1985, 1101–1103.
5. M. Schroter, M. Friedrich, and H.-M. Rein, A generalized integral charge-control relation and its application to compact models for silicon based HBTs, *IEEE Trans. Electron Dev.*, 40, 1993, 2036–2046.
6. H.-M. Rein, H. Stübing, and M. Schroter, Verification of the integral charge-control relation for high-speed bipolar transistors at high current densities, *IEEE Trans. Electron Dev.*, 32, 1985, 1070–1076.
7. M. Schroter, A compact physical large-signal model for high-speed bipolar transistors with special regard to high current densities and two-dimensional effects, PhD thesis (in German), Ruhr-University Bochum, Bochum, Germany, 1988.
8. M. Friedrich and H.-M. Rein, Analytical current–voltage relations for compact SiGe HBT models, *IEEE Trans. Electron Dev.*, Vol. ED-46, 2001, 1384–1401.
9. J.J. Ebers and J.L. Moll, Large-signal behavior of junction transistors, *Proc. IRE*, 42, 1761–1772, 1954.
10. H. Klose and A. Wieder, The transient integral charge-control relation—a novel formulation of the currents in a bipolar transistor, *IEEE Trans. Electron Dev.*, ED-34, 1090–1099, 1987.
11. J.S. Hamel and C.R. Selvakumar, The general transient charge–control relation: a new charge-control relation for semiconductor devices, *IEEE Trans. Electron Dev.*, ED-38, 1467–1476, 1991.

A.4
Sample SiGe HBT Compact Model Parameters

Ramana M. Malladi
IBM Microelectronics

This appendix contains a sample set of compact model parameters for a representative $0.32 \times 16.8\ \mu m^2$ first-generation npn SiGe HBT with a peak f_T of 50 GHz, for each of the three dominant higher-order SiGe compact models available in the public domain and in leading circuit simulators: HICUM, MEXTRAM, and VBIC. Each model was carefully calibrated to a comprehensive set of measured dc and ac data.

TABLE A.4.1 HICUM (v 2.1) SiGe HBT Model Parameters

Group	Name	Parameter Description	Value
Transfer current	c10	Constant for ICCR (= Is.qp0)	8.5×10^{-31} A^2 S
	qp0	Zero-bias hole charge in the base	1.4×10^{-13} C
	ich	High-current correction to account for 2D and 3D effects	10 A
	hjei	BE depletion charge weighting factor for HBTs	2
	hjci	BC depletion charge weighting factor for HBTs	0.1
	mcf	Forward nonideality factor for transfer current for HBTs	1
	hfe	Emitter minority charge weighting factor for HBTs	1
	hfc	Collector minority charge weighting factor for HBTs	0.2
Base–emitter currents	ibeis	Internal BE sauration current	7×10^{-20} A
	mbei	Internal BE non-ideality factor	1.005
	ibeps	Peripheral BE saturation current	7×10^{-22} A
	mbep	Peripheral BE non-ideality factor	1
	ireis	Internal BE saturation current (recombination)	1×10^{-17} A
	mrei	Internal BE non-ideality factor (recombination)	2
	ireps	Peripheral BE saturation current (recombination)	1×10^{-20} A
	mrep	Peripheral BE non-ideality factor (recombination)	2
Base–collector currents	ibcis	Internal BC saturation current	1.5×10^{-18} A
	mbci	Internal BC saturation current ideality factor	1.02
	ibcxs	External BC saturation current	2×10^{-19} A
	mbcx	External BC saturation current ideality factor	2
Base–emitter tunnelling current	ibets	BE tunneling saturation current	1×10^{-21} A
	abet	BE tunneling factor	36
Base–collector avalanche current	favl	Avalanche current factor	19.3 V^{-1}
	qavl	Exponent for avalanche current	150×10^{-15} C
Series resistances—base, emitter, and collector	rbi0	Internal base resistance at zero bias	20 Ω
	rbx	Extrinsic base resistance	6 Ω
	fdqr0	Correction factor for modulation by BE and BC SCR	0.2
	fgeo	Geometry factor for current crowding	0.67
	fqi	Ratio of internal to total minority charge	1.0
	fcrbi	Ratio of shunt capacitance (parallel to Rbi) to total internal capacitance	0
Emitter	re	Emitter series resistance	3 Ω
Collector	rcx	Extrinsic collector series resistance	23 Ω
Substrate transistor	iscs	CS diode saturation current	4×10^{-21} A

	msc	CS diode non-ideality factor	1
	itss	Transfer saturation current of substrate transistor	2.5×10^{-19} A
	msf	Forward non-ideality factor of substrate transfer current	1
	msr	Reverse non-ideality factor of substrate transfer current	1
	tsf	Transit time (forward operation)—substrate	2×10^{-12} sec
Substrate network	rsu	Substrate resistance	50 Ω (layout dependent)
	csu	Substrate capacitance	3×10^{-18} F
Self-heating	rth	Thermal resistance	700 K/W
	cth	Thermal capacitance	350 pJ/K
Base–emitter junction capacitance	cjei0	Internal zero-bias BE depletion capacitance	35×10^{-15} F
	vdei	Internal BE built-in voltage	1.0 V
	zei	Internal BE grading coefficient	0.32
	aljei	Maximum internal depletion capacitance divided by cjei0	2.0
	cjep0	Peripheral zero-bias BE depletion capacitance	5×10^{-15} F
	vdep	Peripheral BE built-in voltage	1.0 V
	zep	Peripheral BE grading coefficient	0.32
	aljep	Maximum peripheral depletion capacitance divided by cjep0	2.2
	ceox	Emitter oxide (overlap) capacitance	18×10^{-15} F
Base–collector junction capacitance	cjci0	Internal zero-bias BC depletion capacitance	7×10^{-15} F
	vdci	Internal BC built-in voltage	0.7 V
	zci	Internal BC grading coefficient	0.3
	vptci	Punch-through voltage of internal BC junction	2.5 V
	cjcx0	External zero-bias BC depletion capacitance	30×10^{-15} F
	vdcx	External BC built-in voltage	0.73 V
	zcx	External BC grading coefficient	0.4
	vptcx	Punch-through voltage of external BC junction	100 V
	ccox	BC overlap capacitance	2.5×10^{-15} F
	fbc	Partitioning factor for cjcx and ccox over rbx	0.8
Collector–substrate junction capacitance	cjs0	Zero-bias CS depletion capacitance	40×10^{-15} F
	vds	CS built-in voltage	0.6 V
	zs	CS grading coefficient	0.3
	vpts	Punch-through voltage of CS junction	1×10^{10} V
Diffusion capacitances/transit times—low currents	t0	Low current forward transit time at $V_{cb}=0$ V	2.6×10^{-12} sec
	dt0h	Time constant for base and BC space charge layer width modulation	0.9×10^{-12} sec
	tbvl	Time constant for modeling carrier jam at low V_{ce}	0.7×10^{-12} sec
High currents	tef0	Neutral emitter storage time	40×10^{-15} sec
	gtfe	Exponent for current dependence of neutral emitter storage time	1.0

(*Continued*)

TABLE A.4.1 HICUM (v 2.1) SiGe HBT Model Parameters (*Continued*)

Group	Name	Parameter Description	Value
	thcs	Saturation time constant at high current densities	25×10^{-12} sec
	alhc	Smoothing factor for current dependence of base and collector transit time	0.53
	fthc	Factor for partitioning this into base and collector portion	0.6
	vces	Internal C-E saturation voltage	0.1 V
	rci0	Internal collector resistance at low electric field	20 Ω
	vlim	Voltage separating ohmic (low field) and saturation velocity (high field) regime	0.7 V
	vpt	Collector punch-through voltage	15 V
	tr	Storage time for inverse operation	20×10^{-12} sec
Non-quasistatic effects	alqf	Factor for additional delay time of minority charge	0.125
	alit	Factor for additional delay time of transfer current	0.45
Noise parameters	kf	Flicker noise factor	22×10^{-6}
	af	Flicker noise exponent factor	2.5
	krbi	Noise factor for internal base resistance	1
Temperature effect parameters	tnom	Measurement temperature	25 C
	vgb	Bandgap voltage	1.17 V
	alb	Temperature coefficient of current gain	6×10^{-3}
	alfav	Temperature coefficient of favl	5×10^{-5} K^{-1}
	alqav	Temperature coefficient of qavl	2×10^{-4} K^{-1}
	zetaci	Temperature coefficient for mobility in epi-collector (i.e., for collector resitance)	1.6
	alvs	Relative temperature coefficient of saturation drift velocity	1×10^{-3} K^{-1}
	alces	Relative temperature coefficient of vces	0.4×10^{-3} K^{-1}
	zetarbi	Temperature coefficient for mobility in internal base (i.e., for internal base resistance)	0.6
	zetarbx	Temperature coefficient for mobility in extrinsic base (i.e., for extrinsic base resistance)	0.2
	zetarcx	Temperature coefficient for mobility in extrinsic collector (i.e., for extrinsic collector resistance)	0.2
	zetare	Temperature coefficient for emitter resistance	0
	alt0	First-order temperature coefficient of t0	1×10^{-3} K^{-1}
	kt0	Second-order temperature coefficient of t0	1×10^{-5} K^{-2}

TABLE A.4.2 MEXTRAM 504 SiGe HBT Model Parameters

Group	Name	Parameter Description	Value
Forward and reverse currents	is	Transistor main saturation current	5×10^{-18} A
	ik	Knee current for high-injection effects in the base	4.5×10^{-2} A
	bf	Ideal forward current gain	95
	ibf	Saturation-current of the non-ideal forward base current	2×10^{-17} A
	mlf	Non-ideality factor of the non-ideal forward base current	1.545
	xibi	Sidewall component of ideal base current	0
	bri	Ideal reverse current gain	3.77
	ibr	Saturation current of the non-ideal reverse base current	1.7×10^{-15} A
	vlr	Cross-over voltage of the non-ideal reverse base current	1×10^{-2} V
	xext	Partitioning factor for the extrinsic region	0.19
Early voltage	ver	Reverse Early voltage	4.8 V
	vef	Forward Early voltage	65 V
Weak avalanche	wavl	Epilayer thickness used in weak-avalanche model	2.44×10^{-7} M
	vavl	Voltage determining curvature of avalanche current	0.63 V
	sfh	Current spreading factor of avalanche model (when exavl $=1$)	1.7
Resistances and quasisaturation	re	Emitter resistance	3.0 Ω
	rbc	Constant part of the base resistance	6 Ω
	rbv	Zer-bias value of the bias-dependent base resistance	20 Ω
	rcc	Constant part of the collector resistance	17 Ω
	rcv	Resistance of the un-modulated epilayer	52 Ω
	scrcv	Space charge resistance of the epilayer	54 Ω
	ihc	Critical current for velocity saturation in the epilayer	3.56×10^{-3} A
	axi	Smoothness parameter for the onset of quasi-saturation	0.21
Base–emitter junction capacitance	cje	Zero-bias emitter–base depletion capacitance	42×10^{-15} F
	vde	Emitter–base diffusion voltage	0.9 V
	pe	Emitter–base grading coefficient	0.23
	xcje	Fraction of the emitter–base depletion capacitance that belongs to the side-wall	0
	cbeo	Emitter–base overlap capacitance	18×10^{-15} F
Base–collector junction capacitance	cjc	Zero–bias collector–base depletion capacitance	7×10^{-15} F
	vdc	Collector–base diffusion voltage	0.75 V
	pc	Colector–base grading coefficient	0.28
	xp	Constant part of cjc	1×10^{-3}

(*Continued*)

TABLE A.4.2 MEXTRAM 504 SiGe HBT Model Parameters (*Continued*)

Group	Name	Parameter Description	Value
	mc	Coefficient for the current modulation of the collector–base depletion capacitance	0.5
	xcjc	Fraction of the collector–base depletion capacitance under the emitter	8.7×10^{-2}
	cbco	Collector–base overlap capacitance	2.5×10^{-15} F
Collector–substrate junction capacitance	cjs	Zero-bias collector–substrate depletion capacitance	45×10^{-15} F
	vds	Collector–substrate diffusion voltage	0.6 V
	ps	Collector–substrate grading coefficient	0.3
	vgs	Bandgap voltage of the substrate	1.17 V
Diffusion capacitances/transit times	mtau	Non-ideality factor for the emitter stored charge	0.388
	taue	Minimum transit time of stored emitter charge	52×10^{-15} sec
	taub	Transit time of stored base charge	1.44×10^{-12} sec
	tepi	Transit time of stored epilayer charge	14.4×10^{-12} sec
	taur	Transit time of reverse extrinsic stored base charge	20×10^{-12} sec
HBT parameters	deg	Bandgap difference over the base	0.03 eV
	xrec	Pre-factor of the recombination part of ideal base current	0
Temperature coeffcients	aqbo	Temperature coefficient of the zero-bias base charge	0.34
	ae	Temperature coefficient of the resistivity of the emitter	0
	ab	Temperature coefficient of the resistivity of the base	1.22
	aepi	Temperature coefficient of the resistivity of the epilayer	1.88
	aex	Temperature coefficient of the resistivity of the extrinsic base	8.7×10^{-7}
	ac	Temperature coefficient of the resistivity of the buried layer	1
	as	Temperature coefficient for Iss and Iks (for a closed buried layer, as=ac and for an open buried layer, as=aepi)	0.76
	dvgbf	Bandgap voltage difference for forward current gain	3.75×10^{-2} V
	dvgbr	Bandgap voltage difference for reverse current gain	4.38×10^{-2} V
	vgb	Bandgap voltage of the base	1.15 V
	vgc	Bandgap voltage of the collector	1.18 V
	vgj	Bandgap voltage: recombination of the emitter–base junction	1.15V
	dvgte	Bandgap voltage difference of emitter stored charge	0.236 V
1/*f* Noise	af	Exponent of the flicker noise	2.5
	kf	Flicker-noise coefficient of the ideal base current	22×10^{-6}
	kfn	Flicker noise coefficient of the non-ideal base current	20×10^{-12}
Substrate transistor	iss	Base-substrate saturation current	2.5×10^{-19} A
	iks	Base-substrate high-injection knee current	50 A
Self-heating network	rth	Thermal resistance	700 K/W
	cth	Thermal capacitance	350 pJ/K

TABLE A.4.3 VBIC SiGe HBT Model Parameters

Group	Name	Parameter Description	Value
Saturation currents and ideality factors	is	Transport saturation current (collector)	4.85×10^{-18} A
	ibei	Ideal base–emitter saturation current	7×10^{-20} A
	iben	Nonideal base–emitter saturation current	1×10^{-15} A
	ibci	Ideal base–collector saturation current	1.5×10^{-18} A
	ibcn	Nonideal base–collector saturation current	1×10^{-34} A
	isp	Parasitic transport saturation current	3×10^{-19} A
	ibcip	Ideal parasitic base–collector saturation current	
	ibcnp	Nonideal parasitic base–collector saturation current	1×10^{-40} A
	ibeip	Ideal parasitic base–emitter saturation current	2.52×10^{-18} A
	ibenp	Nonideal parasitic base–emitter saturation current	1×10^{-28} A
	nf	Forward emission coefficient	1.0003
	nei	Ideal base–emitter emission coefficient	1.026
	nen	Nonideal base–emitter emission coefficient	2.5
	nr	Reverse emission coefficient	1
	nci	Ideal base–collector emission coefficient	1.02
	ncn	Nonideal base–collector emission coefficient	1.00
	nfp	Parasitic forwad emission coefficent	1.00
	ncip	Ideal parasitic base–collector emission coefficient	1.00
	ncnp	Nonideal parasitic base–collector emission coefficient	2
Knee currents	ikf	Forward knee current	4.5×10^{-2} A
	ikr	Reverse knee current	4.8×10^{-3} A
	ikp	Parasitic knee current	10 A
Avalanche breakdown	avc1	Base–collector weak avalanche parameter 1	19.2
	avc2	Base–collector weak avalanche parameter 2	23.6 V
Series resistances	rbi	Intrinsic base resistance	20 Ω
	rbx	Extrinsic base resistance	6 Ω
	rbp	Parasitic base resistance	1 Ω
	re	Emitter resistance	3 Ω
	rcx	Extrinsic collector resistance	23 Ω
	rs	Substrate resistance	50 Ω
Self-heating	rth	Thermal resistance	700 K/W
	cth	Thermal capacitance	350 pJ/W

(*Continued*)

TABLE A.4.3 VBIC SiGe HBT Model Parameters (*Continued*)

Group	Name	Parameter Description	Value
Quasi-saturation parameters	rci	Intrinsic collector resistance	40 Ω
	vo	Epi drift saturation voltage	1×10^{-10} V
	gamm	Epi doping parameter	5×10^{-13} V
	hrcf	High-current RC factor	1×10^{-13}
	qco	Epi-charge parameter	1.4×10^{-15}
Early effect parameters	vef	Forward Early voltage	65 V
	ver	Reverse Early voltage	5.5 V
Base–emitter junction capacitance	cje	Base–emitter zero-bias capacitance	42 fF
	me	Base–emitter grading coefficient	1.0
	pe	Base–emitter built-in potential	0.3 V
	cbeo	Extrinsic base–emitter overlap capacitance	18×10^{-15} F
	fc	Forward bias depletion capacitance limit	0.93
	aje	Base–emitter capacitance switching parameter	−0.1
Base–collector junction capacitance	cjc	Base–collector intrinsic zero-bias capacitance	7×10^{-15} F
	mc	Base–collector grading coefficient	0.3
	pc	Base–collector built-in potential	0.7 V
	cbco	Extrinsic base–collector overlap capacitance	2.5×10^{-15} F
	cjep	Base–collector extrinsic zero bias capacitance	30×10^{-15} F
	ajc	Base–collector capacitance switching parameter	−0.1
Collector–substrate junction capacitance	cjcp	Substrate–collector zero bias capacitance	40×10^{-15} F
	ms	Substrate–collector grading coefficient	0.3
	ps	Substrate–collector built-in potential	0.6 V
	ccso	Fixed collector–substrate capacitance	3e−18 F
	ajs	Substrate–collector capacitance switching parameter	−0.9
Transit times and their bias dependence	tf	Forward transit time	2×10^{-12} sec
	itf	Coefficient of tf dependence of ic	0.32
	vtf	Coefficient of tf dependence of V_{bc}	23.7
	qtf	Variation of tf with base-width modulation	0
	td	Forward excess-phase delay time	0.7×10^{-12} sec
Temperature effect parameters	ea	Activation energy for is	1.17 V
	eaie	Activation energy for ibei	1.17 V
	eaic	Activation energy for ibci and ibeip	1.17 V
	eais	Activation energy for ibcip	1.17 V
	eanc	Activation energy for ibcn/ibenp	1.17 V

eane	Activation energy for iben	1.17 V
eans	Activation energy for ibcnp	1.17 V
xii	Temperature coefficient for ibei, ibci, ibeip, and ibcip	2.0
xin	Temperature coefficient for iben, ibcn, ibenp, and ibcnp	2.0
xis	Temperature coefficient for is	1.9
xre	Temperature coefficient for re	0
xrb	Temperature coefficient for rbi	0
xrc	Temperature coefficient for rc	0
xrs	Temperature coefficient for rs	0
xvo	Temperature coefficient for vo	0
tavc	Temperature coefficient for avc2	250×10^{-6} V

References

1. HICUM bipolar transistor model: http://www.iee.et.tu-dresden.de/iee/eb/comp_mod.html.
2. Mextram bipolar transistor model: http://www.semiconductors.phillips.com/acrobat/other/phillips-models/NLUR2000811_7.pdf.
3. VBIC bipolar transistor model: http://www.designers-guide.org/VBIC/references.html.

Index

A

Abrasive etching method, 12-5
Active millimeter multipliers, designs of, 8-6
Adjacent-channel-leakage ratio (ACLR), 4-17
Adjacent channel selectivity, 16-17
Analog RF circuit, switching speeds, 9-3
Analog VGA low-pass filter, 16-15
Antenna connector, TX Power, 16-17
ASSURA-RF extractor, 16-8
Atmospheric pressure CVD (AP-CVD), 2-3, 7-2
Attenuation constant, 12-7
Avalanche breakdown, 4-9, 4-11, 4-13

B

Ball grid arrays (BGA), 13-2–13-4
Bandgap engineering, 1-3–1-4, 2-2–2-4, 4-1, 5-9, 5-36
Bandpass filters, 11-11–11-16, 16-11
 layout of, 11-13
 tunable filter, 11-11
Bandwidth-limited instability, 9-3
Base transit time, A.2-4
BGA package-based 3D module, 13-4
Bias current
 bias current source, 9-26
 cascode preamplifier, 9-16
 common-emitter preamplifier, 9-15
 differential cascode preamplifier, 9-17
 and gain-bandwidth product for transistors, 9-2
 Gilbert multiplier, 9-23
Biasing network, 11-7
BiCMOS DECT transceiver, 4-19, 4-20
Bipolar gain cell, schematic of, 7-13
Bipolar junction transistor (BJT), 2-4
Bipolar transistor, 16-4, 16-5, 16-7
 high-frequency region of, 9-3
Bluetooth technology, 4-3
Boltzman's constant *(k),* 4-14
Breakdown voltage, 4-8, 4-9, 4-11, 9-13–9-14
Broadband noise, 10-10
Buffers, 10-17–10-18

C

Capacitive loads, 11-7
 computed performance, 11-7, 11-8
Capacitors, 11-12, 16-6
Cascode cell
 equivalent circuit of, 7-14
 gain, feedback and attenuation, 7-13
 output admittance of, 7-14
Cascode LNAs
 differential cascode preamplifier, 9-16–9-17
 impedance matching networks, 9-19
Cascode preamplifier
 bandwidth and reverse isolation, 9-16
 topologies of, 9-15
Cavity-etched patch antenna
 fabrication of, 12-4
 input return of, 12-6
 radiation patterns of, 12-6
CDR-demultiplexer chip, 14-9
Cellular handsets, 4-1–4-4
Characteristic impedance, 11-11
Chemical vapor deposition (CVD), 2-3, 4-8, 11-2
Cherry-Hooper amplifier, 7-11, 14-3, 14-11
Chip micrographs, dynamic FD, 15-5
Class-AB input stage, distortion minimization, 9-22–9-23
Clock and data recovery (CDR), 14-2, 14-5, 14-8
Clock multiplier unit (CMU), 14-11, 14-12
CMOS base process, 16-5
CMOS LC oscillators, 14-5
CMOS LC-VCOs, 14-5, 14-6
CMOS transistor, 16-3
COB module, RF analog behavior, 16-14
Coefficient of thermal expansion (CTE), 13-6
Common-base amplifier
 cascode preamplifier, 9-15
 Gilbert multiplier, 9-23
 transistor noise sources of, 9-25
Common-emitter preamplifier, topologies for, 9-15
Common-mode rejection, differential and symmetric circuit topologies, 9-15
Communication building blocks
 frequency divider and buffers, interfacing of, 10-19

low-noise amplifier and high frequency oscillator, 10-3
Communication system design, 10-3
Conductor shielding, 9-33
Conventional emitter doping (CED) transistors, A.3-3
Coplanar bond-pad arrangement, 7-5–7-6
Coplanar oscillators
designs on HRS, 8-4
layout of, 8-3
passive elements, modeling of, 8-3
phase noise and conversion efficiency of, 8-4
Coplanar transmission lines
attenuation and slowing factor for, 7-4
quality factor of, 8-2
Coplanar waveguide (CPW), 12-6, 12-7, 12-14
Cost-performance, 13-4–13-5
Cutoff frequencies, 11-2, 11-3

D

DC bias lines, 11-19
DC offset variation effect
correction of, 10-23
dynamic, 10-6–10-7
static, 10-6
Demodulator (DEMO), 15-6
Demultiplexer
circuit, block diagram, 14-4
1:16 DEMUX, 15-4
design of, 14-3
Design frequency, 11-17, 11-19
Dielectric constant model, 12-13
Differential cascode preamplifier, single-ended equivalent circuit for, 9-16–9-17
Differential Colpitts oscillator, 9-36, 9-38–9-39
Differential shielding
Q-factor for, 9-34
transmission line, 9-33
Differential topology, 10-3, 10-5–10-6
Differential VCO tuning, 9-36
Digital capacitive loads, electrical equivalent circuit, 11-8
Diodes, 16-6
Direct conversion micromixer IC, 10-22
Direct conversion radio architectures
bipolar transistors and MOS transistors, 10-10
compensation schemes, 10-23
DC offset variation effect, 10-6–10-7
developmental history of, 10-2
differential topology of, 10-3, 10-5–10-6
direct conversion micromixer IC, performance of, 10-21
frequency allocation of, 10-11–10-12
impact of building blocks in, 10-5
integrated direct conversion receiver, performance of, 10-23
mixing technique, 10-10
reverse isolation, 10-10
signal paths, nonidealities in, 10-6
space compaction in, 10-6
static offset corrections, 10-23
system performance impact on device technology, 10-9–10-10
variants of, 10-4
Distributed amplifiers
constant group velocity, 7-12
equivalent circuit of a common-emitter bipolar gain cell in, 7-13
scattering parameters of, 7-15
III–V-based Distributed amplifiers gain-bandwidth, 7-16
Doubler
conversion gain of, 8-6
effects, 4-2
scheme of, 8-6
sensors, lumped elements, 7-9
Double sideband (DSB) noise, 8-7
Down-converter IC
conversion gain of, 7-9
IF output buffer, 7-9
layout of, 7-8–7-9
Down-converter mixer, 10-15
layout, 16-10
schematic, 16-9
Drain transmission line, 7-16
Dual-in-line package (DIP), 13-2, 13-4
Duplexer losses, 16-6
Dynamic frequency divider, 7-8
chip micrographs, 15-5
input sensitivity of, 15-5
Dynamic toggle flip-flop (D-TFF), 15-5

E

ECL logic, 14-7
Electrical design, 13-6
Electronic-toll-collection (ETC) transceiver IC, 15-6
Emitter-coupled logic (ECL), 2-4, 14-3
Emitter follower, nonlinearity effects, 10-18
ESD protection diodes, 16-8
ETC transceiver IC
block diagram, 15-7
receiver transfer characteristics, 15-7

F

Fabricated filters, 11-13
Fabricated monolithic tuners, 11-20
FC module, 16-14
FDD standards, 16-3, 16-4
FEM model, 12-5
Fifth-order low-pass filter, schematic of, 16-11
Filter with MEMS switches, 11-14, 11-15

Finite difference time-domain (FDTD), 12-2
Flicker noise, 9-13
Flip chip
configuration, 13-5
devices, 13-4
technology, 13-4
Four-bit reconfigurable tuner
matched load impedances, 11-18
schematic of, 11-17
Frequency
allocation of, 10-11–10-12
doubler cell
conversion gain of, 7-7
fundamental frequency suppression, 7-8
synthesizers, 4-13
Full-rate (bottom) transmitters, block diagram, 14-10
Full-rate multiplexer architecture, 14-12
Full-rate transmitters
designs, 14-11
implementation, 14-16
measurement results, 14-15
Fully differential distributed amplifier, 7-15

G

GaAs-based devices, 4-3
GaAs MMICs
substrate resistivity in, 7-2
transmission lines, 7-3
Gain switching and current steering approach, 10-15
Gate transmission line, 7-16
Ge
effective mass parameters, A.1-2
energy band structure, A.1-1
properties of, A.1-2
Generalized ICCR (GICCR), A.3-2–A.3-6, A.3-13
Gilbert cell, current sources, 8-7
Gilbert cell mixer
bias circuits, 10-18
buffer circuitry, 10-17–10-18
current source and transconductor for, 10-16
load network, 10-17
micromixer and high impedance inductively degenerated, 10-18–10-19
single-ended interface, 10-15
switching pair of, 10-16–10-17
Gilbert-micromixer, 10-19
Gilbert multiplier
class-AB mixer input stage, 9-23–9-24
impedance load, 9-22
linear range of, 9-23
noise minimization, 9-22
Ground aperture, radiation pattern, 12-16
Ground-coupled noise, 7-5–7-6
Grounded shield, 9-33
GSM handsets, 4-5

H

Half- and full-rate VCOs, phase noise performance, 14-15
Half-rate multiplexer architecture, 14-12
Half-rate transmitters
designs, 14-11
implementation, 14-16
measurement results, 14-15
HBT-based amplifier, design of, 8-6
HBT SiGeC cross section, 16-6
Heterodyne receiver, in cellular handset, 4-4
Heterojunction bipolar transistor (HBT)
charge control relation, A.3-9–A.3-12
nonlinear behavior of, 16-4
HICUM (v 2.1) SiGe HBT model parameters, A.4-2–A.4-4
High-frequency low-noise amplifier, 10-3
High-frequency region, 9-3
High impedance inductively degenerated Gilbertcell mixer, 10-18–10-19
High-resistivity silicon (HRS), substrate
Microstrip/coplanar transmission lines, 8-1
SPDT switches on, 8-7
High-resistivity substrates, 7-3
Hole-perforated dielectric constant, 12-10
Homojunction transistors, charge control relation, A.3-6–A.3-9

I

IC Chipset, 15-3
IC transmission lines, 9-5
Ideal transformer
current and voltage transformations in, 9-10
magnetic coupling strength of, 9-10–9-11
magnetizing inductance of, 9-11
IEEE 802.11a standard, 10-11
IF output buffer, 7-9
Impatt diodes, 8-7
Indium tin oxide (ITO), 11-19
Injection locking, 8-7
Integral charge-control relation (ICCR), A.3-2
hetereojunction transistors, A.3-9–A.3-12
homojunction transistors, A.3-6–A.3-9
Integrated circuit (IC) package, 13-1–13-3
Integrated circuit oscillators, 9-36
Integrated direct conversion receiver, 10-23
Integrated LNA, validation of, 16-14
Integrated receiver IC using Si–SiGe HBTs, 7-8
Intercept point (IIP3), third-order, 9-15, 9-20, 9-21, 9-26
Interconnect wiring, time-of-flight, 9-4
Intermodulation characteristics, 16-17
Intermodulation distortion
fifth-order, 9-16
second-order and third-order, 9-14–9-15

transistor bias current and power supply voltage, 9-15–9-16
Interwinding capacitance
effect of, 9-12–9-13
primary and secondary, 9-11
Isolation constraint, 16-13

J

Johnson limit, 4-9

K

Kroemer's approach, A.2-4

L

Leeson's model, 9-29
Lift-off process, 12-11
Limited-reaction-processing CVD (LRP-CVD), 2-2–2-4
Line width, 12-16
Liquid-crystal polymers (LCP), 13-6
LNA, *see* Low-noise preamplifier
LNA Bottleneck, 4-4
LNA implementation, 16-6
LNA inductors, 16-12
LNA-mixer isolation strategy, 16-12–16-13
Load network, 10-18
Local multiple-distribution services (LMDS), 15-4
Local oscillator, 4-4–4-7, 4-17
Low-emitter concentration (LEC) transistors, A.3-3
Low-loss resonators, 11-6–11-11
Low-noise amplifiers (LNAs), 4-2–4-3, 16-2
circuit for cascode switch gain, 10-13–10-14
gain switching approach, 10-14
input and output matching, 10-14–10-15
input bond-pad and ground-coupled noise, 7-5–7-6
input circuitry of, 7-5–7-6
Low-noise preamplifier
common-emitter preamplifier, 9-15–9-16
gain and third-order intercept point (IIP3), 9-20
intermodulation and harmonic distortions, 9-15–9-16
symmetric inductor, 9-4
Low-pass filter (LPF), 16-10–16-12
Low-pass filter operational amplifier
gain bandwidth product of, 16-11
schematic, 16-11
Low-resistivity substrates, 8-1
Low-temperature Si epi growth techniques, 2-3
LRS, *see* Low-resistivity substrates
Lumped passive elements
magnetically coupled inductors, 7-9
and silicon MMICs designers, 7-2

M

Magnetically coupled inductors, layout of, 7-9
Magnetic coupling
common-emitter transistor amplifier, 9-19
windings, 9-10
Master-slave flip-flops, 14-7
Matching networks (Tuners), 11-16–11-22
Mechanical design, 13-6
Media access control (MAC) functions, 4-7
MEMS devices, 11-1, 11-2
Metal-insulator-metal capacitor
equivalent circuit of, 7-5
parasitic series resistance, 7-4
MEXTRAM 504 model parameters, A.4-5–A.4-6
Micromachining approach, 12-6
Micromixers, 10-18
Microstrip antennas
bandwidth of, 12-6
cavity-etched silicon substrate, 12-2, 12-3
design parameters, 12-3
hole-perforated silicon substrate, 12-10
radiation pattern, 12-2, 12-3
resonant frequency, 12-2
return loss, 12-4, 12-5, 12-8
view of, 12-7
Microstrip-based bandstop filter, 11-11
Microstrip inductor, symmetric
coupled microstrips, winding of, 9-31
lumped equivalent circuit model of, 9-32
physical layouts for differential drive, 9-30
Microstrip-on-silicon transmission line models
circuit components of, 9-7
one-section mode and two-section mode, 9-8
skin effect and attenuation constant, 9-7
Microstrip patch antennas
dielectric constants, 12-5
hole-perforated substrate, 12-10
Microstrip transmission (MST) lines
attenuation and slowing factor for, 7-4
Q-factor, strip width, and frequency, relationship between, 9-6
RLC model for, 9-7
technology, 8-4
Microstrip tuner, double-stub, 11-19
Microstrip VCO, 8-5
Microwave balun structures, 9-11–9-12
Miller capacitance, 9-13, 9-15, 9-19
Millimeter-wave generation
coplanar oscillators
designs on HRS, 8-4
layout of, 8-3
passive elements, modeling of, 8-3
phase noise and conversion efficiency of, 8-4
microstrip oscillators, 8-4
multiplier, 8-6
MIM capacitor, *see* Metal-insulator-metal capacitor
Mixer MMIC

conversion gain of, 8-7
design of, 8-7
M2 MOS transistor, 16-7
Modem, 4-7
Moll-Ross collector current density relation, A.2-3
Moll-Ross relation
assumptions for, A.2-1
generalized, A.2-2
Monolithic microwave integrated circuits (MMICs), 7-1
Monolithic transformer, 9-10
Monolithic tuner with DC bias lines, 11-20
Monopole antenna array
photomicrograph of, 12-18
SEM of, 12-18
Moore's law, 4-1
MOS capacitors, inversion-mode and accumulation-mode, 9-35
MOS substrate modulation, 16-3
MOS transistors, noise characteristics of, 10-10
MOS varactors, Q-factors for, 9-36
Mott diodes, 8-7
Multichip module, 13-4–13-5
Multifilament transformers, 9-11
Multiplexer, 14-2
4:1 Multiplexer, 8-6
block liagram, 14-11, 15-2
outpu waveform of, 15-3
16:1 MUX (multiplexer), 15-4

N

Narrow-band circuits
current sources, 7-7
operating principle of, 7-6
Neutralization
Collector- base feedback capacitance, 9-19
transformer turns ratio, 9-20
Noise factor *(F)*, 4-12, 9-17–9-18
Noise figure (NF), 4-12, 5-1, 16-5
Balun-coupled mixer, 9-25
emitter areas, 9-18
Gilbert multiplier, 9-23
microwave transistor, 9-17
mixer, 9-13
optimization of, 9-17
RF signal and, 9-23–9-24
Nonlinearity effects
emitter follower, 10-18
MOS transistors, 10-10

O

OFDM subcarriers, 10-11
Off-chip interconnections and time delay, 9-4
Ohmic switch
cantilever, 11-4, 11-7
schematic of, 11-2, 11-3
simulated S-parameters, 11-5
Optical data communication, 14-2
Optical fiber, 14-2
Optical-fiber-link communication systems
block diagram, 15-2
IC chipset, 15-1
Optical transmission systems
integrated circuit(IC), 15-1
large-scale ICs, 15-1
Oscillator phase noise
down converted spectrum, 9-27–9-28
in high-quality oscillator, 9-29
Leeson's model, 9-29
Output admittance, 7-14

P

Package design, 13-1, 13-5–13-7
Parallel feedback technique, 7-9, 7-11
Phase detector, 14-8
block diagram, 14-6
three-level gate generating, 14-7
Phase-locked loop (PLL), 4-17, 14-2, 14-5, 14-11, 14-15, 15-4, 15-6
Photodetector, 14-2
Pin diodes, 8-7
Pin grid array (PGA), 13-2
Planar inductors for monolithic circuits, 9-9
Planar transmission lines, 8-2
Plastic leaded chip carrier (PLCC), 13-2
PLL bandwidth design, 14-13
Polyimide cavity
polythickness Values, 12-8
Simulated S_{11}, 12-9
Postmixer amplifier (PMA), 16-2
differential operational amplifier, 16-9
implementation, 16-8
layout, 16-10
Preamplifier stages, downconverter IC, 7-8–7-9
Printed circuit board (PCB), 13-1
Printed wiring board (PWB), 13-1

Q

Quality factor, 11-7

R

Radio-frequency amplifiers
low-noise preamplifier (LNA)
common-emitter preamplifier, 9-15
noise suppression, 9-14
spurious-free dynamic range (SFDR), 9-14
operating bandwidth of, 9-14
Radio-frequency circuit design, 9-1
MIM capacitors, 9-8

monolithic circuit technologies for, 9-5
monolithic transformer fabrication, 9-10
multifilament transformers, 9-11
passive transformer, 9-9–9-10
planar inductors, 9-9
polysilicon resistors and films, 9-8–9-9
resonant-tuned (LC) circuits, 9-9
transmission line interconnections
circuit components, 9-5
semiconducting substrate and skin-effect, 9-5–9-6
shunt parasitics, 9-6–9-7
slow-wave and quasi-TEM modes, 9-5–9-6
transmission line Q-factor, 9-6
Radio-frequency integrated circuits
design for silicon, 9-36
RFIC wireless front-end building blocks, 9-1–9-2
Receiver
broadband noise, 10-10
chip, 14-2–14-14
circuit, in optical link
block design, 14-3–14-4
demultiplexed eye diagram, 14-9
DC blocking capacitors, 10-20
DC offset variation effect, 10-6
front end
characterization of, 10-21
layout symmetry, 10-19
performance of fabricated hardwares, 10-21–10-22
implementation, 16-13
layout symmetry, 10-20
noise floor, 10-7
signal-generation path for, 10-7–10-9
signal path and SNR degradation, 10-6–10-7
Resonant frequency, 11-7, 11-12, 12-8
Resonant tanks, 9-30
Resonant-tuned (LC) circuits, 9-9
Resonator with MEMS switches
performance, 11-9
SEM photo, 11-9
Reverse-biased diode, depletion region width of, 9-35
RF amplifiers, *see* Radio-frequency amplifiers
RF-analog-mixed-signal IC, 4-3
RF and analog blocks, validation, 16-14–16-17
RF circuit design, *see* Radio-frequency circuit design
RF circuit performance, 9-4
RFICs, *see* Radio-frequency integrated circuits
RF low-noise amplifier, design parameters
circuit topologies, 5-3
gain, 5-3–5-4
linearity, 5-4–5-5
noise parameters, 5-2–5-3
power consumption, 5-4–5-5
RF MEMS
switches, 11-2–11-6, 11-7, 11-12, 11-13, 11-16, 11-19
bandwidth, 11-14
center frequency, 11-14
photos of cantilever, 11-4
reliability, 11-5
simulated S-parameters states, 11-6
techniques, 11-2–11-6
schematic of, 11-3
surface micromachining, 11-2
tuners, 11-17
RF Mixer
as analog multiplier and frequency converter, 9-22
Gilbert multiplier, 9-22–9-24
third-order intercept point (IIP3), 9-26
transformer balun-coupled mixer, 9-23–9-25
RF Pads, 16-8
RF power spectrum, 4-3, 9-27
RF system performances
impact on RF circuits, 10-10–10-11
mismatching impact on, 10-10
RF variable-gain amplifier (RF-VGA), 4-16

S

SAW filter, 16-15
Second-order input-referred intercept point (IIP2), 4-12
III–V Semiconductor technologies, 7-2
SFDR, *see* Spurious-free dynamic range
Shockley boundary condition, generelized, A.2-2, A.2-3
Shunt diodes, insertion loss of, 8-7
Shunt parasitics, 9-6–9-7
Si
effective mass parameters, A.1-2
energy band structure, A.1-1
properties of, A.1-2
SiGe alloys, 1-4, 2-1, 5-9, 8-8, 10-2, 12-3
SiGe BiCMOS process, 14-3, 14-10, 14-12, 14-17, 15-4, 15-6
SiGe BiCMOS technology, 4-2, 9-2
SiGe bipolar devices, gain-bandwidth product of, 9-2
SiGe BJT, 9-21
SiGe cascode LNA design
analytical expressions, 5-9–5-13
device model, use of, 5-6–5-8
optimization strategies, 5-5–5-6
simulation in, 5-8–5-9
SiGe:C channel pMOSFET, 2-8
SiGe:C HBT technology, 2-6
SiGeC heterojunction bipolar transistor (HBT), 16-5
SiGe circuits, 13-1
SiGe direct conversion radio
based on 50Ω interfacing impedance, 10-11–10-12
frequency allocation of, 10-11–10-13
IEEE802.11a standards, 10-11
metal layers, design platform, 10-13
SiGe distributed amplifier MMICs, 8-6
SiGe HBT BiCMOS technology, 2-5
SiGe HBTs
development of, 2-3–2-6
graded Ge base profile, A.2-4
high-frequency region of, 9-3

high-power, 2-6
technology, 15-3, 15-5, 16-5
SiGe HFETs, 7-16
SiGe MMICs, 7-2
SiGe-Si strained layer epitaxy development, 2-2–2-3
SiGe-strained Si FETs development, 2-6–2-8
Signal-generation path, transceiver
frequency generation, 10-8
nonlinearity effects, 10-7
VCO pulling effects, 10-9
Signal path
direct conversion radios, 10-4
nonidealities in, 10-6
receiver, 10-6–10-7
signal generation path, 10-7
transmitter, 10-6
Signal-to-noise ratio, 4-4, 4-12, 5-1
Balun-coupled mixer, 9-25
of RF mixers, 9-22–9-24
of transistor, 9-18
Silicidation, of SiGeC, 2-11
Silicon-based micromachined patch
effective dielectric constant *vs.* air gap thickness, 12-3
Silicon-based wired data communications circuits, 14-1
Silicon dies, 13-4
interconnect lengths, 13-5
packaging, 13-4
Silicon dioxide and polyimide, 12-8–12-10
Silicon MOSFET and BJT
gain, flicker noise levels and breakdown voltage, 9-13
Johnson limit, 9-14
Silicon NMOS devices, 9-2
Silicon substrate
attenuation constant, 12-7
CPW dimensions *vs.* characteristic impedances, 12-16
dielectric constant, 12-10
microstrip/coplanar transmission lines, 8-1–8-2
relative permittivity, 12-6
simulated input return loss for, 12-17
thermal conductivity of, 8-2
W-band TSA, 12-12
Silicon wafer, 12-11
configuration of, 12-15
CPW lines, characteristic impedance of, 12-16
H-plane pattern, 12-13
TSA patterns, 12-13, 12-14, 12-15
Si MMICs
SiGe distributed amplifier MMICs, 8-6
spiral inductors and metal–insulator–metal (MIM) capacitors, 7-3
Single-chip transceiver LSI, block diagram of, 15-4
Single-stage wide-band amplifiers, *see* Transimpedance amplifier
Si-SiGe HBT low-noise amplifier
circuit linearity, effect of, 4-12–4-13
noise factor *(F)*, 4-12
Si-SiGe HBT technologies, 7-2
Si-SiGe HBT voltage-controlled oscillators
fully monolithic, 4-15, 4-16
noise factor *(F)*, 4-14
performance issues, 4-13
schematic of, 4-14
Slow-wave and quasi-TEM modes, 9-5–9-6
SNR degradation
of direct conversion receivers, 10-7–10-8
nonideality in receiver signal path, 10-6–10-7
nonideality in transmitter signal path, 10-6
SPDT switches, 8-7
Spiral inductors
equivalent circuit for, 7-4–7-5
ground shields and eddy currents, 7-5
Spurious-free dynamic range, 9-14
Static frequency divider, 7-8
Strained Si CMOS, 1-2, 1-6, 1-8–1-10, 2-11
Strained Si nMOSFET, 2-8
Stub capacitances, 11-16
Substrate current, 9-9
Substrates, semiconducting
propagating waves and, 9-9
resistivity and dielectric properties, 7-3
skin-effect and slow-wave mode of propagation, 9-5–9-6
Superheterodyne receiver
direct conversion frequency scheme and, 10-8
frequency shifting, 10-2
Surface acoustic wave (SAW) filter, 4-4
Surface micromachining, 11-2, 12-2, 12-14
Surface mount technology (SMT), 13-2
Switched gain LNA
layout view, 16-8
simplified schematic, 16-7
Switching pair
device mismatching, 10-17
inputs of, 10-16
Switch lifetime, 11-4, 11-5

T

Tapered slot antenna (TSA), 12-11
Thermal design, 13-6
Thick metal inductors, 16-6
Thin-film microstrip lines
and coplanar waveguides, 7-4
interconnects, 8-2
Third-order input-referred intercept point (IIP3), 4-12, 5-5
Threshold voltage, 16-3
Thru-reflect-line (TRL), 11-12
Time-of-flight delays, 9-3, 9-4
Transceiver LSI, single-chip, 15-3–15-4
Transfer function HBT, 16-4
Transformer balun-coupled mixer
monolithic transformer balun, 9-24
noise degradation, 9-25
transformed source resistance, 9-25

Transformer feedback amplifiers
  circuit parameters, 9-19
  impedance match, 9-20
  unilateralization and neutralization, 9-19–9-20
Transformer, RFIC design
  balun, square symmetric, 9-12
  monolithic transformer fabrication, 9-10
  multifilament transformers, 9-11
  passive transformer, 9-9–9-10
Transient ICCR (TICC), A.3-2
Transimpedance amplifier, 7-9
Transistor noise figure optimization, 9-17
Transmission line effects
  off-chip interconnections, 9-4
  propagation delays, 9-3
Transmission line interconnections
  circuit components, 9-5
  total Q-factor, 9-6
Transmission line Q-factor, 9-6
Transmitter implementations
  full-rate architecture, 14-9–14-10
    Phase noise plots, 14-16
    VCOs used in, 14-13
  half-rate architecture, 14-9–14-10
    phase noise plots, 14-16
    VCOs used in, 14-13
Transmitter (TX), 15-6, 16-12
  blocking characteristics, 16-18
  chip, 14-2–14-14
  signal-generation path for, 10-7–10-9
  signal path and in band SNR degradation, 10-6
  spurious emission, 16-18
Two-pole MEMS filter
  performance, 11-10
  SEM photo, 11-10

U

Ultrafine pitch (UFP) wire bond technology, 13-4
Ultrahigh vacuum chemical vapor deposition (UHV/CVD), 2-3, 4-1
Ultrawide-band MMIC amplifier, 7-16
Unilateralization, 9-19

V

Validation strategy, 16-13–16-14
Varactor diode, 4-15
Varactors, 16-6
  bias voltage across, 9-35
  figure-of-merit Q-factor, 9-34
Varactors diode, 9-35
Variable gain amplifier (VGA), 16-10–16-12
VBIC SiGe HBT model parameters, A.4-7–A.4-9
Voltage controlled oscillator (VCO), 4-13–4-15
  circuit diagram of, 8-5
  low-frequency noise source effects, 9-29
  oscillator phase noise, 9-28–9-29
  phase noise performance of, 14-8
  pulling effects, 10-8–10-9
  resonant tanks, 9-30
  spiral inductors, 9-30–9-31
Volterra-series method, 5-8, 5-10

W

W-band TSA, 12-11, 12-13
W-CDMA handset, 4-7
W-CDMA receiver
  global test, 16-18–16-19
  simulated bit error rate performance, 16-18
W-CDMA SiGe
  direct-conversion receiver chip, 4-17
  heterodyne transmitter chip, 4-18
Wide-band circuits, 7-9
Wideband code-division multiaccess (WCDMA), 15-7, 16-1
Winding, primary and secondary
  self-inductance of, 9-18
Wireless communication systems
  design considerations
    direct upconversion architecture, 4-7
    heterodyne receiver, 4-3–4-4
    homodyne receiver, 4-5
    indirect upconversion transmitter, 4-6
  electronic-toll-collection transceiver IC, 15-6–15-7
  Frequency-Divider ICs, 15-4–15-6
  frequency-divider ICs, 15-4–15-6
Wireless local area networks (WLAN), 15-4
Wireless transceivers design, 16-1
Wireless transreceivers, SiGe HBT based
  avalanche breakdown, 4-9
  base width, 4-9, 4-10
  breakdown voltage, 4-9
  equivalent circuit and cross section of, 4-8
  speed of, 4-10

X

XOR gate, 14-7

Z

Zero-IF WCDMA architecture, 16-2